Unsteady Combustor Physics

Second Edition

Explore a unified treatment of the dynamics of combustor systems, including acoustics, fluid mechanics, and combustion in a single rigorous text. This updated new edition features an expansion of data and experimental material, updates the coverage of flow stability, gives an enhanced treatment of flame dynamics, and addresses the system dynamics of clean energy and propulsion systems used in low emissions systems; it synthesizes the fields of fluid mechanics and combustion into a coherent understanding of the intrinsically unsteady processes in combustors. This is a perfect reference for engineers and researchers in fluid mechanics, combustion, and clean energy.

Tim C. Lieuwen is Regents' Professor and Executive Director of the Strategic Energy Institute at Georgia Tech. He is also the founder and CTO of TurbineLogic, an energy analytics firm. He has authored 4 books and over 350 other publications. Board positions include governing/advisory boards for Oak Ridge National Lab, Pacific Northwest National Lab, and the National Renewable Energy Lab, and appointment by the DOE Secretary to the National Petroleum Counsel. He is an elected member of the National Academy of Engineering, a fellow of ASME and AIAA, and recipient of the AIAA Lawrence Sperry Award and ASME's George Westinghouse Gold Medal.

Unsteady Combustor Physics

Second Edition

TIM C. LIEUWEN
Georgia Institute of Technology

CAMBRIDGE
UNIVERSITY PRESS

University Printing House, Cambridge CB2 8BS, United Kingdom

One Liberty Plaza, 20th Floor, New York, NY 10006, USA

477 Williamstown Road, Port Melbourne, VIC 3207, Australia

314–321, 3rd Floor, Plot 3, Splendor Forum, Jasola District Centre, New Delhi – 110025, India

103 Penang Road, #05–06/07, Visioncrest Commercial, Singapore 238467

Cambridge University Press is part of the University of Cambridge.

It furthers the University's mission by disseminating knowledge in the pursuit of education, learning, and research at the highest international levels of excellence.

www.cambridge.org
Information on this title: www.cambridge.org/9781108841313
DOI: 10.1017/9781108889001

© Tim C. Lieuwen 2012, 2021

This publication is in copyright. Subject to statutory exception
and to the provisions of relevant collective licensing agreements,
no reproduction of any part may take place without the written
permission of Cambridge University Press.

First published 2012

Printed in the United Kingdom by TJ Books Limited, Padstow, Cornwall

A catalogue record for this publication is available from the British Library.

ISBN 978-1-108-84131-3 Hardback

Additional resources for this publication at cambridge.org/9781108841313

Cambridge University Press has no responsibility for the persistence or accuracy of URLs for external or third-party internet websites referred to in this publication and does not guarantee that any content on such websites is, or will remain, accurate or appropriate.

Summary Contents

	Acknowledgments	page xiv
	Introduction	1
	Overview of the Book	7
1	**Basic Equations**	9
2	**Decomposition and Evolution of Disturbances**	27
3	**Hydrodynamic Flow Stability I: Linear Instability**	81
4	**Hydrodynamic Flow Stability II: Common Combustor Flow Fields**	113
5	**Acoustic Wave Propagation I: Basic Concepts**	176
6	**Acoustic Wave Propagation II: Heat Release, Complex Geometry, and Mean Flow Effects**	210
7	**Flame Sheet and Flow Interactions**	261
8	**Ignition**	296
9	**Internal Flame Processes**	321
10	**Flame Stabilization, Flashback, Flameholding, and Blowoff**	379
11	**Forced Response I: Flamelet Dynamics**	406
12	**Forced Response II: Heat Release Dynamics**	463
	Index	511

Detailed Contents

Acknowledgments		*page* xiv
Introduction		1
Updates to the Second Edition		4
Overview of the Book		7

1 Basic Equations — 9

1.1	Thermodynamic Relations in a Multicomponent Perfect Gas	9
1.2	Continuity Equation	10
1.3	Momentum Equation	11
1.4	Species Conservation Equation	12
1.5	Energy Equation	13
1.6	Aside: Discussion of the Vorticity and Circulation Equations	17
1.7	Nomenclature	20
	1.7.1 Latin Alphabet	21
	1.7.2 Greek Alphabet	23
	1.7.3 Subscripts	24
	1.7.4 Superscripts	25
	1.7.5 Other Symbols	25
Exercises		25
References		26

2 Decomposition and Evolution of Disturbances — 27

2.1	Descriptions of Flow Perturbations	27
2.2	Small-Amplitude Propagation in Uniform, Inviscid Flows	29
	2.2.1 Motivating Example: Solutions of the One-Dimensional Linearized Euler Equations	30
	2.2.2 Decomposition Approach	32
	2.2.3 Comments on the Decomposition	36
	2.2.4 Molecular Transport Effects on the Decomposition	38
2.3	Modal Coupling Processes	39
	2.3.1 Coupling through Boundary Conditions	39
	2.3.2 Coupling through Flow Inhomogeneities	40
	2.3.3 Coupling through Nonlinearities	42

	2.4 Energy Density and Energy Flux Associated with Disturbance Fields	44
	2.5 Linear and Nonlinear Stability of Disturbances	48
	2.5.1 Linearly Stable/Unstable Systems	49
	2.5.2 Nonlinearly Unstable Systems	52
	2.5.3 Forced and Limit Cycling Systems	54
	2.6 Complex Phase Space Dynamics	60
	2.7 Decompositions of Data	61
	2.7.1 Prescribed Basis Functions	62
	2.7.2 Empirical Basis Functions	66
	2.8 Aside: Triple Decomposition	71
	2.9 Aside: Effects of Simultaneous Acoustic and Vortical Velocity Disturbances	74
	2.10 Aside: Further Consideration of the Disturbance Energy Equation	75
	Exercises	76
	References	77
3	**Hydrodynamic Flow Stability I: Linear Instability**	**81**
	3.1 Linear Stability Notions	82
	3.2 Global and Convective Instability	84
	3.3 Transient Growth	87
	3.4 Normal Modes in Parallel Flows	88
	3.4.1 Basic Formulation	88
	3.4.2 General Results for Temporal Instability	90
	3.4.3 Revisiting Convective/Absolute Instability for Parallel Flows	94
	3.4.4 Extended Example: Spatial Mixing Layer	95
	3.4.5 Connection between Local and Global Analysis	99
	3.5 Mean Flow Stability	100
	3.6 Aside: Vortex Mutual Induction	103
	3.7 Aside: Receptivity	106
	3.8 Aside: Rayleigh–Taylor and Centrifugal Instability	106
	Exercises	108
	References	110
4	**Hydrodynamic Flow Stability II: Common Combustor Flow Fields**	**113**
	4.1 Free Shear Layers	113
	4.1.1 Flow Stability and Unsteady Structure	116
	4.1.2 Effects of Harmonic Excitation	120
	4.2 Wakes and Bluff Body Flow Fields	123
	4.2.1 Parallel Flow Stability Analysis	125
	4.2.2 Bluff Body Wake	128
	4.2.3 Shear Layer Dynamics in Wake Flows	130
	4.2.4 Effects of Harmonic Excitation	131

4.3	Jets		132
	4.3.1	Parallel Flow Stability Analysis	133
	4.3.2	Constant-Density Jet Dynamics	136
	4.3.3	Effects of Harmonic Excitation	137
	4.3.4	Jets in Cross Flow	138
4.4	Swirling Jets and Wakes		144
	4.4.1	Rotating and Winding Directions of Flow Disturbances	146
	4.4.2	Vortex Breakdown	147
	4.4.3	Swirling Jet and Wake Dynamics	151
	4.4.4	Effects of Harmonic Excitation	153
4.5	Backward-Facing Steps and Cavities		156
	4.5.1	Parallel Flow Stability Analysis	157
	4.5.2	Unsteady Flow Structure	158
4.6	Aside: Boundary Layers		161
4.7	Aside: Confinement Effects and Arrays of Jets/Wakes		163
4.8	Aside: Noncircular Jets		164
Exercises		165
References		165

5 Acoustic Wave Propagation I: Basic Concepts — 176

5.1	Traveling and Standing Waves	177
5.2	Boundary Conditions: Reflection Coefficients and Impedance	181
5.3	Natural Modes of Simple Geometries	187
	5.3.1 One-Dimensional Modes	188
	5.3.2 Multidimensional Rectangular Duct Modes	191
	5.3.3 Circular Duct Modes	192
	5.3.4 Lumped Elements and Helmholtz Resonators	196
	5.3.5 Convective Modes	198
5.4	Forced Oscillations	198
	5.4.1 One-Dimensional Forcing and Resonance	199
	5.4.2 Forced Oscillations in Ducts and Cutoff Modes	200
5.5	Aside: Annular and Sector Circular Geometries	205
5.6	Aside: Natural Modes in Annular Geometries	206
Exercises		208
References		209

6 Acoustic Wave Propagation II: Heat Release, Complex Geometry, and Mean Flow Effects — 210

6.1	Introduction	210
6.2	Mean Flow Effects	213
	6.2.1 Mean Flow Effects on Wave Propagation	213
	6.2.2 Mean Flow Effects on Boundary Conditions	215
	6.2.3 Mean Flow Compressibility Effects and Acoustic/Entropy Coupling	216
	6.2.4 Variable Temperature Effects	217

		6.2.5	Example Problem: Wave Reflection and Transmission through Variable Temperature Region	218
		6.2.6	Example Problem: Natural Frequencies of a Variable Temperature Region	220
	6.3	Variable Area and Complex Geometry Effects		221
		6.3.1	Baseline Results	222
		6.3.2	Nozzles and Diffusers	223
		6.3.3	Wave Refraction and Injector Coupling	225
		6.3.4	Unsteady Vorticity Generation and Acoustic Damping	228
	6.4	Acoustic Damping Processes		231
	6.5	Unsteady Heat Release Effects		233
		6.5.1	Thermoacoustic Stability Model Problem	235
		6.5.2	Further Discussion of Thermoacoustic Instability Trends	239
	6.6	Nonlinear Effects and Limit Cycles		242
		6.6.1	Formulation of Modal and Amplitude Equations	244
		6.6.2	Sources of Nonlinearities	248
	6.7	Aside: Sturm–Liouville Eigenvalue Problems		250
		6.7.1	Orthogonality of Eigenfunctions	251
		6.7.2	Real/Imaginary Characteristics of Eigenvalues	252
		6.7.3	Asymptotic Representation of Eigenfunctions	252
	6.8	Aside: Wave Propagation through Regions with Slowly Varying Properties		252
	6.9	Aside: Approximate Methods for Linearized Jump Conditions across Compact Zones		253
	6.10	Aside: Wave Interactions with Compact Nozzles or Diffusers		255
	Exercises			255
	References			256
7	**Flame Sheet and Flow Interactions**			261
	7.1	Surface Dynamics		261
	7.2	Field Equations for Premixed and Nonpremixed Flames		263
		7.2.1	Premixed Flames	263
		7.2.2	Nonpremixed Flames	264
		7.2.3	Comparison of Premixed and Nonpremixed Flame Evolution Equations	266
	7.3	Jump Conditions		267
		7.3.1	Premixed Jump Conditions	267
		7.3.2	Nonpremixed Jump Conditions	269
		7.3.3	Velocity and Pressure Relations	269
		7.3.4	Vorticity Relations and Vortex–Flame Interactions	274
	7.4	Stretching of Material and Flame Surfaces		282
		7.4.1	Stretching of Material Surfaces	282
		7.4.2	Premixed Flame Stretch	283
		7.4.3	Example Problem: Stretching of Material Line by a Vortex	283

	7.5	Influence of Premixed Flames on the Approach Flow	285
	7.6	Aside: Finite Flame Thickness Effects on Flame Jump Conditions	288
	7.7	Aside: Further Analysis of Vorticity Jump Conditions across Premixed Flames	289
	7.8	Aside: Linearized Analysis of Flow Field Modification by the Flame	291
		Exercises	292
		References	293
8	**Ignition**		296
	8.1	Overview	296
	8.2	Autoignition	298
		8.2.1 Ignition of Homogeneous, Premixed Reactants	298
		8.2.2 Effects of Losses and Flow Inhomogeneity	301
	8.3	Forced Ignition	311
		Exercises	316
		References	317
9	**Internal Flame Processes**		321
	9.1	Premixed Flame Overview	322
		9.1.1 Premixed Flame Structure	322
		9.1.2 Premixed Flame Dependencies	326
	9.2	Premixed Combustion in Inhomogeneous, Autoigniting Mixtures	328
	9.3	Premixed Flame Stretch and Extinction	331
		9.3.1 Overview	332
		9.3.2 Expressions for Flame Stretch	334
		9.3.3 Weak Stretch Effects	335
		9.3.4 Strong Stretch Effects, Consumption and Displacement Speeds, and Extinction	338
	9.4	Premixed Flames: Unsteady Effects	341
	9.5	Nonpremixed Flame Overview	343
	9.6	Finite-Rate Effects in Nonpremixed Flames	345
	9.7	Edge Flames and Flame Spreading	349
		9.7.1 Overview	349
		9.7.2 Buckmaster's Edge Flame Model Problem	351
		9.7.3 Edge Flame Velocities	354
		9.7.4 Conditions at the Flame Edge	357
		9.7.5 Implications on Flame Spread after Ignition	359
	9.8	Intrinsic Flame Instabilities	360
	9.9	Aside: Unsteady Flame Response Effects	363
	9.10	Aside: Flame Extinction by Vortices	366
	9.11	Aside: Wave Speeds of Reaction–Diffusion Equations	367
		Exercises	371
		References	371

10	**Flame Stabilization, Flashback, Flameholding, and Blowoff**	379
	10.1 Flashback and Flameholding	379
	10.1.1 Flame Propagation in the Core Flow	380
	10.1.2 Boundary Layer Flashback	382
	10.2 Flame Stabilization and Blowoff	388
	10.2.1 Basic Effects in Premixed Flames: Kinematic Balance Between Flow and Burning Velocities	390
	10.2.2 Stretch Rates for Shear Layer Stabilized Flames	392
	10.2.3 Product Recirculation Effects on Flame Stabilization and Blowoff	396
	10.2.4 Nonpremixed Flame Liftoff and Blowoff	400
	References	402
11	**Forced Response I: Flamelet Dynamics**	406
	11.1 Overview of Length/Time Scales	406
	11.1.1 Premixed Flame Interactions with Broadband Disturbance Fields	407
	11.1.2 Flame Interactions with Narrowband Velocity Disturbance Fields	412
	11.2 Dynamics of Premixed Flame Sheets	415
	11.2.1 Model Problems for Two-Dimensional Configurations	415
	11.2.2 Linearized Dynamics of Constant Burning Velocity Flames	419
	11.2.3 Nonlinear Flame Front Dynamics	433
	11.3. Dynamics of Nonpremixed Flame Sheets	442
	11.3.1 Example Problem: Mixing Layer	442
	11.3.2 Example Problem: Transient Stagnation Flame	445
	11.3.3 Example Problem: Isothermal Nonpremixed and Premixed Flame Rollup by a Vortex	446
	11.3.4 Example Problem: Harmonic Forcing of a Confined, Overventilated Flame	448
	11.4 Aside: Dissipation and Dispersion of Disturbances on Premixed and Nonpremixed Flames	452
	11.5 Aside: Harmonic Forcing of Turbulent, Premixed Flames	454
	11.6 Aside: Forced Response Effects on Natural Flame Instabilities	456
	Exercises	457
	References	459
12	**Forced Response II: Heat Release Dynamics**	463
	12.1 Overview of Forced Flame Response Mechanisms	463
	12.2 Flame Configuration Effects on Response Sensitivities	469
	12.2.1 Geometry and Flame Area Distribution Effects	470
	12.2.2 Burning Rate Distribution Effects	471

12.3 Harmonic Flame Excitation 472
    12.3.1 Linear Dynamics: Velocity-Coupled Response 472
    12.3.2 Linear Dynamics: Equivalence Ratio Coupling 479
    12.3.3 Nonlinear Dynamics 481
12.4 Broadband Excitation and Turbulent Flame Speeds 488
    12.4.1 Time-Averaged Burning Rates 488
    12.4.2 Fluctuating Burning and Heat Release Rates 493
    12.4.3 Combustion Noise 498
12.5 Aside: Effect of External Forcing on Limit Cycle Oscillations 502
Exercises 503
References 504

Index 511

Acknowledgments

Many individuals must be acknowledged for the completion of this book. First, I am deeply appreciative to my dear wife, Rinda, and daughters Liske, Anneke, Carolina, and Janna Lieuwen for their love, encouragement, and support.

This book would not have been possible without the financial support provided through Joseph Citeno, which got the project kicked off, and the support of Vigor Yang through my department. I am deeply grateful for their support, which made initiating this project possible.

Next, this book would never have been possible without the enormous help provided by my group here at Georgia Tech. They were a great help in pulling together references, performing calculations, critiquing arguments, fleshing out derivations, catching mistakes, and being a general sounding board. Particular thanks go to Mike Aguilar, Alberto Amato, Ianko Chterov, Jack Crawford, Ben Emerson, Chris Foley, Julia Lundrigan, Nick Magina, Mike Malanoski, Andrew Marshall, Jacqueline O'Connor, Shreekrishna, Vishal Acharya Srinivas, Dong-Hyuk Shin, Ryan Sullivan, Prabhakar Venkateswaran, and Ben Wilde. I have been very fortunate to have had such a great team to work with and I thank all of them for their help.

Next, special thanks to Ben Bellows, Enrique Portillo Bilbao, Baki Cetegen, Jeff Cohen, Joel Daou, Catalin Fotache, Fei Han, Santosh Hemchandra, Hong Im, Matthew Juniper, Vince McDonell, Randal McKinney, Venkat Narra, Bobby Noble, Preetham, Rajesh Rajaram, Mike Renfro, Paul Ronney, Dom Santavicca, Thomas Sattelmayer, David Scarborough, Thierry Schuller, Santosh Shanbhogue, Shiva Srinivasan, R.I. Sujith, Sai Kumar Thumuluru, and Qingguo Zhang for their feedback and suggestions on the outline and content. In addition, Siva Harikumar, Faisal Ahmed, and Jordan Blimbaum were a great editorial support team.

In addition, my sincere thanks go to my colleagues and mentors Ben Zinn, Robert Loewy, Lakshmi Sankar, Jeff Jagoda, Jerry Seitzman, Suresh Menon, and Vigor Yang for their help and support.

I am deeply appreciative of the suggestions for second edition contents and comments on the manuscript by many of my colleagues, including Ben Emerson, Vishal Acharya, Jacqueline O'Connor, Santosh Hemchandra, and Hong Im.

I am also very thankful for the help and support of Austin Matthews in helping to pull together this second edition. His carefulness and attention to detail made this a much better book.

Finally, I am very appreciative of the assistance of my group here at Georgia Tech in developing figures, performing calculations, and checking the text. In particular, thanks to Aravind Chandh, Lane Dillon, Chris Douglas, Tony John, Henderson Johnson II, Sriram Kalathoor, Jeong-Won Kim, Vedanth Nair, Shivam Patel, Parth Patki, Sara Schmidheiser, and Sukruth Somappa.

Introduction

This book is about unsteady combusting flows, with a particular emphasis on the system dynamics that occur at the intersection of the combustion, fluid mechanics, and acoustic disciplines – i.e., on *combustor* physics. In other words, this is not a combustion book – rather, it treats the interactions of flames with unsteady flow processes that control the behavior of combustor systems. While numerous topics in reactive flow dynamics are "unsteady" (e.g., internal combustion engines, detonations, flame flickering in buoyancy-dominated flows, thermoacoustic instabilities), this text specifically focuses on unsteady combustor issues in high Reynolds number, gas-phase flows. This book is written for individuals with a background in fluid mechanics and combustion (it does not presuppose a background in acoustics), and is organized to synthesize these fields into a coherent understanding of the intrinsically unsteady processes in combustors.

This book follows several texts or monographs which have treated related topics – including Toong's *Combustion Dynamics* [1], Crocco and Cheng's *Theory of Combustion Instability in Liquid Propellant Rocket Motors* [2], *Liquid Propellant Rocket Combustion Instability* by Harrje and Reardon [3], Putnam's *Combustion Driven Oscillations in Industry* [4], Fred Culick's *Unsteady Motions in Combustion Chambers for Propulsion Systems* [5], or Lieuwen and Yang's edited *Combustion Instabilities in Gas Turbine Engines* [6]. Similarly, several dedicated texts on turbulent combustion have been written, including Peters [7] and Lipatnikov [8].

Unsteady combustor processes define many of the most important considerations associated with modern combustor design. These unsteady processes include *transient, time harmonic,* and *stochastic* processes. For example, ignition, flame blowoff and flashback are *transient* combustor issues that often define the range of fuel/air ratios or velocities over which a combustor can operate. As we discuss in this book, these transient processes involve the coupling of chemical kinetics, mass and energy transport, flame propagation in high shear flow regions, hydrodynamic flow stability, and interaction of flame-induced dilatation on the flow field – much more than a simple balance of flame speed and flow velocity.

Similarly, combustion instabilities are a *time-harmonic* unsteady combustor issue where the unsteady heat release excites natural acoustic modes of the combustion chamber. These instabilities cause such severe vibrations in the system that they can impose additional constraints on where combustor systems can be operated. The acoustic oscillations associated with these instabilities are controlled by the entire combustor system; i.e., they involve the natural acoustics of the coupled plenum, fuel delivery system, combustor, and turbine transition section. Moreover, these acoustic oscillations often excite natural hydrodynamic instabilities of the flow, which then wrinkle the flame front and cause modulation of the heat release rate. As such, combustion instability problems involve the coupling of acoustics, flame dynamics, and hydrodynamic flow stability.

Turbulent combustion itself is an intrinsically unsteady problem involving *stochastic* fluctuations that are both stationary (such as turbulent velocity fluctuations) and nonstationary (such as turbulent flame brush development in attached flames). Problems such as turbulent combustion noise generation require an understanding of the broadband fluctuations in heat release induced by the turbulent flow, as well as the conversion of these fluctuations into propagating sound waves. Moreover, the turbulent combustion problem is a good example for a wider motivation of this book – many time-averaged characteristics of combustor systems cannot be understood without understanding their unsteady features. For example, the turbulent flame speed, related to the time-averaged consumption rate of fuel, can be one to two orders of magnitude larger than the laminar flame speed, precisely because of the effect of unsteadiness on the time-averaged burning rate. In turn, crucial issues such as flame spreading angle and flame length, which then directly feed into basic design considerations such as locations of high combustor wall heat transfer, or combustor length requirements, are then directly controlled by unsteadiness.

Even in nonreacting flows, intrinsically unsteady flow dynamics control many time-averaged flow features. For example, it became clear a few decades ago that turbulent mixing layers did not simply consist of broadband turbulent fluctuations, but were, rather, dominated by quasi-periodic structures. Understanding the dynamics of these large-scale structures has played a key role in our understanding of the time-averaged features of shear layers, such as growth rates, mixing rates, or exothermicity effects. Additionally, this understanding has been indispensable in understanding intrinsically unsteady problems, such as how shear layers respond to external forcing.

Similarly, many of the flow fields in combustor geometries are controlled by hydrodynamic flow instabilities and unsteady large-scale structures that, in turn, are also profoundly influenced by combustion-induced heat release. It is well known that the instantaneous and time-averaged flame shapes and recirculating flow fields in many combustor geometries often bear little resemblance to each other, with the instantaneous flow field exhibiting substantially more flow structures and asymmetry. Flows with high levels of swirl are a good example of this, as shown by the comparison of time-averaged (a) and instantaneous (b–d) streamlines in Figure I.1. Understanding such features as recirculation zone lengths and flow topology, and how these features are influenced by exothermicity or operational conditions, necessarily requires an understanding of the dynamic flow features. To summarize, continued progress in predicting steady-state combustor processes will come from a fuller understanding of their time dynamics.

Modern computations and diagnostics have revolutionized our understanding of the spatiotemporal dynamics of flames since the publication of Markstein's *Nonsteady*

Introduction

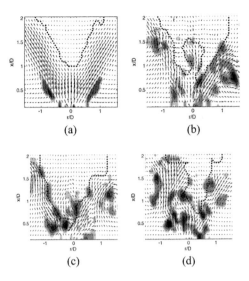

Figure I.1 (a) Time-averaged and (b–d) instantaneous flow field in a swirling combustor flow. Dashed line denotes isocontour of zero axial velocity and shaded regions denote vorticity values. Image courtesy of M. Aguilar, M. Malanoski, and J. O'Connor.

Flame Propagation [9]. Indeed, massive improvements in computational power and techniques for experimental characterization of the spatial features of reacting flows has led to a paradigm shift in recent decades in our understanding of turbulent flame processes. For example, well-stirred reactors once served as a widely accepted physical model used to describe certain types of flames, using insight based on line-of-sight flame imaging, such as shown in the top three images taken from a swirling flow in Figure I.2. These descriptions suggest that the combustion zone is essentially a homogeneous, distributed reaction zone due to the vigorous stirring in the vortex breakdown region. Well-stirred reactor models formed an important conceptual picture of the flow for subsequent modeling work, such as to model blowoff limits or pollutant formation rates. However, modern diagnostics, as illustrated by the planar cuts through the same flame that are shown in the bottom series of images in Figure I.2, show a completely different picture. These images show a thin, but highly corrugated, flame sheet. This flame sheet is not distributed, but a thin region that is so wrinkled in all three spatial dimensions that a line-of-sight image suggests a homogeneous reaction volume.

Such comparisons of the instantaneous versus time-averaged flow field and flame, or the line-of-sight versus planar images, suggest that many exciting advances still lie in front of this community. These observations – that a better understanding of temporal combustor dynamics will lead to improved understanding of both its time-averaged and unsteady features – serve as a key motivator for this book. I hope that it will provide a useful resource for the next generation of scientists and engineers working in the field, grappling with some of the most challenging combustion and combustor problems yet faced by workers in this difficult yet rewarding field.

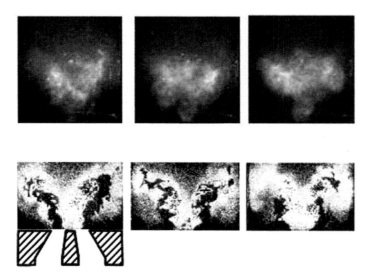

Figure I.2 Line-of-sight (top) and planar (bottom) OH-PLIF images of turbulent, swirling flame [10]. Images courtesy of B. Bellows.

Updates to the Second Edition

It is hard to believe that 10 years have gone by since we started this project. Many of the motivators and drivers of this book remain the same, but much has changed as well. From a societal point of view, the march toward decarbonization is accelerating, motivating topics like hydrogen combustion or combustor operability limits of alternative fuels. The commercial space market has taken off and combustion instabilities remain a key risk for rocket development. Major interest has developed in rotating detonation engines, where the wave dynamics in annular passages and coupled injector dynamics mirror many similar combustion instability topics. There is a resurgence of interest in data-driven approaches for the analysis of complex data sets, active control, or prediction of future or unmeasured system behaviors. Significant developments have also taken place in hydrodynamic stability, particularly reacting flows.

With this in mind, this book has been refreshed and updated. New sections have been added or major updates made in the following sections:

- Vorticity/circulation dynamics, Section 1.6.
- Exact solution of one-dimensional, linearized Navier–Stokes equations, motivating the canonical decomposition into entropy, acoustic, and vortical disturbances, Section 2.2.1.
- New example problem on randomly forced nonlinear oscillators in Section 2.5.3.3.
- Phase space dynamics of nonlinear systems, Section 2.6.

- Decompositions of data, including the Fourier transform, wavelets, partial orthogonal decomposition (POD), spectral POD, and dynamic mode decomposition, Section 2.7.
- New material and complete reorganization of Chapter 3, organized around different approaches for analyzing hydrodynamic flow stability, including a new section on mean flow stability theory and limit cycle amplitudes of globally unstable flows.
- Section 3.8 on instabilities in rotating and density stratified flows.
- Reacting jets in cross flow in Section 4.3.4.2.
- Stability of confined and multielement canonical flows in Section 4.7.
- Acoustic wave dynamics in annular passages in Section 5.6.
- Acoustic–entropy mode coupling in Section 6.2.3.
- Acoustic wave interactions with injectors in Section 6.3.3.
- Generalized discussion of surface dynamics, including constant property, passive scalar, and propagating surfaces in Chapter 7. Also reorganized material from Chapter 11 on premixed and nonpremixed flame surface dynamics into this chapter.
- Autoignition waves in inhomogeneous mixtures in Sections 8.2.2.4 and 9.2.
- Wave solutions of reaction–diffusion equations in Section 9.11, showing how fundamentally different types of wave solutions are possible depending on the shape of the reaction rate curve.
- Flame propagation in flows with deterministic velocity disturbances in Section 11.2.3.3, showing two fundamentally different regimes where flame propagation speed is controlled by localized points or the entire velocity field.
- Flame position, heat release response, and sound radiation from flames disturbed by three-dimensional disturbances in Sections 11.2.2.4, 12.3.1.3, and 12.4.3.2.
- Harmonic forcing effects of turbulent flames in Section 11.5.
- Flame configuration effects on its sensitivity to disturbances in Section 12.2.
- Heat release response of nonpremixed flames to harmonic forcing in Section 12.3.1.2.

References

[1] Toong T.Y., *Combustion Dynamics: The Dynamics of Chemically Reacting Fluids*. 1983, McGraw-Hill.
[2] Crocco L. and Cheng S.I., *Theory of Combustion Instability in Liquid Propellant Rocket Motors*. AGARDograph No. 8. 1956, Butterworths Scientific Publications.
[3] Harrje D.T. and Reardon F.H., *Liquid Propellant Rocket Combustion Instability*. 1972, NASA.
[4] Putnam A.A., *Combustion-Driven Oscillations in Industry*. Fuel and Energy Science Series. 1971, American Elsevier Publishing Company, Inc.
[5] Culick F.E.C., *Unsteady Motions in Combustion Chambers for Propulsion Systems*. 2006, RTO/NATO.
[6] Lieuwen T.C. and Yang V., eds. *Combustion Instabilities in Gas Turbine Engines: Operational Experience, Fundamental Mechanisms, and Modeling*. Progress in Astronautics and Aeronautics Vol. 210. 2005, AIAA.

[7] Peters N., *Turbulent Combustion*. 1st ed. 2000, Cambridge: Cambridge University Press.
[8] Lipatnikov A., *Fundamentals of Premixed Turbulent Combustion*. 2012, CRC Press.
[9] Markstein G.H., *Nonsteady Flame Propagation*. 1964, Oxford: Pergamon Press.
[10] Bellows B.D., Bobba M.K., Seitzman J.M., and Lieuwen T., Nonlinear flame transfer function characteristics in a swirl-stabilized combustor. *Journal of Engineering for Gas Turbines and Power*, 2007, **129**(4): pp. 954–961.

Overview of the Book

This section previews the structure and content of this book and provides suggestions for how readers of different backgrounds can use it most effectively. The bulk of Chapter 1 is dedicated to reviewing the basic equations to be used in this text. Then, the remainder of the book is divided into three main sections: Chapters 2–6, 7–9, and 10–12. The first section, Chapters 2–6, discusses flow disturbances in combustors. Chapter 2 details how different types of disturbances arise and propagate in inhomogeneous, reacting combustor environments. By introducing the decomposition of flow disturbances into acoustic, vortical, and entropy disturbances, this chapter sets the stage for Chapters 3–6 which delve into the dynamics of disturbances in inhomogeneous environments in more detail. Specifically, Chapters 3 and 4 focus on the evolution of vortical disturbances in combustor environments. Chapter 3 provides a general overview of hydrodynamic stability theory and details some general features controlling the conditions under which flows are unstable. Chapter 4 then details specific canonical flow configurations that are particularly relevant to combustor environments, such as shear layers, wakes, and swirling jets. This chapter also discusses effects of flow inhomogeneity and acoustic forcing effects on flow instabilities.

Chapters 5 and 6 treat acoustic wave propagation in combustor environments. Chapter 5 provides a general introduction to acoustic wave propagation, boundary conditions, and natural acoustic modes. Chapter 6 then provides additional treatment of the effects of heat release, mean flow, and complex geometries on sound waves. This chapter also includes an extensive discussion of thermoacoustic instabilities.

The second section of the book, Chapters 7–9, incorporates reacting flow phenomena and kinetics. Chapter 7 introduces the surface dynamics of propagating surfaces (a model for premixed flames), constant-property surfaces (nonpremixed flames), and material surfaces (nondiffusive passive scalars), illustrating their similarities and distinctives. It also discusses how flames influence the bulk flow field, but does not treat internal flame processes explicitly. Rather, it focuses on the influence of a flame on pressure, entropy, vorticity, and velocity fields. Chapter 8 then treats auto- and forced ignition. Chapter 9 covers flames, first reviewing premixed and nonpremixed fundamentals, then moving on to more complex topics such as flame stretch, flame extinction, and edge flames.

The third section of the book, Chapters 10–12, treats transient (in addition to the ignition processes discussed in Chapter 8) and time-harmonic combustor phenomenon. Chapter 10 focuses on the transient, unsteady combustor issues of blowoff, flashback, and flame stabilization in general. Chapters 11 and 12 then focus on forced flame dynamics and discuss the interactions of these nominal flame dynamics with narrowband and broadband (turbulent) acoustic and vortical forcing.

The text is intended to be accessible to the new reader who has taken an introductory graduate course in fluid mechanics and had an undergraduate exposure to combustion. Expanded discussions of various topics are also included in the "Asides." While the book has been organized to be read through in the order the material is presented, there are several topical groupings of material that readers using this text for reference will find useful. Readers particularly interested in hydrodynamic stability or large-scale structures in combustor flows can start with Chapter 2 to understand, first, the more general context of disturbance propagation modes. They can then proceed to Chapters 3 and 4. Similarly, readers interested in acoustic phenomena can start with Chapter 2 and then proceed to Chapters 5 and 6. Those interested in thermoacoustics will also want to read Chapters 11 and 12 on forced flame response. Finally, those interested in flame stabilization, blowoff, and flashback phenomena can work through the material in Chapters 7, 9, and 10. In addition, readers specifically interested in topics outside of the scope of this text, such as detonations and/or two-phase combustor physics, will find several of these topical groupings, such as hydrodynamic stability, thermoacoustics, or flame stabilization, to be useful introductions to foundational issues controlling dynamics of other flows.

1 Basic Equations

1.1 Thermodynamic Relations in a Multicomponent Perfect Gas

This chapter presents the key equations for a multicomponent, chemically reacting perfect gas which will be used in this text [1]. These equations describe the thermodynamic relationships between state variables in a perfect gas, such as the interrelationship between pressure, density, and entropy. They also describe the physical laws of conservation of mass, which relates the density and velocity, the momentum equation, which relates the velocity and pressure, and the energy equation, which relates the internal and kinetic energy of the flow to work and heat transfer to the fluid.

This chapter's primary purpose is to compile in one place the key equations to be used throughout the text, and assumes that the reader has some prior familiarity with them. A number of references are provided to readers for further details and derivations of these equations. It is not necessary to follow the derivations to understand the subsequent chapters, although an understanding of the physics embodied in each equation is critical. For these reasons, discussions of various terms in these expressions are included in this chapter.

We will use the following perfect gas equations of state:

$$p = \frac{\rho \mathscr{R}_u T}{\overline{MW}}, \tag{1.1}$$

$$de = Td\mathscr{S} + p\frac{d\rho}{\rho^2} + \sum_{i=1}^{N} \frac{\mu_i}{MW_i} dY_i, \tag{1.2}$$

where $e = \sum_{i=1}^{N} Y_i e_i$, $\mathscr{S} = \sum_{i=1}^{N} Y_i \mathscr{S}_i$, and $\mu_i = \left.\frac{\partial E}{\partial n_i}\right|_{\mathscr{S},\rho,n_{j\neq i}}$ is the chemical potential of species i. All variables are defined in the nomenclature in Section 1.7. The mixture-averaged molecular weight is given by

$$\overline{MW} = \sum_{i=1}^{N} X_i \cdot MW_i, \tag{1.3}$$

where the mole fractions X_i are related to the mass fractions Y_i by

$$X_i = \frac{Y_i \overline{MW}}{MW_i}. \tag{1.4}$$

The enthalpy is written as

$$h = \sum_{i=1}^{N} Y_i h_i = \sum_{i=1}^{N} Y_i \int_{T_{ref}}^{T} c_{p,i}(T^*)dT^* + \sum_{i=1}^{N} h_{f,i}^0 Y_i = \underbrace{\int_{T_{ref}}^{T} c_p(T^*)dT^*}_{h_{sens}} + \underbrace{\sum_{i=1}^{N} h_{f,i}^0 Y_i}_{h_{chem}}.$$

(1.5)

The first and second terms on the right-hand side of this expression are the sensible enthalpy, h_{sens}, and chemical enthalpy, h_{chem}, of the system. The internal energy and enthalpy are related as follows:

$$e = h - \frac{p}{\rho} = \int_{T_{ref}}^{T} c_p(T^*)dT^* - \frac{\mathscr{R}_u T}{MW} + \sum_{i=1}^{N} h_{f,i}^0 Y_i = \sum_{i=1}^{N} Y_i e_i. \quad (1.6)$$

Additionally, the "stagnation" or "total" enthalpy and internal energy, defined as the enthalpy or internal energy of the flow when adiabatically brought to rest, are given by

$$h_T = h + |\vec{u}|^2/2; \quad e_T = e + |\vec{u}|^2/2. \quad (1.7)$$

The rest of the chapter gives an overview of the conservation of mass, momentum, and energy equations, while also providing evolution equations for other important quantities such as fluid dilatation, entropy, vorticity, and kinetic energy.

1.2 Continuity Equation

The continuity equation is given by

$$\frac{\partial \rho}{\partial t} + \nabla \cdot (\rho \vec{u}) = 0. \quad (1.8)$$

Rewriting this equation in terms of the substantial derivative,

$$\frac{D}{Dt}(\ ) = \frac{\partial (\ )}{\partial t} + \vec{u} \cdot \nabla (\ ), \quad (1.9)$$

it can be cast in the alternative form

$$\frac{1}{\rho} \frac{D\rho}{Dt} + \nabla \cdot \vec{u} = 0. \quad (1.10)$$

Physically, this equation states that if one follows a given (i.e., fixed mass) packet of fluid, the normalized time rate of change of its density is equal to the negative of the local divergence of the velocity. The local velocity divergence is, itself, directly proportional to the rate of change of volume of a fluid element; i.e., its "dilatation rate," given by the symbol $\Lambda = \nabla \cdot \vec{u}$. Moreover, the dilatation rate is also equal to the instantaneous flux of fluid out of a differential volume element of space. This can be seen by integrating the dilatation rate over a volume and utilizing Gauss's divergence theorem:

$$\iiint \Lambda dV = \iint \vec{u} \cdot \vec{n} \, dA. \tag{1.11}$$

The resulting surface integral equals the instantaneous volume flux of fluid through the control surface.

1.3 Momentum Equation

The momentum equation is given by

$$\frac{\partial \vec{u}}{\partial t} + \vec{u} \cdot \nabla \vec{u} = -\frac{\nabla p}{\rho} + \frac{\nabla \cdot \underline{\underline{\tau}}}{\rho} + \sum_{i=1}^{N} Y_i \vec{F}_i, \tag{1.12}$$

where the viscous stress tensor is given by

$$\underline{\underline{\tau}} = \mu_\lambda (\nabla \cdot \vec{u}) \underline{\underline{\delta}} + 2\mu \underline{\underline{S}} \tag{1.13}$$

and μ, μ_λ are the first and second coefficients of viscosity, $\underline{\underline{\delta}}$ is the Kronecker delta unit tensor, and $\underline{\underline{S}}$ is the symmetric "strain rate tensor," given by

$$\underline{\underline{S}} = \frac{1}{2}\left[\nabla \vec{u} + (\nabla \vec{u})^T\right]. \tag{1.14}$$

The momentum equation is an expression of Newton's second law, stating that the net acceleration of a fixed fluid element ($D\vec{u}/Dt$) equals the force per unit mass exerted on it. The force terms on the right-hand side denote surface forces due to pressure and viscous stress, and body forces. We will write three useful rearrangements of the momentum equation next.

First, we can write a vector equation for the flow vorticity $\vec{\Omega} = \nabla \times \vec{u}$ by taking the curl of Eq. (1.12):

$$\nabla \times \left(\frac{\partial \vec{u}}{\partial t}\right) + \nabla \times (\vec{u} \cdot \nabla \vec{u}) = -\nabla \times \left(\frac{\nabla p}{\rho}\right) + \nabla \times \left(\frac{\nabla \cdot \underline{\underline{\tau}}}{\rho}\right) + \nabla \times \left(\sum_{i=1}^{N} Y_i \vec{F}_i\right). \tag{1.15}$$

Expanding further,

$$\frac{\partial}{\partial t}(\nabla \times \vec{u}) + \nabla \times \left(\nabla \left(\frac{1}{2}|\vec{u}|^2\right) + (\vec{\Omega} \times \vec{u})\right)$$
$$= -\left(\frac{1}{\rho} \nabla \times \nabla p - \frac{\nabla \rho \times \nabla p}{\rho^2}\right) + \nabla \times \left(\frac{\nabla \cdot \underline{\underline{\tau}}}{\rho}\right) + \nabla \times \left(\sum_{i=1}^{N} Y_i \vec{F}_i\right). \tag{1.16}$$

Noting that $\nabla \cdot \vec{\Omega} = 0$ and expanding $\nabla \times (\vec{\Omega} \times \vec{u}) = (\nabla \cdot \vec{u} + \vec{u} \cdot \nabla)\vec{\Omega} - (\vec{\Omega} \cdot \nabla)\vec{u}$, the vorticity equation can be written as follows:

$$\frac{D\vec{\Omega}}{Dt} = -(\nabla \cdot \vec{u})\vec{\Omega} + (\vec{\Omega} \cdot \nabla)\vec{u} + \frac{\nabla \rho \times \nabla p}{\rho^2} + \nabla \times \left(\frac{\nabla \cdot \underline{\underline{\tau}}}{\rho}\right) + \nabla \times \left(\sum_{i=1}^{N} Y_i \vec{F}_i\right). \tag{1.17}$$

Section 1.6 discusses the terms in this equation.

We next write a scalar equation for the evolution of the fluid dilatation Λ by taking the divergence of Eq. (1.12):

$$\rho \frac{D\Lambda}{Dt} = -\rho \nabla \vec{u} : \nabla \vec{u} - \nabla^2 p + \frac{\nabla \rho \cdot \nabla p}{\rho} + \nabla \cdot (\nabla \cdot \underline{\underline{\tau}}) - \frac{\nabla \rho \cdot (\nabla \cdot \underline{\underline{\tau}})}{\rho} + \rho \sum_{i=1}^{N} \nabla \cdot (Y_i \vec{F}_i). \tag{1.18}$$

The double dot product appearing in the first term on the right-hand side is expressed in tensor notation as

$$\rho \nabla \vec{u} : \nabla \vec{u} = \rho \frac{\partial u_I}{\partial x_J} \frac{\partial u_J}{\partial x_I}. \tag{1.19}$$

The last equation to be developed from the momentum equation relates to the kinetic energy per unit mass, $|\vec{u}|^2/2$. This is obtained by taking the dot product of the velocity vector $\vec{u}$ with Eq. (1.12). This leads to the scalar equation

$$\rho \frac{D}{Dt}\left(\frac{\vec{u} \cdot \vec{u}}{2}\right) = -(\vec{u} \cdot \nabla)p + \vec{u} \cdot (\nabla \cdot \underline{\underline{\tau}}) + \rho \left(\sum_{i=1}^{N} Y_i \vec{u} \cdot \vec{F}_i\right). \tag{1.20}$$

We will utilize all of these momentum equation variants in the following sections and subsequent chapters.

1.4 Species Conservation Equation

The species conservation equation is given by

$$\frac{\partial (\rho Y_i)}{\partial t} + \nabla \cdot (\rho Y_i \vec{u}_i) = \dot{w}_i. \tag{1.21}$$

Note the distinction between the mass-averaged fluid velocity, $\vec{u}$, and the velocity of a given species, $\vec{u}_i$. Their difference is the diffusion velocity, $\vec{u}_i - \vec{u} = \vec{u}_{D,i}$. This species conservation equation states that the time rate of change of a fixed mass of a given element of species i equals the rate of production or consumption by chemical reaction, $\dot{w}_i$.

It is typically more useful to replace the species velocities $\vec{u}_i$ in this expression with the bulk gas velocity, $\vec{u}$, and diffusion velocity, $\vec{u}_{D,i}$. This leads to

$$\rho \frac{DY_i}{Dt} = \dot{w}_i - \nabla \cdot (\rho Y_i \vec{u}_{D,i}). \tag{1.22}$$

These diffusion velocities are described implicitly by the following equation:

$$\nabla X_i = \sum_{j=1}^{N} \left(\frac{X_i X_j}{\mathcal{D}_{ij}}\right)\left(\vec{u}_{D,j} - \vec{u}_{D,i}\right) + (Y_i - X_i)\left(\frac{\nabla p}{p}\right)$$
$$+ \sum_{j=1}^{N} \left[\left(\frac{X_i X_j}{\rho \mathcal{D}_{ij}}\right)\left(\frac{\mathcal{D}_{T,j}}{Y_j} - \frac{\mathcal{D}_{T,i}}{Y_i}\right)\right]\left(\frac{\nabla T}{T}\right) + \frac{\rho}{p}\sum_{j=1}^{N} Y_i Y_j \left(\vec{F}_i - \vec{F}_j\right). \quad (1.23)$$

The reader is referred to Refs. [1, 2] for discussion of the terms in this expression, which describe the diffusion of mass by gradients in concentration, pressure, and temperature, as well as a species-dependent body force term. Equation (1.23) leads to the familiar Fickian diffusion expression below for a binary mixture when Soret, pressure, and body force terms are neglected:

$$\vec{u}_{D,i} = -\mathcal{D}_i \nabla \ln Y_i. \quad (1.24)$$

This expression can be used to recast the species conservation equation as the following unsteady convection–reaction–diffusion equation:

$$\rho \frac{DY_i}{Dt} = \dot{w}_i + \nabla \cdot (\rho \mathcal{D}_i \nabla Y_i). \quad (1.25)$$

The species equations can also be recast into "conserved scalar" equations that are source free. The key idea is that although atoms may move from one compound to another during chemical reactions, they themselves are "conserved" and not created or destroyed. For example, one can define conserved scalars based on each particular atom, such as the hydrogen atom, H, and combine the different H-containing species equations (e.g., H_2, H_2O) to arrive at a source- and chemistry-free equation [2, 3]. This approach is particularly useful in a system with a single fuel and single oxidizer stream with the same mass diffusivities, where the mixture fraction, Z, is commonly defined as the mass fraction of material originating from the fuel jet. The governing equation for the mixture fraction is given by (see Exercise 1.8):

$$\rho \frac{\partial Z}{\partial t} + \rho \vec{u} \cdot \nabla Z - \nabla \cdot (\rho \mathcal{D} \nabla Z) = 0. \quad (1.26)$$

We will return to this equation in discussion of constant-property surfaces in Chapter 7, and in the treatment of nonpremixed flames in Chapter 9. This equation is generally valid for premixed systems as well, but is less useful as Z is constant everywhere in a perfectly premixed system.

1.5 Energy Equation

The energy equation is given by

$$\rho \frac{De_T}{Dt} = -\nabla \cdot \vec{q} - \nabla \cdot (p\vec{u}) + \nabla \cdot (\vec{u} \cdot \underline{\underline{\tau}}) + \rho \sum_{i=1}^{N} Y_i (\vec{u} + \vec{u}_{D,i}) \cdot \vec{F}_i. \quad (1.27)$$

This is a statement of the first law of thermodynamics – the time rate of change of internal and kinetic energy per unit mass of a given fluid element (the left-hand side) equals the rate of heat transfer, $\nabla \cdot \vec{q}$, minus the rate of work out (the latter three terms on the right-hand side). There are three work terms, relating to work on the fluid at the surfaces by pressure forces, $\nabla \cdot (p\vec{u})$, and viscous forces, $\nabla \cdot (\vec{u} \cdot \underline{\underline{\tau}})$, and by body forces, $\sum_{i=1}^{N}(\vec{u} + \vec{u}_{D,i}) \cdot (\rho_i \vec{F}_i)$. The heat flux vector $\vec{q}$ is given by

$$\vec{q} = -k_T \nabla T + \rho \sum_{i=1}^{N} h_i Y_i \vec{u}_{D,i} + \mathcal{R}_u T \sum_{i=1}^{N} \sum_{j=1}^{N} \left(\frac{X_j \mathcal{D}_{T,i}}{W_i \mathcal{D}_{ij}} \right) (\vec{u}_{D,i} - \vec{u}_{D,j}) + \vec{q}_{Rad}. \quad (1.28)$$

See Ref. [1] for a discussion of these terms, which describe heat transfer by conduction, mass diffusion, multicomponent mass diffusion due to concentration gradients, and radiation, respectively.

Alternatively, the energy equation can be written in terms of total enthalpy, $h_T = e_T + (p/\rho)$, by moving the pressure work term from the right-hand side of Eq. (1.27) to the left:

$$\rho \frac{Dh_T}{Dt} = -\nabla \cdot \vec{q} + \frac{\partial p}{\partial t} + \nabla \cdot (\vec{u} \cdot \underline{\underline{\tau}}) + \rho \sum_{i=1}^{N} Y_i (\vec{u} + \vec{u}_{D,i}) \cdot \vec{F}_i. \quad (1.29)$$

Equations for the internal energy and enthalpy can be obtained by subtracting the kinetic energy equation, Eq. (1.20), leading to

$$\rho \frac{De}{Dt} = -\nabla \cdot \vec{q} - p(\nabla \cdot \vec{u}) + \underline{\underline{\tau}} : (\nabla \vec{u}) + \rho \sum_{i=1}^{N} Y_i \vec{u}_{D,i} \cdot \vec{F}_i, \quad (1.30)$$

$$\rho \frac{Dh}{Dt} = -\nabla \cdot \vec{q} + \frac{Dp}{Dt} + \underline{\underline{\tau}} : (\nabla \vec{u}) + \rho \sum_{i=1}^{N} Y_i \vec{u}_{D,i} \cdot \vec{F}_i. \quad (1.31)$$

It is useful to explicitly bring out the chemical component of the enthalpy. This is done by first writing an evolution equation for the chemical enthalpy of the system using the species conservation equation, Eq. (1.22), as follows:

$$\frac{Dh_{chem}}{Dt} = \sum_{i=1}^{N} h_{f,i}^0 \frac{DY_i}{Dt} = \frac{1}{\rho} \sum_{i=1}^{N} h_{f,i}^0 \dot{w}_i - \frac{1}{\rho} \nabla \cdot \left(\rho \sum_{i=1}^{N} h_{f,i}^0 Y_i \vec{u}_{D,i} \right). \quad (1.32)$$

This can be further simplified to yield

$$\rho \frac{Dh_{chem}}{Dt} = -\dot{q} - \nabla \cdot \left(\rho \sum_{i=1}^{N} h_{f,i}^0 Y_i \vec{u}_{D,i} \right), \quad (1.33)$$

where the chemical source term $\dot{q}$ is the volumetric heat release rate due to combustion, given by

$$\dot{q} = -\sum_{i=1}^{N} h_{f,i}^0 \dot{w}_i. \quad (1.34)$$

1.5 Energy Equation

This chemical enthalpy equation, Eq. (1.33), is then subtracted from Eq. (1.31) to yield the following equation for sensible enthalpy:

$$\rho \frac{Dh_{sens}}{Dt} = \dot{q} - \nabla \cdot \vec{q} + \frac{Dp}{Dt} + \underline{\underline{\tau}} : (\nabla \vec{u}) + \nabla \cdot \left(\rho \sum_{i=1}^{N} h_{f,i}^{0} Y_i \vec{u}_{D,i} \right) + \rho \sum_{i=1}^{N} Y_i \vec{u}_{D,i} \cdot \vec{F}_i.$$

(1.35)

The energy equation can also be written in terms of entropy, $\mathscr{s}$. From Eq. (1.2),

$$T \frac{D\mathscr{s}}{Dt} = \frac{De}{Dt} - \frac{p}{\rho^2} \frac{D\rho}{Dt} - \sum_{i=1}^{N} \frac{\mu_i}{MW_i} \frac{DY_i}{Dt}. \tag{1.36}$$

Substituting Eqs. (1.30) and (1.25) into the above expression yields

$$\rho T \frac{D\mathscr{s}}{Dt} = -\nabla \cdot \vec{q} + \underline{\underline{\tau}} : (\nabla \vec{u}) - \rho \sum_{i=1}^{N} \mathscr{D}_i \nabla Y_i \cdot \vec{F}_i - \sum_{i=1}^{N} \frac{\mu_i}{MW_i} (\dot{w}_i + \nabla \cdot (\rho \mathscr{D}_i \nabla Y_i)).$$

(1.37)

This equation shows how the entropy of a given mass of fluid is altered by mass, momentum, and energy diffusion, as well as body forces and chemical reaction.

Finally, the energy equation is written in terms of temperature and pressure. This can be done by starting from the equation for sensible enthalpy and, noting that $dh_{sens} = c_p dT$, rewriting Eq. (1.35) as

$$\rho \frac{Dh_{sens}}{Dt} - \frac{Dp}{Dt} = \dot{q} - \nabla \cdot \vec{q} + \underline{\underline{\tau}} : (\nabla \vec{u}) + \nabla \cdot \left(\rho \sum_{i=1}^{N} h_{f,i}^{0} Y_i \vec{u}_{D,i} \right) + \rho \sum_{i=1}^{N} Y_i \vec{u}_{D,i} \cdot \vec{F}_i.$$

(1.38)

The left-hand side of Eq. (1.38) can be expressed as the substantial derivative of either temperature or pressure. For example, writing changes in temperature as

$$dT = \left(\frac{\partial T}{\partial p} \right)_{\rho, Y_i} dp + \left(\frac{\partial T}{\partial \rho} \right)_{p, Y_i} d\rho + \sum_{i=1}^{N} \left(\frac{\partial T}{\partial Y_i} \right)_{p, \rho, Y_{j \neq i}} dY_i, \tag{1.39}$$

the substantial derivative of temperature may be expressed as

$$\frac{DT}{Dt} = \frac{T}{p} \frac{Dp}{Dt} - \frac{T}{\rho} \frac{D\rho}{Dt} - T \sum_{i=1}^{N} \frac{\overline{MW}}{MW_i} \frac{DY_i}{Dt}, \tag{1.40}$$

so that

$$\rho \frac{Dh_{sens}}{Dt} - \frac{Dp}{Dt} = \rho c_p \frac{DT}{Dt} - \frac{Dp}{Dt}$$

$$= \rho c_p T \left(\frac{1}{\gamma p} \frac{Dp}{Dt} + \nabla \cdot \vec{u} - \frac{\overline{MW}}{\rho} \sum_{i=1}^{N} \frac{\dot{w}_i}{MW_i} - \frac{\overline{MW}}{\rho} \sum_{i=1}^{N} \frac{\nabla \cdot (\rho \mathscr{D}_i \nabla Y_i)}{MW_i} \right)$$

$$= \rho c_p T \left(\frac{1}{\gamma T} \frac{DT}{Dt} + \frac{\gamma - 1}{\gamma} \left(\nabla \cdot \vec{u} - \frac{\overline{MW}}{\rho} \sum_{i=1}^{N} \frac{\dot{w}_i}{MW_i} - \frac{\overline{MW}}{\rho} \sum_{i=1}^{N} \frac{\nabla \cdot (\rho \mathscr{D}_i \nabla Y_i)}{MW_i} \right) \right).$$

(1.41)

Hence, the energy equation may be written in terms of the temperature as

$$\frac{1}{T}\frac{DT}{Dt}+(\gamma-1)\nabla\cdot\vec{u}=\left\{\begin{array}{l}\dfrac{\dot{q}}{\rho c_v T}+(\gamma-1)\dfrac{\dot{n}}{n}+(\gamma-1)\dfrac{1}{n}\sum_{i=1}^{N}\dfrac{\nabla\cdot(\rho\mathscr{D}_i\nabla Y_i)}{MW_i}\\ +\dfrac{1}{\rho c_v T}\left(-\nabla\cdot\vec{q}+\underline{\underline{\tau}}:(\nabla\vec{u})+\nabla\cdot\left(\rho\sum_{i=1}^{N}h_{f,i}^0 Y_i\vec{u}_{D,i}\right)+\rho\sum_{i=1}^{N}Y_i\vec{u}_{D,i}\cdot\vec{F}_i\right)\end{array}\right.,$$

(1.42)

or for pressure as

$$\frac{1}{\gamma p}\frac{Dp}{Dt}+\nabla\cdot\vec{u}=\left\{\begin{array}{l}\dfrac{\dot{q}}{\rho c_p T}+\dfrac{\dot{n}}{n}+\dfrac{1}{n}\sum_{i=1}^{N}\dfrac{\nabla\cdot(\rho\mathscr{D}_i\nabla Y_i)}{MW_i}\\ +\dfrac{1}{\rho c_p T}\left(-\nabla\cdot\vec{q}+\underline{\underline{\tau}}:(\nabla\vec{u})+\nabla\cdot\left(\rho\sum_{i=1}^{N}h_{f,i}^0 Y_i\vec{u}_{D,i}\right)+\rho\sum_{i=1}^{N}Y_i\vec{u}_{D,i}\cdot\vec{F}_i\right)\end{array}\right.,$$

(1.43)

where $\dot{n}$ describes the time rate of change of the number of moles of the gas and is defined as

$$\dot{n}=\sum_i\frac{\dot{w}_i}{MW_i},\quad n=\frac{\rho}{\overline{MW}}.$$

(1.44)

Note that

$$-\frac{1}{\overline{MW}}\frac{D}{Dt}\overline{MW}=\frac{\dot{n}}{n},$$

(1.45)

so that $\dot{n}=0$ if the average molecular weight is constant.

Neglecting molecular transport and body forces, Eq. (1.43) is

$$\frac{1}{\gamma p}\frac{Dp}{Dt}+\nabla\cdot\vec{u}=\frac{\dot{q}}{\rho c_p T}+\frac{\dot{n}}{n}.$$

(1.46)

The two terms on the right-hand side of this expression are source terms that describe volume production due to chemical reactions – the first because of unsteady heat release and the second due to changes in the number of moles of the gas. To illustrate the relative magnitudes of these two source terms, $\dot{q}$ and $\dot{n}$, consider the ratio of product to reactant gas volume, assuming constant pressure combustion:

$$\frac{\text{Mole Production}+\text{Heat Release}}{\text{Heat Release Alone}}=\frac{\left(\dfrac{T^b}{T^u}\cdot\dfrac{\overline{MW}^u}{\overline{MW}^b}-1\right)}{\left(\dfrac{T^b}{T^u}-1\right)},$$

(1.47)

where the superscripts u and b denote the reactant (unburned) and product (burned) values. When fuels are burned in air, the $\dot{n}$ term is small relative to the $\dot{q}$ term, since the reactive species are strongly diluted in inert nitrogen. However, there are applications, most notably oxycombustion, where the molecular weight change between reactants and products is far more prominent. An equilibrium calculation of stoichiometric methane–air and methane–oxygen combustion shows that this ratio equals 1.01 and 1.27 for the air and oxygen systems at 1 bar, respectively. Consequently, for air-breathing systems the molar production term can be neglected, as it will be in the remainder of this text. In this case, Eq. (1.46) may be written as

$$\frac{Dp}{Dt} = -\gamma p (\nabla \cdot \vec{u}) + (\gamma - 1)\dot{q}. \tag{1.48}$$

1.6 Aside: Discussion of the Vorticity and Circulation Equations

This section discusses the various terms in the vorticity equation, reproduced below:

$$\frac{D\vec{\Omega}}{Dt} = (\vec{\Omega} \cdot \nabla)\vec{u} - \vec{\Omega}(\nabla \cdot \vec{u}) - \frac{\nabla p \times \nabla \rho}{\rho^2} + \nabla \times \left(\frac{\nabla \cdot \tau}{\rho}\right) + \sum_{i=1}^{N} \nabla \times \left(Y_i \vec{F}_i\right). \tag{1.49}$$

The left-hand side of this equation physically describes the time rate of change of vorticity of a fixed fluid element. The right-hand side describes vorticity source or sink terms. The first term, $(\vec{\Omega} \cdot \nabla)\vec{u}$, is the vortex stretching and bending term. This term is intrinsically three-dimensional and, therefore, is identically zero in a two-dimensional flow. There are two processes described in this term, as illustrated in Figure 1.1. The first process is the increase or decrease of vorticity by vortex stretching or contraction, respectively. For example, consider a vortex tube oriented in the axial direction, Ω_x. If this axial flow is accelerating, $\partial u_x / \partial x > 0$, then the tube is stretched, causing an increase in axial vorticity, i.e., $D\Omega_x / Dt = \Omega_x \partial u_x / \partial x > 0$.

The second phenomenon is the bending of a vortex tube originally inclined in another dimension by the local flow. For example, a vortex tube that is initially

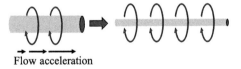

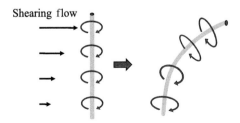

Figure 1.1 Illustration of stretching (top) and bending (bottom) effects on a vortex tube.

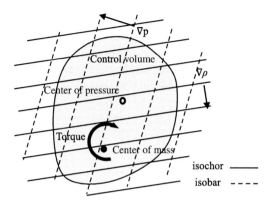

Figure 1.2 Illustration of processes by which a misaligned pressure and density gradient lead to torque on a fluid element, and thus vorticity production.

oriented vertically, i.e., Ω_y, is rotated toward the x-axis in an axially shearing flow $\partial u_x/\partial y > 0$. This rotation of the vortex tube causes it to be partially oriented in the axial direction, inducing a Ω_x component, i.e., $D\Omega_x/Dt = \Omega_y \partial u_x/\partial y$.

Returning to the full vorticity equation, the second term, $(\nabla \cdot \vec{u})\vec{\Omega}$, describes flow dilatation impacts on vorticity. It is only nonzero in compressible flows. Positive dilatation, i.e., expansion in cross-sectional area of a vortex tube, leads to a reduction in vorticity. This effect is analogous to a spinning skater who extends their arms outward, leading to a slowing in angular velocity, or vice versa. This term has important damping influences on vorticity as it propagates through the flame.

The third term, $\frac{\nabla p \times \nabla \rho}{\rho^2}$, describes vorticity production via the baroclinic mechanism, which occurs when the density and pressure gradients are misaligned. This term is identically zero in fluids where $\rho = \rho(p)$, e.g., in isentropic flows of a perfect gas. The torque induced on the flow by a misaligned pressure and density gradient can be understood from Figure 1.2. Consider the shaded control volume in a field with a spatially varying pressure and density. The pressure gradient induces a net force on the control volume, which acts through the center of pressure, shown in the figure. Because of the density gradient, the center of mass is displaced from the center of pressure. If the pressure and density gradients are not aligned (i.e., $\nabla \rho \times \nabla p \neq 0$), then the force acting at the center of pressure induces a torque on the fluid element about this center of mass, creating vorticity.

Pressure only enters the vorticity equation through this baroclinic term. This reflects the fact that only when the pressure and density gradients are misaligned can the normal pressure forces exert a torque on the flow.

It is also important to note that vorticity can only be generated at no-slip boundaries, by baroclinic torque, or through body forces. All of the other terms in Eq. (1.49) describe the amplification, stretching, bending, or diffusion of vorticity that already exists in the flow. Thus, large-scale vortical structures which play such a key role in Chapters 3 and 4 do not arise "from nothing" in flows without vorticity sources, but from the complex reorganization of vorticity that enters the flow from boundary layers (e.g., in jets, wakes, etc.). An enormous range of possible flow structures can arise from this vorticity, depending on the characteristics of the specific flow in which it arises.

More insights into sources of new vorticity, relative to reorganization/amplification of existing vorticity, can be obtained by looking at the circulation. The circulation, Γ, is defined as the surface integral of the vorticity or, equivalently, as the line integral of the tangential velocity along the bounding surface, i.e.,

$$\Gamma = \int_A \vec{\Omega} \cdot d\vec{A} = \oint_{S_1} \vec{u} \cdot d\vec{l} \ . \tag{1.50}$$

An equation for the circulation is

$$\frac{D\Gamma}{Dt} = -\int_A \frac{1}{\rho^2} (\nabla p \times \nabla \rho) \cdot d\vec{A} + \int_A \nabla \times \left(\frac{\nabla \cdot \underline{\underline{\tau}}}{\rho} \right) \cdot d\vec{A} + \sum_{i=1}^N \int_A \nabla \times (Y_i \vec{F}_i) \cdot d\vec{A} \ . \tag{1.51}$$

Comparing this equation and the vorticity equation shows that, while stretching or dilatation act as sources/sinks of vorticity, they do not appear as sources of circulation. Baroclinic torque, diffusion, and nonconservative body forces are circulation sources/sinks.

In addition, the manner in which vorticity induces fluid motion can be inferred from Eq. (1.50). To illustrate, Figure 1.3 illustrates a two-dimensional region where vorticity is confined to a region bounded by S_1. Outside of S_1, the flow is irrotational. Equation (1.50) can then be manipulated to yield the relation

$$\int_1 \Omega_z dA = \oint_2 \vec{u} \cdot d\vec{l} \ . \tag{1.52}$$

In cases where the outer surface is much farther from the vorticity-containing region, the value of u_θ along the circular bounding surface 2 will become progressively more uniform, so that the line integral can be evaluated to yield

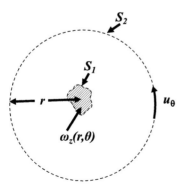

Figure 1.3 Illustration of velocity induced by a patch of vorticity.

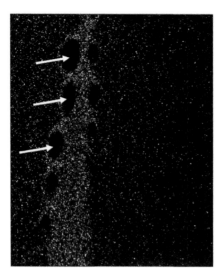

Figure 1.4 Mie scattering image of a nonreacting, $J = 15$ jet in cross flow. Dark regions indicated by arrows are regions of intense vorticity where seed particles are centrifuged out. Image courtesy of Vedanth Nair.

$$u_\theta = \frac{1}{2\pi r} \int_1 \Omega_z dA. \tag{1.53}$$

This equation shows that the induced velocity, u_θ, is proportional to the integral of the vorticity. This induced motion has important practical consequences on the stability of flows with multiple vorticity-containing regions, such as in wake flows where two vortex sheets are convected from the bluff body as the boundary layer separates. These vortex sheets mutually induce motion on each other, a fundamental driver of the von Kármán vortex street behind bluff bodies, as further discussed in Section 3.2.

This relation can also be useful for inferring average vorticity values from velocity measurements around a bounding surface. For example, in particle image velocimetry (PIV) measurements, there are often no seed particles in the vortex cores due to the intense azimuthal velocities, such as shown in Figure 1.4. With no seed particles present, the local velocity, velocity gradient, or vorticity cannot be measured in the core region where the arrows point. However, by integrating the tangential velocity along a bounding surface where sufficient seed is present, the average vorticity can be determined from Eq. (1.52).

1.7 Nomenclature

This section details the nomenclature used in the text. Maintaining a consistent nomenclature across the whole text is challenging given the different uses of common symbols across the combustion, acoustics, and hydrodynamic stability communities. This uniformity of definition admittedly also makes the nomenclature complicated, and the reader is encouraged to spend a few minutes understanding a few basic items.

All gas velocities are given by u. Subscripts such as x, y, and z, or r and θ, are used to denote the specific scalar component of the vector. Superscripts u and b are used to denote the value just up- or downstream of the premixed flame. Numbered subscripts are used to indicate perturbation quantities, using expansions detailed in Eq. (2.2). For example, $u^u_{x,1}$ denotes the linear perturbation expansion of the axial velocity component just upstream of the flame.

For flame-based quantities, subscripts c and d are used to denote consumption- and displacement-based velocities, and the superscript 0 to denote unstretched values. Thus, s^b_d denotes the laminar displacement speed of the flame with respect to the burned gases. Other nomenclature questions can be addressed by referencing the list that follows.

1.7.1 Latin Alphabet

a	Radius
A	Cross-sectional area
c, c_0, c_{ph}	Speed of sound; phase speed
c_p, c_v	Specific heat
$\tilde{e}$	Reaction progress variable
D	Diameter
Dr	Dispersion relationship
Da	Damköhler number
$\mathscr{D}_{ij}, \mathscr{D}_i, \mathscr{D}_{T,i}$	Diffusion coefficients
e, e_i	Internal energy per mass; internal energy of species i per mass
$\vec{e}_x, \vec{e}_y, \vec{e}_z$	Unit vector
E	Energy
$\vec{F}_i$	Body force per unit mass acting on species i
f, f_0, f_{BVK}, f_{KH}	Frequency
F	Arbitrary function; transfer function
g	Spatial gradient of flame position or velocity (g_u)
$\mathscr{G}$	Green's function
G	Premixed flame level set function
$h, h_i, h^o_{f,i}, h_{chem}$	Enthalpy per mass
h_R	Heat release per unit mass of reactant consumed
H	Geometric scale
He	Helmholtz number, ratio of geometric length scale to acoustic wavelength
$\vec{I}$	Acoustic intensity
J_m	Bessel function of the first kind
k_c	Normalized disturbance convection velocity, $u_c/u_{x,0}$
k_T	Thermal conductivity

Ka	Karlovitz number
L	Length scale
L_F	Flame length
L_{11}	Integral flow length scale
L_η	Kolmogorov flow length scale
Le	Lewis number
$\dot{m}$	Mass flow rate
$\dot{m}'', \dot{m}_F''$	Mass flux; burning mass flux of flame
M	Mach number
m	Mode number
Ma	Markstein number
$MW_i, \overline{MW}$	Molecular weight
n	Unsteady heat release gain factor
$\vec{n}$	Unit normal vector
$\dot{n}$	Molar production rate
n	Number of moles, normal direction, axial acoustic mode number
p	Pressure
Pr	Prandtl number
$\vec{q}, \vec{q}_{Rad}$	Heat flux
$\dot{q}$	Chemical heat release rate per unit volume or flamelet surface area
Q	Heat release per unit mass of fuel reacted
$\dot{Q}$	Spatially integrated heat release rate
r	Radial location
r_{corr}	Correlation function or correlation coefficient
R	Acoustic wave reflection coefficient
$\mathcal{R}_{corr}$	Coherence
$Re, Re_D, Re_x, Re_\delta$	Reynolds number
$Real(X)$	Real part of X
$\mathcal{R}, \mathcal{R}_u$	Gas constant, universal gas constant
s, s_i	Entropy per unit mass; entropy of species i per unit mass
s_c, s_d, s^u, s^b	Flame speed, speed of isosurface with respect to flow (s_d)
$\underline{\underline{S}}$	Strain rate tensor, see Eq. (1.14)
Sa	Sankaran parameter, $s_d \lvert \nabla \tau_{ign} \rvert$
SR	Spin ratio, defined in Eq. (5.120)
St	Strouhal number, fL/u
St_2	$St/\cos^2(\theta)$, used in Chapter 12
St_δ	Strouhal number based on shear layer thickness, $f\delta/u_x$
St_F	Strouhal number based on internal flame processes, $f\delta_F/s^{u,0}$
St_W	Strouhal number based on burner width, fW_{II}/u_x, used for nonpremixed flame discussions in Chapter 11
S_v, S_m	Swirl number
$\vec{t}$	Unit tangential vector
T	Temperature

$\mathcal{T}$	Period of oscillations $(1/f)$
T	Acoustic wave transmission coefficient
$u_x, u_y, u_z,$ $u_r, u_\theta, u_n{}^u, u_n{}^b$	Scalar gas velocity component
$\vec{u}, \vec{u}_\xi, \vec{u}_i, \vec{u}_{D,i},$ $\vec{u}^u, \vec{u}^b$	Fluid velocity vector
$u_c, u_{c,F}$	Phase speed of velocity; flame wrinkle disturbance
V	Volume
$\vec{v}_F$	Velocity of flame surface
$\dot{w}_i$	Mass-based production rate of species i by chemical reactions
W	Duct height
$W(y)$	Flame area distribution weighting factor used in Chapter 12
X_i	Mole fraction of species i
Y_i	Mass fraction of species i
Y_m	Bessel function of the second kind
Z	Acoustic impedance
$\mathcal{Z}$	Mixture fraction
Ze	Zeldovich number

1.7.2 Greek Alphabet

α	Thermal diffusivity, $k_T/\rho c_p$
β	Parameter
χ	Flow velocity ratio (Chapters 3 and 4) or scalar dissipation rate (Chapter 9)
δ	Shear layer thickness or Dirac delta function
$\overline{\overline{\delta}}$	Kronecker delta unit tensor
δ_{99}	Boundary layer thickness
δ_F	Flame thickness
δ_k	Critical ignition kernel size
δ_M	Markstein length
δ_T	Turbulent flame brush thickness
$\Delta_x, \Delta_y, \Delta_z, \Delta_t$	Spatial/temporal separation used for correlation calculations
ε	Small parameter used for perturbation expansions
$\mathcal{E}$	Turbulent dissipation rate
ϕ	Equivalence ratio
Φ	Source terms
γ	Specific heats ratio
η	Modal amplitude
φ_{Ox}	Stoichiometric mass ratio of oxidizer to fuel
κ	Flame stretch rate
$\lambda_\Lambda, \lambda_c, \lambda_f$	Acoustic wave length; convective wave length; flame sheet wrinkling wavelength

Λ	Dilatation rate, $\nabla \cdot \vec{u}$
μ	First coefficient of viscosity
μ_i	Chemical potential of species i
μ_λ	Second coefficient of viscosity
ν_ν	Dynamic viscosity, μ/ρ
θ	Angle or phase
$\theta_i, \theta_r, \theta_{nom}$	Incident, refraction, and nominal angle
θ_{pq}	Phase difference between pressure and heat release
ρ	Density
σ_T	Temperature ratio
σ_ρ	Density ratio
τ	Time scale
τ_t	Separation time
τ_{ign}	Autoignition time
$\underline{\underline{\tau}}$	Shear stress
ω	Angular frequency
$\dot{\omega}$	Molar-based production rate of species i by chemical reactions
$\vec{\Omega}$	Vorticity
ξ	Interface position, usually referring to flame
ψ	Stream function; spatial mode shape
ζ	Decay, damping coefficient

1.7.3 Subscripts

$0, 1, 2, \ldots$	Perturbations
A	Reference position or value
B	Reference position or value
c	Flame speed: consumption-based definition, such as, s_c^u; flow velocity: disturbance convection velocity, u_c
$curv$	Curvature-induced flame stretch
D	Displacement-based velocity definition, such as displacement speed
η	Associated with Kolmogorov scales
Λ	Acoustic mode
i	Imaginary
$lean$	Lean mixture
MR	Most reacting
n	Normal component
NL	Nonlinear part
p	Potential mode
θ	Azimuthal vector component (polar coordinates)
r	Real part, or radial vector component (for polar coordinates)
s	Entropy mode

S	Hydrodynamic strain component
Sens	Sensible
t	Tangential component
T	Stagnation
ξ	At the flame front
Ω	Vortical mode
x, y, z	Velocity coordinates in rectangular coordinate system

1.7.4 Superscripts

0	Unstretched value of flame quantity
b	Value of quantity at flame sheet on burned side
u	Value of quantity at flame sheet on unburned side

1.7.5 Other Symbols

( )′	Fluctuating quantity
( )	Time average of quantity
[ ]	Jump across the flame/interface
(^)	Fourier transform of variable

Exercises

(*Solutions are available at* cambridge.org/9781108841313)

1.1. A flow is isentropic if the entropy of a given mass of fluid does not change, i.e., $Ds/Dt = 0$. Work out the conditions that are required for this to be true.

1.2. Can a flow be isentropic, i.e., $Ds/Dt = 0$, and still have entropy fluctuations?

1.3. Work out the conditions for the stagnation enthalpy of a given mass of fluid to remain constant, i.e., $Dh_T/Dt = 0$.

1.4. The preceding problems worked out the conditions for the stagnation enthalpy and entropy to remain constant *along* a pathline. Crocco's equation [4, 5] is a useful expression providing insight into the relationship between flow and thermodynamic properties *across* streamlines (which is the same as a pathline only in a steady flow). Moreover, it provides an enlightening reformulation of the momentum equation into a form illustrating the relationship between vorticity, stagnation enthalpy, and entropy. Starting with Eq. (1.12), neglect body forces and molecular transport, and substitute in the state relation

$$dh = Tds + \frac{dp}{\rho} + \sum_{i=1}^{N} \frac{\mu_i}{MW_i} dY_i \qquad (1.60)$$

to prove the following relation:

$$\frac{\partial \vec{u}}{\partial t} + \nabla h_T - T\nabla\mathscr{s} - \vec{u} \times \vec{\Omega} - \sum_{i=1}^{} \frac{\mu_i}{MW_i} \nabla Y_i = 0. \qquad (1.61)$$

1.5. *Implications of Crocco's equation.* Crocco's equation provides important insights into the relationships between flow and thermodynamic properties between adjacent streamlines. Assuming a two-dimensional steady flow with no chemical reaction and molecular transport, and that h_T is constant along a streamline, show that Crocco's equation may be written in the following scalar form along a streamline (where n is the coordinate locally perpendicular to the streamline):

$$\frac{dh_T}{dn} = |\vec{u}||\vec{\Omega}| + T\frac{d\mathscr{s}}{dn}. \qquad (1.67)$$

1.6. *Implications of Crocco's equation.* Following on the development from the preceding problem, assume that h_T is constant everywhere. Discuss the physical implications of Crocco's equation.

1.7. *Quasi-one-dimensional forms.* Derive approximate mass, momentum, and energy equations for a quasi-one-dimensional flow in a channel with slowly varying cross-sectional area $A(x)$. Neglect molecular transport, body forces, and chemical reactions. You can do this by developing differential expressions for the conservation laws, but note that although the normal velocity is zero along the side walls, the pressure is not.

1.8. *Mixture fraction equation, Z.* The mixture fraction, Z, is defined as the mass fraction of material originating from the fuel jet, $Z \equiv Y_{Fuel} + \left(\frac{1}{\varphi_{Ox}+1}\right) Y_{Prod}$. Derive Eq. (1.26) assuming that only fuel, oxidizer, and product species exist and that all species have equal diffusivities.

References

[1] Williams F.A., *Combustion Theory: The Fundamental Theory of Chemically Reacting Flow Systems*. 1994, Perseus Books.
[2] Poinsot T. and Veynante D., *Theoretical and Numerical Combustion*. 2nd ed. 2005, R.T. Edwards, Inc.
[3] Peters N., *Turbulent Combustion*. 1st ed. 2000, Cambridge: Cambridge University Press.
[4] Wu C.S. and Hayes W.D., Crocco's vorticity law in a non-uniform material. *Quarterly Journal of Applied Mathematics*, 1958, **16**: pp. 81–82.
[5] Currie I.G., *Fundamental Mechanics of Fluids*. 3rd ed. Mechanical Engineering. 2002, Marcel Dekker.

2 Decomposition and Evolution of Disturbances

A key focus of this text is to relate the manner in which fluctuations in flow or thermodynamic variables propagate and interact in combustion systems. In this chapter, we demonstrate that combustor disturbances can be decomposed into three canonical types of fluctuations – acoustic, entropy, and vorticity disturbances. This decomposition is highly illustrative in understanding the spatial/temporal dynamics of combustor disturbances [1]. For example, the velocity field can be decomposed into acoustic fluctuations, which propagate at the speed of sound with respect to the flow, and vorticity fluctuations, which are advected by the flow. This decomposition is important because, as shown in Chapters 11 and 12, two velocity disturbances of the same magnitude can lead to very different influences on the flame, depending on their phase speeds and space–time correlation. Section 2.9 further emphasizes how this decomposition provides insight into behavior measured in a harmonically oscillating flow field.

This chapter is organized in the following manner. Section 2.1 introduces the basic approach for analyzing disturbances, and illustrates the formal process of perturbation expansions used throughout the text. Section 2.2 then considers small-amplitude disturbance propagation in homogeneous flows. This limit is helpful for understanding key aspects of the problem, as the disturbance modes do not interact and are not excited. Section 2.3 closely follows this material by showing how these disturbance modes are excited and how they interact with each other. Section 2.4 then considers the energy density and energy flux associated with these disturbances.

Section 2.5 moves from specific analysis of gas dynamic disturbance to a more general overview of linear and nonlinear stability concepts. This and subsequent sections pull together a number of results associated with linear and nonlinear systems, the effects of forcing on self-excited systems, and effects of interactions between multiple self-excited oscillators. These results have been selected from more general nonlinear dynamics treatments for their relevance to important unsteady combustor phenomena, which will be referred to throughout the text.

2.1 Descriptions of Flow Perturbations

In this section we introduce the concepts used to quantitatively analyze fluctuations. In order to illustrate these definitions, consider the following example problem, motivated by the isentropic pressure–density relationship, describing the time variation of the pressure, $p(t)$:

$$p(t) = \mathcal{B}_1(1 + \varepsilon \sin(\omega t))^\gamma. \tag{2.1}$$

We will denote the "base" or "nominal" flow as the value of the quantity in the absence of perturbations. This quantity is given the subscript 0, so in this case $p_0 = \mathcal{B}_1$. It is often useful to expand the quantity about this base state in a Taylor series:

$$p = p_0 + p_1 + p_2 + \cdots, \tag{2.2}$$

where, by assumption, $p_1 \ll p_0, p_2 \ll p_1$, and so forth.[1] Thus, for this problem we can determine the form of these various orders of approximations by expanding Eq. (2.1) to yield

$$\frac{p_1(t)}{p_0} = \varepsilon \gamma \sin(\omega t), \tag{2.3}$$

$$\frac{p_2(t)}{p_0} = \frac{\varepsilon^2 \gamma(\gamma-1)\sin^2(\omega t)}{2} = \varepsilon^2 \frac{\gamma(\gamma-1)}{4}(1 - \cos(2\omega t)), \tag{2.4}$$

$$\frac{p_3(t)}{p_0} = \frac{\varepsilon^3 \gamma(\gamma-1)(\gamma-2)\sin^3(\omega t)}{6} = \varepsilon^3 \frac{\gamma(\gamma-1)(\gamma-2)}{24}(3\sin\omega t - \sin 3\omega t). \tag{2.5}$$

Thus, the leading-order term, p_1, is linear in disturbance amplitude, ε, and oscillates with the same frequency as the disturbance. Approximating $p(t) - p_0$ by this first-order term is referred to as the "linear approximation." The second-order term is quadratic in disturbance amplitude, ε^2, and has two different terms, one that does not vary in time and the other that oscillates at 2ω. The former term describes the change in time-averaged pressure by the disturbance, and the latter describes the introduction of higher harmonics of the disturbance frequency by nonlinearities. The third-order term is cubic in disturbance amplitude, ε^3, and has one term that oscillates at ω and the other at 3ω. The former term describes the alteration of the amplitude of oscillations at the disturbance frequency by nonlinear processes, and the latter describes the introduction of an additional higher harmonic frequency by nonlinearities.

Several patterns are evident: terms of even order have time-invariant terms and higher harmonics with frequencies that are even multiples of the disturbance frequency. Odd-order terms alter the amplitude of oscillations at the disturbance frequency, and also introduce higher harmonics with frequencies that are odd multiples of the disturbance frequency. As shown in Exercise 2.1, if the disturbance

---

[1] A note on conventions: perturbation expansions of this form more typically use $p = p_0 + \varepsilon p_1 + \varepsilon^2 p_2 + \cdots$, where p_1 and p_2 are $O(1)$ quantities. While this expansion is very natural for the problems described in this section, we have used the form in Eq. (2.2) here as it is more convenient for other analyses and we prefer to use a single, consistent expansion approach throughout the text.

consists of multiple frequencies, ω_1 and ω_2, nonlinearities also generate sum and difference frequencies, such as $\omega_1 - \omega_2$, $\omega_1 + \omega_2$, $2\omega_1 - \omega_2$, and so on.

We will denote the time average of a quantity with an overbar, e.g., $\bar{p}$. Fluctuations of quantities about the time average are denoted with $(\ )'$:

$$p'(t) = p(t) - \bar{p}. \tag{2.6}$$

In many problems, flow fluctuations can be further decomposed into random and deterministic or quasi-deterministic fluctuations. For example, coherent flow structures often consist of quasi-deterministic flow features, superposed upon a background of fine-scale turbulence. Similarly, pressure fluctuations in thermoacoustically unstable combustors are superpositions of broadband turbulent flow and combustion noise, and near perfect tones at one or more frequencies. A "triple decomposition" can be used to differentiate these different types of fluctuations, as described in Section 2.8.

Note that the time-averaged value is not equal to its base or nominal value; i.e., $\bar{p} \neq p_0$. Rather, $\bar{p}$ and p_0 are only equivalent to first order in perturbation amplitude. For instance, in this example problem,

$$\frac{\bar{p}}{p_0} = 1 + \frac{\varepsilon^2 \gamma(\gamma - 1)}{4} + \frac{\varepsilon^4 \gamma(\gamma - 1)(\gamma - 2)(\gamma - 3)}{64} + O(\varepsilon^6), \tag{2.7}$$

$$\frac{p'(t)}{p_0} = \left(\varepsilon\gamma + \frac{\varepsilon^3 \gamma(\gamma - 1)(\gamma - 2)}{8}\right) \sin(\omega t) - \frac{\varepsilon^2 \gamma(\gamma - 1)}{4} \cos(2\omega t) \\ - \frac{\varepsilon^3 \gamma(\gamma - 1)(\gamma - 2)}{24} \sin(3\omega t). \tag{2.8}$$

In general, the base value, $(\ )_0$, of a quantity is not experimentally accessible, but its time average is (similarly, the $(\ )_1$, $(\ )_2$, ... perturbation terms are not experimentally accessible in general). For example, in a turbulent flow, the base flow consists of the laminar flow that would exist at those same conditions. However, since the flow is unstable, it is impossible to observe what this base flow would look like. Nonetheless, expanding about a base flow is useful for approximate analytical techniques to determine whether such a flow is linearly stable. Moreover, analysts often use the formulae developed from such expansions and apply them to turbulent flows by replacing the base flow values by the time-averaged ones. We will discuss this specifically in Sections 3.5 and 11.5.

With this distinction in mind, theoretical expressions developed from perturbation expansions will be presented in this text as p_0, p_1, p_2, and so on. In contrast, measurements or computations are reported in terms of $\bar{p}$ and p'.

2.2 Small-Amplitude Propagation in Uniform, Inviscid Flows [2]

In this section we analyze the solution characteristics of the linearized Navier–Stokes and energy equations, by assuming infinitesimal perturbations superposed upon a spatially homogeneous background flow. Each variable is written as

$$p(\vec{x},t) = p_0 + p_1(\vec{x},t),$$
$$\rho(\vec{x},t) = \rho_0 + \rho_1(\vec{x},t), \qquad (2.9)$$
$$\vec{u}(\vec{x},t) = \vec{u}_0 + \vec{u}_1(\vec{x},t).$$

We will also assume (1) that the gas is nonreacting and calorically perfect, implying that the specific heats are constants, and (2) viscous and thermal transport are negligible. The effects of this latter assumption are discussed in Section 2.2.4.

2.2.1 Motivating Example: Solutions of the One-Dimensional Linearized Euler Equations

Before considering the problem more generally, it is helpful to consider harmonically oscillating solutions of the one-dimensional, linearized Euler equations; i.e., assume that the base quantities are uniform, that $u_{y,0} = 0$, and that all disturbances are only functions of x. This problem enables simple solutions, showing that the linearized Euler equations naturally admit a superposition of acoustic, entropy, and vortical disturbances.

Without proof, the linearized equations to be solved are (details on the derivation can be obtained by taking the one-dimensional versions of the equations developed in Section 2.2.2):

continuity:
$$\frac{\partial \rho_1}{\partial t} + u_{x,0}\frac{\partial \rho_1}{\partial x} + \rho_0\frac{\partial u_{x,1}}{\partial x} = 0, \qquad (2.10)$$

x-momentum:
$$\frac{\partial u_{x,1}}{\partial t} + u_{x,0}\frac{\partial u_{x,1}}{\partial x} = -\frac{1}{\rho_0}\frac{\partial p_1}{\partial x}, \qquad (2.11)$$

y-momentum:
$$\frac{\partial u_{y,1}}{\partial t} + u_{x,0}\frac{\partial u_{y,1}}{\partial x} = 0, \qquad (2.12)$$

energy:
$$\frac{\partial p_1}{\partial t} + u_{x,0}\frac{\partial p_1}{\partial x} = -\gamma p_0 \frac{\partial u_{x,1}}{\partial x}. \qquad (2.13)$$

Assume the disturbances to be of the form $p_1(x,t) = \hat{p}_1(x,\omega)e^{-i\omega t}$, $u_{y,1}(x,t) = \hat{u}_{y,1}(x,\omega)e^{-i\omega t}$, and so forth. Here, the hatted quantities denote the Fourier transform (discussed further in Section 2.7.1). Note that the governing equation for $\hat{u}_{y,1}$ is decoupled from the others, with a solution given by

$$\hat{u}_{y,1}(x) = A_\Omega \exp\left(\frac{i\omega x}{u_{x,0}}\right), \qquad (2.14)$$

where A_Ω is the integration constant, to be discussed further below. With some manipulation, the coupled system governing $\hat{\rho}_1$, $\hat{u}_{x,1}$, and $\hat{p}_1$ can be expressed in matrix form as

2.2 Small-Amplitude Propagation: Uniform Inviscid Flows

$$\begin{bmatrix} \dfrac{d\hat{p}_1}{dx} \\ \dfrac{d\hat{u}_{x,1}}{dx} \\ \dfrac{d\hat{p}_1}{dx} \end{bmatrix} = \begin{bmatrix} \dfrac{i\omega}{u_{x,0}} & \dfrac{i\omega\rho_0}{\left(c_0^2 - u_{x,0}^2\right)} & \dfrac{i\omega}{u_{x,0}\left(u_{x,0}^2 - c_0^2\right)} \\ 0 & \dfrac{i\omega}{u_{x,0}\left(u_{x,0}^2 - c_0^2\right)} & \dfrac{i\omega}{\rho_0\left(c_0^2 - u_{x,0}^2\right)} \\ 0 & \dfrac{c_0^2 i\omega\rho_0}{\left(c_0^2 - u_{x,0}^2\right)} & \dfrac{i\omega u_{x,0}}{\left(u_{x,0}^2 - c_0^2\right)} \end{bmatrix} \begin{bmatrix} \hat{p}_1 \\ \hat{u}_{x,1} \\ \hat{p}_1 \end{bmatrix}. \quad (2.15)$$

This equation is (for unknowns in vector $[v]$) of the form

$$\frac{d}{dx}[v] = [B][v], \quad (2.16)$$

whose solution is given by

$$[v] = \sum_{j=1}^{n} C_j e^{\lambda_j x} [v_j], \quad (2.17)$$

where λ_j are the eigenvalues of $[B]$ and $[v_j]$ are the corresponding eigenvectors. Solving for the eigenvectors and eigenvalues, we can write the solution as

$$\begin{bmatrix} \hat{p}_1(x) \\ \hat{u}_{x,1}(x) \\ \hat{p}_1(x) \end{bmatrix} = A_p \begin{bmatrix} \dfrac{1}{c_0^2} \\ \dfrac{1}{\rho_0 c_0} \\ 1 \end{bmatrix} e^{\left(\frac{i\omega x}{c_0 + u_{x,0}}\right)} + A_s \begin{bmatrix} 1 \\ 0 \\ 0 \end{bmatrix} e^{\left(\frac{i\omega x}{u_{x,0}}\right)} + B_p \begin{bmatrix} \dfrac{1}{c_0^2} \\ -\dfrac{1}{\rho_0 c_0} \\ 1 \end{bmatrix} e^{-\left(\frac{i\omega x}{c_0 - u_{x,0}}\right)}.$$

(2.18)

Now, using these solutions, we can also calculate the temperature, entropy, vorticity, and dilatational fluctuations, writing all these solutions in matrix form as

$$\begin{bmatrix} \hat{p}_1(x) \\ \hat{\rho}_1(x) \\ \hat{T}_1(x) \\ \hat{u}_{x,1}(x) \\ \hat{u}_{y,1}(x) \\ \hat{s}_1(x) \\ \hat{\Omega}_1(x) \\ \hat{\Lambda}_1(x) \end{bmatrix} = \begin{bmatrix} 1 & 1 & 0 & 0 \\ \dfrac{1}{c_0^2} & \dfrac{1}{c_0^2} & 1 & 0 \\ \dfrac{\gamma-1}{\gamma p_0}T_0 & \dfrac{\gamma-1}{\gamma p_0}T_0 & -\dfrac{T_0}{\rho_0} & 0 \\ \dfrac{1}{\rho_0 c_0} & -\dfrac{1}{\rho_0 c_0} & 0 & 0 \\ 0 & 0 & 0 & 1 \\ 0 & 0 & -\dfrac{\gamma \mathcal{R}}{(\gamma-1)\rho_0} & 0 \\ 0 & 0 & 0 & \dfrac{i\omega}{u_{x,0}} \\ \dfrac{i\omega}{\rho_0 c_0(c_0 + u_{x,0})} & \dfrac{i\omega}{\rho_0 c_0(c_0 - u_{x,0})} & 0 & 0 \end{bmatrix} \begin{bmatrix} A_p e^{\left(\frac{i\omega x}{c_0+u_{x,0}}\right)} \\ B_p e^{-\left(\frac{i\omega x}{c_0-u_{x,0}}\right)} \\ A_s e^{\left(\frac{i\omega x}{u_{x,0}}\right)} \\ A_\Omega e^{\left(\frac{i\omega x}{u_{x,0}}\right)} \end{bmatrix}.$$

(2.19)

Consider first the terms multiplied by the coefficients A_p and B_p. The terms in the complex exponentials show that these describe disturbances moving at speeds of $c_0 + u_{x,0}$ and $c_0 - u_{x,0}$, respectively. As such, these are acoustic disturbances propagating downstream and upstream, respectively. They are the only terms influencing $\hat{\Lambda}_1$, showing that they are compressional disturbances, and pressure, $\hat{p}_1$, showing the direct association of acoustic and pressure disturbances. They also influence the fluctuating density, $\hat{\rho}_1$, temperature, $\hat{T}_1$, and axial velocity, $\hat{u}_{x,1}$, and have no influence on the entropy, $\hat{s}_1$, or vortical disturbance, $\hat{\Omega}_1$.

Next, consider the terms multiplied by the term A_s. The terms in the complex exponentials show that these describe disturbances moving at a speed of $u_{x,0}$. This is the only term influencing $\hat{s}_1$, showing that it is directly associated with entropy fluctuations. The fluctuating density and temperature also include a term associated with it. As such, both acoustic and entropy disturbances influence $\hat{\rho}_1$ and $\hat{T}_1$. Also, this term has no influence on the unsteady velocity $\hat{u}_{x,1}$ or $\hat{u}_{y,1}$ (and, therefore, the dilatation $\hat{\Lambda}_1$ or vorticity $\hat{\Omega}_1$), or on the pressure, $\hat{p}_1$.

Finally, consider the terms multiplied by A_Ω. The terms in the complex exponentials show that these describe disturbances moving at a speed of $u_{x,0}$. This is the only term influencing $\hat{\Omega}_1$, showing that it is directly associated with vortical fluctuations. The fluctuating transverse velocity includes a term associated with it. Ordinarily, the axial velocity would as well, but it is suppressed because of the assumption of no variations in y. This term has no influence on the pressure, dilatation, entropy, density, or temperature fluctuations.

To summarize, this solution shows that there are three independent types of disturbances, acoustic (i.e., dilatational or pressure disturbances), entropy, and vorticity. Acoustic disturbances propagate with the speed of sound, with a correction due to mean flow advection, and entropy and vorticity disturbances move with the flow. Finally, the acoustic, vortical, and entropy disturbances are independent of each other. The acoustic and vortical disturbances induce fluid motion; acoustic and entropy disturbances induce fluctuations in thermodynamic variables.

With this motiving example, the next section considers this problem in more generality.

2.2.2 Decomposition Approach

Having motivated the basic approach and decomposition for a one-dimensional problem, this section analyzes the dynamics of the first-order fluctuations more generally in a field of uniform base variables. The fluctuating variables are functions of both space and time. We start with the continuity equation, and write the derivatives in density in terms of derivatives in entropy and pressure, utilizing the thermodynamic relation Eq. (1.36):

$$-\frac{1}{\rho}\frac{D\rho}{Dt} = \frac{1}{c_p}\frac{Ds}{Dt} - \frac{1}{\gamma p}\frac{Dp}{Dt} = \nabla \cdot \vec{u}. \qquad (2.20)$$

2.2 Small-Amplitude Propagation: Uniform Inviscid Flows

Note that Ds/Dt is identically zero because of the neglect of molecular transport terms and chemical reaction (see Eq. (1.37)), yielding a simplified equation relating pressure disturbances and flow dilatation:

$$-\frac{1}{\gamma p}\frac{Dp}{Dt} = \nabla \cdot \vec{u}. \tag{2.21}$$

Expanding each variable into base and fluctuating components yields

$$-\frac{1}{\gamma(p_0 + p_1)}\left(\frac{\partial p_1}{\partial t} + (\vec{u}_0 + \vec{u}_1)\cdot\nabla p_1\right) = \nabla\cdot\vec{u}_1. \tag{2.22}$$

This equation can be manipulated to yield

$$-\frac{1}{\gamma}\left(\frac{\partial p_1}{\partial t} + \vec{u}_0\cdot\nabla p_1 + \vec{u}_1\cdot\nabla p_1\right) = p_0\nabla\cdot\vec{u}_1 + p_1\nabla\cdot\vec{u}_1. \tag{2.23}$$

Retaining only first-order terms leads to the linearized equations:

$$-\frac{1}{\gamma p_0}\frac{D_0 p_1}{Dt} = \nabla\cdot\vec{u}_1 \equiv \Lambda_1, \tag{2.24}$$

where we use the following notation for the substantial derivative, D_0/Dt:

$$\frac{D_0}{Dt}(\ ) = \frac{\partial}{\partial t}(\ ) + \vec{u}_0\cdot\nabla(\ ); \tag{2.25}$$

this physically describes the time rate of change of a quantity in a reference frame moving with the base flow. In the same way, the linearized form of the momentum equation, Eq. (1.12), is

$$\rho_0\frac{D_0\vec{u}_1}{Dt} = -\nabla p_1. \tag{2.26}$$

Finally, the linearized energy equation, written in terms of the entropy, is

$$\frac{D_0 s_1}{Dt} = 0. \tag{2.27}$$

We next consider the linearized equations for the vorticity, $\vec{\Omega}_1 = \nabla\times\vec{u}_1$, and dilatation, $\Lambda_1 = \nabla\cdot\vec{u}_1$. These can be obtained by either linearizing the explicit expressions from Eqs. (1.49) and (1.18) in the previous chapter or by taking the curl and divergence, respectively, of Eq. (2.26). This leads to

$$\frac{D_0\vec{\Omega}_1}{Dt} = 0, \tag{2.28}$$

$$\rho_0\frac{D_0\Lambda_1}{Dt} + \nabla^2 p_1 = 0. \tag{2.29}$$

Finally, an equation for the unsteady pressure itself can be obtained by taking the substantial derivative, D_0/Dt, of Eq. (2.24) and substituting this into Eq. (2.29) to yield the advected wave equation,

$$\frac{D_0^2 p_1}{Dt^2} - c_0^2 \nabla^2 p_1 = 0, \tag{2.30}$$

where c_0 is the speed of sound:

$$c_0 = \sqrt{\gamma p_0/\rho_0} = \sqrt{\gamma \mathscr{R} T_0}. \tag{2.31}$$

The wave equation can equivalently be written as

$$\frac{\partial^2 p_1}{\partial t^2} + 2\vec{u}_0 \cdot \nabla \frac{\partial p_1}{\partial t} + \vec{u}_0 \cdot \nabla (\vec{u}_0 \cdot \nabla p_1) - c_0^2 \nabla^2 p_1 = 0. \tag{2.32}$$

For low Mach number flows the $\vec{u}_0$-containing terms are negligible, leaving the familiar wave equation:

$$\frac{\partial^2 p_1}{\partial t^2} - c_0^2 \nabla^2 p_1 = 0. \tag{2.33}$$

Summarizing, we have the following sets of equations describing vorticity, acoustic, and entropy fluctuations:

Vorticity:

$$\frac{D_0 \vec{\Omega}_1}{Dt} = 0; \tag{2.34}$$

Acoustic:

$$\frac{D_0^2 p_1}{Dt^2} - c_0^2 \nabla^2 p_1 = 0, \tag{2.35}$$

$$\frac{D_0^2 \Lambda_1}{Dt^2} - c_0^2 \nabla^2 \Lambda_1 = 0; \tag{2.36}$$

Entropy:

$$\frac{D_0 \mathscr{s}_1}{Dt} = 0. \tag{2.37}$$

Note that pressure and dilatational disturbances are interchangeable, as they are related through Eq. (2.24).

We will now decompose these equations into three subsystems. The basic idea is to write each fluctuation as a superposition of three more "fundamental" fluctuations, each of which is induced by the canonical acoustic, vortical, and entropy fluctuations. These canonical fluctuations are indicated by the subscripts Λ, Ω, and $\mathscr{s}$, respectively. Clearly, what is actually measured in an experiment or computed in a calculation is the total fluctuation in, for example, velocity. We will show, however, that this total fluctuation can be more easily interpreted when it is understood that it is actually a superposition of more fundamental quantities. Some examples of attempts to decompose measured/computed observables into these canonical modes can be found in the literature [1, 3].

As such, we write:

$$\vec{\Omega}_1 = \vec{\Omega}_{1\Lambda} + \vec{\Omega}_{1s} + \vec{\Omega}_{1\Omega},$$
$$s_1 = s_{1\Lambda} + s_{1s} + s_{1\Omega}, \quad (2.38)$$
$$p_1 = p_{1\Lambda} + p_{1s} + p_{1\Omega}.$$

To illustrate, $\vec{\Omega}_{1s}$ denotes vorticity fluctuations induced by entropy mode fluctuations. Similarly, $p_{1\Omega}$ denotes pressure fluctuations associated with disturbances in the vorticity mode, and so forth. These expressions can then be substituted into Eqs. (2.27), (2.28), and (2.30). Because these dynamical equations are linear, they can then similarly be decomposed into subsystems of equations showing the influence of each type of disturbance mode on the actual flow and thermodynamic variables. In reality, this decomposition could potentially be performed in a variety of ways, but following Chu and Kovásznay [2] and results from Section 2.2.1, the choice used here is motivated by the desire for simplicity and clarity.

Oscillations Associated with Vorticity Mode

$$\frac{D_0 \vec{\Omega}_{1\Omega}}{Dt} = 0 \quad (2.39)$$

$$p_{1\Omega} = s_{1\Omega} = T_{1\Omega} = \rho_{1\Omega} = 0 \quad (2.40)$$

$$\frac{D_0 \vec{u}_{1\Omega}}{Dt} = 0 \quad (2.41)$$

Equation (2.39) for $\vec{\Omega}_{1\Omega}$ shows that vorticity fluctuations are convected by the mean flow. The pressure equation, $p_{1\Omega} = 0$, shows that vorticity fluctuations induce no fluctuations in pressure within this linear approximation. As discussed later, vorticity fluctuations do excite pressure oscillations when higher-order terms are included. For example, prediction of jet noise, which is intrinsically associated with the generation of propagating pressure disturbances by vortical flow disturbances, requires inclusion of second-order (i.e., nonlinear) terms. Similarly, Eq. (2.40) shows that vorticity fluctuations induce no fluctuations in entropy, density, temperature, or dilatation.

Oscillations Associated with Acoustic Mode

$$\frac{D_0^2 p_{1\Lambda}}{Dt^2} - c_0^2 \nabla^2 p_{1\Lambda} = 0 \quad (2.42)$$

$$\vec{\Omega}_{1\Lambda} = s_{1\Lambda} = 0 \quad (2.43)$$

$$\rho_{1\Lambda} = \frac{p_{1\Lambda}}{c_0^2} = \frac{\rho_0 c_p}{c_0^2} T_{1\Lambda} \quad (2.44)$$

$$\rho_0 \frac{D\vec{u}_{1\Lambda}}{Dt} = -\nabla p_{1\Lambda}. \tag{2.45}$$

These equations show that the propagation of acoustic pressure disturbances (as well as density, velocity, and temperature) are described by a wave equation. The density, temperature, and pressure fluctuations are locally and algebraically related through their isentropic relations, as expressed in Eq. (2.44). Equation (2.43) also shows that acoustic disturbances do not excite vorticity or entropy disturbances.

Although labeled "acoustic," these disturbances may occur even if the flow is incompressible [4]. To make this more explicit, we can define a potential mode, p, that is a special case of the acoustic mode. Such a potential mode would satisfy the homogeneous relations

$$\vec{\Omega}_{1p} = s_{1p} = p_{1p} = \nabla \cdot \vec{u}_{1p} = 0. \tag{2.46}$$

Two examples where this potential mode would arise are geometries with slowly moving boundaries or mass sources.

Oscillations Associated with Entropy Mode

$$\frac{D_0 s_{1s}}{Dt} = 0 \tag{2.47}$$

$$p_{1s} = \vec{\Omega}_{1s} = \vec{u}_{1s} = 0 \tag{2.48}$$

$$\rho_{1s} = -\frac{\rho_0}{c_p} s_{1s} = -\frac{\rho_0}{T_0} T_{1s}. \tag{2.49}$$

These expressions show that entropy fluctuations, s_{1s}, are described by a convective equation. They also show that, within the linear approximation, entropy oscillations do not excite vorticity, velocity, pressure, or dilatational disturbances, as shown by Eq. (2.48). They do excite density and temperature perturbations; see Eq. (2.49).

2.2.3 Comments on the Decomposition

In this section we present some general conclusions from these relations. First, in a homogeneous, uniform flow, these three disturbance modes propagate independently in the linear approximation. In other words, the three modes are decoupled within the approximations of this analysis – vortical, entropy, and acoustically induced fluctuations are completely independent of each other. For example, velocity fluctuations induced by vorticity and acoustic disturbances, $\vec{u}_{1\Omega}$ and $\vec{u}_{1\Lambda}$, are independent of each other and each propagates as if the other were not there. Moreover, there are no sources or sinks of any of these disturbance modes. Once created, they propagate with constant amplitude. In the next section we will show how these results change as a result of boundaries, flow inhomogeneities, and nonlinearities.

Second, acoustic disturbances propagate with respect to the flow at a characteristic velocity equal to the speed of sound. This can be seen from Eq. (2.42). In contrast, vorticity and entropy disturbances are convected at the bulk flow velocity, $\vec{u}_0$. Consequently, in low Mach number flows, these disturbances have substantially different length scales. Acoustic properties vary over an acoustic length scale, given by $\lambda_\Lambda = c_0/f$, while entropy and vorticity modes vary over a convective length scale, given by $\lambda_c = u_0/f$. Thus, the entropy and vortical mode "wavelength" is shorter than the acoustic wavelength by a factor equal to the mean flow Mach number, $\lambda_c/\lambda_\Lambda = u_0/c_0 = M$. This can have important implications on unsteady flow interactions with flames, nozzles, and so on. For example, a flame whose length, L_F, is short relative to an acoustic wavelength, $L_F \ll \lambda_\Lambda$, may be of the same order as, or longer than, a convective wavelength. Thus, a convected disturbance, such as an equivalence ratio oscillation, may have substantial spatial variation along the flame front that results in heat release disturbances generated at different points of the flame being out of phase with each other. As shown in Chapter 12, a Strouhal number, defined as $St = fL_F/u_0$, is a key parameter that affects the flame response to perturbations; note that St is proportional to the ratio of the flame length and convective wavelength, $St = L_F/\lambda_c$. A flame whose length is much less than an acoustic/convective wavelength is referred to as acoustically/convectively *compact*. See also Section 2.9 for a discussion of how these differences in propagation velocity influence the characteristics of the total perturbation velocity in systems where vortical and acoustic disturbances coexist.

Third, entropy and vorticity disturbances propagate with the mean flow and diffuse from regions of high to low concentration. In contrast, acoustic disturbances, being true waves, reflect off boundaries, are refracted at property changes, and diffract around obstacles. For example, the reflection/refraction of acoustic waves from multidimensional flame fronts generally results in a complex, multidimensional acoustic field in the vicinity of the flame, as shown in Figure 2.1.

The refraction of sound also has important influences on the acoustic sensitivities of burners. For example, consider burners in an annular combustion chamber where

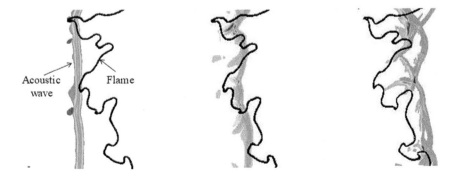

Figure 2.1 Computed image of the instantaneous pressure field and flame front of a sound wave incident on a turbulent flame from the left at three successive time instants [5]. Images courtesy of D. Thévenin.

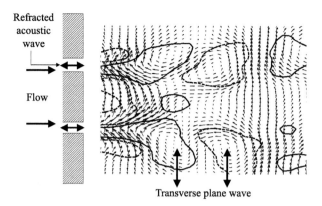

Figure 2.2 Illustration of acoustic refraction effects in a transversely forced, swirling jet. Contour lines indicate fluctuating vorticity. Data courtesy of J. O'Connor.

sound waves propagate azimuthally around the annulus (these modes are discussed more fully in Chapter 5). These acoustic disturbances diffract "around the corner" into the burners, and can introduce very strong axial velocity disturbances as well, as discussed in Section 6.3.3. To illustrate, Figure 2.2 shows this effect in velocity data from a transversely excited annular jet; the flow is from left to right, while the transverse acoustic field fluctuates up and down. The presence of the transverse acoustics is particularly obvious at the top and bottom edges of the figure, while right at the nozzle exit the fluctuation has a significant axial component. This is caused by sound diffraction around the edge of the nozzle. In addition, strong vortical velocity fluctuations excited by the sound waves are evident and indicated by the contour lines.

2.2.4 Molecular Transport Effects on the Decomposition

This section provides a brief discussion of the effects of molecular transport, neglected in the previous discussion [2, 6]. Most obviously, molecular transport leads to diffusion of vorticity and entropy, effects that can be seen explicitly in Eqs. (1.17) and (1.37). These terms have been dropped from all equations in the previous section. Second, molecular energy transport leads to an entropic velocity component, $\vec{u}_{1,4}$. This can be seen from Eq. (2.20), which shows that the velocity divergence has a component arising from $D\phi/Dt$ that is, in turn, related to molecular transport terms through Eq. (1.37). This also implies that the pressure and dilatational fluctuations are not one-to-one related when molecular transport effects are included. Third, the acoustic wave equation has a dissipation term. This dissipation term leads to entropy excitation and causes the acoustic and entropy modes to be coupled, even within the assumptions of linear, homogeneous mean flow.

Molecular transport effects on acoustic disturbances are negligible, except for very high frequencies where the acoustic wavelength is of the order of the molecular mean free path, and in boundary layers. Similarly, their effects on vorticity and entropy disturbances are negligible in high Reynolds number flows, except in boundary layers.

2.3 Modal Coupling Processes

Section 2.2 showed that the small-amplitude canonical disturbance modes propagate independently within the fluid domain in a homogeneous, inviscid flow. This section shows how these modes couple with each other at boundaries, in regions of flow inhomogeneity, and through nonlinearities. The primary focus of this section is on introducing why modal coupling occurs – most examples and calculations of modal coupling are introduced in their relevant sections throughout the rest of this text.

2.3.1 Coupling through Boundary Conditions

Disturbance modes couple at boundaries because boundary conditions apply to the total value of the fluctuation itself, and not its components. For example, a velocity boundary condition applies to $\vec{u}_{1\Lambda} + \vec{u}_{1\Omega}$, and not $\vec{u}_{1\Lambda}$ and $\vec{u}_{1\Omega}$ separately. The same idea holds for thermal boundary conditions, which apply to the sum, $T_{1\sigma} + T_{1\Lambda}$.

To illustrate, consider Figure 2.3, which shows an acoustic wave impinging obliquely on a rigid wall at an angle θ_i in a low Mach number cross flow. Several boundary conditions must be applied at the wall. First, the normal velocity is zero at the wall, $\vec{u}_1 \cdot \vec{n} = 0$. If this is the only boundary condition applied, then this leads to the prediction that an acoustic wave of equal strength to the incident one must be reflected at the wall, whose reflection angle equals the incident angle, θ_i. The second boundary condition is the no-slip condition, $(\vec{u}_1 \times \vec{n}) \times \vec{n} = 0$. In order to satisfy this, the incident acoustic wave must excite a vorticity disturbance at the wall whose transverse magnitude is equal and of opposite sign to that associated with the transverse components of the incident and reflected acoustic waves. This is a transfer of energy from the acoustic to the vortical modes, and results in damping of the acoustic mode and excitation of the vortical mode. Thus, the reflected acoustic wave has a magnitude less than the incident wave. Moreover, the reflected acoustic wave

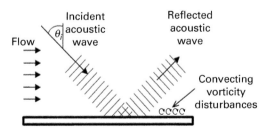

Figure 2.3 Acoustic wave impinging obliquely on wall, exciting vorticity.

propagates away at the speed of sound, while the excited vortical wave convects and diffuses downstream.

Similarly, if we assume that the fluctuating temperature at the boundary is zero, then $T_1 = 0 = T_{1s} + T_{1\Lambda}$ implies that the incident acoustic wave excites an entropy disturbance. Again, this results in damping of the acoustic mode and excitation of the entropy mode. For example, assuming a no-slip and no fluctuating temperature boundary condition, the fraction of the incident acoustic energy flux that is converted into vorticity/entropy disturbances is given by [6]

$$1 - |R|^2 = \frac{4\overline{\mathcal{N}}\sqrt{2}\cos\theta_i}{\left(\sqrt{2}\cos\theta_i + \overline{\mathcal{N}}\right)^2 + \overline{\mathcal{N}}^2}, \tag{2.50}$$

where

$$\overline{\mathcal{N}} = \sqrt{\frac{\omega v_\nu}{c_0^2}\left(\sin^2\theta_i + \frac{\gamma - 1}{\sqrt{Pr}}\right)} \tag{2.51}$$

and R is the reflection coefficient, which equals the ratio of the amplitudes of the reflected and incident acoustic waves. Note the $\omega^{1/2}$ frequency scaling.

Indeed, it turns out that, except for very long range acoustic propagation, acoustic damping at system boundaries dominates the actual damping of acoustic energy by viscous and molecular transport processes within the fluid itself.

2.3.2 Coupling through Flow Inhomogeneities

In a flow with mean property gradients, such as shear flows with varying mean velocities, or with gradients in mean temperature, the disturbance modes couple. Thus, fluctuations in one mode continuously excite others. This is a complex subject for which there is extensive literature, particularly in aeroacoustics [7, 8]. We will illustrate the key ideas next, by dropping the requirement from Section 2.1 that the base flow variables are spatially constant, now allowing them to vary spatially, i.e., $p_0 = p_0(x, y, z)$.

For the entropy and vorticity equations, we can think of the previously neglected nonhomogeneous terms as "source terms" on the right-hand side of their propagation equations, Eqs. (2.39) and (2.47). To illustrate, consider the full vorticity equation, Eq. (1.49), and linearize it about the base flow, neglecting body forces and molecular transport:

$$\frac{D_0\vec{\Omega}_1}{Dt} = -\vec{u}_1\cdot\nabla\vec{\Omega}_0 + \vec{\Omega}_0\cdot\nabla\vec{u}_1 + \vec{\Omega}_1\cdot\nabla\vec{u}_0 - \vec{\Omega}_0\nabla\cdot\vec{u}_1 - \vec{\Omega}_1\nabla\cdot\vec{u}_0 + \frac{\nabla p_1 \times \nabla p_0}{\rho_0^2} + \frac{\nabla\rho_0 \times \nabla p_1}{\rho_0^2}. \tag{2.52}$$

Section 1.6 provides a physical interpretation of these various terms. Consider first the terms involving dot products between velocity and vorticity gradients. The u_1 term in this equation has an acoustic and vortical component, $\vec{u}_1 = \vec{u}_{1\Lambda} + \vec{u}_{1\Omega}$. This equation shows that an acoustic wave propagating through a steady, shearing flow with nonzero

$\vec{\Omega}_0$ continuously excites vorticity through a convective $(\vec{u}_1 \cdot \nabla \vec{\Omega}_0)$, vortex stretching $(\vec{\Omega}_0 \cdot \nabla \vec{u}_1)$, and dilatational $(\vec{\Omega}_0 \nabla \cdot \vec{u}_1)$ mechanism. In other words, the acoustic disturbance is a vorticity source in an inhomogeneous flow. Turn next to the pressure gradient and density gradient terms. The $\frac{\nabla \rho_0 \times \nabla p_1}{\rho_0^2}$ term shows that an acoustic pressure fluctuation propagating at an angle to a density gradient, such as a flame, generates vorticity through the baroclinic mechanism [9]. The other term, $\frac{\nabla \rho_1 \times \nabla p_0}{\rho_0^2}$, shows that a density fluctuation, which could arise from an entropy or acoustic wave, propagating at an angle through a mean pressure gradient also excites vorticity. An illustration of this would be an entropy disturbance propagating through a shock wave.

Another important example for combustor applications is the acoustic–entropy coupling that occurs in regions of mean flow acceleration [10, 11]. This coupling can be understood purely from a one-dimensional analysis. In essence, if an entropy disturbance is accelerated or decelerated (such as when it passes through the combustor exit nozzle), this generates an acoustic disturbance. This can be seen by linearizing the momentum equation, Eq. (1.12):

$$\rho_0 \frac{\partial \vec{u}_1}{\partial t} + \rho_0 \vec{u}_0 \cdot \nabla \vec{u}_1 + \rho_0 \vec{u}_1 \cdot \nabla \vec{u}_0 + \rho_1 \vec{u}_0 \cdot \nabla \vec{u}_0 = -\nabla p_1. \qquad (2.53)$$

An entropy disturbance, s_{1d}, convecting through the nozzle is also associated with a density disturbance, $\rho_{1d} = -\frac{\rho_0}{c_p} s_{1d}$. This entropy-induced density fluctuation appears in the term $\rho_1 \vec{u}_0 \cdot \nabla \vec{u}_0$ on the left-hand side of the equation, which describes the acceleration of a density disturbance by the mean flow. This density fluctuation induces an oscillatory pressure drop across the nozzle in an accelerating or decelerating mean flow. This compensating unsteady pressure gradient (the right-hand side of this equation) leads to up- and downstream-propagating acoustic waves. Such sound generation by density disturbances (referred to as "indirect sound generation") has received extensive discussion for its potential role in low-frequency oscillations in combustors [12]. This point will be considered further in Section 6.3.2.

Another important example of acoustic–vortical coupling in combustor environments occurs when sound waves propagate through swirl vanes. The axially oscillating sound field excites axial vorticity and azimuthal velocity fluctuations. Although the ratio of axial and azimuthal velocities, and therefore the swirl number, at the swirler exit may stay constant in time, the swirl number oscillates downstream because the axial oscillations propagate with the sound speed, while the azimuthal velocity disturbances propagate at the flow speed [13–15].

The above examples for vorticity and entropy modes are fairly straightforward and can be handled by simple inspection. This is not generally the case for acoustic waves, due to their wave nature. In brief, the problem is that some inhomogeneous terms lead to coupling/damping of sound waves ("sound sources or sinks"), while other terms describe the convection and refraction of sound waves by mean flow or temperature gradients ("sound propagation terms") [16]. As such, distinguishing between acoustic source and propagation terms can be difficult. A simple example of this is the following wave equation, derived in Section 6.2.4, in the presence of mean temperature gradients with no flow:

$$\nabla^2 p_1 + \frac{1}{T_0}\nabla T_0 \cdot \nabla p_1 - \frac{1}{c_0^2}\frac{\partial^2 p_1}{\partial t^2} = 0. \qquad (2.54)$$

It will be shown in Chapter 6 that the inhomogeneous temperature term does not lead to sound amplification or damping; it simply leads to sound wave reflection and refraction.

2.3.3 Coupling through Nonlinearities

In this section we show that nonlinear terms lead to coupling and excitation of disturbance modes, even in a homogeneous flow. To introduce the ideas, we revisit the expansions initiated in Section 2.1, but include higher-order terms:

$$\begin{aligned} p &= p_0 + p_1 + p_2 + \cdots, \\ \vec{\Omega} &= \vec{\Omega}_0 + \vec{\Omega}_1 + \vec{\Omega}_2 + \cdots, \\ \vec{u} &= \vec{u}_0 + \vec{u}_1 + \vec{u}_2 + \cdots, \\ \rho &= \rho_0 + \rho_1 + \rho_2 + \cdots, \end{aligned} \qquad (2.55)$$

where, for example, p_1 is $O(\varepsilon)$, p_2 $O(\varepsilon^2)$, and so forth. In order to simplify the presentation, we will again assume a homogeneous flow, i.e., that p_0 and ρ_0 are constants. These expansions can be substituted into the general governing equations and matched at each order of ε, i.e., terms of $O(1)$, $O(\varepsilon)$, $O(\varepsilon^2)$, and so on. The linearized equations considered in Section 2.1 were, in essence, obtained by matching terms of $O(\varepsilon)$. In this section we will consider the higher-order terms by explicitly considering $O(\varepsilon^2)$ terms and their impacts.

To illustrate the basic idea, consider the continuity equation, Eq. (1.8). Substituting in the expansions and equating terms of equal order in ε leads to the following system of equations to $O(\varepsilon^2)$:

$$O(1) : \nabla \cdot (\rho_0 \vec{u}_0) = 0 \qquad (2.56)$$

$$O(\varepsilon) : \frac{\partial \rho_1}{\partial t} + \nabla \cdot (\rho_0 \vec{u}_1 + \rho_1 \vec{u}_0) = 0 \qquad (2.57)$$

$$O(\varepsilon^2) : \frac{\partial \rho_2}{\partial t} + \nabla \cdot (\rho_0 \vec{u}_2 + \rho_2 \vec{u}_0) = -\nabla \cdot (\rho_1 \vec{u}_1) \qquad (2.58)$$

Equation (2.58) shows that $O(\varepsilon)$ fluctuations in density and velocity, the right-hand side of Eq. (2.58), act as "source terms" to the second-order corrections. Clearly, the significance of these "source" terms grows quadratically with disturbance amplitude. Similarly, products like $\rho_1 \vec{u}_2$ act as "source terms" to the third-order equations that are not shown above.

The vorticity equation, Eq. (1.49), can be similarly expanded. Matching terms of $O(\varepsilon^2)$:

$$\rho_0 \frac{D_0 \vec{\Omega}_2}{Dt} = -\rho_0 \vec{\Omega}_1 \nabla \cdot \vec{u}_1 + \rho_0 \vec{\Omega}_1 \cdot \nabla \vec{u}_1 - \rho_0 \vec{u}_1 \cdot \nabla \vec{\Omega}_1 - \frac{\nabla p_1 \times \nabla \rho_1}{\rho_0}. \qquad (2.59)$$

2.3 Modal Coupling Processes

Expanding the source terms by writing them explicitly in terms of their entropy, vorticity, and acoustic components:

$$\rho_0 \frac{D_0 \vec{\Omega}_2}{Dt} = -\rho_0 \vec{\Omega}_{1\Omega} \nabla \cdot \vec{u}_{1\Lambda} + \rho_0 \vec{\Omega}_{1\Omega} \cdot \nabla (\vec{u}_{1\Lambda} + \vec{u}_{1\Omega})$$
$$- \rho_0 (\vec{u}_{1\Lambda} + \vec{u}_{1\Omega}) \cdot \nabla \vec{\Omega}_{1\Omega} - \frac{\nabla p_{1\Lambda} \times \nabla \rho_{1s}}{\rho_0}. \tag{2.60}$$

From this expression, various types of terms can be recognized. Some terms describe the nonlinear interactions of vorticity upon itself (e.g., $\vec{\Omega}_{1\Omega} \cdot \nabla \vec{u}_{1\Omega}$), a well-known phenomenon in incompressible turbulence. Other terms, $-\vec{\Omega}_{1\Omega} \nabla \cdot \vec{u}_{1\Lambda} + \vec{\Omega}_{1\Omega} \cdot \nabla \vec{u}_{1\Lambda} - \vec{u}_{1\Lambda} \cdot \nabla \vec{\Omega}_{1\Omega}$, describe nonlinear interactions of sound waves and vorticity in exciting vorticity. The key point to note here is that nonlinear interactions occur because of finite-amplitude modal disturbances. While acoustic and vortical disturbances propagate independently at $O(\varepsilon)$, they do not at $O(\varepsilon^2)$; rather, they interact and modify each other.

Similarly, the second-order pressure equation can be obtained by taking the divergence of the second-order momentum equation. Analysis of the wave equation must be handled with more care because of the potential for nonlinear interactions to be responsible for both generating sound (i.e., a true sound source) and/or scattering sound that already exists [17]. We can, however, recognize various types of terms that arise. For example, the term $\nabla \vec{u}_{1\Omega} : \nabla \vec{u}_{1\Omega}$ appears and describes nonlinear interactions between vorticity disturbances. This is a very important term for the field of jet noise in aeroacoustics, and plays a key role in sound generation by vortical flows, such as turbulent jets. This term must clearly be a source of sound, as it is nonzero even if there are no first-order acoustic fluctuations. In contrast, a sound scattering term implies that the sound already existed, but that it was simply redirected by nonlinear interactions, such as with a vortex [18–20]. For example, terms like $\nabla \vec{u}_{1\Lambda} : \nabla \vec{u}_{1\Omega}$ describe the scattering of sound by vortical disturbances, which essentially acts as an oscillatory convective component to the sound field.

Although we will not present the full details here, we can summarize the different types of second-order sources of sound, vorticity, and entropy in Table 2.1 [2]. The leftmost column lists the type of interaction, while the topmost horizontal row lists the effect of this modal interaction. A "—" means there is no interaction to the given mode at $O(\varepsilon^2)$ if molecular transport effects are neglected.

We conclude this section by noting that the approach described here is quite general and can be used as a point of departure for a variety of different specialties. For example, the study of nonlinear interactions of vorticity is a restatement of the entire incompressible turbulence problem. The interactions between entropy and vorticity lead to many problems in gas dynamics, shocks, and high Mach number turbulence. The scattering and excitation of sound by vorticity–vorticity interactions, sound–vorticity, entropy–sound, and others forms the basis for the field of aeroacoustics and aerodynamic generation of sound. The rest of this book will use this approach specifically for understanding various unsteady features most relevant to combustor systems.

Table 2.1 Summary of nonlinear sound, vorticity, and entropy sources [2].

	Sound source/scattering	Vorticity source	Entropy source
Sound–Sound	Steepening and self-scattering $\nabla \vec{u}_{1\Lambda} : \nabla \vec{u}_{1\Lambda} + c_0^2 \nabla^2 p_{1\Lambda}^2$ $+ \frac{\gamma - 1}{2} \frac{\partial}{\partial t^2} p_{1\Lambda}^2$	—	—
Vorticity–Vorticity	Generation $\nabla \vec{u}_{1\Omega} : \nabla \vec{u}_{1\Omega}$	Convection/stretching/bending $\vec{\Omega}_{1\Omega} \cdot \nabla \vec{u}_{1\Omega} - \vec{u}_{1\Omega} \cdot \nabla \vec{\Omega}_{1\Omega}$	—
Entropy–Entropy	—	—	—
Sound–Vorticity	Scattering $2\nabla \vec{u}_{1\Lambda} : \nabla \vec{u}_{1\Omega}$	Convection/stretching/bending/dilatation $-\vec{\Omega}_{1\Omega} \nabla \cdot \vec{u}_{1\Lambda} + \vec{\Omega}_{1\Omega} \cdot \nabla \vec{u}_{1\Lambda} - \vec{u}_{1\Lambda} \cdot \nabla \vec{\Omega}_{1\Omega}$	—
Sound–Entropy	Scattering $\frac{\partial}{\partial t} \nabla(\mathit{s}_s \vec{u}_{1\Lambda})$	Generation $-\frac{\nabla p_{1\Lambda} \times \nabla \rho_{1s}}{\rho_0^2}$	Convection $-\vec{u}_{1\Lambda} \cdot \nabla \mathit{s}_{1s}$
Vorticity–Entropy	—	—	Convection $-\vec{u}_{1\Omega} \cdot \nabla \mathit{s}_{1s}$

2.4 Energy Density and Energy Flux Associated with Disturbance Fields

In this section we discuss the energy density, E, flux vector, $\vec{I}$, and source terms, Φ, associated with disturbance fields. Although the time average of the disturbance fields may be zero, they nonetheless contain nonzero time-averaged energy, and also propagate energy whose time average is nonzero. As a simple example, the kinetic energy per unit volume is $E_{kin} = \frac{1}{2}\rho(\vec{u}_1 \cdot \vec{u}_1)$, a quantity whose time average is clearly nonzero even if the time average of the fluctuating velocity is zero.

The key objective here is to analyze conservation equations of the form

$$\frac{\partial E}{\partial t} + \nabla \cdot \vec{I} = \Phi. \tag{2.61}$$

This equation can be integrated over a control volume to obtain

$$\frac{d}{dt} \iiint_V E \, dV + \oiint_A \vec{I} \cdot \vec{n} \, dA = \iiint_V \Phi \, dV, \tag{2.62}$$

which states that the sum of the time rate of change of energy in the control volume and the net flux of energy out of the control surface equals the net rate of production of energy in the control volume. From the general energy, Eq. (1.27), we can recognize the general forms of these three terms, such as $E = \rho(e + |\vec{u}|^2/2)$.

We next consider manipulation of this equation in order to develop expressions for the energy and energy flux associated with fluctuations. In a general inhomogeneous flow, the simultaneous presence and interaction of the various disturbance modes makes the development and interpretation of such an equation surprisingly subtle and difficult. We will not delve into these difficulties in this text, but refer the reader to Section 2.10 and key references [21–29].

In order to fix some ideas we will start with a simplified example: an acoustic energy equation, derived by assuming that entropy and vorticity fluctuations are zero, that the mean velocity is zero, and that there are no gradients in all other mean quantities. Furthermore, the combustion process is assumed to be isomolar; see Eq. (1.45). The acoustic energy equation comes from multiplying the linearized energy equation, Eq. (1.48), by $p_1/\gamma p_0$ and adding it to the dot product of the linearized momentum equation, Eq. (1.12), with $\vec{u}_1$. This yields

$$\frac{\partial}{\partial t}\left(\frac{1}{2}\rho_0(\vec{u}_1 \cdot \vec{u}_1) + \frac{1}{2}\frac{p_1^2}{\rho_0 c_0^2}\right) + \nabla \cdot (p_1 \vec{u}_1) = \frac{(\gamma - 1)}{\gamma p_0} p_1 \dot{q}_1. \tag{2.63}$$

From this expression, we can recognize the following acoustic energy density, flux, and source terms:

$$E_\Lambda = \frac{1}{2}\rho_0(\vec{u}_1 \cdot \vec{u}_1) + \frac{1}{2}\frac{p_1^2}{\rho_0 c_0^2}, \tag{2.64}$$

$$\vec{I}_\Lambda = p_1 \vec{u}_1, \tag{2.65}$$

$$\Phi_\Lambda = \frac{(\gamma - 1)}{\gamma p_0} p_1 \dot{q}_1. \tag{2.66}$$

The energy density, E_Λ, is a linear superposition of the kinetic energy associated with unsteady motions, and internal energy associated with the isentropic compression of an elastic gas. The flux term, $p_1 \vec{u}_1$, reflects the familiar pdV done by pressure forces on a system. The reader is referred to Section 2.10 for further discussion of these terms and their relation to the general energy, Eq. (1.27).

Next, look at the source term, Φ_Λ. This term shows that unsteady heat release can add or remove energy from the acoustic field, depending on its phasing with the acoustic pressure. Specifically, unsteady heat release acts as an acoustic energy source when the time average of the product of the fluctuating pressure and heat release is greater than zero, referred to as *Rayleigh's criterion* [30], which should not be confused with the criterion of the same name used in hydrodynamic stability problems to be discussed in Chapter 3. For a harmonically oscillating field, this shows that unsteady heat addition locally adds energy to the acoustic field when the magnitude of the phase between the pressure and heat release oscillations, θ_{pq}, is less than 90 degrees (i.e., $0 < |\theta_{pq}| < 90$). Conversely, when these oscillations are out of phase (i.e., $90 < |\theta_{pq}| < 180$), the heat addition oscillations damp the acoustic field. To illustrate this energy source term, Figure 2.4 plots the dependence of the thermoacoustic instability amplitude on the measured pressure–heat release phase, measured in a combustor with variable geometry over a range of operating conditions [31]. This graph clearly illustrates that the highest pressure amplitudes are observed at conditions

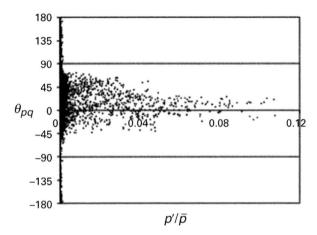

Figure 2.4 Measured dependence of combustion instability pressure amplitude on pressure–heat release (or, more precisely, the OH* chemiluminescence) phase [31]. Data courtesy of K. Kim and D. Santavicca.

where the pressure and heat release are in phase. It also illustrates that the pressure amplitudes are at background noise levels for $|\theta_{pq}|$ values greater than about 60 or 70 degrees. The fact that this threshold value is less than 90 degrees is probably due to acoustic damping in the combustor. We will discuss the factors influencing the phase between pressure and heat release more fully in Chapters 6 and 12.

The physical reason for this energy exchange follows from determining the conditions under which the unsteady heat release performs work on the gas. Heat release at constant pressure results in gas expansion, analogous to blowing up a balloon. Rayleigh's criterion states that the unsteady heat release performs work on the gas when this expansion occurs in phase with the pressure. This is analogous to the "pdV work" in the energy flux term, where an unsteady force (i.e., the pressure) must be in phase with the velocity (i.e., the gas dilatation rate) if net work is to be performed.

Equation (2.63) can be readily generalized to include a uniform mean flow, $\vec{u}_0$, as shown in Exercise 2.3. This leads to the following terms [6, 32]:

$$E_\Lambda = \frac{1}{2}\rho_0(\vec{u}_1 \cdot \vec{u}_1) + \frac{1}{2}\frac{p_1^2}{\rho_0 c_0^2} + \rho_1 \vec{u}_0 \cdot \vec{u}_1, \tag{2.67}$$

$$\vec{I}_\Lambda = (p_1 + \rho_0 \vec{u}_0 \cdot \vec{u}_1)\left(\vec{u}_1 + \frac{p_1}{\rho_0 c_0^2}\vec{u}_0\right) = p_1\vec{u}_1 + \rho_0\vec{u}_1(\vec{u}_0 \cdot \vec{u}_1) + \frac{p_1^2}{\rho_0 c_0^2}\vec{u}_0 + \frac{p_1 \vec{u}_0}{c_0^2}(\vec{u}_0 \cdot \vec{u}_1), \tag{2.68}$$

$$\Phi_\Lambda = \frac{(\gamma - 1)}{\gamma p_0}p_1 \dot{q}_1. \tag{2.69}$$

Note that the energy flux contains several new terms describing the convection of acoustic energy by the mean flow. In many practical cases these convective acoustic flux terms are very significant because the time average of p_1 and u_1 is small; see Exercise 2.4.

2.4 Disturbance Fields: Energy Density and Flux

Having considered the simple case of a uniform flow with only acoustic disturbances present, we now consider the general case of an inhomogeneous field with acoustic, entropy, and vortical modes present. Expressions for the energy density, flux, and source terms are [25, 33]:

$$E = \frac{p_1^2}{2\rho_0 c_0^2} + \frac{1}{2}\rho_0(\vec{u}_1 \cdot \vec{u}_1) + \rho_1 \vec{u}_0 \cdot \vec{u}_1 + \frac{\rho_0 T_0 s_1^2}{2c_p}, \quad (2.70)$$

$$\vec{I} = (p_1 + \rho_0 \vec{u}_0 \cdot \vec{u}_1)\left(\vec{u}_1 + \frac{p_1}{\rho_0}\vec{u}_0\right) + \rho_0 \vec{u}_0 T_1 s_1, \quad (2.71)$$

$$\Phi = \rho_0 \vec{u}_0 \cdot (\vec{\Omega}_1 \times \vec{u}_1) + \rho_1 \vec{u}_1 \cdot (\vec{\Omega}_0 \times \vec{u}_0)$$
$$- s_1(\rho_0 \vec{u}_1 + \rho_1 \vec{u}_0) \cdot \nabla T_0 + s_1 \rho_0 \vec{u}_0 \cdot \nabla T_1 + \left(\frac{\dot{q}_1 T_1}{T_0} - \frac{\dot{q}_0 T_1^2}{T_0^2}\right). \quad (2.72)$$

The fluctuations in this equation describe the superposition of acoustic, entropy, and vortical disturbances. Given that energy is an additive quantity, a natural question to consider is whether one can decompose the energy density, flux, and source terms into their contributions to each mode and, moreover, write three systems of equations for each mode, where

$$\begin{aligned} E &= E_\Lambda + E_s + E_\Omega, \\ \vec{I} &= \vec{I}_\Lambda + \vec{I}_s + \vec{I}_\Omega, \\ \Phi &= \Phi_\Lambda + \Phi_s + \Phi_\Omega. \end{aligned} \quad (2.73)$$

There is some ambiguity in this decomposition because of the various products. For example, does the term $\vec{u}_\Lambda \cdot \vec{u}_\Omega$ describe kinetic energy associated with the acoustic or vortical modes? Similar terms can be found in the flux and source terms. This question has been addressed in a limited manner in the technical literature [24, 26]. However, we will not consider it further, except for some speculations on which types of disturbance modes the various source terms below deposit energy into.

We will again discuss various terms in these expressions. Starting with the energy density, note the presence of a new term proportional to the square of the fluctuating entropy, $\frac{\rho_0 T_0 s_1^2}{2c_p}$. Recall that the fluctuating pressure term describes the work done in gas compression. Because the entropy disturbance occurs at constant volume, this term describes changes in internal energy associated with fluctuations in temperature. Similarly, the flux term contains the additional term $\rho_0 \vec{u}_0 T_1 s_1$, which is only nonzero in the presence of a mean flow. This term describes convection of energy associated with nonisentropic density fluctuations. Finally, a variety of new source terms are present. Starting with the unsteady heat release term, $\frac{\dot{q}_1 T_1}{T_0}$, note the difference from the earlier development, Eq. (2.63), where the product $\frac{(\gamma-1)}{\gamma p_0} p_1 \dot{q}_1$ appears. If the temperature fluctuations are isentropic, then $T_\Lambda = \frac{p_\Lambda}{\rho_0 c_p}$ and the two terms are identical. As such, the best interpretation of this result is that $\frac{(\gamma-1)}{\gamma p_0} p_1 \dot{q}_1$ represents the rate of energy addition to the acoustic field by the unsteady heat release, while $\frac{\dot{q}_1 T_1}{T_0}$ represents the rate of energy addition to the total disturbance field [23]. Note also that this term remains nonzero and does not change form for a one-dimensional, irrotational flow, suggesting that this total fluctuating energy is deposited into the acoustic and entropy modes, but not into the vortical mode.

Returning to the source terms, note that the first two terms involve the Coriolis term, $\rho_0 \vec{u}_0 \cdot (\vec{\Omega}_1 \times \vec{u}_1) + \rho_1 \vec{u}_1 \cdot (\vec{\Omega}_0 \times \vec{u}_0)$. The first term remains unchanged for purely incompressible flows, and describes increases in vortical energy associated with positive $\vec{\Omega}_1 \times \vec{u}_1$. Treating the Coriolis term as a pseudo-force, then both terms show that a positive dot product of the velocity with this pseudo-force leads to transfer of energy to unsteady vortical motions.

Finally, the last term contains an expression which is of similar form to the Rayleigh source term, with interchanges in the mean and fluctuating terms. This term describes energy input into the disturbance field by temperature fluctuations in heat release regions.

2.5 Linear and Nonlinear Stability of Disturbances

The disturbances which have been analyzed in this chapter arise because of underlying instabilities, either in the local flow profile (such as the hydrodynamic instabilities treated in Chapter 3) or in the coupled flame–combustor acoustic systems (such as thermoacoustic instabilities). This section provides a more general background to stability concepts by considering the time evolution of a disturbance of amplitude $\mathcal{A}(t)$ given by the first-order model

$$\frac{d\mathcal{A}(t)}{dt} = F_A - F_D, \qquad (2.74)$$

where F_A and F_D denote processes responsible for amplification and damping of the disturbance, respectively; see Figure 2.5. Thus, the amplitude of the oscillations grows if the rate of amplification exceeds dissipation. The primary attention of these next two

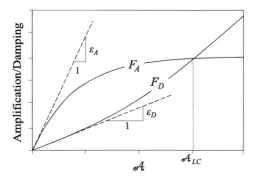

Figure 2.5 Illustration of the amplitude dependence of the amplification and damping terms, F_A and F_D.

sections is on the stability of the fixed point $\mathcal{A}(t) = 0$ to infinitesimal (Section 2.5.1) and finite-amplitude (Section 2.5.2) disturbances. If the fixed point is unstable, the system evolves to a new state, such as a limit cycle. While we discuss the stability characteristics of these more complicated system orbits in a cursory manner, the reader is referred to specialized treatments for more details on this topic.

2.5.1 Linearly Stable/Unstable Systems

In a linearly stable/unstable system, infinitesimally small disturbances decay/grow, respectively. An example of a linearly unstable situation is a pen perfectly balanced on its point on a flat desk, where any small disturbance causes it to fall away from this unstable equilibrium point.

To illustrate these points, we can expand the functions F_A and F_D around their $\mathcal{A} = 0$ values in a Taylor series:

$$F_A = \varepsilon_A \mathcal{A} + F_{A,NL}, \tag{2.75}$$

$$F_D = \varepsilon_D \mathcal{A} + F_{D,NL}, \tag{2.76}$$

where the subscript NL denotes the nonlinear part of F_A and F_D, and ε is given by

$$\varepsilon_A = \frac{\partial F_A}{\partial \mathcal{A}}\bigg|_{\mathcal{A}=0}. \tag{2.77}$$

The amplification and damping curves intersect at the origin, indicating that a zero-amplitude oscillation is a potential equilibrium point. For the example shown in Figure 2.5, however, this equilibrium point is unstable, since $\varepsilon_A > \varepsilon_D$ and any small disturbance that moves the system away from the origin produces a condition in which F_A is larger than F_D, resulting in further growth of the disturbance. Because these two curves diverge near the origin, their difference increases with amplitude, implying that $d\mathcal{A}(t)/dt$ increases with amplitude. In contrast, if $\varepsilon_A < \varepsilon_D$, the $\mathcal{A} = 0$ equilibrium point is stable to small disturbances and all fluctuations of $\mathcal{A}$ about this 0 value are damped. In this case, the $\mathcal{A} = 0$ point is an example of an "attractor" in that disturbances are drawn toward it. These results can also be seen mathematically by noting that the small-amplitude solution of Eq. (2.74) is

Linearized solution: $\quad \mathcal{A}_1(t) = \mathcal{A}(t=0) \exp\left((\varepsilon_A - \varepsilon_D)t\right).$ (2.78)

This linearized solution may be a reasonable approximation to the system dynamics if the system is linearly stable (unless the initial excitation $\mathcal{A}(t=0)$ is large, or there is large transient disturbance growth before decay, a phenomenon that occurs with multiple degree of freedom systems that have nonorthogonal eigenmodes; see Section 3.1). However, it clearly is only valid for some small time interval when the system is unstable, as disturbance amplitudes cannot increase indefinitely. In this situation, the amplitude dependence of system amplification/damping is needed to describe the system dynamics. Figure 2.5 describes a situation where F_A saturates and F_D increases linearly with the amplitude $\mathcal{A}$, resulting in an intersection of the two curves at the amplitude $\mathcal{A}_{LC}$. In practice, this saturation in F_A could be due to either nonlinearities in the gain, or phase characteristics of processes leading to amplification. For example, in combustion instability problems where the acoustic energy source term is given by Eq. (2.69), this can occur due to an amplitude-dependent relation between either the amplitude or the phase of p_1 and $\dot{q}_1$ (a topic discussed extensively in Chapter 12). Returning to Figure 2.5, note that this steady-state amplitude, $\mathcal{A}_{LC}$, is stable; i.e., a perturbation of the amplitude to the left (right) of this intersection point causes F_A to become larger (smaller) than F_D, causing the amplitude to increase (decrease) and return to its value $\mathcal{A}_{LC}$. Thus, for the configuration shown in Figure 2.5, there are two possible steady-state points, $\mathcal{A} = 0$ and $\mathcal{A} = \mathcal{A}_{LC}$. $\mathcal{A} = 0$ is a "repelling" fixed point and $\mathcal{A} = \mathcal{A}_{LC}$ an attractor.

The physical situation described by these solutions depends on the definition of $\mathcal{A}(t)$ for the specific problem of interest. Returning to the expansion of Eq. (2.55), if $\mathcal{A}(t)$ is defined, for example, by the expression

$$p(t) - p_0 = \mathcal{A}(t), \qquad (2.79)$$

then in this case the problem depicted in Figure 2.5 illustrates a situation where $p = p_0$ is a repelling point and the system will instead have the steady-state value of $p = p_0 + \mathcal{A}_{LC}$, a steady state that is stable to small disturbances. In many other problems, $\mathcal{A}(t)$ is used to describe the amplitude of a fluctuating disturbance, for example

$$p(t) - p_0 = \mathcal{A}(t) \cos(\omega t). \qquad (2.80)$$

For this situation, Figure 2.5 illustrates a situation where the steady-state fixed point $p = p_0$ is a repelling point and the system instead oscillates harmonically at a limit cycle amplitude $\mathcal{A}_{LC}$, $p(t) - p_0 = \mathcal{A}_{LC} \cos(\omega t)$. This limit cycle is stable – in other words, the system will be pulled back into this attracting, periodic orbit even when it is slightly perturbed. Figure 2.6 shows a phase portrait illustrating the instability of the fixed point and stability of the limit cycle that would occur for a situation as sketched in Figure 2.5.

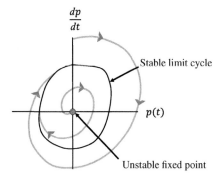

Figure 2.6 Phase portrait of stable limit cycle oscillation where the solid line denotes the limit cycle orbit. The light solid lines denote trajectories of orbits initially inside and outside the limit cycle, showing their attraction toward the stable limit cycle.

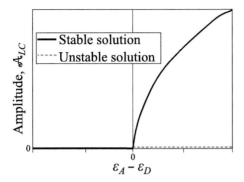

Figure 2.7 Supercritical bifurcation diagram.

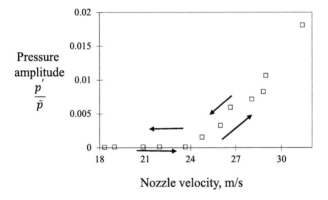

Figure 2.8 Measured pressure data illustrating supercritical bifurcation at a combustion nozzle velocity of 23.5 m/s. Data obtained by author.

We will assume that $\mathcal{A}(t)$ denotes the amplitude of a periodic disturbance, see Eq. (2.80), for the rest of the discussion in this chapter. Next, consider a situation where some combustor parameter is systematically varied in such a way that ε_A increases while ε_D remains constant; see Figure 2.5. For $\varepsilon_A < \varepsilon_D$, $\mathcal{A} = 0$ is the stable solution. However, when $\varepsilon_A > \varepsilon_D$, the solution $\mathcal{A} = 0$ becomes unstable, causing the amplitude of the disturbance to increase towards a new stable, limit cycle, equilibrium. The $\varepsilon_A = \varepsilon_D$ condition separates two regions of fundamentally different dynamics and is referred to as a *supercritical bifurcation point*. These ideas are illustrated in the bifurcation diagram in Figure 2.7, which shows the dependence of the amplitude, $\mathcal{A}_{LC}$, on $\varepsilon_A - \varepsilon_D$.

Figure 2.7 shows that as $\varepsilon_A - \varepsilon_D$ becomes positive, the $\mathcal{A}_{LC} = 0$ steady-state solution becomes linearly unstable and a new stable limit cycle solution arises. While $\varepsilon_A - \varepsilon_D$ is used as the stability parameter for Figure 2.7, it could be replaced in practice by any more directly measurable parameter that controls the system's stability; e.g., for the thermoacoustic stability problems described in Chapter 6, it could be the nozzle velocity, as shown in Figure 2.8 [34]. This data shows a smooth, monotonic dependence of the combustor pressure amplitude on the nozzle velocity, which is indicative of a supercritical bifurcation in this combustor.

2.5.2 Nonlinearly Unstable Systems

In a nonlinearly unstable system, small-amplitude disturbances decay while disturbances with amplitudes exceeding a critical value grow. This type of instability is referred to as *subcritical*. A simple example of a nonlinearly unstable system is shown in Figure 2.9, which shows a ball in a depression on the top of a hill. When pushed, this ball returns to its equilibrium point as long as it is subjected to disturbances with amplitudes that do not get it over the side walls of the depression. However, for a sufficiently large disturbance amplitude, the ball will roll out of the depression and down the hill. Although not shown, it will then reach some new steady state or oscillatory equilibrium.

Such behavior occurs frequently in the hydrodynamic stability of shear flows without inflection points (see Section 3.3), high activation energy kinetics (see Section 8.2.2.1), and thermoacoustic instabilities in combustors, to name a few. It has historically been referred to as "triggering" in the context of combustion instabilities in rockets [35].

Figure 2.10 provides an example of the amplitude dependence of F_A and F_D that produces the previously discussed behavior. In this case, the system has three

Figure 2.9 Example of a nonlinearly unstable system.

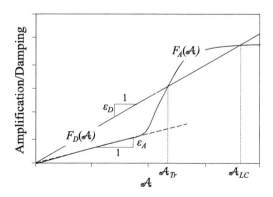

Figure 2.10 Illustration of the amplitude dependence of the amplification and damping terms, F_A and F_D, that leads to hysteresis and "triggering."

2.5 Linear and Nonlinear Stability of Disturbances

equilibrium points where the amplification and dissipation curves intersect. Specifically, the dissipation exceeds the amplification when $\mathcal{A} < \mathcal{A}_{Tr}$, indicating that $\mathcal{A} = 0$ is a stable fixed point, as all disturbances in the range $0 < \mathcal{A} < \mathcal{A}_{Tr}$ decay to $\mathcal{A} = 0$. The next equilibrium amplitude where the amplification and damping curves intersect is at $\mathcal{A} = \mathcal{A}_{Tr}$. This is an unstable equilibrium point because any disturbance that shifts the system from this point continues to increase in time. The third equilibrium point, $\mathcal{A} = \mathcal{A}_{LC}$, is also stable. Thus, in such a system all disturbances with amplitudes $\mathcal{A} < \mathcal{A}_{Tr}$ return to the solution $\mathcal{A} = 0$ (subject to the caveats in the next paragraph), and disturbances with amplitudes $\mathcal{A} > \mathcal{A}_{Tr}$ grow until their amplitude attains the value $\mathcal{A} = \mathcal{A}_{LC}$. Consequently, two stable solutions exist at this operating condition. The one observed at any point in time will depend on the history of the system. An example of such a system's time evolution in the phase plane is shown in Figure 2.11.

In almost all real problems, the system response consists of multiple modes, whose superposition describes the net spatiotemporal dynamics of the system pressure or velocity. If these modes are not orthogonal, or "nonnormal," then even if these modes have amplitudes that individually decay, their temporal superposition can lead to initial growth of the total disturbance magnitude before eventual decay. If this is the case, an initial disturbance whose $t = 0$ amplitude $\mathcal{A}$ is less than $\mathcal{A}_{Tr}$ will initially grow. This is a purely linear process. If the disturbance amplitude then grows to the point where it exceeds $\mathcal{A}_{Tr}$, nonlinear processes can take over and drive the system towards the $\mathcal{A} = \mathcal{A}_{LC}$ point. Thus, $\mathcal{A}(t=0)$ can still be less than $\mathcal{A}_{Tr}$ for "triggering" to occur. These points are discussed further in Sections 3.1 and 3.3.

A typical bifurcation diagram of this type of system is shown in Figure 2.12. It shows that for $\varepsilon_A < \varepsilon_D$, $\mathcal{A} = 0$ and $\mathcal{A} = \mathcal{A}_{LC}$ are stable solutions, as noted above. For $\varepsilon_A > \varepsilon_D$, the $\mathcal{A} = 0$ solution is unstable, and only a single stable solution is present. In this case,

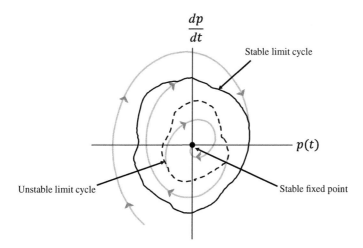

Figure 2.11 Phase portrait showing evolution of system states for a system with stable fixed point, unstable limit cycle, and stable limit cycle.

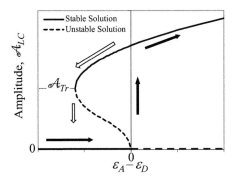

Figure 2.12. Subcritical bifurcation diagram showing hysteretic behavior.

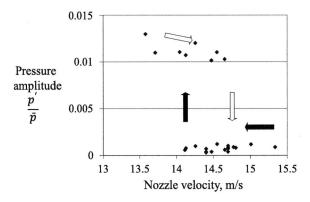

Figure 2.13 Experimental data showing evidence of a subcritical bifurcation. Data obtained by author.

if a system parameter is monotonically increased to change the sign of $\varepsilon_A - \varepsilon_D$ from negative to positive, the system's amplitude jumps discontinuously from $\mathcal{A} = 0$ to $\mathcal{A} = \mathcal{A}_{LC}$ at $\varepsilon_A - \varepsilon_D = 0$. Hysteresis is also present in the system, as illustrated by the fact that if the system parameter is subsequently decreased, the system's steady-state amplitude decreases as it follows the stable branch on top, even for a range of $\varepsilon_A < \varepsilon_D$ values, before it discontinuously "jumps" to the $\mathcal{A} = 0$ solution. Thermoacoustic instability data exhibiting such behavior is presented in Figure 2.13 [34].

2.5.3 Forced and Limit Cycling Systems

This section further discusses the characteristics of systems that are externally forced, as well as those that display intrinsic oscillations in the absence of forcing because of instability. In practice, the spectra of a lightly damped system that is externally excited by broadband noise, and a limit cycling system with additive noise, may not be distinguishable – they both exhibit a narrowband response at the system natural frequencies. The question of how to distinguish between these two possibilities appears routinely in interpretation of data from thermoacoustically unstable combustors [36]. This is an important question, as the approaches needed to model the characteristics of these systems, such as the amplitude of oscillations, are fundamentally different.

This section considers several examples to illustrate different facets of this problem. It starts by reviewing the forcing of lightly damped, linear systems in Section 2.5.3.1, then proceeds to limit cycle oscillations of linearly unstable systems in Section 2.5.3.2. Section 2.5.3.3 treats an important generalization of these two problems – the response of a limit cycling system to external forcing. Finally, Section 2.5.3.4 briefly treats nonlinear interactions between two self-excited oscillators. These are fascinating topics, of which we can only scratch the surface; the reader is referred to specialized texts on nonlinear oscillations for more detailed discussions [37].

2.5.3.1 Example: Forced Response of Lightly Damped, Linear Systems

In this section we briefly review the response of a lightly damped, second-order system to forcing. These response characteristics can be understood from the following model equation for a linear, second-order oscillator [37]:

$$\frac{d^2 x}{dt^2} + 2\zeta \omega_0 \frac{dx}{dt} + \omega_0^2 x = F(t), \tag{2.81}$$

where ζ denotes the damping parameter, and $F(t)$ an arbitrary forcing function. It is most convenient to revert to the frequency domain and write the solution for the Fourier transform of $x(t)$, given by $\hat{x}(\omega)$, as

$$|\hat{x}(\omega)| = \frac{|\hat{F}(\omega)|}{\omega_0^2 \sqrt{\left(1 - \left(\frac{\omega}{\omega_0}\right)^2\right)^2 + 4\zeta^2 \left(\frac{\omega}{\omega_0}\right)^2}}. \tag{2.82}$$

Assuming a broadband excitation source, $\hat{F}(\omega)$, the magnitude of $\hat{x}(\omega)$ peaks at ω_0 (with an $O(\zeta^2)$ correction). This peak in the frequency domain can be quite narrow. One measure of this width is to define it via the half-maxima, $\Delta \omega$, which for a lightly damped system is

$$\frac{\Delta \omega}{\omega_0} = 2\zeta + O(\zeta^2). \tag{2.83}$$

This expression shows that stable systems exhibit a very narrowband response when $\zeta \ll 1$, making it difficult to differentiate them from a system exhibiting intrinsic, self-excited oscillations. However, a key difference from the self-excited system is that the system requires a persistent excitation source for the oscillations. These oscillations are dissipated at a rate proportional to the damping time scale, $1/(\omega_0 \zeta)$. As such, oscillations excited at some time $t = 0$ will make negligible contribution to the disturbance amplitude at times $t \gg 1/(\omega_0 \zeta) = t_2$. Moreover, oscillations in some time interval around t_2 will be poorly correlated with those around the time interval at $t = 0$. It therefore follows that the damping coefficient is directly related to the time interval over which oscillations are correlated. This point can be quantified using the autocorrelation function,

$$G(\tau) = \frac{\langle x(t)x(t+\tau)\rangle}{\langle x^2(t)\rangle}. \qquad (2.84)$$

Assuming that the power spectrum of the forcing term $\hat{F}(\omega)$ is approximately constant over the oscillator bandwidth, it can be shown that $G(\tau)$ is given by [38]

$$G(\tau) = e^{-\omega_0 \zeta \tau} \cos(\omega_0 \tau) + O(\zeta). \qquad (2.85)$$

This expression shows that the autocorrelation coefficient oscillates at the natural frequency with increasing time delay and decays at a rate given by the damping coefficient. This decay in autocorrelation is a diagnostic for differentiating self-excited from lightly damped, noise-driven systems [39, 40].

2.5.3.2 Example: Limit Cycling Systems

The following model equation generalizes Eq. (2.81) to include nonlinear terms and is useful for illustrating a nonlinear, second-order oscillator that displays intrinsic oscillations [37]:

$$\frac{d^2 x}{dt^2} + 2\zeta \omega_0 (1 - \beta x^2) \frac{dx}{dt} + \omega_0^2 x = 0. \qquad (2.86)$$

The $x = 0$ equilibrium point is linearly unstable when $\zeta < 0$ and the system evolves to a limit cycle orbit at an amplitude where the time average of the linear negative damping and nonlinear positive damping terms balance, similar to Figure 2.6. Exercise 2.6 derives the following limit cycle solution amplitude for small values of ζ [37]:

$$x(t) = \frac{2}{\sqrt{\beta}} \cos(\omega_0 t), \qquad (2.87)$$

i.e., $A_{LC} = 2/\sqrt{\beta}$. These are "self-excited" oscillations, as they are present even in the absence of external forcing.

2.5.3.3 Example: Forced Response of Limit Cycling Systems

This section considers an example problem where forced oscillations are present in a system exhibiting intrinsic oscillations. We will consider two cases: single-frequency excitation and random excitation. The single-frequency excitation problem occurs, for example, in situations where acoustic oscillations act as background forcing to inherent fluid mechanic instabilities. The noise-driven problem approximately models the high degree of freedom turbulent fluctuations that are present in high Reynolds number systems.

In either case, because of nonlinearities, the effect of forced oscillations during limit cycle conditions is not simply additive. Rather, strong nonlinear interactions occur between the limit cycle oscillations and the forced oscillations, which are typically at different frequencies.

These nonlinear interactions can be illustrated by adding forcing terms to Eq. (2.86):

2.5 Linear and Nonlinear Stability of Disturbances

$$\frac{d^2x}{dt^2} + 2\zeta\omega_0\left(1 + F_1(t) - \beta x^2\right)\frac{dx}{dt} + \omega_0^2 x = F_2(t). \tag{2.88}$$

The term $F_2(t)$ is referred to as "additive" forcing, while disturbances in the parameters ζ and β are referred to as "parametric" disturbances. Here, $F_1(t)$ is a parametric disturbance in the damping rate ζ.

Consider first a single-frequency, additive forcing example, where $F_2(t) = \mathcal{A}_f \cos \omega_f t$ and $F_1(t) = 0$. Assuming ω_f is sufficiently far from ω_0, and that $\omega_0 \zeta \ll 1$, an approximate solution can be written as the superposition of the natural and forced oscillations:

$$x(t) = \mathcal{A}(t) \cos(\omega_0 t) + \frac{\mathcal{A}_f}{\omega_0^2 - \omega_f^2} \cos \omega_f t, \tag{2.89}$$

where the steady-state amplitude of $\mathcal{A}(t)$ is

$$\mathcal{A}_{LC} = \frac{2}{\sqrt{\beta}}\sqrt{1 - \frac{\beta \mathcal{A}_f^2}{2\left(\omega_0^2 - \omega_f^2\right)^2}}. \tag{2.90}$$

This dependence of $\mathcal{A}_{LC}$ is plotted in Figure 2.14 as a function of the forcing amplitude, $\mathcal{A}_f$. It shows the monotonic reduction in limit cycle amplitude, $\mathcal{A}_{LC}$, until its amplitude reaches zero, as the forcing amplitude $\mathcal{A}_f$ is increased. This behavior is referred to as "frequency locking" or "entrainment." Such behavior is observed in both hydrodynamic and thermoacoustic instability problems and is referred to in Sections 3.3, 4.2.4, and 12.5.

The reduction in amplitude of the self-excited, limit cycle oscillations with increased forcing amplitude is associated with the saturating nature of the nonlinearity. To understand why it occurs, consider a simplified example with a "hard saturation," where the amplitude of oscillations of $x(t)$ above some threshold cannot exceed $\mathcal{A}_{total}$, which is a constant. Write $\mathcal{A}_{total} = \mathcal{A}_{LC} + \mathcal{A}_f$ (a simplification because of the inherent excitation of harmonics, but nonetheless useful for the purposes of this discussion). If $\mathcal{A}_{total}$ is constant because of the assumed hard saturating nonlinearity, then increasing $\mathcal{A}_f$ implies that $\mathcal{A}_{LC}$ must decrease.

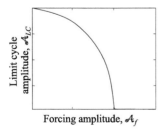

Figure 2.14 Demonstration of frequency locking, showing the dependence of limit cycle amplitude, $\mathcal{A}_{LC}$, on forcing amplitude, $\mathcal{A}_f$, for the example problem.

Consider next a noise-driven example where $F_1(t) = \tilde{\varepsilon}_1 \sigma_1(t)$, $F_2(t) = \varepsilon_2 \sigma_2(t)$, $\sigma_i(t)$ is zero-mean, Gaussian white noise, $\langle \sigma_i(t)\sigma_j(t')\rangle = \delta_{ij}\delta(t-t')$, and $\tilde{\varepsilon}_1$ and ε_2 are the root mean square (RMS) of the noise terms F_1 and F_2 respectively. Defining an instantaneous amplitude and phase by $x(t) = A(t)\sin(\omega_0 t + \theta(t))$, the stationary distribution of the probability density function of the amplitude $A(t)$ is given by (see Exercise 2.7) [41]:

$$p_{st}(A) = NA z^{\left\{-\frac{2}{3} - \frac{8\varsigma\omega_0}{3\varepsilon_1^2}\left(1 + \frac{\beta\varepsilon_2^2}{3\omega_0^2\varepsilon_1^2}\right)\right\}} \exp\left(\frac{-2\beta\varsigma\omega_0 A^2}{3\varepsilon_1^2}\right), \qquad (2.91)$$

where $z = \frac{3\varepsilon_1^2 A^2}{16} + \frac{\varepsilon_2^2}{4\omega_0^2}$, N is the normalization constant, and $\varepsilon_1 = -2\varsigma\omega_0\tilde{\varepsilon}_1$. The stationary probability density function (PDF) of the phase, $\theta(t)$, is uniform; i.e., all phase values are equally probable.

Figure 2.15 plots this stationary amplitude distribution for several parametric forcing intensities ε_1. The vertical line denotes the steady-state amplitude of the forced deterministic solution. The figure demonstrates several effects of noise which in probability space are referred to as "diffusion" and "drift." Noise broadens the distribution of amplitudes, a manifestation of "diffusion." Parametric noise also shifts the average amplitude, a manifestation of "noise-induced drift," and increases the tails of the PDFs. Finally, noise can alter the stability characteristics of the system, referred to as "noise-induced transitions."

The right image in Figure 2.15 plots the stationary amplitude distribution for several additive noise intensities ε_2. "Diffusion," i.e., the broadening of the amplitude, is clearly evident in this case as well.

2.5.3.4 Nonlinear Interactions between Multiple Oscillators

The previous section showed that external excitation, even additive forcing, alters the nature of self-excited limit cycle oscillations because of system nonlinearities. In this

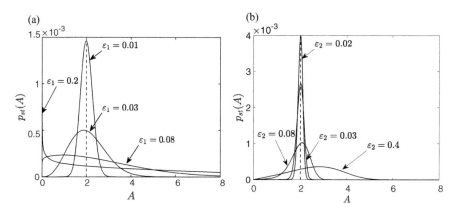

Figure 2.15 Stationary probability distribution of (a) parametrically forced, nonlinear oscillator ($\varsigma\omega_0 = -5 \times 10^{-4}$, $\beta = 1$, $\varepsilon_2 = 0$) and (b) additively forced, nonlinear oscillator ($\varsigma\omega_0 = -5 \times 10^{-4}$, $\beta = 1$, $\varepsilon_1 = 0$). In both cases, the vertical dashed line denotes the deterministic amplitude without forcing, Eq. (2.87).

section we show that similar interactions occur when a system has multiple self-excited oscillators that nonlinearly couple. Important examples of this are unstable combustors with multiple natural acoustic modes that may be simultaneously excited. For this problem, a linearized analysis would suggest that oscillations associated with each oscillator should be present. However, nonlinear interactions may lead to very different results, as discussed next. To illustrate, consider the following example with two nonlinearly coupled oscillators:

$$\frac{d^2\eta_a}{dt^2} + 2\zeta_a\omega_a \mathcal{D}(\eta_a,\eta_b)\frac{d\eta_a}{dt} + \omega_a^2\eta_a = 0,$$
$$\frac{d^2\eta_b}{dt^2} + 2\zeta_b\omega_b \mathcal{D}(\eta_a,\eta_b)\frac{d\eta_b}{dt} + \omega_b^2\eta_b = 0,$$
(2.92)

where

$$\mathcal{D}(\eta_a,\eta_b) = 1 - \beta(\eta_a^2 + \eta_b^2)$$
(2.93)

is a nonlinear damping coefficient that couples the two modes when $\beta \neq 0$. A typical computed solution is plotted in Figure 2.16 for parameter values where both modes are linearly unstable, i.e., $\zeta_a, \zeta_b < 0$. As such, both $\mathcal{A}_a$ and $\mathcal{A}_b$ initially grow exponentially with time, as nonlinear effects are negligible. However, for normalized time values greater than about 20, nonlinear interactions noticeably influence the results. Beyond a dimensionless time of 40, the amplitude of $\mathcal{A}_b$ begins to drop and eventually goes essentially to zero. As such, although both oscillators are linearly unstable, *only oscillator a is observed at the limit cycle*. In other words, a linear analysis could

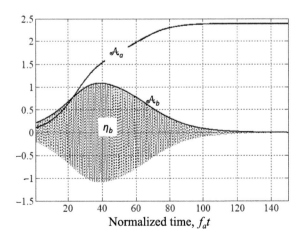

Figure 2.16 Time variation of modal amplitudes, showing nonlinear modal interactions of two oscillators described by Eqs. (2.92) and (2.93), where the solid line denotes amplitude and the dashed line denotes instantaneous value ($2\zeta_a\omega_a = -0.13$, $2\zeta_b\omega_b = -0.20$, $\beta = 0.7$, $\omega_a = 2\pi$ and $\omega_b = 6\pi$). Calculation courtesy of Shreekrishna.

correctly predict that two modes are simultaneously linearly unstable, but not be able to describe the fact that only one of them is experimentally observed.

2.6 Complex Phase Space Dynamics

Figures 2.6 and 2.11 summarize potential topological features of phase space orbits for second-order systems. Indeed, for second-order nonlinear differential equations with constant coefficients, a small set of behaviors are possible: stable/unstable fixed points or stable/unstable limit cycles. Limit cycles are closed trajectories in phase space; they can be stable (orbits spiral into them) or unstable (orbits spiral away from them). Each limit cycle divides the phase space into two regions, inside and outside the closed orbit.

For continuous systems, much more complex dynamics are possible at third order or higher [42]. The transition from second to third order or higher allows completely new behaviors. One can envision arbitrarily complex phase orbits in three dimensions, where nonintersecting lines can trace out complex orbits that intertangle and wrap around each other without ever intersecting, much like a tangled ball of string. Limit cycles can still occur, although they may exhibit multiple periodicities before closing their loops. Bounded orbits are possible that never intersect, though this cannot occur in a two-dimensional phase plane. Such nonintersecting orbits can occur with systems that exhibit oscillations at two different frequencies that are not integer multiples, termed "quasi-periodic." In addition, "chaotic" orbits are possible, where orbits that are bounded (i.e., not exponentially diverging) do not intersect, and where two adjacent initial conditions diverge exponentially (although never getting too far apart because the orbits are bounded). The rate of divergence of two closely spaced initial conditions is quantified via the "Lyapunov exponent." Because of the inherent uncertainty in any measurement or computer resolution, this divergence of initial trajectories implies that one loses the ability to predict future system behavior after a certain time. Any small variation/uncertainty in initial conditions eventually leads to completely different time variation of the system for two calculations. In other words, even though the system is deterministic (i.e., its initial conditions uniquely prescribe its future state for all future time), it is unpredictable.

Examples of three-dimensional phase orbits for a "period-1," "period-2," and chaotic orbit are shown in Figure 2.17. Shown below each phase orbit is a "Poincaré section," which reduces the dimensionality of the representation by one, and shows the points of intersection of these orbits with a cut-plane. In this representation, a period-1 limit cycle becomes a single point, i.e., a fixed point. A period-2 orbit manifests as two dots, and the chaotic orbit as a structured sea of points.

In addition, intermittent orbits can occur. Intermittency occurs when the system switches back and forth between two qualitatively different behaviors, even in the absence of background noise or variation in system parameters. For example, the system can switch between almost periodic behavior and chaotic behavior, with the average time that it spends in each basin of attraction depending on some system

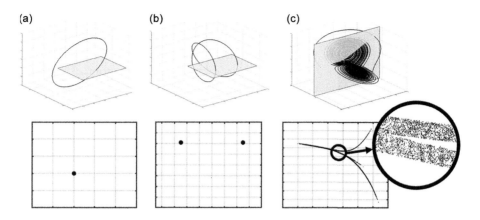

Figure 2.17 Phase orbits for (a) period-1 and (b) period-2 limit cycles, and (c) chaotic orbits (using Lorentz equations for this example [42]). A corresponding Poincaré section for each case is shown below each image.

parameter (such as the coefficient multiplying a nonlinearity). Intermittency and chaotic behavior have been characterized in detail in combustion applications [43].

2.7 Decompositions of Data

High Reynolds number reacting flows often consist of a superposition of high degree of freedom, turbulent flow disturbances, narrowband acoustic disturbances, and quasi-coherent flow disturbances associated with spatially or globally unstable hydrodynamic disturbances. Given the massive size of the space–time resolved data coming from computations and experiments, a fundamental question arises around how to develop reduced-order models, extract insights into dominant flow/flame features, filter out less important features, and generally extract insight into controlling processes. One way to do this is to express these data as a superposition of simpler functions. For reduced-order modeling, one can do a Galerkin projection of these modes onto a given partial differential equation to derive a set of ordinary differential equations for the time coefficients; see, e.g., Section 6.6.1, where natural acoustic mode shapes are projected onto the Euler equations to derive a set of equations for the nonlinear evolution of acoustic modes.

A common linear approach for such decompositions is to expand the data as a superposition of basis functions (or "modes"):

$$u(\vec{x}, t) = \sum_m \beta_m F_m(\vec{x}, t), \tag{2.94}$$

where any complete, orthogonal set of basis functions can exactly represent the original data or any other function. So what should these functions be? There are a variety of possible approaches, and no "right answer." For example, there are numerous sets of complete, orthogonal basis functions, such as polynomials (t^m), harmonic functions ($\sin(m\omega t)$), weighted Bessel functions ($J_m(t)$), a large class of other solutions of second-order ordinary differential equations (see Section 6.7), or a host of other functions. However, a *truncated* superposition of such basis functions will necessarily provide varying degrees of fidelity for the original data; more significantly, it may not reflect the underlying functional dependence of the phenomenon. For example, while a sinusoidal function can certainly be represented by an infinite series of polynomials, truncation of this summation can lead to significant distortions in the reconstructed data and will extrapolate poorly outside the fit region. Moreover, a sinusoidally oscillating function is most naturally represented by a sinusoidally oscillating basis function. This is a critical issue, as one would like to represent the data by a small number of coefficients and space/time mode shapes that concentrate the energy or other metric of interest into a small number of modes. Superposition of these modes will then ideally provide a key summary of the dominant flow features, with substantial reduction in independent numbers. This naturally leads to the question of the most useful basis functions, F_m.

This is a major topic, and the intention of this section is to briefly introduce a few key linear approaches. The reader is referred to the excellent references that are dedicated to these topics more generally [44, 45]. In addition, nonlinear decompositions are not covered here, but references on this topic exist as well [46]. One way to differentiate the myriad different possibilities for F_m is to organize them by those which are imposed on the data, either based on (i) some theoretical model (e.g., using Bessel functions for decomposing acoustic data from cylindrical geometries, see Section 5.3.3, or other linearized solutions of the Navier–Stokes equations) or a useful set of orthogonal functions (e.g., Fourier series) – we will refer to these as "prescribed basis functions"; or (ii) to use the data itself to extract a set of basis functions – these are "empirical basis functions." These are both discussed further in the following sections.

2.7.1 Prescribed Basis Functions

For the treatment of imposed basis functions, this section will consider Fourier series in most detail, closely following Mallat [44]. The discussion below will treat one dimension, using time, but can also be generalized to multiple dimensions. The Fourier integrals are defined as

$$\hat{f}(\omega) = \int_{-\infty}^{+\infty} f(t) e^{-i\omega t} dt, \qquad (2.95)$$

$$f(t) = \frac{1}{2\pi} \int_{-\infty}^{+\infty} \hat{f}(\omega) e^{i\omega t} d\omega. \qquad (2.96)$$

Examples of temporal Fourier series are shown in Figures 4.17 and 6.24. The magnitude of $\hat{f}(\omega)$ provides a measure of the amplitude of oscillations at that given frequency. How natural such a representation of the data is depends on the problem; a function that is naturally described by a truncated polynomial series like $F = a + bt + ct^2$ is clearly better described by polynomial basis functions. However, there is a large class of problems where such representations are very useful. First, complex sinusoidals, $e^{i\omega t}$, are the eigenvectors of time-invariant linear systems [44], implying that they are natural basis functions for linearized systems; they are even used in situations with strong nonlinearities. For example, spatial Fourier-based representations are useful for describing the turbulence cascade, i.e., that turbulent kinetic energy scales as $k^{-5/3}$ in the inertial subrange.

Consider next the relationship between $f(t)$ and $\hat{f}(\omega)$ as summarized in Table 2.2, where * denotes the convolution operator:

$$f(t) * g(t) = \int_{-\infty}^{\infty} f(t')g(t-t')dt' = \int_{-\infty}^{\infty} g(t')f(t-t')dt'. \tag{2.97}$$

An additional inequality, referred to as the time–frequency uncertainty principle, is

$$\sigma_t^2 \sigma_\omega^2 \geq \frac{1}{4}, \tag{2.98}$$

where σ_t and σ_ω denote the variance of the temporal signal and its Fourier transform. This inequality provides insights into the relationship between the temporal localization of a signal and the range of frequencies over which its Fourier transform is distributed, and is illustrated in Figure 2.18. This figure plots the time variation of a signal and its Fourier transform. The area of the box, given by the product $\sigma_t \sigma_w$, cannot be arbitrarily small but must satisfy the inequality in Eq. (2.98). Decreasing the width of the box causes its height to increase, and vice versa. For example, a Dirac delta function, $\delta(t - t_0)$, is the extreme example of a temporally localized function; in contrast, its Fourier transform, $e^{-i\omega t_0}$, has a uniform magnitude over all frequencies. While this frequency distribution goes all the way to infinity for the Dirac delta

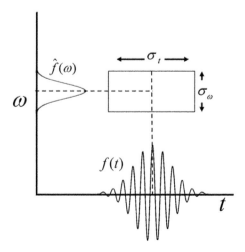

Figure 2.18 Illustration of the time–frequency uncertainty principle in Eq. (2.98). Adapted from Mallat [44].

Table 2.2 Summary of the properties of the Fourier transform [44].

Property	Function	Fourier transform
Convolution	$f_1(t) * f_2(t)$	$\hat{f}_1(\omega)\hat{f}_2(\omega)$
Multiplication	$f_1(t)f_2(t)$	$\frac{1}{2\pi}\hat{f}_1(\omega) * \hat{f}_2(\omega)$
Translation	$f(t-u)$	$e^{-iu\omega}\hat{f}(\omega)$
Modulation	$e^{i\xi t}f(t)$	$\hat{f}(\omega - \xi)$
Scaling	$f(t/s)$	$|s|\hat{f}(s\omega)$
Time derivatives	df/dt	$-i\omega\hat{f}(\omega)$

function, an approximation of the delta function with temporal width τ will have uniform frequency content up to frequencies of $\sim 1/\tau$. The inverse of this example is a harmonically oscillating signal at ω_0 that exists for $-\infty < t < \infty$: its Fourier transform is a Dirac delta function, $\delta(\omega - \omega_0)$; i.e., all of its frequency content is concentrated at a single frequency. Truncation of the signal to a temporal duration T (as must be the case for all real sampled data), causes broadening of its spectral content to a frequency range $1/T$.

Referring to Table 2.2, insights into these behaviors are also evident in the "scaling" property of the Fourier transform. Compressing the signal in time (i.e., increasing the decay rate of an exponential, or decreasing the period of oscillations) expands the signal in the frequency dimension, i.e., causing it to be represented by larger frequencies.

Consider the other properties in Table 2.2. The "translation" property shows that shifting a signal in time or space does not alter the magnitude of the Fourier transform; i.e., the magnitude of the Fourier transform of $f(t)$ and $f(t-t_0)$ is the same. Rather, a phase shift is added to the Fourier transform, so that the phase of the Fourier transform between the two signals differs by $e^{-i\omega t_0}$. The analog of this property is "modulation" – shifting the spectral content of a signal from $\hat{f}(\omega)$ to $\hat{f}(\omega - \omega_0)$ is equivalent to multiplying the time domain signal by $e^{i\omega_0 t}$.

The "convolution" property provides insight into how a transfer function acting on an input signal alters its time domain input. For example, an excitation signal with a spectral content of $\hat{f}_1(\omega)$, multiplied by some transfer function $\hat{f}_2(\omega)$, causes the time domain version of the output to equal the temporal convolution of these two signals. The inverse of this property is "multiplication," where the product of two signals equals the convolution of their Fourier transforms in the frequency domain. This property provides insight into how nonlinear interactions between two quantities cause the spectra to evolve. An example calculation simulating a convective non-linearity is included in Exercise 2.5.

The "derivative" property shows that the Fourier transform of the derivative of a quantity shifts the phase of the Fourier transform at all frequencies by 90 degrees, and its magnitude by $\omega |\hat{f}(\omega)|$.

Finally, we have Parseval's formula:

$$\int_{-\infty}^{+\infty} |f(t)|^2 dt = \frac{1}{2\pi} \int_{-\infty}^{+\infty} |\hat{f}(\omega)|^2 d\omega. \qquad (2.99)$$

This relation shows the equivalence of "energy" in the time and frequency domain. In other words, the Fourier transform is a nonlocal transformation: a change in $f(t)$ in some small interval around t_0 affects $\hat{f}(\omega)$ for all ω, and vice versa. Nonetheless, certain integral measures of both signals are preserved and related. For example, given a time domain waveform with a given RMS (the left-hand side of Eq. (2.99)), very narrow bandwidth signals will necessarily have high Fourier transform magnitudes, limiting to a delta function for a pure tone such that the right-hand side of Eq. (2.99) remains fixed. In comparing the spectra of two time domain signals with the same RMS, the one with more spectrally broadband content will necessarily have lower magnitudes in the frequency domain, as the same "energy" is spread over a broader range of frequencies. Finally, Parseval's formula shows how each component of the spectrum of a signal contributes to the RMS of its time domain equivalent.

Consider next the spatial Fourier transform, which, in most respects, mirrors the spatial transform properties. One important case is decomposition of azimuthal dependencies for problems with circular coordinate systems, i.e., where $f = f(\theta)$:

$$\hat{f}_m = \frac{1}{2\pi} \int_0^{2\pi} f(\theta) e^{-im\theta} d\theta. \qquad (2.100)$$

In this case, the index m is not continuous but discrete. This is a common decomposition for characterizing the dominant eigenmodes of circular jets/wake flows, and will be used in Chapter 4. In this case, $m = 0$ denotes an axisymmetric disturbance and $|m| \neq 0$ denotes azimuthally periodic disturbances, such as helical or other three-dimensional disturbances.

Wavelets are a generalization of the Fourier transform. Rather than taking the inner product of the original function with $e^{-i\omega t}$, an alternative basis function $\psi(t)$ with a more restricted temporal domain is used (e.g., $e^{-i\omega t} e^{-\sigma t^2}$). Then, by considering the inner product of this basis function that is translated in time and rescaled as $\psi((t - t_0)/a)$, where a is an "inverse frequency," as shown in Figure 2.19, the wavelet transform is then a function of t_0 and a:

$$\hat{f}(t_0, a) = \int_t f(t) \psi\left(\frac{t - t_0}{a}\right) dt. \qquad (2.101)$$

An example application area where wavelets are useful are nonstationary data, whose spectra or statistics vary in time. There are a variety of other wavelet applications and potential wavelet basis functions; the reader is referred to Mallat's text [44] for coverage of this topic.

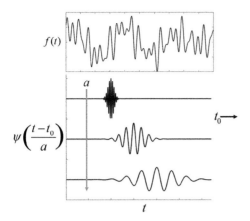

Figure 2.19 Graphical illustration of wavelet operation, showing the wavelet basis function $\psi(t)$ and the translation and rescaling operations.

2.7.2 Empirical Basis Functions

2.7.2.1 Proper Orthogonal Decomposition: One-Dimensional Data

As noted at the beginning of this section, different types of orthogonal, complete sets of basis functions will have different rates of convergence in terms of the number of terms to include in the summation in Eq. (2.94) to capture a given data set. This point naturally motivates the question of whether one can determine an "optimal" set of orthogonal basis functions, where the first j modes (for arbitrary j) more accurately represent the underlying data than all other potential orthogonal sets of basis functions. The answer to this question is yes, and the nature of these modes is the focus of this subsection. There are a number of good references to proper orthogonal decomposition (POD, also referred to as Karhunen–Loeve decomposition) to which the reader is referred [44, 47]; this section will be light on mathematical rigor and primarily focus on what the decomposition is doing, starting with one-dimensional data.

The basic approach is sketched out here for a zero-mean, one-dimensional function defined over some time interval [0, 1], where the basis functions are orthogonal and normalized such that $\int_0^1 F_n(t) F_m(t) dt = \delta_{nm}$. It can, however, be generalized in a straightforward way to an arbitrary number of dimensions, a topic that is touched on in the next section. As such, write Eq. (2.94) as

$$u(t) = \sum_{m=1}^{\infty} \beta_m F_m(t). \tag{2.102}$$

The data $u(t)$ may be random, and so the functions β_m would be as well, but $F_m(t)$ is not. Define $u_j(t)$ based on truncation of this summation to an arbitrary j, and the error in the estimate as $e_j(t) = u(t) - u_j(t)$:

$$u_j(t) = \sum_{m=1}^{j} \beta_m F_m(t). \tag{2.103}$$

Next, we must quantify the approach that will be used to quantify the error that is to be minimized. The POD modes are the ones that minimize the following L_2 inner product for arbitrary j:

$$\mathcal{E}_j = \int_0^1 e_j^2(t)dt. \qquad (2.104)$$

Stated differently, the POD modes are the ones where the variance of $u_j(t)$ is closest to the variance of the original signal, $u(t)$, for any value of j. This problem can be solved using the calculus of variations (i.e., find the functions F_m that minimize $\mathcal{E}_j$) and, without proof, are given by the solution to [47]:

$$\int_0^1 \langle u(t)u(t')\rangle F_m(t')dt' = \lambda_m F_m(t). \qquad (2.105)$$

This equation shows that the POD basis functions are eigenfunctions of an integral equation with a kernel given by the autocorrelation function, $\langle u(t)u(t')\rangle$, whose eigenvalues are λ_m. In the discrete case (as it will always be for decomposing data or computations), the POD becomes equivalent to principal component analysis (PCA), where the eigenvectors are the PCA axes.

A few observations on the decomposition. First, consider how the variance (or "energy") in the original signal is distributed among the different modes. Defining the "energy" as

$$E \equiv \int_0^1 \langle u^2(t)\rangle dt, \qquad (2.106)$$

it can be shown that

$$E = \sum_{m=1}^\infty \langle \beta_m^2\rangle = \sum_{m=1}^\infty \lambda_m. \qquad (2.107)$$

Thus, the mth eigenvalue, λ_m, equals the "energy" associated with that mode, and the overall "energy" of the signal equals the sum of the energy in each mode. This property is reminiscent of Parseval's formula, Eq. (2.99). Note that in defining this set of basis functions as optimal, we have assumed that the eigenvalues are ordered by magnitude.

Second, it should be emphasized that, in determining these basis functions, it is necessary to prescribe the approach for quantifying "goodness of fit." The POD modes are optimal with respect to the L_2 inner product, but one could certainly define other distance metrics and would have a different set of optimal basis functions. Because this inner product is proportional to the squared distance of the estimate from the real function, it will necessarily have sensitivity to outliers. This is analogous to the sensitivity of regression approaches to outliers when fitting data using mean-squared minimization approaches to fitting a line or polynomial through data.

Third, it is important to recognize that the appearance of a POD mode exhibiting coherent structure does not imply an underlying pattern in the data. In fact, under certain homogeneity/stationarity assumptions in space–time, the POD of a random signal converges to Fourier modes [44]. In such cases, the POD basis yields sinusoidal mode shapes which may appear to show underlying patterns in the data, even though such shapes are simply the optimal way for the decomposition to capture random variation in the signal.

2.7.2.2 Proper Orthogonal Decomposition: Spatiotemporal Data

While the above example has been worked out for one-dimensional data, this section discusses generalizations to multidimensional data, in particular where a quantity is determined as a function of both space and time. For example, Figure 2.20 shows a sequence of images of a bluff body stabilized flame that is axially forced at a frequency of f_f.

One POD approach is to explicitly separate the space and time dependence:

$$u(\vec{x}, t) = \sum_{m=1}^{\infty} \beta_m(t) F_m(\vec{x}). \tag{2.108}$$

In this formulation, the coefficients are functions of time, and the modes are only functions of space. This approach is useful for reduced-order modeling; e.g., spatial POD modes can be projected onto the Navier–Stokes equations in order to develop a system of nonlinear ordinary differential equations for the temporal coefficients [47]. A typical example showing the application of this decomposition to the images in Figure 2.20 is shown in Figure 2.21, showing POD spatial modes 1, 2, and 7, the spectra of their time coefficients, and the relative energy in each POD mode.

The original data in Figure 2.20 clearly show the rollup of the flame by a series of convecting vortical structures. It can be seen how a superposition of the spatial modes in Figure 2.21 reproduces this pattern; increasing the number of modes captures the underlying data with increasing fidelity. The energy convergence plot shows that this data set is well captured by a small number of modes: four modes capture more than half of the variance in the original signal. In many practical cases, however, many more modes are needed to capture a comparable level of "energy." For example,

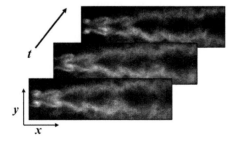

Figure 2.20 Successive snapshots of chemiluminescence images from reacting bluff body. Images courtesy of Ben Emerson [48].

2.7 Decompositions of Data

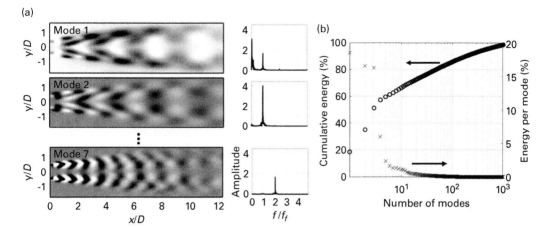

Figure 2.21 POD decomposition of mean-subtracted chemiluminescence data from reacting bluff body showing (a) POD modes 1, 2, and 7 and corresponding spectra of the time coefficient $\beta_m(t)$, and (b) the relative energy in each POD mode. Images courtesy of Ben Emerson and Hanna Ek.

velocity data from a high Reynolds number swirl flow show that about 40 modes are required to capture half the variance [49].

While this decomposition approach is quite common, note that the spatial mode shapes do not depend on the time evolution of the system because they are only a function of the spatial correlation function, $\langle u(\vec{x}_1)u(\vec{x}_2)\rangle$, as can be inferred from Eq. (2.105). In other words, the temporal ordering of the realizations has no influence on the mode shapes, a point that must be kept in mind in cases where disturbances have coupled space–time coherence, clearly evident in this example data set in Figure 2.20. For example, information on a given disturbance will be distributed across multiple modes for structures with spatiotemporal coherence [50], such as a convecting disturbance described by the function $u(x,t) = u\left(t - \frac{x}{u_c}\right)$. For example, the convecting velocity disturbance $u(x,t) = \sin\left(\omega\left(t - \frac{x}{u_c}\right)\right)$ must be written as the superposition of two "modes" if written as a product of only space and only time, i.e., $\sin\left(\omega\left(t - \frac{x}{u_c}\right)\right) = \sin(\omega t)\cos\left(\omega\frac{x}{u_c}\right) - \cos(\omega t)\sin\left(\omega\frac{x}{u_c}\right)$. This shows that "mode 2" is correlated with "mode 1" at a different spatial point and later time. For the data shown in Figure 2.21, mode 1 is such a pair to the illustrated mode 2, and together they describe a convecting disturbance.

An alternative way to manage spatiotemporal data is to define modes that depend on both space and time, i.e., where the space–time correlation matrix, $\langle u(\vec{x}_1, t_1)u(\vec{x}_2, t_2)\rangle$, is used to define the mode shapes. "Spectral POD" (SPOD) [50, 51] does this by taking the temporal Fourier transform of the data and expanding the modes as

$$\hat{u}(\vec{x},\omega) = \sum_{m=1}^{\infty} \beta_m(\omega) \tilde{F}_m(\vec{x},\omega). \tag{2.109}$$

This approach leads to a series of discrete, orthogonal spatial modes, each of which is distributed over a range of frequencies. The relationships between $\tilde{F}_m$ and F_m are presented elsewhere [50], but are complex and not localized; the mth mode of F_m is potentially a function of every single SPOD mode, $\tilde{F}_m$, as well as integrated over all frequencies. This is also a manifestation of the nonlocal relationship between a signal's time and frequency domain characteristics discussed in Section 2.7.1. This point could be seen in a very simple fashion in the $\sin\left(\omega\left(t - \frac{x}{u_c}\right)\right)$ example.

Motivated by these issues, another approach, also called "spectral POD" despite being different, has been developed that relies on prefiltering the correlation matrix before solving the eigenvalue problem [52]. The reader is referred to the references for a discussion of the interrelationships between these different approaches [53].

2.7.2.3 Linear Operators and Dynamic Mode Decomposition [54, 55]

An alternative approach for empirically decomposing data is to determine a linear, time-invariant operator (e.g., ordinary differential equations are one example of a linear operation on some function) that can reproduce the measured time variation of the data. Rather than attempting to construct basis functions based on their "closeness" to the data, as POD does, this approach expands data into summations of basis functions that capture the dynamics associated with this linear operator. This dynamic mode decomposition (DMD) is an approach for determining these eigenmodes for the discrete time case. It can be thought of as a best-fit linear dynamical system that advances measurements forward in time. As in the previous section, this section will be light on mathematical details and focus on explanations of what this decomposition is doing; details can be found in other references.

Consider the time snapshots of planar data shown in Figure 2.20. Each time snapshot of planar data consists of a matrix of elements associated with the resolution of the camera or computational grid, denoted here as $X(t)$. Subsequent snapshots are then denoted by matrices $X(t + j\Delta t)$, where Δt is the sampling interval. The DMD starts by determining an estimated best-fit linear operator, A, to take data at time t and predict the data matrix at $t + \Delta t$; i.e., it estimates

$$X(t + \Delta t) = AX(t). \tag{2.110}$$

Because the eigenvalues of linear, time-invariant systems are complex exponentials, the eigenvalues of such an operator are complex exponentials. Thus, each eigenmode oscillates at a single frequency and temporally grows or decays at a given exponential rate. Thus, the DMD decomposes the data into a series of spatial modes F_m (not necessarily orthogonal), each modulated by a complex exponential, ω_m:

$$u(\vec{x},t) = \text{Real}\left(\sum_{m=1}^{j} F_m(\vec{x}) e^{-i(\omega_{r,m} + i\omega_{i,m})t}\right). \tag{2.111}$$

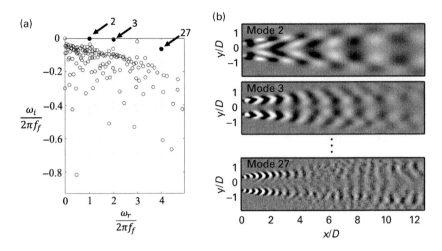

Figure 2.22 Illustration of (a) DMD spectrum and (b) spatial modes associated with representative points for the data set shown in Figure 2.20. Images courtesy of Ben Emerson and Hanna Ek.

Thus, DMD is a best-fit linear model to take the data snapshots from time t to $t + \Delta t$. The superposition of the first j leading eigenvalues/eigenvectors provides an approximation to predict future behavior of the system from a current-time data set.

DMD modes for the same data set shown in Figure 2.20 are plotted in Figure 2.22. The left plot shows the oscillatory frequency and decay rate of these modes, and the right plot shows their spatial structure. The spatial modes are shown for three DMD modes, associated with the indicated frequency/decay points on the left. The left plot shows both discrete points and clouds of points that appear to be representing continuous parts of the spectra. For example, there are four discrete points on or near the $\omega_i = 0$ axis, describing periodic oscillations at the forcing frequency and its harmonics. Presumably, each of these is a component of a limit cycle oscillation. The associated spatial modes show spatially periodic patterns. Note how spatially well defined they are for the first four diameters downstream. Further downstream, these same patterns are evident, but are increasingly diffuse. This reflects the growing presence of phase noise in the convecting structures (see Section 2.8) and their progressive reduction in amplitude.

The relationship between DMD and POD is discussed in more detailed references [50, 52].

2.8 Aside: Triple Decomposition

The triple decomposition is used to differentiate the time average, $\overline{(\ )}$, random, $(\ )'$, and deterministic, $\langle (\ )' \rangle$, components of a random variable [56]. It is useful in situations where the flow disturbances are composed of both random and deterministic fluctuations, such as during thermoacoustic instabilities or coherent vortical structures in high

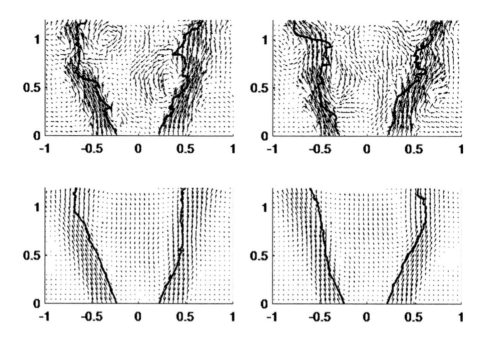

Figure 2.23 Overlay of instantaneous flame edge and velocity vectors (top), and phase-averaged velocity and flame position (bottom) [57].

Reynolds number flows. For example, Figure 2.23 shows images of a forced swirling flame at two phases of the forcing cycle. The top images show instantaneous snapshots and the bottom show phase averages. The coherent pattern in flame undulation is much clearer in the phase averaging, while the instantaneous images reveal the much broader range of scales in the flow and flame wrinkling that are present.

To illustrate the procedure, consider the decomposition on the pressure:

$$p(\vec{x},t) = \bar{p}(\vec{x}) + \langle p'(\vec{x},t) \rangle + \breve{p}'(\vec{x},t). \tag{2.112}$$

The time average is defined in the conventional way as

$$\bar{p}(\vec{x},t) = \frac{1}{T}\int_0^T p(\vec{x},t)dt \tag{2.113}$$

and the fluctuations as

$$p'(\vec{x},t) = p(\vec{x},t) - \bar{p}(\vec{x}) = \langle p'(\vec{x},t) \rangle + \breve{p}'(\vec{x},t). \tag{2.114}$$

The ensemble average, $\langle \ \rangle$, describes the average of a variable over a number of repeated trials. To illustrate, let $p(\vec{x},t)^{(n)}$ denote the measured value of $p(\vec{x},t)$ in the nth experiment. The ensemble average is defined by [58]:

$$\langle p'(\vec{x},t) \rangle = \frac{1}{N}\sum_{n=1}^{N}\left(p(\vec{x},t)^{(n)} - \bar{p}(\vec{x})\right). \tag{2.115}$$

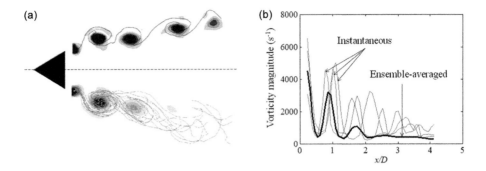

Figure 2.24 (a) Top: Instantaneous vorticity contours and corresponding flame sheet location. Bottom: Ensemble-averaged vorticity contours and representative realizations of the corresponding flame sheet locations. (b) Comparison of instantaneous and ensemble-averaged axial vorticity variation. Data courtesy of S. Shanbhogue.

In other words, the time average and ensemble average are not generally equal. For example, the ensemble average of $\sin(\omega t)$ is $\sin(\omega t)$, while the time average of $\sin(\omega t)$ is zero. In contrast, both the time and ensemble average of a zero time average random variable tends to zero as N and $T \to \infty$.

While the triple decomposition is a conceptually useful way to decompose spatio-temporally complex flows with both broadband and narrowband spectral features, its practical utility in highly turbulent flows is complicated by inherent space–time variability in quasi-deterministic features ("phase jitter"). Moreover, quasi-deterministic structures can be spatially smeared and reduced in magnitude by the ensemble averaging operation. For example, Figure 2.24(a) compares instantaneous (top) and ensemble-averaged (bottom) vorticity field measurements of a harmonically forced, bluff body stabilized flame. Also shown is the corresponding instantaneous flame position for the top image, and an overlay of several instantaneous realizations of flame position on the bottom. These same data are quantitatively illustrated in Figure 2.24(b), which plots the axial vorticity distribution at the vortex core of three instantaneous realizations and the ensemble-averaged vorticity. For this example, there is no temporal phase jitter because of the phase-locked reference, but there is spatial jitter, as the vortex positions are influenced by both the average flow and the random turbulent fluctuations. This spatial phase jitter grows downstream, so that the instantaneous and ensemble-averaged vorticity fields are similar near $x = 0$, but diverge increasingly downstream. Moreover, the data show only three structures in the ensemble-averaged image due to the smearing effect of averaging, while at least five are clearly shown in the instantaneous image. As such, quantities such as coherent vorticity magnitudes or spatial decay rates can be quite different based on the analysis method. For example, Figure 2.24(b) shows a factor of ~2.9 difference in peak vorticity for the structure at $x/D \sim 1.8$ [59]. Similar effects can be seen in the velocity data in Figure 2.23 – clear concentrations of rotating velocity structures are evident in the instantaneous images, but are much more diffuse in the phase-averaged images.

2.9 Aside: Effects of Simultaneous Acoustic and Vortical Velocity Disturbances

Velocity measurements at a given point measure the total unsteady velocity, and thus cannot differentiate between acoustic- and vortical-induced velocities. However, their different characteristics can be inferred from measurement of the spatiotemporal distribution of the velocity field. For example, the substantially different propagation speeds of acoustic and vortical disturbances can lead to interference patterns in the unsteady flow velocity. To illustrate, consider the following model problem where a harmonically oscillating acoustic plane wave and vortical disturbance are propagating in the same direction, x:

$$u_1(x,t) = u_{1,\Lambda} + u_{1,\Omega} = \mathcal{A}_\Lambda \cos(\omega(t - x/c_0)) + \mathcal{A}_\Omega \cos(\omega(t - x/u_{x,0})). \quad (2.116)$$

For the simple case of $\mathcal{A} = \mathcal{A}_\Lambda = \mathcal{A}_\Omega$, this expression can be rewritten using trigonometric identities as

$$u_1(x,t) = 2\mathcal{A}\cos\left(\frac{c_0 - u_{x,0}}{2c_0 u_{x,0}}\omega x\right) \cos\left(\omega t - \frac{c_0 + u_{x,0}}{2c_0 u_{x,0}}\omega x\right). \quad (2.117)$$

The resulting disturbance field magnitude has some analogies to the acoustic standing waves discussed extensively in Chapter 5. This equation shows that the velocity field oscillates harmonically at each point as $\cos(\omega t)$. The amplitude of these oscillations varies spatially as $\cos\left(\frac{c_0 - u_{x,0}}{2c_0 u_{x,0}}\omega x\right)$, due to interference. This type of behavior is often observed in acoustically excited combustors, where acoustic and vortical disturbances coexist, such as shown in Figure 2.2. To illustrate this point quantitatively, Figure 2.25 illustrates data obtained from a bluff body stabilized reacting flow. The symbols denote the experimentally measured velocity amplitude and phase. The line represents a fit to these data using $\mathcal{A}_\Lambda/\mathcal{A}_\Omega = 0.6$. Note also the lines on the amplitude graph denoted "acoustic only" and "vortical only" – these lines show the axial variation of

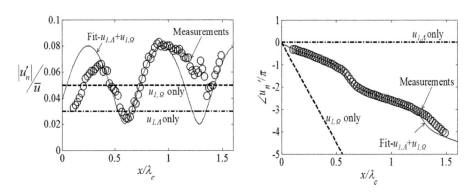

Figure 2.25 Measured PIV data showing interference effects in unsteady velocity due to superposition of a convecting vortical disturbance and a transverse acoustic wave. Image courtesy of V. Acharya, D.H. Shin, and B. Emerson.

amplitude of the acoustic and vortical waves separately. This plot shows that the superposition of two disturbances with spatially constant amplitudes still leads to an undulating overall disturbance field magnitude because of their differing phase speeds.

2.10 Aside: Further Consideration of the Disturbance Energy Equation [21]

This section considers the disturbance energy equations in more detail. There are a few points relating to Eq. (2.63) which require some care. First, note that this equation is second order in fluctuations (e.g., $\frac{1}{2}\rho|\vec{u}|^2$), but was derived from linearized equations, accurate only to first order. This raises the question of whether this equation represents a consistent ordering and inclusion of terms. Second, the relationship of this equation to the full energy equation, Eq. (1.27), which includes all nonlinear and higher-order terms, is not immediately clear.

To answer these questions, we consider the relationship between the full energy equation expanded to second order in perturbation amplitude, and Eq. (2.63). We will assume a spatially homogeneous medium, no mean flow, and a calorically perfect gas for this development; see Myers [21] for a more general analysis. Consider first the internal energy term in Eq. (1.27). The deviation of the instantaneous internal energy from its value in the absence of fluctuations is given by $(\rho e) - (\rho e)|_0$. We can expand this term in a Taylor series to second order to obtain

$$\rho e - (\rho e)_0 = \left[\frac{\partial(\rho e)}{\partial \rho}\right]_0 (\rho - \rho_0) + \frac{1}{2}\left[\frac{\partial^2(\rho e)}{\partial \rho^2}\right]_0 (\rho - \rho_0)^2 + O((\rho - \rho_0)^3). \quad (2.118)$$

The derivatives can be evaluated as

$$\left[\frac{\partial(\rho e)}{\partial \rho}\right]_0 = h_0, \quad (2.119)$$

$$\left[\frac{\partial^2(\rho e)}{\partial \rho^2}\right]_0 = \frac{c_0^2}{\rho_0}. \quad (2.120)$$

Pulling these together, this shows that the total energy density associated with fluctuations, expanded to second order, is

$$\rho\left(e + \frac{|\vec{u}|^2}{2}\right) - \rho\left(e + \frac{|\vec{u}|^2}{2}\right)\bigg|_0 = h_0(\rho - \rho_0) + \frac{1}{2}\frac{c_0^2}{\rho_0}(\rho - \rho_0)^2 + \rho_0\frac{|\vec{u}|^2}{2}$$

$$= h_0(\rho - \rho_0) + \frac{1}{2}\frac{1}{\rho_0 c_0^2}(p - p_0)^2 + \rho_0\frac{|\vec{u}|^2}{2}. \quad (2.121)$$

The latter two terms are identical to the energy density terms in Eq. (2.63). However, there is also clearly a nonzero term that is first order in $\rho - \rho_0$. We will next use this result for expanding the left-hand side of the full energy equation, Eq. (1.27), to second order to yield

$$\frac{\partial}{\partial t}\left[h_0(\rho-\rho_0)+\frac{1}{2}\frac{1}{\rho_0 c_0^2}(p-p_0)^2+\rho_0\frac{|\vec{u}|^2}{2}\right]+\nabla\cdot([\rho_0 e_0+h_0(\rho-\rho_0)+p]\vec{u}).\tag{2.122}$$

Utilizing the continuity equation, this can be written as

$$\frac{\partial}{\partial t}\left[\frac{1}{2}\frac{p_1^2}{\rho_0 c_0^2}+\rho_0\frac{|\vec{u}_1|^2}{2}\right]+\nabla\cdot(p_1\vec{u}_1).\tag{2.123}$$

Note that this is identical to that derived earlier in Eq. (2.63), showing that Eq. (2.63), derived from the linearized, first-order momentum and energy equations, is exact to second order. However, Eq. (2.123) also shows that the terms in this expression that were interpreted as energy density and energy flux are actually not complete: they are missing the terms in Eq. (2.121) that are linear in $\rho-\rho_0$. These terms have a zero time average and so do not affect calculations of time-averaged energy density, but do influence instantaneous values.

Exercises

2.1 Consider the following generalization of Eq. (2.1) to a case with multiple frequencies:

$$p(t)=B_1(1+\varepsilon_1\sin(\omega_1 t)+\varepsilon_2\sin(\omega_2 t))^{\gamma}.\tag{2.124}$$

Derive expressions for $p(t)$ to $O(\varepsilon^2)$ and discuss the frequency components that appear at each order.

2.2. Derive the second-order entropy source terms in Table 2.1.

2.3. Derive the expressions for acoustic energy density and flux from Eqs. (2.67) and (2.68).

2.4. Consider the acoustic energy flux term from Eq. (2.68) for a purely one-dimensional flow. Assuming that the unsteady pressure and velocity are given by $p_1=|p_1|\cos(\omega t)$ and $u_1=\frac{|p_1|\mathcal{A}}{\rho_0 c_0}\cos(\omega t+\theta)$, determine the time-averaged energy flux. Develop a relationship for the relative contributions of the purely acoustic (i.e., the terms that persist with no flow) and convective terms in terms of $\mathcal{A}$, θ, and the Mach number, M.

2.5. Consider two time-varying functions, $F_1(t)$ and $F_2(t)$. Using the derivative and multiplication properties of the Fourier transform (see Table 2.2), derive an expression for the Fourier transform of the product $F_1(t)dF_2(t)/dt$. Simplify this expression for sinusoidally varying functions, oscillating at ω_1 and ω_2.

2.6. Derive an expression for the limit cycle amplitude for the dynamical system given in Eq. (2.86).

2.7. In this exercise we will derive the stationary probability distribution of the stochastically forced nonlinear oscillator in Eq. (2.88). Consider a stochastic differential equation given by

$$\dot{x} = f(x) + g(x)\sigma(t), \qquad (2.137)$$

where $\sigma(t)$ is Gaussian white noise with $\langle \sigma(t) \rangle = 0$ and $\langle \sigma(t)\sigma(t') \rangle = \delta(t-t')$. Then, the probability density $p(x,t)$ satisfies the Fokker–Planck equation [41]:

$$\frac{\partial p}{\partial t} = -\frac{\partial}{\partial x}(fp) + \frac{1}{2}\frac{\partial^2}{\partial x^2}(g^2 p), \qquad (2.138)$$

where the first and second terms on the right-hand side are denoted as the "drift" and "diffusion" terms, respectively. It can be shown that the stochastically averaged form of Eq. (2.88) is given by the following stochastic differential equation:

$$\frac{dA}{dt} = \varsigma\omega_0\left(\frac{\beta}{4}A^3 - A\right) + \frac{5}{16}\varepsilon_1^2 A + \frac{1}{4A\omega_0^2}\varepsilon_2^2 + \sqrt{\frac{3}{8}\varepsilon_1^2 A^2 + \frac{\varepsilon_2^2}{2\omega_0^2}}\sigma(t). \qquad (2.139)$$

Derive the expression for the stationary probability distribution of the amplitude given in Eq. (2.91).

References

1. Muthukrishnan M., Strahle W.C. and Neale D.H., Separation of hydrodynamic, entropy, and combustion noise in a gas turbine combustor. *AIAA Journal*, 1978, **16**(4): pp. 320–327.
2. Chu B.T. and Kovásznay L.S.G., Non-linear interactions in a viscous heat-conducting compressible gas. *Journal of Fluid Mechanics*, 1957, **3**(5): pp. 494–514.
3. Jou W. and Menon S., Modes of oscillation in a nonreacting ramjet combustor flow. *Journal of Propulsion and Power*, 1990, **6**(5): pp. 535–543.
4. Goldstein M., Characteristics of the unsteady motion on transversely sheared mean flows. *Journal of Fluid Mechanics*, 1978, **84**(2): pp. 305–329.
5. Laverdant A. and Thévenin D., Interaction of a Gaussian acoustic wave with a turbulent premixed flame. *Combustion and Flame*, 2003, **134**(1–2): pp. 11–19.
6. Pierce A.D., *Acoustics: An Introduction to its Physical Principles and Applications*. 1989, Melville: Acoustical Society of America. 678.
7. Goldstein M.E., *Aeroacoustics*. 1976, New York: McGraw-Hill.
8. Lighthill M.J., On the energy scattered from the interaction of turbulence with sound or shock waves. *Mathematical Proceedings of the Cambridge Philosophical Society*, 1952, **49**: pp. 531–551.
9. Lieuwen T., Theoretical investigation of unsteady flow interactions with a premixed planar flame. *Journal of Fluid Mechanics*, 2001, **435**: pp. 289–303.
10. Marble F.E. and Candel S.M., Acoustic disturbance from gas non-uniformities. *Journal of Sound and Vibration*, 1977, **55**(2): pp. 225–243.
11. Cuadra E., Acoustic wave generation by entropy discontinuities flowing past an area change. *The Journal of the Acoustical Society of America*, 1966, **42**(4): pp. 725–732.

12. Cumpsty N.A. and Marble F.E., Core noise from gas turbine engines. *Journal of Sound and Vibration*, 1977, **54**(2): pp. 297–309.
13. Komarek T. and Polifke W., Impact of swirl fluctuations on the flame response of a perfectly premixed swirl burner. *Journal of Engineering for Gas Turbines and Power*, 2010, **132**(6): pp. 061503(1)–061503(7).
14. Palies P., Durox D., Schuller T., and Candel S., Modeling of premixed swirling flames transfer functions. *Proceedings of the Combustion Institute*, 2011, **33**(2): pp. 2967–2974.
15. Acharya V. and Lieuwen T., Effect of azimuthal flow fluctuations on flow and flame dynamics of axisymmetric swirling flames. *Physics of Fluids*, 2015, **27**(10).
16. Mohring W., On energy, group velocity and small damping of sound waves in ducts with shear flow. *Journal of Sound and Vibration*, 1973, **29**(1): pp. 93–101.
17. Howe M.S., *Acoustics of Fluid-Structure Interactions*. 1st ed. Cambridge Monographs on Mechanics. 1998, Cambridge: Cambridge University Press, p. 572.
18. O'Shea S., Sound scattering by a potential vortex. *Journal of Sound and Vibration*, 1975, **43**(1): pp. 109–116.
19. Chih-Ming H. and Kovasznay L.S.G., Propagation of a coherent acoustic wave through a turbulent shear flow. *Journal of the Acoustical Society of America*, 1975, **60**(1): pp. 40–45.
20. George J. and Sujith R.I., Emergence of acoustic waves from vorticity fluctuations: Impact of non-normality. *Physical Review E*, 2009, **80**(4): pp. 046321:1–6.
21. Myers M.K., Transport of energy by disturbances in arbitrary steady flows. *Journal of Fluid Mechanics*, 1991, **226**: pp. 383–400.
22. Morfey C.L., Acoustic energy in non-uniform flows. *Journal of Sound and Vibration*, 1971, **14**(2): pp. 159–170.
23. Nicoud F. and Poinsot T., Thermoacoustic instabilities: Should the Rayleigh criterion be extended to include entropy changes? *Combustion and Flame*, 2005, **142**: pp. 153–159.
24. Jenvey P.L., The sound power from turbulence: A theory of the exchange of energy between the acoustic and non-acoustic fields. *Journal of Sound and Vibration*, 1989, **121**(1): pp. 37–66.
25. Myers M.K., An exact energy corollary for homentropic flow. *Journal of Sound and Vibration*, 1986, **109**(2): pp. 277–284.
26. Doak P.E., Momentum potential theory of energy flux carried by momentum fluctuations. *Journal of Sound and Vibration*, 1989, **131**(1): pp. 67–90.
27. Chu B.T., On the energy transfer to small disturbances in fluid flow (Part I). *Acta Mechanica*, 1965, **1**(3): pp. 215–234.
28. Chu B.T. and Apfel R.E., Are acoustic intensity and potential energy density first or second order quantities? *American Journal of Physics*, 1983, **51**(10): pp. 916–918.
29. George J. and Sujith R.I., On Chu's disturbance energy. *Journal of Sound and Vibration*, 2011, **330**: pp. 5280–5291.
30. Rayleigh J.W.S., *The Theory of Sound*, Vol. 2. 1896, New York: Macmillan and Co.
31. Kim K.T., Lee J.G., Quay B., and Santavicca D., The dynamic response of turbulent dihedral V flames: An amplification mechanism of swirling flames. *Combustion Science and Technology*, 2011, **183**(2): pp. 163–179.
32. Cantrell R.H. and Hart R.W., Interactions between sound and flow in acoustic cavities: Mass, momentum, and energy considerations. *The Journal of the Acoustical Society of America*, 1964, **36**(4): pp. 697–706.
33. Karimi N., Brear M.J., and Moase W.H., Acoustic and disturbance energy analysis of a flow with heat communication. *Journal of Fluid Mechanics*, 2008, **597**: pp. 67–98.

34. Lieuwen T.C., Experimental investigation of limit-cycle oscillations in an unstable gas turbine combustor. *Journal of Propulsion and Power*, 2002, **18**(1): pp. 61–67.
35. Oefelein J.C. and Yang V., Comprehensive review of liquid-propellant combustion instabilities in F-1 engines. *Journal of Propulsion and Power*, 1993, **9**(5): pp. 657–677.
36. Cohen J.M., Proscia W., and Delaat J., Characterization and control of aeroengine combustion instability: Pratt & Whitney and NASA experience, in *Combustion Instabilities in Gas Turbine Engines. Operational Experience, Fundamental Mechanisms, and Modeling*, T.C. Lieuwen and V. Yang, eds. 2005, AIAA, pp. 113–144.
37. Nayfeh A.H. and Mook D.T., *Nonlinear Oscillations*. 1995, Wiley.
38. Gardiner C.W., *Handbook of Stochastic Methods*. 1997, Springer: New York.
39. Lieuwen T.C., Statistical characteristics of pressure oscillations in a premixed combustor. *Journal of Sound and Vibration*, 2003, **260**(1): pp. 3–17.
40. Lieuwen T.C., Online combustor stability margin assessment using dynamic pressure measurements. *Journal of Engineering for Gas Turbines and Power*, 2005, **127**(3): pp. 478–482.
41. Roberts J.B. and Spanos, P.D., Stochastic averaging: An approximate method of solving random vibration problems. *International Journal of Non-Linear Mechanics*, 1986, **21**(2): pp. 111–134.
42. Hilborn R.C., *Chaos and Nonlinear Dynamics: An Introduction for Scientists and Engineers*. 1994, New York: Oxford University Press.
43. Nair V., Thampi G., and Sujith R.I., Intermittency route to thermoacoustic instability in turbulent combustors. *Journal of Fluid Mechanics*, 2014, **756**: pp. 470–487.
44. Mallat S.G. and Mallat, C., *A Wavelet Tour of Signal Processing*. 2nd ed. Wavelet Analysis and Its Applications. 1999, Elsevier Science & Technology.
45. Courant R. and Hilbert D., *Methods of Mathematical Physics*. 1989, John Wiley & Sons.
46. Petropulu A.P., *Higher-Order Spectral Analysis*. 2000, CRC Press LLC.
47. Holmes P., Lumley J.L., Berkooz G., and Rowley C.W., *Turbulence, Coherent Structures, Dynamical Systems and Symmetry*. 1996, New York: Cambridge University Press.
48. Emerson B., *Dynamical Characteristics of Reacting Bluff Body Wakes*. 2013, Georgia Institute of Technology.
49. Oberleithner K., Sieber M., Nayeri C.N., Paschereit C.O., Petz C., Hege H.-C., Noack B.R., and Wygnanski I., Three-dimensional coherent structures in a swirling jet undergoing vortex breakdown: Stability analysis and empirical mode construction. *Journal of Fluid Mechanics*, 2011, **679**: pp. 383–414.
50. Towne A., Schmidt O.T., and Colonius T., Spectral proper orthogonal decomposition and its relationship to dynamic mode decomposition and resolvent analysis. *Journal of Fluid Mechanics*, 2018, **847**: pp. 821–867.
51. Lumley J.L., The structure of inhomogeneous turbulent flows, in *Atmospheric Turbulence and Radio Wave Propagation*, A.M. Yaglom and V.I. Tatarsky, eds. 1967, Nauka, pp. 166–176.
52. Sieber M., Paschereit C., and Oberleithner K., Spectral proper orthogonal decomposition. *Journal of Fluid Mechanics*, 2016, **792**: pp. 798–828.
53. Sieber M., Paschereit C., and Oberleithner K., On the nature of spectral proper orthogonal decomposition and related modal decompositions. 2017, arXiv:1712.08054.
54. Kutz J.N., Brunton S.L., Brunton B.W., and Proctor J.L, *Dynamic Mode Decomposition: Data-Driven Modeling of Complex Systems*. 2016, Society for Industrial and Applied Mathematics.

55. Schmid P.J., Dynamic mode decomposition of numerical and experimental data. *Journal of Fluid Mechanics*, 2010, **656**(1): pp. 5–28.
56. Hussain A.K.M.F. and Reynolds W.C., The mechanics of an organized wave in turbulent shear flow. *Journal of Fluid Mechanics*, 1970, **41**(2): pp. 241–258.
57. Smith T.E., Chterev I.P., Emerson B.L., Noble D.R., and Lieuwen T.C., Comparison of single- and multinozzle reacting swirl flow dynamics. *Journal of Propulsion and Power*, 2017, **34**(2).
58. Pope S.B., *Turbulent Flows*. 2000, Cambridge University Press.
59. Shanbhogue S.J., Seelhorst M., and Lieuwen T., Vortex phase-jitter in acoustically excited bluff body flames. *International Journal of Spray and Combustion Dynamics*, 2009, **1**(3): pp. 365–387.

3 Hydrodynamic Flow Stability I: Linear Instability

In Chapter 2 we showed that flow disturbances can be decomposed into vorticity, entropy, and dilatational/acoustic fluctuations. The next two chapters focus on the evolution of vorticity in flows, and how vorticity in one region of the flow interacts with other regions of vorticity to influence hydrodynamic flow stability, leading to self-organization into concentrated regions of vorticity and flow rotation. Such large-scale structures, embedded on a background of acoustic waves and broadband, smaller-scale turbulence, dominate the unsteady flow fields in combustors. These large-scale structures play important roles in processes such as combustion instabilities, mixing and entrainment, flashback, and blowoff. For example, we will discuss vortex–flame interactions repeatedly in discussions of combustion instabilities in later chapters.

The interactions of vortical regions are strongly influenced by combustion, through gas expansion and baroclinic effects. In addition, density gradients also introduce mechanisms of instability even in the absence of vorticity, as discussed in Section 3.8.

High Reynolds number flows are effectively inviscid outside the boundary layer. Vorticity in the flow largely originates from boundary layers in approach flow passages or other walls. Free shear layers arise at points of boundary layer separation, initiating a sequence of large-scale flow instabilities as this vorticity is then stretched and amplified by the base flow.

The instabilities that are focused on in the next two chapters play important roles in unsteady combustor dynamics. Many of them also manifest themselves in a variety of other instances, including in spectacular, large-scale fashion in nature. To illustrate, Figure 3.1(a) shows cloud patterns showing the Kelvin–Helmholtz instability, discussed in Sections 3.4.4 and 4.1. Figure 3.1(b) illustrates the Bénard/von Kármán instability over an island visualized from space, to be discussed in Section 4.2.

In discussing these flow instabilities, it is important to distinguish between two types of "instabilities" or "transitions" – flow instabilities with a small number of degrees of freedom, such as the Kelvin–Helmholtz instability [3], and the actual transition to turbulence, which is characterized by disturbances with a large number of degrees of freedom manifested over a broad range of spatiotemporal scales. The use of the word "instability" in this text will always refer to low-dimensional flow instabilities.

This chapter introduces basic linear stability concepts. The focus here is on general mechanisms of stability of shear flows, effects of density gradients, factors influencing

 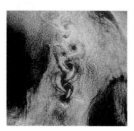

Figure 3.1 Images showing two flow instabilities in the atmosphere visualized by cloud patterns. These same two instabilities also play a prominent role in unsteady combustor processes [1, 2].

the growth rate and phase speed of hydrodynamic disturbances, and convective/absolute instability concepts. Chapter 4 then focuses on vortical structures in specific flow configurations, including the effects of heat release and external forcing.

3.1 Linear Stability Notions

The objective of this section is to further expand the linear stability concepts initiated in Section 2.5 to spatiotemporally developing flows with many degrees of freedom. The goal of linear stability analysis is to determine the evolution of small perturbations about the base state. In other words, starting with the continuity, momentum, and energy equations, Eqs. (1.10), (1.12), and (1.27), we add a small perturbation to the base state and develop linearized equations, as was done in Section 2.2.1, and determine whether this disturbance grows or decays. However, the phrase "whether this disturbance grows or decays" can be interpreted in multiple ways, with potentially different conclusions in spatiotemporally evolving systems with more than one degree of freedom. A convenient way to classify different behaviors where disturbance growth could occur is shown in Table 3.1.

Consider first the top row of Table 3.1, and the question "Does a disturbance, when measured at a fixed position, grow or decay?" The solution, $F(t)$, to an nth-order linear system of ordinary differential equations (assuming no degenerate eigenvalues) is a superposition of n eigenvectors multiplied by complex exponentials of the form $e^{-i\omega_{r,m}t}e^{\omega_{i,m}t}$, where m varies from 1 to n. Clearly, if any of these eigenvalues is a growing exponential, $\omega_{i,m} > 0$, then the solution is linearly unstable and grows in time. This linearly unstable system is referred to as "globally unstable," the top right scenario in Table 3.1, and is discussed further in Section 3.2. In this case, small system disturbances will grow and nonlinear effects must be considered to determine the subsequent system dynamics, as detailed in Section 2.5. Consider next the situation where all of the eigenvalues are stable, $\omega_{i,m} < 0$. In this case, the disturbance will decay given sufficient time ($t \to \infty$), i.e., it is linearly stable. However, additional possibilities are present for short time intervals ($t = 0^+$). To illustrate, consider the superposition of two eigenvectors oriented at an angle θ with respect to each other,

3.1 Linear Stability Notions

Table 3.1 Different approaches for classifying linear stability of a spatiotemporally evolving system.

	Short time ($t = 0^+$)	Long time ($t \to \infty$)
Measurement at a fixed position	Transient growth	Globally unstable
Measurement at a moving position	Transient growth	Convectively unstable

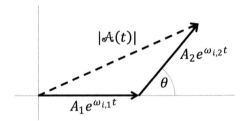

Figure 3.2 Vector superposition of two components, each with exponentially decaying magnitudes.

each with stable eigenvalues, as depicted in Figure 3.2. The magnitude of the two modes is given by $A_1 e^{\omega_{i,1} t}$ and $A_2 e^{\omega_{i,2} t}$, where A_1 and A_2 are arbitrary and set by the initial conditions.

The magnitude of their vector sum is given by

$$|\mathcal{A}(t)| = \sqrt{(A_1 e^{\omega_{i,1} t} + A_2 e^{\omega_{i,2} t} \cos \theta)^2 + (A_2 e^{\omega_{i,2} t} \sin \theta)^2}. \quad (3.1)$$

If $\omega_{i,1}$ and $\omega_{i,2}$ are both negative, $|\mathcal{A}(t)|$ will exponentially decay in time over a long time window. But are there other possibilities for $|\mathcal{A}(t)|$ over short time intervals, the top left scenario of Table 3.1? Insight into $|\mathcal{A}(t)|$ over short times can be obtained from

$$\left[\frac{1}{|\mathcal{A}(t)|} \frac{d|\mathcal{A}(t)|}{dt} \right]_{t=0} = \frac{A_1 A_2 \cos\theta (\omega_{i,1} + \omega_{i,2}) + A_1^2 \omega_{i,1} + A_2^2 \omega_{i,2}}{A_1^2 + 2 A_1 A_2 \cos\theta + A_2^2}. \quad (3.2)$$

If the two modes are orthogonal, $\theta = 90°$, then this equation shows that $|\mathcal{A}(t)|$ decreases immediately at time $t = 0$. However, if the modes are not orthogonal, then there are combinations of initial conditions and θ values where $\mathcal{A}(t)$ initially increases, even as it eventually asymptotically decays to zero at a rate given by the slowest-decaying, stable eigenmode. These scenarios can be seen from Figure 3.3, which plots examples of $|\mathcal{A}(t)|$ for different values of θ.

For the orthogonal, $\theta = 90°$, case, $|\mathcal{A}(t)|$ monotonically decays. However, $|\mathcal{A}(t)|/|\mathcal{A}(t = 0)|$ grows at time zero, reaches a peak, and then decays for the $\theta = 130°$ and $170°$ cases. Moreover, one can readily experiment with different values of parameters in Eq. (3.1) and show that $|\mathcal{A}(t)|$ can achieve arbitrarily large peak values, depending on parameter choice.

This example shows that a system with nonorthogonal but individually stable eigenvalues can exhibit transient growth, potentially very large transient growth.

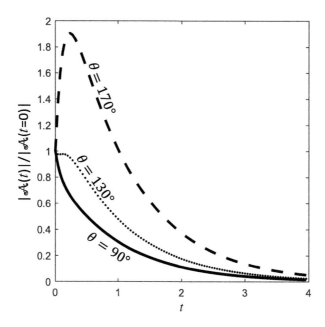

Figure 3.3 Time dependence of vector superposition of the two components sketched in Figure 3.2 for $A_1/A_2 = 1.2$, $\omega_{i,1} = -1$, $\omega_{i,2} = -8$.

Considering a situation such as shown in Figure 2.10, this shows that a very small perturbation that is well below A_{Tr} could grow to values that exceed A_{Tr} and lead to nonlinear instability. We will consider these types of systems in Section 3.3.

Returning to Table 3.1, consider next the bottom right box, labeled "Convectively unstable." To understand this situation, consider the following example of a convecting wave packet:

$$A(x,t) = t \sin(\omega(t - x/u))e^{-(t-x/u)^2}. \tag{3.3}$$

In a moving coordinate system, where $t - x/u$ is constant, the function grows in time. In contrast, in a stationary coordinate system, A will ultimately decrease in time. This is illustrated in Figure 3.4. While disturbances at a fixed position will ultimately decay to zero, this result shows that such a system acts as an amplifier to persistent disturbances. We will consider these type of systems in Section 3.2.

3.2 Global and Convective Instability [4, 5]

Starting with the linearized equations (Section 2.2.1) and boundary conditions, consider the following initial value problem: given some base flow where all fluctuation amplitudes are identically zero at $t = 0$, calculate the spatiotemporal evolution of the flow disturbances in response to the impulsive flow excitation $\delta(\vec{x} - \vec{x}_0)\delta(t)$, where δ

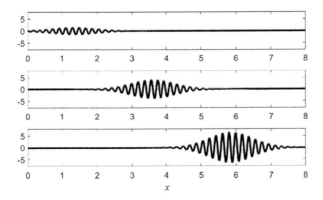

Figure 3.4 Illustration of the function in Eq. (3.3) whose amplitude ultimately decays in time when measured at a fixed location, but that grows continuously as it convects downstream.

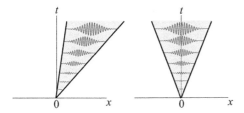

Figure 3.5 Evolution of a disturbance initiated at $(x_0, t_0) = (0,0)$ in a convectively unstable (left) and globally unstable (right) flow. Adapted from Blackburn et al. [6].

denotes the Dirac delta function. The spatiotemporal evolution of the flow disturbances can then be recognized as the Green's function, $\mathcal{G}_u(\vec{x}, \vec{x}_0, t)$.

In a *convectively unstable* flow, this impulsive disturbance grows as it propagates downstream, such as shown in Figure 3.4. However, the disturbance propagates out of the flow domain, so that the disturbance field in the domain returns to zero after some time, since the excitation source is impulsive. In a *globally unstable* flow, the impulsive disturbance grows exponentially in time at the point of impulse, so that it eventually exists everywhere in the domain. These points are also illustrated in Figure 3.5. A flow will be globally unstable if any of the eigenvalues of the linearized solution correspond to an exponentially growing disturbance. It is globally stable if all of these eigenvalues correspond to decaying disturbances.

An important special case that we will consider in detail in Section 3.4 is a "parallel flow," where the mean velocity is one dimensional, does not evolve axially, and is only a function of one variable, i.e., $u_{x,0}(y)$. In this case, the globally unstable flow is referred to as *absolutely unstable*.

To summarize, amplified disturbances in a convectively unstable flow are swept downstream so that the response at any fixed spatial point eventually tends toward zero, as shown in the example in Eq. (3.3). In contrast, perturbations in globally unstable flows increase in amplitude at a fixed spatial location, even as the actual fluid is convected downstream. In a flow without any axial variation in properties, the distinction between absolute and convective instability is simply one of reference frame. However, this distinction has important physical significance in any real system where Galilean invariance does not hold (and, therefore, the system does not strictly satisfy the parallel flow assumption), such as in the downstream evolution of a mixing layer flowing over a splitter plate.

A "convectively unstable" flow region acts an *amplifier*, so that disturbances created at some point grow in amplitude as they convect out of the system. However, the oscillations are not self-excited, but require a continuous disturbance source to persist. In any real situation there are a variety of background disturbances, such as broadband turbulence, that provide this continuous source of excitation. The flow then selectively filters these background disturbances; e.g., in convectively unstable shear flows, the spectrum shows a distributed peak around a frequency associated with the most amplified mode, a frequency that can often be predicted from linear instability theory. These systems generally show a strong response when excited by external disturbances, such as harmonic acoustic waves.

A globally unstable flow is an *oscillator* – it is self-excited, does not require external disturbances to persist, and oscillates temporally at the global mode frequency, ω_{Global} [5, 7, 8]. For example, in the bluff body wake, $\omega_{Global} = 2\pi(St)u_0/D$, where $St \sim 0.22$ for cylinders, and D is the bluff body diameter; see Section 4.2.2. In such a self-excited system, the amplitude grows before saturating into a limit cycle oscillation or more complex phase orbit (see Section 2.6) [9]. Moreover, the spectrum of oscillations shows one or more narrowband peaks associated with this natural frequency, such as shown in Figure 3.6. When externally forced at another frequency, the limit cycle behavior can remain independent of the external forcing at low forcing amplitudes, and change in amplitude and/or frequency

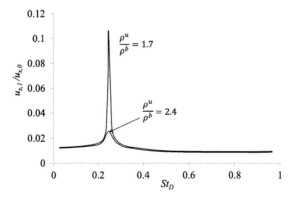

Figure 3.6 Unsteady velocity spectrum in reacting bluff body wake showing transition to absolute instability at low flame density ratios [10]. See also Section 4.2.1. Image courtesy of B. Emerson.

at high forcing amplitude. For example, "lock-in" refers to a phenomenon where oscillations at the global mode frequency disappear altogether with sufficient forcing amplitude. An example of this behavior is discussed in the context of Figure 4.19, and in the model problem presented in Section 2.5.3.3.

Extrapolating from the linear theory to nonlinear systems, in a convectively unstable flow there is a monotonic relationship between forcing amplitude and system response, while in a globally unstable system the limit cycle oscillations may be independent of external forcing, except at high forcing levels. These distinctions are important in combustion instability problems, where different types of hydrodynamic instabilities may be present that interact with acoustic waves in the overall system feedback mechanism. In one case, low-amplitude acoustic excitation will induce a proportional response, while in the other, it may not.

3.3 Transient Growth

Returning to Table 3.1, a large class of flows are linearly, globally stable, but exhibit transient growth [11]. As demonstrated by the simple problem in Eq. (3.1), the superposition of modal solutions can lead to short-time behavior that cannot be understood from considering each stable mode separately, and is qualitatively different from its long-time solution. For example, viscous stability theory shows that a Poiseuille flow becomes globally unstable at a Reynolds number of 5772. In other words, all eigenmodes are damped for $Re < 5772$. On the other hand, disturbances only monotonically decay in time for $Re < 50$. For intermediate Reynolds numbers, $50 < Re < 5772$, some disturbances initially grow and can be amplified by several orders of magnitude before decaying [12]. Similar behavior can be seen in Figure 3.7 for a backward-facing step – only below $Re \sim 58$ do disturbances decay

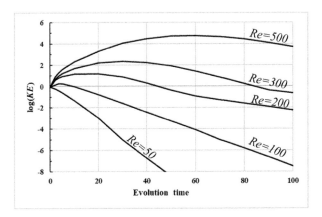

Figure 3.7 Computed ratio of disturbance kinetic energy (normalized by its initial value) for backward-facing step flow at $Re = 50, 100, 200, 300,$ and 500. Adapted from Blackburn et al. [6].

monotonically. Above this value, even for Reynolds numbers where the flow is globally stable, the system undergoes transient amplification. These curves show that the "optimum" disturbance amplification – i.e., the initial condition that leads to the maximum amplification at that Reynolds number – have values $\sim 10^5$.

Thus, even though the system is globally stable, linear calculations quickly lose accuracy and nonlinear effects must be accounted for to understand the system dynamics. This also explains why the stability boundaries of such flows are a strong function of background noise levels. The term "bypass transition" is used to describe such behaviors, defined by Schmid and Henningson [13] as "transition emanating from non-modal growth mechanisms." Recent work in the specific context of combustion instabilities has also emphasized these nonnormal effects [14, 15].

3.4 Normal Modes in Parallel Flows

3.4.1 Basic Formulation [4, 5, 16–19]

While the previous discussion is useful for qualitatively understanding the types of behaviors that can be exhibited, a general analysis of specific problems is not generally possible and most results are computational or experimental. In order to gain insight through theory, this section considers some significant simplifications – analysis of inviscid, incompressible, parallel (i.e., negligible base transverse velocity and negligible variation in the base axial velocity in the axial direction, x) flows. In particular, we will focus on analyzing fundamental features in the flow that cause it to be stable or unstable to small, harmonic disturbances. Of course, the prior section has demonstrated the shortcomings of considering only modal disturbances. However, for most applications of interest to this book, the Reynolds numbers are high enough and the flow exhibits inflection points, and so they are always unstable. In this case, quantities of interest are the temporal and/or spatial growth/decay rate of perturbations, whether the instability is convective or global, the most amplified frequency of periodic perturbations, and the propagation speed of flow disturbances.

To address these questions, we consider a flow with a base axial velocity field of $u_{x,0}(y)$, where $u_{y,0} = 0$. We specifically evaluate the dynamics of small-amplitude, two-dimensional, incompressible perturbations on this base flow (see Exercises 3.1 and 3.2 for the treatment of three-dimensional disturbances). The leading-order terms for the flow perturbations, obtained from Eqs. (1.10) and (1.12), are [16]:

$$\frac{\partial u_{x,1}}{\partial x} + \frac{\partial u_{y,1}}{\partial y} = 0, \qquad (3.4)$$

$$\frac{\partial u_{x,1}}{\partial t} + u_{x,0}\frac{\partial u_{x,1}}{\partial x} + u_{y,1}\frac{\partial u_{x,0}}{\partial y} + \frac{1}{\rho_0}\frac{\partial p_1}{\partial x} = 0, \qquad (3.5)$$

$$\frac{\partial u_{y,1}}{\partial t} + u_{x,0}\frac{\partial u_{y,1}}{\partial x} + \frac{1}{\rho_0}\frac{\partial p_1}{\partial y} = 0. \qquad (3.6)$$

We will seek "normal mode" solutions of the form:

$$u_{x,1}(x,y,t) = Real\{\hat{u}_{x,1}(y)e^{ikx}e^{-i\omega t}\}, \tag{3.7}$$

$$u_{y,1}(x,y,t) = Real\{\hat{u}_{y,1}(y)e^{ikx}e^{-i\omega t}\}, \tag{3.8}$$

$$p_1(x,y,t) = Real\{\hat{p}_1(y)e^{ikx}e^{-i\omega t}\}, \tag{3.9}$$

$$\psi_1(x,y,t) = Real\{\hat{\psi}_1(y)e^{ikx}e^{-i\omega t}\}. \tag{3.10}$$

It is important to note that this assumed form of x dependency (which, in effect, implies either a periodic or infinite domain) can qualitatively influence the solution characteristics, beyond the parallel flow assumptions on the base quantities. For example, imposing a boundary condition on disturbances at a fixed axial location qualitatively influences the stability characteristics of parallel, swirling flows [20]. Note that $\hat{u}$, k, and ω are complex quantities in general, i.e., $k = k_r + ik_i$ and $\omega = \omega_r + i\omega_i$. We will use essentially the same assumed disturbance fields in Chapter 11 to analyze the response of flames to these hydrodynamic flow disturbances.

Define the complex phase velocity as

$$c_{ph} = \omega/k. \tag{3.11}$$

Vortical hydrodynamic disturbances are generally dispersive, i.e., their propagation speed varies with frequency, $c_{ph} = c_{ph}(\omega)$. In contrast, the propagation speed of acoustic waves is essentially constant for all frequencies of interest to combustion problems.

Substituting these normal mode solutions into Eqs. (3.4), (3.5), and (3.6) leads to:

$$ik\hat{u}_{x,1} + \frac{\partial \hat{u}_{y,1}}{\partial y} = 0, \tag{3.12}$$

$$ik(u_{x,0} - c_{ph})\hat{u}_{x,1} + \hat{u}_{y,1}\frac{\partial u_{x,0}}{\partial y} + \frac{ik}{\rho_0}\hat{p}_1 = 0, \tag{3.13}$$

$$ik(u_{x,0} - c_{ph})\hat{u}_{y,1} + \frac{1}{\rho_0}\frac{\partial \hat{p}_1}{\partial y} = 0. \tag{3.14}$$

Define the stream function $\hat{\psi}_1$ by the relations

$$\hat{u}_{x,1} = \frac{\partial \hat{\psi}_1}{\partial y}, \hat{u}_{y,1} = -\frac{\partial \hat{\psi}_1}{\partial x}. \tag{3.15}$$

Equations (3.12)–(3.14) can be combined into the following expression, referred to as the Rayleigh equation [5]:

$$\frac{\partial^2 \hat{\psi}_1}{\partial y^2} - k^2 \hat{\psi}_1 - \frac{(\partial^2 u_{x,0}/\partial y^2)}{u_{x,0} - c_{ph}}\hat{\psi}_1 = 0. \tag{3.16}$$

These equations, coupled with a set of boundary conditions, define an eigenvalue problem – i.e., nontrivial solutions only exist for certain combinations of the complex wavenumber and frequency that satisfy a dispersion relation given by the function Dr:

$$Dr(k, \omega) = 0. \tag{3.17}$$

Section 3.4.4 provides a specific example showing the derivation of such a dispersion relation. The associated spatial distributions of the unsteady velocity and pressure are controlled by eigenmodes. For the *temporal stability* problem, one calculates the complex frequency (i.e., frequency, ω_r, and temporal growth rate, ω_i) that satisfies the governing equations for each given *real* wavenumber, k_r. In contrast, *spatial stability* problems assume a given *real* temporal frequency, ω_r, and determine the spatial growth rate, k_i, and wavenumber, k_r. Thus, k_i and ω_i are set to zero in the temporal and spatial stability problems, respectively. Section 3.4.3 provides an extended example to illustrate the application of both approaches.

It should be emphasized that temporal and spatial growth rates are not related to each other through the base flow velocity, $u_{x,0}$; their relationship is more complicated. However, near neutral stability points, i.e., where k_i or $\omega_i \approx 0$, a useful relationship, referred to as "Gaster's transformation," can be applied [21]. This relationship states that the real part of the frequency of the temporal and spatial modes is the same, and that their growth rates are related by

$$\omega_i|_{Temporal\ problem} \approx -c_g k_i|_{Spatial\ problem}, \tag{3.18}$$

where the group velocity, c_g, is calculated from the dispersion relation, $Dr(k, \omega)$, between ω and k (see, e.g., Eq. (3.38)):

$$c_g = Real\left\{\frac{\partial \omega}{\partial k}\right\}. \tag{3.19}$$

Exercise 3.6 provides an example application of this transformation.

3.4.2 General Results for Temporal Instability

This section discusses several general conclusions regarding the *temporal stability* problem that can be derived from these equations. There is no analogous section in this chapter on the *spatial stability* problem, as a general body of related results does not exist.

3.4.2.1 Necessary Conditions for Temporal Instability [16]

One of the most important general results for temporal flow stability (see Exercise 3.3) is Rayleigh's inflection point theorem, which states that a necessary, but not sufficient, condition for temporal instability in inviscid flows is that the base velocity profile, $u_{x,0}(y)$, has an inflection point, i.e., $d^2 u_{x,0}/dy^2 = 0$, somewhere in the flow domain. For the parallel base flow profiles considered here, note that the vorticity is given by $\Omega_{0,z} = du_{x,0}/dy$, and so this criterion is equivalent to stating that $d\Omega_{0,z}/dy = 0$

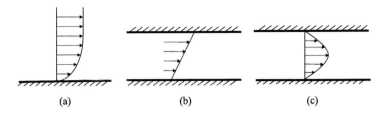

Figure 3.8 Examples of inviscid, linearly stable flows corresponding to (a) zero pressure gradient boundary layer, (b) Couette, and (c) Poiseuille flows.

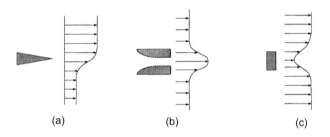

Figure 3.9 Examples of flow that can be destabilized through inviscid, inertial processes: (a) mixing layers, (b) jets, and (c) wakes.

somewhere in the flow. This criterion implies that the vorticity must have a local maximum or minimum for instability to occur; it can be further strengthened [22] to show that the baseline vorticity must have a maximum (thus, a vorticity minimum is not destabilizing). These criteria are highly significant in that we can make immediate inferences on the stability of certain types of flow fields.

As shown in Figure 3.8, many bounded shear flows such as boundary layers over a flat plate, Couette, or Poiseuille flow do not have velocity inflection points and, as such, are inviscidly, linearly stable. However, such flows clearly develop instabilities and transition to turbulence above some Reynolds number. Several subtle, interacting effects must be considered when analyzing such flows. First, viscous effects can be destabilizing under certain circumstances. It can be shown that in order for small-amplitude disturbances to grow, the flow must have a mean velocity gradient, and the correlation of the axial and transverse fluctuations must be nonzero. In the absence of viscosity, this correlation is zero in inviscidly stable flows. However, viscosity influences this phasing between $u_{x,1}$ and $u_{y,1}$, causing energy flow into the disturbance field in some cases [5, 13, 16, 18]. The growth rates of the normal modes of such viscously excited disturbances are small relative to that of the more powerful inviscid mechanisms. Indeed, nonmodal, transient growth processes generally dominate the stability of such flows, as detailed in Section 3.3.

In contrast, Figure 3.9 shows a number of flow configurations that may be destabilized through purely inertial, inviscid mechanisms. These include most free shear flows, such as mixing layers, jets, and wakes. The physical mechanism for instability follows from the fact that perturbing the vorticity in one region induces a perturbation in velocity

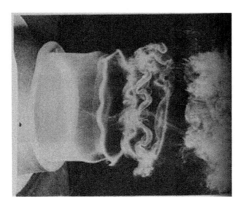

Figure 3.10 Smoke visualization of a jet flow showing flow rollup into vortex rings which subsequently become unstable to azimuthal disturbances. Reproduced from Michalke [25].

in another region; see, for example, Eq. (1.53). Perturbing the velocity, and therefore location of the vorticity, in the other region in turn induces a perturbation in velocity (and therefore location) of the initially perturbed vortex. This feedback loop can be destabilizing, as shown explicitly for several point vorticity array examples in Section 3.6.

Rayleigh's inflection point theorem can be generalized to flows with transverse density variations; i.e., a necessary condition for temporal instability is that $\frac{d}{dy}\left(\rho_0(y)\frac{du_{x,0}(y)}{dy}\right) = \frac{d}{dy}(\rho_0(y)\Omega_{z,0}(y)) = 0$ somewhere in the flow domain [23]. Depending on the details of the density profile (e.g., the width and location of the density gradient with respect to the vorticity), this density-weighted derivative can have multiple local or spatially shifted maxima; see Exercise 3.9 and the discussion around Figure 4.2.

In most instances, the primary instabilities in a flow evolve spatiotemporally and, in turn, develop additional secondary instabilities. These secondary instabilities can be important in their own right, as discussed next. They can lead to further instabilities in the flow which interact nonlinearly and result in rapid disorganization of the initially well-defined, coherent vortical structures into much more complicated motions, or even broadband turbulence. For example, in a spatially developing shear layer, three-dimensional structures, associated with streamwise-oriented vortices, appear downstream of the initially largely two-dimensional flow field. A second example is the occurrence of symmetric ring structures in unstable jets which develop azimuthal instabilities, the "Widnall instability" [24], leading to azimuthal, periodic bending of the vortex, such as shown in Figure 3.10.

3.4.2.2 Growth Rate and Propagation Speed Bounds [16]

This section presents several results associated with the temporal instability growth rate and disturbance phase speed. Both quantities also have important influences on

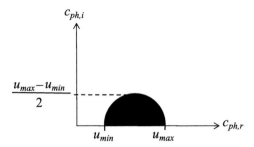

Figure 3.11 Illustration of bounds on instability wave phase speed, $c_{ph,r}$, and growth rate, $c_{ph,i}$. The hashed area denotes the region of permissible solutions for disturbance phase speed and growth rate.

the spatiotemporal dynamics of flames perturbed by flow instabilities, as detailed in Chapter 11. While these quantities can be calculated for a given flow profile $u_{x,0}(y)$ and boundary conditions, it is helpful to look at general bounds on these quantities. Since k is assumed real for this temporal stability analysis, Eq. (3.7) can be written as

$$u_{x,1}(x,y,t) \propto e^{ikx} e^{-ikc_{ph}t} = e^{ikx} e^{-ikc_{ph,r}t} e^{kc_{ph,i}t}. \tag{3.20}$$

Thus, $c_{ph,r}$ and $kc_{ph,i}$ denote the propagation speed and exponential growth rate of the disturbance mode, respectively. Moreover, $\lambda = 2\pi/k$ denotes the spatial wavelength of the disturbance, and ω_r its angular frequency.

The expression below, referred to as Howard's semicircle theorem and illustrated in Figure 3.11, can be derived from Rayleigh's equation and provides a bound on propagation velocities and growth rates for all amplified disturbances where $\omega_i/k = c_{ph,i} > 0$ [26]:

$$\left(c_{ph,r} - \frac{u_{min} + u_{max}}{2}\right)^2 + c_{ph,i}^2 \leq \left(\frac{u_{max} - u_{min}}{2}\right)^2, \tag{3.21}$$

where u_{max} and u_{min} denote the maximum and minimum values of $u_{x,0}(y)$.

Several points are evident from this plot. First, the phase speed, $c_{ph,r}$, of the disturbance is bounded as $u_{min} < c_{ph,r} < u_{max}$. It also shows that the fastest possible growing disturbances propagate at the average flow speed, $(u_{min} + u_{max})/2$. We should emphasize that this result only holds for temporal stability theory and does not apply to spatially growing disturbances.

Second, the maximum possible growth rate is bounded by $\omega_i/k = c_{ph,i} < (u_{max} - u_{min})/2$. This implies that faster-growing instabilities are possible in flows with higher velocity gradients and that the upper bound on maximum instability growth rate increases linearly with k. In other words, upper bounds on growth rates increase as the disturbance length scale decreases. The dependence of the maximum instability growth rate on flow shear is even more explicit in the following additional inequality [27]:

$$\omega_i = kc_{ph,i} \leq \frac{1}{2} \max \left|\frac{\partial u_{x,0}(y)}{\partial y}\right|. \tag{3.22}$$

This shows that upper bounds on instability growth rates are directly proportional to the maximum base flow velocity gradient. Exercises 3.5 and 4.5 compare these bounds with calculated $c_{ph,i}$ vs $c_{ph,r}$ relationships for several problems.

3.4.3 Revisiting Convective/Absolute Instability for Parallel Flows

This section revisits the convective instability discussion from Section 3.2 within the simplifications of a parallel flow. Determining whether a flow is convectively or absolutely unstable can be done by evaluation of the dispersion relation. We must first consider whether the flow is linearly stable or unstable. If the flow is linearly unstable, then the instability can be classified as absolute or convective. The first question can be addressed by considering the temporal stability problem: calculate ω_i for all real k values using the dispersion relation. If the maximum value of all temporal instability growth rates are negative, $\omega_{i,max} < 0$, the flow is linearly stable. If $\omega_{i,max} > 0$, the flow is linearly unstable.

For linearly unstable flows, the absolute/convective determination is obtained by solving for the (k_0, ω_0) combination where

$$Dr(k_0, \omega_0) = 0 \quad \text{and} \quad \frac{\partial Dr}{\partial k}(k_0, \omega_0) = 0. \tag{3.23}$$

Physically, this corresponds to the disturbance with zero group velocity, c_g. If the imaginary part of the frequency associated with this zero group velocity disturbance, denoted as $\omega_{0,i}$, is greater than or less than zero, then the flow is absolutely unstable and convectively unstable, respectively. These stability criteria are summarized in Table 3.2.

Figure 3.12 illustrates these criteria graphically [5]. Consider the temporal instability growth rate of a disturbance as measured by an observer moving along the flow at $v_{observer}$. Three notional plots of the dependence of ω_i on $v_{observer}$, corresponding to a stable, convectively unstable, and absolutely unstable flow, are indicated. The value of ω_i is negative for all $v_{observer}$ values for the linearly stable flow. If there exists some observer reference frame velocity, $v_{observer}$, where $\omega_i > 0$, then the flow is linearly unstable. The flow is then convectively/absolutely unstable for $\omega_i < 0$ and $\omega_i > 0$ at $v_{observer} = 0$, respectively.

Table 3.2 Criteria relating linearly stable, absolutely unstable, and convectively unstable conditions to temporal instability growth rates (see Figure 3.12) [5].

$\omega_{i,max} > 0$	$\omega_{0,i} > 0$	Absolutely unstable
$\omega_{i,max} > 0$	$\omega_{0,i} < 0$	Convectively unstable
$\omega_{i,max} < 0$	$\omega_{0,i} < 0$	Linearly stable

3.4.4 Extended Example: Spatial Mixing Layer [5, 28]

This section provides an example of spatial and temporal stability analyses by presenting a parallel-flow analysis of a piecewise linear shear layer of thickness δ, shown in Figure 3.13. The velocity profile is given by

$$u_{x,0}(y) = \begin{cases} u_{a,0} & y > \delta/2, \\ \dfrac{u_{a,0} + u_{b,0}}{2} + (u_{a,0} - u_{b,0})\dfrac{y}{\delta} & |y| \leq \delta/2, \\ u_{b,0} & y < -\delta/2. \end{cases} \quad (3.24)$$

The effects of density gradients are further considered in Exercise 3.7. In each region, Rayleigh's equation, Eq. (3.16), reduces to the form

$$\frac{d^2 \hat{\psi}_1}{dy^2} - k^2 \hat{\psi}_1 = 0. \quad (3.25)$$

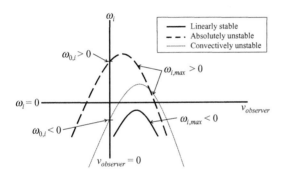

Figure 3.12 Comparison of temporal growth rate of disturbance as a function of observer velocity, $v_{observer}$, for linearly stable, absolutely unstable, and convectively unstable flows [5].

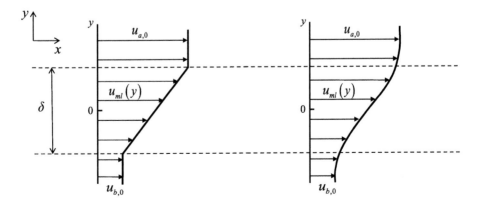

Figure 3.13 Profile of piecewise linear (left) and hyperbolic mixing layer (right) used for stability calculations.

The relationship between the stream function and the pressure can be determined from the linearized x-momentum equation, which for a parallel flow is given by

$$\frac{\partial u_{x,1}}{\partial t} + u_{x,0}\frac{\partial u_{x,1}}{\partial x} + u_{y,1}\frac{\partial u_{x,0}}{\partial y} = -\frac{1}{\rho_0}\frac{\partial p_1}{\partial x}. \tag{3.26}$$

Substituting the stream functions and assumed disturbance forms, Eq. (3.9), into this expression yields

$$-i\omega\frac{\partial \hat{\psi}_1}{\partial y} + iku_{x,0}\frac{\partial \hat{\psi}_1}{\partial y} - ik\hat{\psi}_1\frac{\partial u_{x,0}}{\partial y} = -\frac{ik}{\rho_0}\hat{p}_1. \tag{3.27}$$

Finally, using the relation from Eq. (3.11) and rearranging leads to

$$\hat{p}_1(y) = \rho_0 \frac{du_{x,0}(y)}{dy}\hat{\psi}_1 - \rho_0[u_{x,0}(y) - c_{ph}]\frac{d\hat{\psi}_1}{dy}. \tag{3.28}$$

A general solution of Eq. (3.25) is given by $\hat{\psi}_1 = \mathcal{A}e^{-ky} + \mathcal{B}e^{ky}$, so that the solution in each region is given by

$$\hat{\psi}_{a,1}(y) = \mathcal{A}_a e^{-ky} + \mathcal{B}_a e^{ky}, \qquad y > \delta/2, \tag{3.29}$$

$$\hat{\psi}_{b,1}(y) = \mathcal{A}_b e^{-ky} + \mathcal{B}_b e^{ky}, \qquad y < -\delta/2, \tag{3.30}$$

$$\hat{\psi}_{ml,1}(y) = \mathcal{A}_{ml} e^{-ky} + \mathcal{B}_{ml} e^{ky}, \qquad |y| \leq \delta/2. \tag{3.31}$$

Since the disturbances decay as $y \to \pm\infty$, the coefficients $\mathcal{B}_a$ and $\mathcal{A}_b$ equal zero. These solutions must be matched at the two interfaces, $y = \pm\delta/2$, using continuity of particle displacement and pressure, which take the form (see Exercise 3.4);

$$\left[\frac{\hat{\psi}_1(y)}{u_{x,0}(y) - c_{ph}}\right] = 0, \tag{3.32}$$

$$[\hat{p}_1(y)] = 0, \tag{3.33}$$

where the bracket denotes the jump in values at the respective adjoining edges of the interface. Applying these matching conditions at the two interfaces leads to the following four equations:

$$\mathcal{A}_a e^{-k\delta/2} = \mathcal{A}_{ml} e^{-k\delta/2} + \mathcal{B}_{ml} e^{k\delta/2}, \tag{3.34}$$

$$\mathcal{B}_b e^{-k\delta/2} = \mathcal{A}_{ml} e^{k\delta/2} + \mathcal{B}_{ml} e^{-k\delta/2}, \tag{3.35}$$

$$-k(u_{a,0} - c_{ph})\mathcal{A}_a e^{-k\delta/2} = k(u_{a,0} - c_{ph})\left(-\mathcal{A}_{ml} e^{-k\delta/2} + \mathcal{B}_{ml} e^{k\delta/2}\right) \\ - \frac{\Delta u}{\delta}\left(\mathcal{A}_{ml} e^{-k\delta/2} + \mathcal{B}_{ml} e^{k\delta/2}\right), \tag{3.36}$$

$$k(u_{b,0} - c_{ph})\mathcal{B}_b e^{-k\delta/2} = k(u_{b,0} - c_{ph})\left(-\mathcal{A}_{ml} e^{k\delta/2} + \mathcal{B}_{ml} e^{-k\delta/2}\right) \\ - \frac{\Delta u}{\delta}\left(\mathcal{A}_{ml} e^{k\delta/2} + \mathcal{B}_{ml} e^{-k\delta/2}\right). \tag{3.37}$$

These equations can be combined to yield the dispersion relation

$$Dr = (k\delta)^2 \left(\frac{c_{ph}}{u_{av}} - 1\right)^2 - \chi^2\left[(k\delta - 1)^2 - e^{-2k\delta}\right] = 0, \quad (3.38)$$

where $u_{av} = \frac{u_{a,0}+u_{b,0}}{2}$, $\chi = \Delta u/(2u_{av})$, and $\Delta u = u_{a,0} - u_{b,0}$. Equation (3.38) can be used to study the temporal and spatial stability characteristics of this flow, as well as the convective and absolute stability boundaries.

Starting with temporal stability, we can write the following explicit expression for the complex phase speed, $c_{ph} \equiv \omega/k$:

$$c_{ph} = u_{av} \pm \frac{\Delta u}{2(k\delta)}\left\{[(k\delta) - 1]^2 - e^{-2(k\delta)}\right\}^{1/2}. \quad (3.39)$$

Using this expression, the calculated dependence of the normalized disturbance growth rate, $\omega_i = c_{ph,i}k$, and disturbance phase speed, $c_{ph,r}$, on the dimensionless shear layer thickness, $k\delta$, is plotted in Figure 3.14. Note that there are two temporal stability branches.

Several trends can be deduced from these figures. First, Figure 3.14(a) shows that the flow is linearly unstable, as there is a range of wavenumbers where $\omega_i > 0$ for one of the solution branches, implying that disturbances grow in time. The nondimensionalized threshold wavenumber, $k_{thresh}\delta \approx 1.28$, divides regions of temporally growing and neutrally stable disturbance length scales. Specifically, disturbance length scales shorter than $\lambda_{thresh} = 2\pi/k_{thresh} \approx 4.9\delta$ are neutrally stable; all longer wavelength disturbances are unstable. Figure 3.14(b) shows that the long-wavelength, temporally growing disturbances propagate nondispersively (i.e., with phase speeds that are independent of the disturbance length scale) with a phase speed equal to the average flow velocity, $c_{ph,r} = u_{av}$. The short-wavelength, neutrally stable disturbances propagate dispersively, and have two solutions that converge to $u_{b,0}$ and $u_{a,0}$ in the high-k

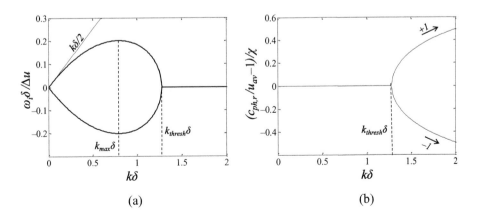

Figure 3.14 Temporal stability calculations showing the dependence of (a) instability growth rate and (b) propagation speed on the disturbance wavenumber.

limit. These phase speed ranges are the same as the general bounds described in the context of Eq. (3.21).

The wavenumber of maximum temporal instability amplification occurs at $k_{max}\delta \approx 0.8$. The associated temporal growth rate is $\omega_i \approx 0.2\Delta u/\delta$. These results can be compared to the two upper bounds in Eqs. (3.21) and (3.22), which are $\omega_i \delta/\Delta u \leq k\delta/2$ and $\omega_i \delta/\Delta u \leq 1/2$, respectively, the former being indicated in the figure.

Consider next a spatial stability analysis. In this case, an explicit expression for k in terms of the real frequency, ω, cannot be determined because Eq. (3.38) is transcendental. Hence, this equation must be solved numerically (or approximated using Gaster's transformation, see Exercise 3.6). The computed dependence of the spatial growth rate and disturbance convection speed on the nondimensional frequency are plotted in Figure 3.15(a) and (b), respectively. Note that the low-frequency disturbances below some threshold value, ω_{thresh}, are amplified spatially, and that higher-frequency ones are neutrally stable. The frequency of maximum amplification occurs at $\omega\delta/u_{av} \approx 0.82$ for $\chi = 0.4$. Larger maximum amplification rates occur with higher velocity ratios, χ.

These threshold and maximum amplification frequencies are both functions of the velocity ratio χ, as plotted in Figure 3.15(c). The frequency corresponding to

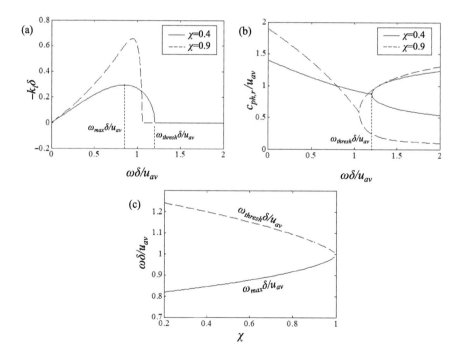

Figure 3.15 Spatial stability calculations showing dependence of (a) instability growth rate and (b) propagation speed on disturbance frequency. (c) Dependence of the threshold angular frequency and frequency of peak amplification on χ.

maximum spatial growth rate is a weak function of χ and has a value of $\omega\delta/u_{av} \sim 0.8 - 1.0$. This corresponds to a Strouhal number of $St_\delta = f\delta/u_{av} \sim 0.13 - 0.16$.

Having demonstrated that this flow is linearly unstable, we next consider the convective/absolute character of the instability. Equation (3.23) shows that the boundary between convective and absolute instability corresponds to the saddle point in the complex k plane, $\partial(\omega\delta/u_{av})/\partial(k\delta) = 0$, where $\omega_i = 0$. Differentiating Eq. (3.38) and setting it to zero leads to

$$\frac{\omega_0\delta}{u_{av}} = k_0\delta - \chi^2\left[(k_0\delta - 1) + e^{-2(k_0\delta)}\right]. \tag{3.40}$$

Substituting the expression for ω from Eq. (3.38) into the above leads to the following transcendental expression:

$$\left\{k_0\delta - 1 + e^{-2(k_0\delta)}\right\}^2 \chi^2 - \left\{[k_0\delta - 1]^2 - e^{-2(k_0\delta)}\right\} = 0. \tag{3.41}$$

This equation must be solved for k_0, whose value is then substituted into the dispersion relation to solve for ω_0. As discussed in the references, this procedure leads to the conclusion that $\omega_{0,i} = 0$ when $\chi = 1$ (assuming, without loss of generality, that $u_{a,0} > u_{b,0}$). Thus, the flow is convectively and absolutely unstable when $\chi < 1$ and $\chi > 1$, respectively. The value $\chi = 1$ corresponds to $u_{b,0} = 0$. Values where $\chi > 1$ correspond to those where there is backflow, $u_{b,0} < 0$; i.e., the lower stream is flowing in the opposite direction to the upper stream. This problem illustrates the significant dependence of absolute stability boundaries on reverse or recirculating flow – similar results will be shown later for wake and backward-facing step flows. As long as both streams are convecting in the same direction, the flow is convectively unstable. This particular example becomes absolutely unstable in the presence of any reverse flow.

It is interesting to compare these analytical results to computations for more realistic mixing layer profiles, such as the continuous hyperbolic tangent profile that is shown in Figure 3.13(b) [29, 30]. This profile gives qualitatively similar results, except that the absolute/convective stability boundary occurs at $\chi = 1.3$; i.e., the flow remains convectively unstable even with some backflow. Moreover, the peak spatial amplification rate of the spatial stability problem occurs at $\omega\delta/u_{av} \sim 0.2$ or $St_\delta = f\delta/u_{av} \sim 0.032$ and is a weak function of χ. This value agrees well with experimental data and is often used to estimate shear layer growth rates in practice.

3.4.5 Connection between Local and Global Analysis [4, 5, 7, 28]

In reality, all flows evolve axially, even if weakly, and have axial boundary conditions on disturbances. A natural question is whether parallel flow concepts can shed light on the global stability of a given flow. The concepts described in earlier sections form a useful starting point if the flow is weakly nonparallel, meaning that the spatial length scale of streamwise variation is small relative to transverse variations, and disturbances are spatially periodic. In these instances, the previously defined dispersion relation, $Dr(k,\omega)$, can be generalized to a function of the axial coordinate, $Dr(k,\omega,x)$, using the local velocity profile at that point, $u_0(x,y)$. Two types of calculation are of interest. First, a local analysis can be performed to determine whether the flow is linearly stable or (convectively or absolutely) unstable at each axial location. To determine the *local* stability of the nonparallel flow at an axial station, x_{local}, a parallel stability analysis is implemented using the local velocity profile, $u_0(x = x_{local}, y)$. Second, the *global* stability characteristics of the flow must be determined. The upper bound on the global mode instability growth rate, $\omega_{Global,i}$, can be shown to equal the maximum local temporal growth rate over all real k and all axial positions, x; i.e.,

$$\omega_{Global,i} \leq \max(\omega_{0,i}). \tag{3.42}$$

This expression implies that a necessary condition for global instability is the existence of a region of local absolute instability in the flow. It also implies that flows that are locally convectively unstable but locally absolutely stable at all spatial points are globally stable – i.e., they will not exhibit intrinsic oscillations, although they will amplify disturbances. The reader is referred to the references for more details on approaches for relating local and global stability concepts, such as the global mode frequency or growth rate.

It should be emphasized that the two restrictions for these relations noted at the beginning of the section, relating to weakly nonparallel and boundary conditions for disturbances, can materially alter the relationship between local and global instability if not satisfied. In particular, it is known that boundary conditions in high-swirl flows can materially alter the flow global stability. For example, neutrally stably swirling flow profiles in an infinite domain become globally unstable when disturbance boundary conditions are applied at a fixed axial location [20].

3.5 Mean Flow Stability [31–35]

This section further expands the concepts from Section 2.1 to analysis of flow stability of unsteady flows, such as turbulent flows. Recall the distinction between the zeroth-order base quantity (e.g., p_0) and the time average ($\bar{p}$), noted for an example problem in Eq. (2.7). Similarly, there are distinctions between the first-order perturbation (p_1) and general disturbance (p'); see also Eq. (2.8). First-order expansions about base flows are a formally accurate approach for analysis of its stability to disturbances. For example, viscous formulations predict the Reynolds numbers at which laminar wake flows become unstable ($Re_D = 46$) [33]. However, the high Reynolds number flows of interest to this text are already turbulent and, therefore, unstable. In this case, analyzing the stability of a laminar base flow at these conditions is analyzing a flow that does not exist. We will consider three related questions in this section. First, can one use the time-averaged, instead of the base, flow to analyze the dominant instability mode shapes, frequencies, and growth rates of, for example, a turbulent jet? Second, essentially all shear flows encountered in combustor environments (jets, wakes, cavity flows, etc.), in addition to being turbulent, exhibit low-dimensional instabilities and large-scale coherent structures. This implies that the system is observed under conditions where it may be executing low-dimensional limit cycle oscillations or strong nonlinearities are present. Do linear instability concepts have any relevance to observations of such flows? For example, can linear predictions of, for example, global mode frequency be used to predict limit cycle frequencies? All of these questions relate in one way or another to whether one can use the stability equations derived earlier and expand about the time-averaged flow, rather than the base flow. We will refer to "linearized" expansions about the time-averaged flow as "mean flow stability analysis." An example of such an expansion is shown below, showing the general time-averaged form (i.e., nonlinear perturbation terms are retained; if they are not, the time-averaged solution trivially reverts to the base flow) and linearized perturbation form of the inviscid, constant-density, momentum equation, Eq. (1.12):

$$\overline{\vec{u}} \cdot \nabla \overline{\vec{u}} = -\nabla \bar{p}/\bar{\rho} - \nabla \cdot (\overline{\vec{u}' \vec{u}'}), \tag{3.43}$$

$$\frac{\partial \vec{u}'}{\partial t} + \vec{u}' \cdot \nabla \overline{\vec{u}} + \overline{\vec{u}} \cdot \nabla \vec{u}' = -\nabla p'/\bar{\rho}. \tag{3.44}$$

Third, related questions also arise around the utilization of quasi-linear theory to predict other nonlinear behavior, such as limit cycle amplitudes. For example, several studies have suggested that the growing disturbances in a globally unstable flow saturate at the amplitude where the growth rate of a disturbance linearized about the mean flow becomes zero – we will call this the "marginal mean flow stability hypothesis" [33, 35, 36]. In other words, the growing disturbance distorts the mean flow field, see Eq. (3.43); that, in turn, alters the growth rate of disturbances on the mean field, described by Eq. (3.44) (in general, the velocity fluctuations consist of both coherent, $\langle \tilde{\vec{u}}' \rangle$, and incoherent, $\langle \vec{u}' \rangle$, oscillations, as detailed in Section 2.8 on the triple decomposition). The marginal mean flow stability theory essentially solves Eqs. (3.43) and (3.44) iteratively, solving for the disturbance amplitude where the instability growth rate becomes zero. Figure 3.16 shows an example of such calculations, plotting the predicted growth rate as a function of disturbance amplitude, calculated using the mean flow at that disturbance amplitude. Also shown in the figure are plots of $\nabla \cdot (\overline{\vec{u}' \vec{u}'})$, showing how this input to Eq. (3.43) varies with disturbance amplitude. Note that $\omega_i = 0$ at $\mathscr{A}^2 = 0.9$ for this calculation.

Formally, the mean flow stability approach cannot be generally true, although, as we discuss in this section, it may be a close approximation in many important cases. The reason that it is not generally true can be seen by considering the following "linearization" of a term such as the convective nonlinearity, $\vec{u} \cdot \nabla \vec{u}'$. While this term may appear to be first order in perturbation amplitude, it actually contains a variety of higher-order terms, as can be seen by referencing the expansions in Eqs. (2.2)–(2.5).

For example, the mean velocity is influenced by the product $\overline{\vec{u}'\,\vec{u}'}$, as shown in Eq. (3.43). Similarly, the perturbation velocity, $\vec{u}'$, is influenced by linear system dynamics, but also higher-order interactions with nonlinear harmonics. As such, while $\overline{\vec{u}} \cdot \nabla \vec{u}'$ appears linear, it implicitly incorporates some, but not all, nonlinear effects. Great care must be exercised in consistently retaining nonlinear terms, as opposed to retaining some and discarding others. Of course, such a consistent ordering of terms can and has been remedied by a more formal approach [32].

With this caution in mind, mean flow stability analysis actually works quite well for several important shear flows, and the conditions when and why it works have also been worked out. To illustrate, Figure 3.17 compares results from a calculation

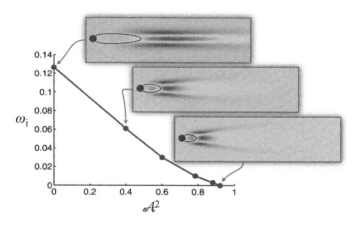

Figure 3.16 Example of the calculation approach for a $Re = 100$ flow, showing growth rate as a function of disturbance amplitude. Inset images show the divergence of Reynolds stress in the axial direction as a colormap; solid lines show the boundary of the recirculation zone. Adapted from Mantic-Lugo et al. [34].

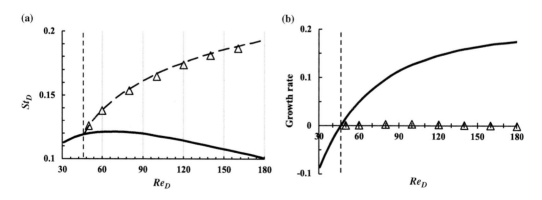

Figure 3.17 Comparison of frequency (a) and growth rate (b) of flow over a cylinder. The dashed curve is the Strouhal number curve-fitted to data from numerical simulation. The solid curves and triangles are computed from a linear stability analysis, using base flow and mean flow, respectively. The vertical dashed lines indicate the bifurcation Reynolds number. Adapted from Barkley [33].

involving a linear expansion about the base flow and time-averaged flow, showing the frequency and growth rate predicted for a flow over a cylinder as a function of Reynolds number.

As expected, the base flow expansion (solid line) approach accurately predicts the Reynolds number at which instability occurs. However, it leads to highly erroneous frequency predictions at Reynolds numbers beyond this bifurcation point, because the flow is unstable and the approximate linearization procedure is invalid. For example, the base flow predicted frequency is off by a factor of nearly 2 at $Re_D = 180$. Also shown in the figure are predictions based on the time-averaged flow profile (the triangles) – note that the frequency prediction is quite accurate and that the predicted instability growth rate is near zero. The latter result is an empirical demonstration of the marginal mean flow stability hypothesis.

Mean flow stability analysis as posed by Eqs. (3.43) and (3.44) captures nonlinear interactions of a single mode with itself (e.g., $\overline{\vec{u}' \vec{u}'}$). However, interactions of higher-order terms also influence the problem. For example, consider the time dynamics of a mode that oscillates as $\sin(\omega t)$. Nonlinear terms generate harmonics, such as $\sin(2\omega t)$. However, nonlinear operations on this harmonic influence both the mean flow (e.g., $\sin^2(2\omega t) = (1 + \sin(4\omega t))/2$) and oscillations at ω (e.g., the product $\sin(\omega t)\sin(2\omega t)$ has oscillations at both ω and 3ω). This suggests that harmonics must be very small relative to the fundamental frequency for mean flow stability approaches to be accurate [31, 37, 38]. While it is not clear how one can generally determine *a priori* if such a criterion will hold, one can analyze their relative magnitudes for a given flow [37] or do an *a posteriori* evaluation based on, for example, measured/computed data [31, 39]. Moreover, the mean flow stability approach can be generalized at the next order to include interactions at ω with 2ω, and self-interactions of 2ω with itself. Successively higher-order corrections can be included by inclusion of higher-order harmonics as well [32].

The wake profile is an example of a configuration where mean flow stability analysis is very accurate, as might be expected from Figure 3.17. This point has also been demonstrated by a formal expansion that rigorously keeps and orders terms of different asymptotic orders [37]. Other studies also suggest that mean flow stability analysis gives accurate predictions for turbulent, swirling jets [31, 39, 40]. However, mean flow stability analysis leads to erroneous results for cavity flows [37], as the cavity flow problem requires inclusion of nonlinear interactions with higher-order harmonics [32].

3.6 Aside: Vortex Mutual Induction

As detailed in Section 3.4.2, the stability of inviscid flows is controlled by the distribution of vorticity. A single point vortex in a medium with no rectilinear flow induces flow rotation, but remains stationary. However, multiple point vortices induce fluid motion upon each other, causing them to move. For example, two counter-rotating vortices of equal strength mutually propel each other in rectilinear motion, as

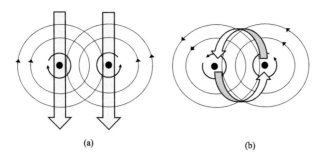

Figure 3.18 Illustration of vortex motion induced by pairs of (a) counter-rotating and (b) co-rotating vortices.

Figure 3.19 Time evolution of a single line of regularly spaced, discrete vortices rotating in the clockwise direction. Adapted from Rosenhead [41].

illustrated in Figure 3.18(a). Two co-rotating vortices of equal strength circle around each other, as illustrated in Figure 3.18(b).

As mentioned in Section 3.4.2, the base vorticity profile plays a key role in determining the necessary conditions for instability. Linear, inviscid instabilities are directly related to the fact that disturbances to unstable vorticity profiles lead to mutual induction of vorticity-containing regions in ways that cause the disturbances to further grow. This can be illustrated by considering the periodic layer of point vortices shown in Figure 3.19 that are initially distributed along a straight line. Consider the evolution of the vortices within the first disturbance wavelength that are disturbed at time $t = 0$. Note how the vortex group that is perturbed downward induces a flow field that is both upward and rightward on the group perturbed upward, and vice versa. This induced motion causes these point vortices to move closer toward each other, leading to an even larger amplitude of the induced flow field disturbances. The initially equally spaced vortices tend to cluster in some regions and spread apart in others. The resulting flow motion evolves in a manner quite reminiscent of the shear layers discussed in Section 4.1. For example, the regions of concentrated and spread out vortices resemble the vortex core and braid, respectively.

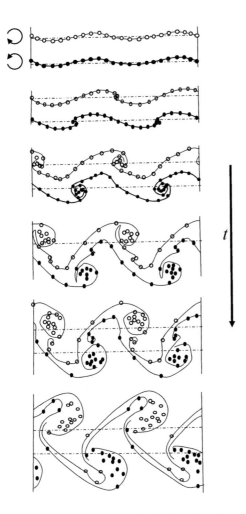

Figure 3.20 Time evolution of a double line of oppositely signed, regularly spaced, discrete vortices. Adapted from Abernathy and Kronauer [42].

In many flows, such as jets and wakes, there are two or more spatially separated sheets of vorticity. In these cases, not only does each vortex sheet induce motion upon itself, but also upon the other vortex sheet. To obtain some insight into the more complex patterns of motion that are then possible, consider the two lines of regularly spaced, counter-rotating vortices shown in Figure 3.20 [42].

This array is initially perturbed in an antisymmetric manner (referred to as a sinuous mode in the context of flow instabilities, to be distinguished from a symmetric, "varicose" mode). Initially, each layer evolves roughly independent of the other, in a manner reminiscent of a single layer. However, as vortices are displaced toward each other, mutual interaction takes place in an accelerating manner. This leads to sweeping of some vortices into concentrations of oppositely signed vortices. Like-signed vortices also accumulate into concentrated regions.

3.7 Aside: Receptivity

All real environments are subjected to various background disturbances. The term "receptivity" is used to describe the mechanisms through which external excitation leads to modal instability waves. Acoustic, vorticity, and entropy disturbances in the free stream excite instabilities in the boundary layer. However, this interaction process cannot be generally described by parallel flow theory because the wavelength and propagation speed of acoustic waves is significantly larger than the length and velocity scales in the boundary layer [13]. Moreover, as noted in Section 2.2, acoustic and vortical disturbances do not linearly interact in uniform flows. Because the dominant instabilities in the boundary layer are of small scale, there must exist a mechanism for energy transfer from the disturbance field. In low Mach number, constant-density flows, this energy transfer process generally occurs in regions of rapid streamwise variations in mean flow properties (which, again, is where acoustic and vortical disturbances interact, see Section 2.3.2), such as at the exit of a splitter plate where the mean flow must rapidly adjust to the change in free stream conditions. Furthermore, in the case of acoustic wave excitation, sharp edges or abrupt changes in geometry lead to much shorter length scale, acoustic disturbance adjustment zones.

The mathematical description of how these acoustic waves lead to excitation of vortical instability waves generally utilizes an "unsteady Kutta condition." The key idea is that at sharp edges, the inviscid acoustic pressure and velocity solutions exhibit a singularity and must excite a vortical disturbance in order to keep the flow disturbances bounded [43]. Elucidating the mechanisms through which flow instabilities are excited by external disturbances is important in understanding the processes through which acoustic waves interact with flow instabilities. Different sections of the next chapter specifically discuss the response of hydrodynamic flow disturbances to acoustic disturbances where data is available.

This scale separation issue does not occur in high-speed boundary layers where vortical disturbances may move at the same speed or even faster than acoustic waves. Moreover, while the stability of low-speed boundary layers is essentially a problem of vorticity dynamics, the coupling and interactions between acoustic, entropy, and vorticity modes profoundly influences the stability of supersonic and hypersonic boundary layers [44].

3.8 Aside: Rayleigh–Taylor and Centrifugal Instability

While this chapter has primarily focused on instability due to distributions of vorticity in the flow, there are other mechanisms of instability as well. This section highlights those occurring in variable-density and swirling flows. Accelerating a density gradient, such as due to gravity, a shock wave, or harmonic oscillations, can be destabilizing, even in the absence of shear in the nominal flow. The simplest example of such an instability is placing a heavier fluid above a lighter one. The temporal instability growth rate is given by [45]:

3.8 Rayleigh–Taylor and Centrifugal Instability

Figure 3.21 Planar laser-induced fluorescence (PLIF) image showing the evolution of Rayleigh–Taylor instability in miscible liquids. Figure courtesy of Devesh Ranjan.

$$\omega_i = \sqrt{kF \frac{(\sigma_\rho - 1)}{(\sigma_\rho + 1)}}, \qquad (3.45)$$

where σ_ρ, k, and F denote the density ratio, spatial wavenumber, and force per unit mass (e.g., the gravitational field), respectively. For example, Figure 3.21 shows an instability of such an interface with flow so that the instability develops from left to right. Note the distinctive shape of the disturbances in the initial development side of the region on the left – the heavy fluid coming down is referred to as a "spike" and the upward motion of the light fluid is the "bubble," an interfacial disturbance shape that is quite distinct from the rollup in Figure 4.4.

Similarly, shock wave propagation through a density gradient leads to instability of the interface, and is referred to as the Richtmyer–Meshkov instability [46, 47]. Premixed flames also divide regions of high and low density ratio. The effect of a steady gravity field on the stability of a flat flame is noted in Section 9.9. During acoustic instabilities, the harmonically varying acceleration field can also stabilize or destabilize the flame, as discussed in Section 11.6.

Additional centrifugal instability mechanisms exist for rotating flows, such as flows with swirl as considered in Section 4.4. An important result for constant-density flows without radial flow components in periodic/infinite axial domains is that a necessary and sufficient condition for stability to axisymmetric ($m = 0$) disturbances is [48]:

$$\frac{d}{dr}(ru_{\theta,0})^2 > 0. \qquad (3.46)$$

This result shows that rotating flows where the velocity decreases with increasing radial distance, or even flows where it increases more slowly than r, are unstable to axisymmetric disturbances. Examples where this can occur are in the outer boundary layer of swirling pipe flows (see Section 4.4) or in flows along curved walls (see Section 4.6). Note also that swirling flows satisfying this criteria can be unstable in finite axial domains [20].

Exercises

3.1. Squire's transformation [5]. In Section 3.1, we assumed two-dimensional disturbances in order to simplify the analysis. In this exercise we will show that the three-dimensional problem can be cast into an equivalent two-dimensional one. Replace the assumed form of the solution by the following, which allows for three-dimensional disturbances:

$$u_1(x, y, z, t) = \text{Real}\left(\hat{u}(y)e^{ik_x x}e^{ik_z z}e^{-i\omega t}\right). \tag{3.47}$$

Using Squire's transformation,

$$\tilde{k}^2 = k_x^2 + k_z^2, \tag{3.48}$$

$$\tilde{k}\tilde{u}_{x,1} = k_x \hat{u}_{x,1} + k_z \hat{u}_{z,1}, \tag{3.49}$$

show that the full three-dimensional problem can be cast into the two-dimensional form analyzed in this chapter.

3.2. This problem further considers the relationship between the stability of parallel flows to two-dimensional and three-dimensional disturbances using Squire's transformation. Consider a three-dimensional mode (k_x, k_z, ω) with a temporal growth rate of ω_i. Show that there exists a two-dimensional mode, $\tilde{k}$, with a larger temporal growth rate and discuss the implications [5]. Hint: Start with the general form of a dispersion relation in three dimensions, $Dr(k, \omega)$, apply Squire's transformation below, and observe the effect on the temporal growth rate. Squire's transformation is:

$$\tilde{c}_{ph} = c_{ph}, \tag{3.77}$$

$$\tilde{k}^2 = k_x^2 + k_z^2, \tag{3.78}$$

$$\tilde{k}\tilde{u}_{x,1} = k_x \hat{u}_{x,1} + k_z \hat{u}_{z,1}, \tag{3.79}$$

$$\tilde{u}_{y,1} = \hat{u}_{y,1}, \tag{3.80}$$

$$\frac{\tilde{p}_1}{\tilde{k}} = \frac{\hat{p}_1}{k_x}. \tag{3.81}$$

3.3. Rayleigh's inflection point theorem is an important result that provides necessary conditions for temporal stability [16]. We will prove this theorem in this exercise. The basic outline of the proof is as follows: multiply Rayleigh's equation by the complex conjugate of the stream function, $\hat{\psi}_1^*$, and integrate over the domain $y_1 \leq y \leq y_2$. Assume that the perturbations vanish at the boundaries, i.e., $\hat{\psi}_1(y_1) = \hat{\psi}_1(y_2) = 0$. Separate the resulting expression into its real and imaginary parts. Obtain Rayleigh's theorem using the imaginary part of the expression.

3.4. In this exercise we derive matching conditions for the use of piecewise continuous flow profiles; particle displacement and pressure must be matched across the weak discontinuities. Derive the continuity of particle displacement and pressure conditions shown in Eqs. (3.32) and (3.33). Hint: Note that the following equation relates the particle location and velocity,

$$u_{y,1}(x,y,t) = \left[\frac{\partial}{\partial t} + u_{x,0}(y)\frac{\partial}{\partial x} + u_{x,1}(x,y,t)\frac{\partial}{\partial x}\right]\xi, \qquad (3.96)$$

at the interface, which is defined by the expression $y = y_0 + \xi(x,t)$.

3.5. The worked problem in Section 3.4.4 provides an explicit relationship between $c_{ph,i}$ and $c_{ph,r}$, which can be used to compare the growth rate and phase speed bounds in Howard's semicircle theorem. Plot $c_{ph,i}$ against $c_{ph,r}$ using $k\delta$ as a parameter for the temporal stability problem. Overlay the inequality from Eq. (3.21) and compare the actual temporal stability characteristics to these bounds.

3.6. Spatial stability of shear layer model problem. Gaster's transformation, Eq. (3.18), provides a useful way to approximately calculate the spatial growth rate based on explicit expressions for the temporal growth rate. Use this transformation to derive an approximate formula for the spatial instability growth rate from the temporal growth rate in Eq. (3.39). Overlay this curve on the exact stability growth rate curve.

3.7. Consider a generalized analysis of the shear layer problem in Section 3.4.3 by allowing the two fluid stream densities to differ, so that the densities are given by ρ_a and ρ_b. Derive the following dispersion relation for the case of $\delta = 0$:

$$-\frac{\left[u_{a,0} - \frac{\omega}{k}\right]^2}{\left[u_{b,0} - \frac{\omega}{k}\right]^2} = \frac{\rho_{b,0}}{\rho_{a,0}}. \qquad (3.113)$$

3.8. Consider the dispersion relation from the previous problem. Plot $c_{ph,r}$ as a function of density ratio for several velocity ratios.

3.9. As noted in Section 3.4.2.1, Rayleigh's inflection point criterion for temporal instability of parallel shear flows takes the following form for flows with density gradients:

$$\frac{d}{dy}\left(\rho_0(y)\frac{du_{x,0}(y)}{dy}\right) = \frac{d}{dy}(\rho_0(y)\Omega_{z,0}(y)) = 0. \qquad (3.120)$$

Depending on the details of the density profile (e.g., the width and location of the density gradient with respect to the vorticity), this density-weighted derivative can have multiple local or spatially shifted maxima. Sketch the $\rho_0(y)\Omega_{z,0}(y)$ profile for shear layers located near premixed flames (where the density monotonically varies between its high and low values) and nonpremixed flame (where it has a minimum between the fuel and oxidizer values), with an offset between the density layer and shear layer, δ. Plot a total of six $\rho_0(y)\Omega_{z,0}(y)$ profiles, corresponding to two density profiles and three offset values ($\delta < 0$, $= 0$, and > 0 offsets). Comment on the number of local maxima, their locations, and the difference between the premixed and nonpremixed density profile results.

References

1. Alfons M., Survey on jet instability theory. *Progress in Aerospace Sciences*, 1984, **21**(0): pp. 159–199.
2. Kabiraj L., Sujith R.I., and Wahi P., Bifurcations of self-excited ducted laminar premixed flames. *Journal of Engineering for Gas Turbines and Power*, 2012, **134**: 031502.
3. Rhode M., Rollins R., Markworth A., Edwards K., Nguyen K., Daw C., and Thomas J., Controlling chaos in a model of thermal pulse combustion. *Journal of Applied Physics*, 1995, **78**(4): pp. 2224–2232.
4. Huerre P. and Monkewitz P.A., Local and global instabilities in spatially developing flows. *Annual Review of Fluid Mechanics*, 1990, **22**(1): pp. 473–537.
5. Godréche C. and Manneville P., *Hydrodynamics and Nonlinear Instabilities*. 1998, Cambridge University Press.
6. Blackburn H., Barkley D., and Sherwin S., Convective instability and transient growth in flow over a backward-facing step. *Journal of Fluid Mechanics*, 2008, **603**: pp. 271–304.
7. Monkewitz P., Huerre P., and Chomaz J., Global linear stability analysis of weakly non-parallel shear flows. *Journal of Fluid Mechanics*, 1993, **251**: pp. 1–20.
8. Chomaz J., Huerre P., and Redekopp L., A frequency selection criterion in spatially developing flows. *Studies in Applied Mathematics*, 1991, **84**: pp. 119–144.
9. Yildirim B. and Agrawal A., Full-field measurements of self-excited oscillations in momentum-dominated helium jets. *Experiments in Fluids*, 2005, **38**(2): pp. 161–173.
10. Emerson B., Lundrigan J., O'Connor J., Noble D., and Lieuwen T., Dependence of the bluff body wake structure on flame temperature ratio, in *Proceedings of 49th AIAA Aerospace Sciences Meeting including the New Horizons Forum and Aerospace Exposition*. 2011, Orlando, FL: AIAA.
11. Schmid P.J., Nonmodal stability theory. *Annual Review of Fluid Mechanics*, 2007, **39**: pp. 129–162.
12. Busse F., Bounds on the transport of mass and momentum by turbulent flow between parallel plates. *Journal of Applied Mathematics and Physics (ZAMP)*, 1969, **20**(1): pp. 1–14.
13. Schmid P.J. and Henningson D.S., *Stability and Transition in Shear Flows*, J.E. Marsden and L. Sirovich, eds. Vol. 142. 2001, New York: Springer.
14. Balasubramanian K. and Sujith R., Non-normality and nonlinearity in combustion–acoustic interaction in diffusion flames. *Journal of Fluid Mechanics*, 2008, **594**: pp. 29–57.
15. Kulkarni R., Balasubramanian K., and Sujith R., Non-normality and its consequences in active control of thermoacoustic instabilities. *Journal of Fluid Mechanics*, 2011, **670**: pp. 130–149.
16. Criminale W.O., Jackson T.L., and Joslin R.D., *Theory and Computation of Hydrodynamic Stability*. 2003, Cambridge University Press.
17. Bagheri S., Schlatter P., Schmid P.J., and Henningson D.S., Global stability of a jet in crossflow. *Journal of Fluid Mechanics*, 2009, **624**: pp. 33–44.
18. Betchov R. and Criminale W.O.J., *Stability of Parallel Flows*. Applied Mathematics and Mechanics, F.N. Frenkiel and G. Temple, eds. Vol. 10. 1967, Academic Press.
19. Ho C. and Huerre P., Perturbed free shear layers. *Annual Review of Fluid Mechanics*, 1984, **16**(1): pp. 365–422.
20. Wang S. and Rusak Z., The dynamics of a swirling flow in a pipe and transition to axisymmetric vortex breakdown. *Journal of Fluid Mechanics*, 1997, **340**: pp. 177–223.

21. Gaster M., A note on the relation between temporally-increasing and spatially-increasing disturbances in hydrodynamic stability. *Journal of Fluid Mechanics*, 1962, **14**(02): pp. 222–224.
22. Fjortoft R., Application of integral theorems in deriving criteria of stability for laminar flows and for the baroclinic circular vortex. *Geofysiske Publikasjoner*, 1950, **17**(6): p. 1.
23. Menkes J., On the stability of a shear layer. *Journal of Fluid Mechanics*, 1959, **6**(4): pp. 518–522.
24. Widnall S.E., Bliss D.B., and Tsai C.-Y., The instability of short waves on a vortex ring. *Journal of Fluid Mechanics*, 1974, **66**(1): pp. 35–47.
25. Michalke A., Survey on jet instability theory. *Progress in Aerospace Sciences*, 1984, **21**: pp. 159–199.
26. Howard L., Note on a paper of John W. Miles. *Journal of Fluid Mechanics*, 1961, **10**(04): pp. 509–512.
27. Høiland E., *On Two-Dimensional Perturbation of Linear Flow*. Vol. 18. 1953, Grøndahl & søns boktr., I kommisjon hos J. Dybwad.
28. Balsa T., On the spatial instability of piecewise linear free shear layers. *Journal of Fluid Mechanics*, 1987, **174**: pp. 553–563.
29. Michalke A., On the inviscid instability of the hyperbolictangent velocity profile. *Journal of Fluid Mechanics*, 1964, **19**(04): pp. 543–556.
30. Monkewitz P. and Huerre P., Influence of the velocity ratio on the spatial instability of mixing layers. *Physics of Fluids*, 1982, **25**: pp. 1137–1143.
31. Tammisola O.J. and Juniper M.P., Coherent structures in a swirl injector at Re = 4800 by nonlinear simulations and linear global modes. *Journal of Fluid Mechanics*, 2016, **792**: pp. 620–657.
32. Meliga P., Harmonics generation and the mechanics of saturation in flow over an open cavity: a second-order self-consistent description. *Journal of Fluid Mechanics*, 2017, **826**: pp. 503–521.
33. Barkley D., Linear analysis of the cylinder wake mean flow. *Europhysics Letters*, 2006, **75**(5): pp. 750–756.
34. Mantic-Lugo V., Arratia C., and Gallaire F., A self-consistent model for the saturation dynamics of the vortex shedding around the mean flow in the unstable cylinder wake. *Physics of Fluids*, 2015, **27**(7): p. 074103.
35. Mezic I., Analysis of fluid flows via spectral properties of the Koopman operator. *Annual Review of Fluid Mechanics*, 2013, **45**: pp. 357–378.
36. Williamson C.H.K., Defining a universal and continuous Strouhal-Reynolds number relationship for the laminar vortex shedding of a circular cylinder. *Physics of Fluids*, 1988, **31**: p. 2742.
37. Sipp D. and Lebedev A., Global stability of base and mean flows: a general approach and its applications to cylinder and open cavity flows. *Journal of Fluid Mechanics*, 2007, **593**: pp. 333–358.
38. Turton S.E., Tuckerman L.S., and Barkley D., Prediction of frequencies in thermosolutal convection from mean flows. *Physics Review E*, 2015, **91**(4): p. 043009.
39. Manoharan K., Frederick M., Clees S., O'Connor J., and Hemchandra S., A weakly nonlinear analysis of the precessing vortex core oscillation in a variable swirl turbulent round jet. *Journal of Fluid Mechanics*, 2019, **884**, A29.
40. Oberleithner K., Sieber M., Nayeri C.N., Paschereit C.O., Petz C., Hege H.C., Noack B.R., and Wygnanski I., Three-dimensional coherent structures in a swirling jet undergoing

vortex breakdown: stability analysis and empirical mode construction. *Journal of Fluid Mechanics*, 2011., **679**: pp. 383–414.
41. Rosenhead L., The formation of vortices from a surface of discontinuity. *Proceedings of the Royal Society of London*, 1931, **134**(823): pp. 170–192.
42. Abernathy F.H. and Kronauer R.E., The formation of vortex sheets. *Journal of Fluid Mechanics*, 1962, **13**(1): pp. 1–20.
43. Crighton D., The Kutta condition in unsteady flow. *Annual Review of Fluid Mechanics*, 1985, **17**(1): pp. 411–445.
44. Mack L.M., Review of linear compressible stability theory, in *Stability of Time Dependent and Spatially Varying Flows*, D.L. Dwoyer and M.Y. Hussaini, eds. 1987, Springer, pp. 164–187.
45. Drazin P.G. and Reid W.H., *Hydrodynamic Stability*. 2nd ed. 2004, Cambridge University Press.
46. Brouillette M., The Richtmyer–Meshkov instability. *Annual Review of Fluid Mechanics*, 2002, **34**(1): pp. 445–468.
47. Mohaghar M., Carter J., Musci B., Reilly D., McFarland J., and Ranjan D., Evaluation of turbulent mixing transition in a shock-driven variable-density flow. *Journal of Fluid Mechanics*, 2017, **831**: pp. 779–825.
48. Drazin P. and Reid W., *Hydrodynamic Stability*. 1st ed. 1981, Cambridge University Press.
49. Day M.J., Mansour N.N., and Reynolds W.C., Nonlinear stability and structure of compressible reacting mixing layers. *Journal of Fluid Mechanics*, 2001, **446**: pp. 375–408.

4 Hydrodynamic Flow Stability II: Common Combustor Flow Fields

This chapter continues the treatment initiated in Chapter 3, focusing on specific flow fields. Hydrodynamic flow stability is a large, rich field and this chapter can only provide a brief introduction to the many fascinating instabilities that arise [1]. For these reasons, attention is specifically focused on high Reynolds number flows and several specific flow configurations of particular significance in combustor systems, including shear layers, wakes, jets, and backward-facing steps.

The vorticity that controls the hydrodynamic stability features of these flows originates largely from the boundary layers in approach flow passages or other walls. The separating boundary layer characteristics serve, then, as an important initial condition for the flows of interest to this chapter. Discussion of the stability and coherent structures present in boundary layers is outside the scope of this book, but several important characteristics of boundary layers are summarized in Section 4.6.

This chapter starts with a discussion of free shear layers in Section 4.1, the most fundamental hydrodynamic instability of interest to practical combustors. It then considers more complex flows that largely involve interactions of multiple free shear layers or of shear layers with walls. For example, two-dimensional wakes (Section 4.2) and jets (Section 4.3) are equivalent to two free shear layers separated by some distance, a, of oppositely signed vorticity. It is recommended that readers using this book as a reference focus on Section 4.1, and then use the remaining sections as references for other specific flow configurations as required.

4.1 Free Shear Layers

Free shear layers are regions of differing velocities between two fluid streams. They are termed "free" because the flow shear is not associated with a wall region, such as is encountered in boundary layers or pipe flows. However, confinement exerts important influences on shear flow stability characteristics [2]; see also Section 4.7. Figure 4.1 illustrates several examples of free shear layers. The dashed lines partition the different regions in the flow by outlining the shear layers, i.e., the regions where velocity gradients exist. Figure 4.1(a) shows a mixing layer between two merging streams of initially different velocities. Figure 4.1(b) shows a free jet, which consists of a potential core near the jet exit that divides two shear layers in the two-dimensional case. As discussed in the second example of Section 3.6, although the vorticity in

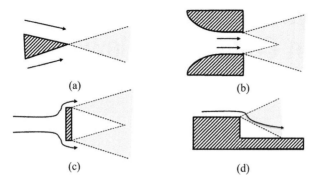

Figure 4.1 Illustration of (a) splitter plate, (b) jet, (c) wake, and (d) backward-facing step flows in which free shear layers occur. Shear regions are shaded in gray.

these two shear layers is spatially separated, they interact because of the flows they induce; this interaction leads to larger-scale jet instabilities. Similarly, Figure 4.1(c) shows the two free shear layers in a bluff body flow, and Figure 4.1(d) shows the flow over a backward-facing step. Note the importance of the separating boundary layer in all of these cases in controlling the "initial conditions" of the free shear flow. In the region immediately downstream of the separation point or splitter plate, the flow enters a complicated adjustment region [3] where it transitions from a boundary layer to a free shear layer, with rapid axial and transverse flow variations. The rest of this section focuses on the flow field beyond this adjustment region where the axial spatial gradients are small relative to those in the transverse direction.

In many combusting flows, the two streams also consist of fluids with different compositions and densities and, as such, shear layers are often interchangeably referred to as mixing layers. For example, in Figure 4.1(a), the top stream could consist of fuel, and the bottom of oxidizer. In this case, a nonpremixed flame resides in the shear layer, which is the region of highest temperatures and lowest densities. For this nonpremixed flame example, the flame would serve as an interface between the regions containing fuel and those containing oxidizer. Note that velocity and species interfaces between the two streams need not exactly coincide, but are generally closely related. For example, in premixed flamelets, the product/reactant interface is very sharp on an instantaneous basis, while that of the velocity can be more diffuse because it is controlled by pressure gradients.

In this same configuration shown in Figure 4.1(a), the top configuration could consist of premixed reactants, and the bottom of hot products. This configuration simulates a flame stabilized in a shear layer with flow recirculation, such as a backward-facing step (shown in Figure 4.1(d)). In this case, a premixed flame would reside in the shear layer, and serve as the instantaneous dividing point between cold reactants and hot products. The resulting temperature and density profile is consequently quite different from the nonpremixed flame example. This point is illustrated in Figure 4.2, which plots the notional velocity profile through a shear layer, as well as the density profile for a premixed and nonpremixed flame, illustrating the

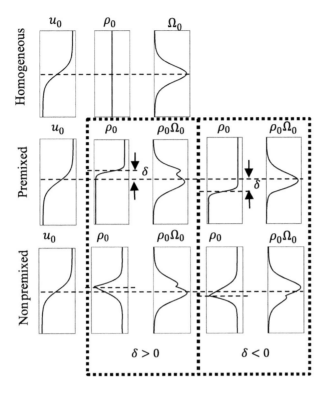

Figure 4.2 Schematic of the spanwise variation of base flow velocity u_0, density ρ_0, and density-weighted vorticity $\rho_0 \Omega_0$ for three cases – homogeneous mixing layer; mixing layer with density jump modeling a premixed flame with an offset δ; mixing layer with density variation modeling a nonpremixed flame with an offset. Image courtesy of Vedanth Nair.

monotonically varying density profile through the premixed flame, and the minimum in density for a nonpremixed flame. The offset between the velocity and density gradient is indicated by δ, and curves are shown for $\delta < 0$ and > 0. Finally, the figure plots the density-weighted vorticity profile for these different configurations (nonreacting, premixed flame with $\delta < 0$ and > 0, nonpremixed flame for $\delta < 0$ and > 0). As noted in Section 3.4.2, the existence of local maxima in these density-weighted vorticity profiles is a necessary condition for temporal instability in parallel flows. Depending on the details of the density profile (e.g., the width and location of the density gradient with respect to the vorticity), this density-weighted derivative can have multiple local or spatially shifted maxima; see Exercise 3.9 for further discussion.

The mixing layers draw ambient fluid from either stream into the mixing region, a process referred to as "entrainment." In nonreacting flows or nonpremixed flames, these two streams then mix in the shear layer. The entrainment process is not symmetric – e.g., for a constant-density shear layer, more fluid is entrained from the high-velocity stream than the low-velocity one [4]. This asymmetry is further increased when the density of the low-speed stream is higher than that of the high-speed stream, and vice versa [4, 5]. This asymmetry has important implications on heat release rates in nonpremixed flames, depending on whether the low- or high-speed side of the shear layer contains the deficient reactant.

The mixing layer evolves axially in a self-similar manner. For example, the scaled mean velocity profiles at every axial location collapse onto a single curve when plotted as a function of $y/\delta(x)$, where $\delta(x)$ is the local shear layer thickness. However, the similarity constants, and therefore the shear layer growth rates, do not have universal values. For example, the axial development of the mixing layer is a strong function of the laminar or turbulent state of the separating boundary layer [6]. This reflects the important role of the large-scale structures in the shear layer, arising from the convectively unstable base flow. Given the sensitivity of convectively unstable flows to external disturbances, the initial conditions matter and are not "forgotten" by the flow [4].

4.1.1 Flow Stability and Unsteady Structure

It is clear from the worked example in Section 3.4.3 that a shear layer is unstable – this instability is referred to as the *Kelvin–Helmholtz instability*, and is one of the key instability mechanisms in shear flows. Although the analysis in Section 3.4.3 was performed assuming a quiescent base flow, $u_{x,0}(y)$, the same basic instability dominates free shear flows at high Reynolds numbers where the base flow is highly turbulent [6] (initially laminar mixing layers transition to turbulence in the Reynolds number range $Re = \bar{u}_x \delta / v_v = 3000$–$5000$ [6]). Indeed, with some adjustments, the same analytical approaches are quite successful in matching data using experimentally measured mean velocity profiles, $\bar{u}_x(y)$; see also Section 3.5. In these cases, the flow consists of quasi-deterministic[1] large-scale structures associated with the underlying inviscid instability of the mean velocity profile $\bar{u}_x(y)$. The flow also contains smaller-scale, broadband turbulent flow fluctuations, as illustrated in Figure 4.3. Studies have clearly shown the persistence of these large-scale structures at Reynolds numbers based on the momentum thickness exceeding 10^6 [7]. The fine-scale turbulence can be

Figure 4.3 Schlieren image of a high Reynolds number shear flow showing large-scale vortical structures and fine-scale broadband turbulence, reproduced from Roshko [9].

---

[1] The term "quasi" is used to indicate the existence of a significant level of spatiotemporal "jitter" in vortex location and pairing dynamics; see also the discussion around Figure 2.24.

thought of as an added "eddy viscosity" acting to dissipate these quasi-deterministic unsteady motions, a familiar concept from Reynolds-averaged analyses of the mean flow [8].

The most amplified disturbance frequency in a laminar and a turbulent mixing layer is $St_\delta = f\delta/u_{x,0} \sim 0.032$ and $St_\delta = f\delta/\bar{u}_x \sim 0.044$, respectively, where δ is the local shear layer thickness [6]. This result varies little with backflow, as discussed at the end of Section 3.4.3. The manifestation of this primary flow instability is the appearance of an essentially two-dimensional disturbance at the interface between the two streams. This disturbance initially appears as small-amplitude oscillations on the interface, then grows and causes the interface to roll up. This rolling action pulls fluid from one side of the shear layer to the other, leading to an undulating interface between the fluid streams that were initially separated; see Figure 4.4. Even in a laminar flow, there is not a real "boundary" because of diffusive transport across this interface, but this is a helpful way to visualize the flow topology.

Vorticity accumulates in the vortex cores. The region between successive vortex cores is referred to as the "braids" where there is strong stretching of the interface between the two fluids, a topic treated in Section 7.4. Figure 4.5 illustrates the flow field in the braid region in a reference frame moving with the vortex cores. Flow is entrained from each stream into the braid regions, which contain stagnation points. At the stagnation point, the flow is divided into counterflowing streams that are drawn into the compressive cores [4]. The extensional and compressive straining in the braid and cores, respectively, can also be anticipated from the distance between the discrete

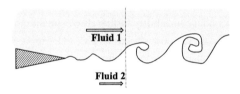

Figure 4.4 Topology of the initial growth and rollup of spanwise flow structures in a shear layer (the cross section of the flow at the vertical cut denoted by the dashed line is shown in Figure 4.6(b)).

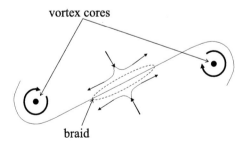

Figure 4.5 Flow structure in a coordinate system moving with the vortices, showing detail of the braid region between two vortex cores, with strong extensional straining of the flow [4].

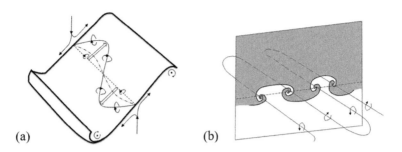

Figure 4.6 Schematics showing (a) the evolution of a perturbed vortex line, showing the conversion of spanwise vorticity to streamwise vorticity, and (b) the streamwise vorticity influence on the mixing layer, as visualized by a planar cut through the mixing layer, see Figure 4.4, following Lasheras et al. [10].

vortices in the vortex array simulation shown in Figure 3.19. Scalar mixing studies show that this entrained flow consists of irrotational fluid from either side of the shear layer, but remains largely unmixed while the structure retains a distinct identity [7]. Once mixing begins, it occurs very rapidly, a phenomenon referred to as the "mixing transition" [10]. This mixing transition is related to the onset of the secondary, streamwise structures (either directly, or due to the smaller-scale structures that appear after this secondary instability) discussed next. Thus, the processes of entrainment and fine-scale mixing are quite distinct [4].

These two-dimensional flow disturbances are subject to secondary instabilities that substantially alter its downstream evolution. Although initially two dimensional, the flow develops a three-dimensional, secondary instability farther downstream that leads to the appearance of streamwise, counter-rotating flow structures [10, 11]. Figure 4.6 illustrates the effect of a small spanwise disturbance on a vortex tube that is initially coincident with the stagnation point shown in Figure 4.5. It is evident that the tube will be pulled towards the adjacent vortex cores, causing it to be rotated into a direction along the flow.

A similar streamwise vortex develops near walls, for different reasons; this is an important effect to be aware of in confined experiments, such as when a nominally two-dimensional bluff body is confined to a finite-width channel [12, 13].

Another important secondary instability is vortex pairing. Adjacent vortices pair as they convect downstream, as shown in Figure 4.7, resulting in the initial vorticity being continually redistributed into larger vortices. The vortex pairing process also leads to a doubling in vortex strength and a doubling and halving in vortex passage wavelength and frequency, respectively. This causes the successive generation of subharmonics in the velocity spectrum with each successive pairing event downstream. Vortex pairing is also a mechanism for growth of the shear layer thickness; the shear layer thickness stays approximately constant between pairing events and grows abruptly after pairing. In a harmonically forced flow, the pairing location can be spatially localized, causing the shear layer thickness to grow in a step-like fashion. In general, however, the space–time location of merging is somewhat random, so that the time-averaged shear layer grows linearly with axial direction.

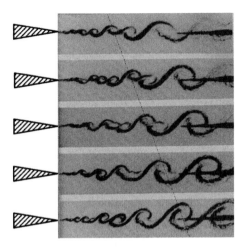

Figure 4.7 Sequence of photographs showing vortex pairing (indicated by the dashed line), adapted from Roshko [14].

As the shear layer width grows, the local growth rate of the most amplified, fundamental mode decreases toward zero, even as that of the subharmonic increases. This can be understood from the growth rate plot in Figure 3.15 by considering the shear layer thickness, δ, to be a function of x. The vortex pairing is observed to occur at the point where the subharmonic growth rate reaches a maximum. However, the distance required for each pairing event increases with each subsequent pairing, since the growth rate of the instability drops as the shear layer thickens. This can also be seen from Figure 3.15, showing that k_i in the spatial growth rate, $\exp(k_i x)$, scales as $k_i \sim 1/\delta$.

Heat release has a significant influence on both spatial amplification rates and the most amplified frequencies of shear layers [15] – the most pronounced effect, which appears to be quite general for a broad range of shear flows, is to cause a decrease in spatial growth rates. Indeed, heat release in zero pressure gradient shear layers decreases mass entrainment into the mixing layer, and therefore product formation rates, and causes a slower rate of shear layer growth [4, 16]. For example, a non-premixed flame residing in a shear layer would cause high temperatures at the centerline that decreases to the ambient value in the two fluid streams. For example, a calculation using a temperature ratio of four and velocity ratio $\chi = 0.2$ (see Eq. (4.1)) shows a threefold decrease in spatial growth rate relative to the constant temperature case [17]. The most amplified frequency also decreases. Different sensitivities are predicted for temperature profiles more representative of a premixed flame stabilized in a mixing layer. Thus, the effect of heat release on the shear layer is quite different for premixed and nonpremixed flames [18], as also discussed in the context of Figure 4.2.

Moreover, the organized vortical structures in the shear layer can propagate along and possibly pass through the flame. As such, the discussion of vortex–flame interactions in Section 7.3.4 directly applies to this configuration. For weak vortices where $u_\theta/s^u \ll 1$, vorticity is damped due to gas expansion across the flame. However, the situation is more complex when $u_\theta/s^u > 1$, leading to rotation and distortion of the flame front and significant changes in the flow field due to baroclinic torque.

The presence of pressure gradients in the flow leads to interactions of shear and baroclinically generated vorticity. For example, consider a problem with the splitter plate configuration in Figure 4.1(a), with co-flowing reactants on top and products on the bottom. A favorable/adverse pressure gradient in the flow direction leads to counterclockwise/clockwise baroclinic vorticity production by the flame. Depending on whether the reactant or product stream is moving faster, this baroclinic vorticity can then reinforce or damp the shear-generated vorticity. If the baroclinic vorticity is of opposite sign to the shear-generated vorticity, heat release can then lead to a complete sign reversal in vorticity direction far downstream of the bluff body, an effect seen in confined bluff body stabilized flames; see Figure 4.12 and Section 4.2.3.

4.1.2 Effects of Harmonic Excitation

Shear layers respond strongly to harmonic excitation, as a result of their convectively unstable nature [19]. This point will be returned to in later chapters, as this leads to important linear mechanisms for self-excited oscillations in combustion chambers; there, acoustic waves excite shear layer disturbances that, in turn, excite the flame. Key to this feedback loop is the monotonic input–output relationship between acoustic waves and the vortical disturbances arising from the shear layer. For example, studies have shown that mixing layers respond at the excitation frequency when forced with disturbances whose amplitudes are three or four orders of magnitude less than the mean velocity. In contrast, such a relationship does not hold for absolutely unstable flows, implying substantially different dynamics.

For many problems of interest in combustors, the frequency of external excitation, f_f, is lower than the most amplified frequency of the unforced shear layer, f_m. To illustrate, assuming a mean velocity of 50 m/s and a shear layer thickness of 1 mm, the most amplified frequency is 1600 Hz, assuming $St_\delta = 0.032$. As another example, a detailed computation of a commercial nozzle showed shear layer frequencies of 13 kHz [20]. Consequently, the rest of this section considers this $f_f < f_m$ case [21].

A measured dependence of the shear layer response frequency, f_r, on forcing frequency, f_f, is plotted in Figure 4.8. The plot shows that the response frequency equals the forcing frequency in the $f_m/2 < f_f < f_m$ range. For forcing frequencies less than half the most-amplified frequency, the very near-field shear layer rollup and development occurs at *approximately* the same frequency as in the unforced case, with the response frequency ranging from $f_r = 0.5f_m$ to $f_r = f_m$. However, these vortices then interact and merge in integer multiples in such a way that the downstream vortex passage frequency matches f_f. For example, in the forcing frequency range of $f_m/3 < f_f < f_m/2$, the response frequency is exactly double that of f_f, but is also close to f_m; i.e., $f_r = 0.66f_m$ to $f_r = 1f_m$. In other words, the excitation frequency

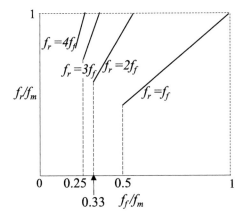

Figure 4.8 Dependence of shear layer response frequency, f_r, on excitation frequency, f_f. Adapted from Ho and Huang [21].

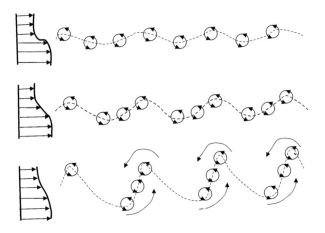

Figure 4.9 Illustration of a forced mixing layer in the forcing frequency range $f_m/4 < f_f < f_m/3$, showing the merging of three sequential vortices. Adapted from Ho and Huang [21].

equals the first subharmonic of the response frequency. In this case, every two vortices merge, leading to a vortex passage frequency equal to that of the excitation frequency. Vortex merging occurs due to the same mutual induction process previously discussed – the amplification of subharmonic oscillations associated with the imposed forcing displaces the vortices transversely, causing their induced flow field to further draw them towards each other.

Similarly, if the forcing frequency falls in the range $f_m/4 < f_f < f_m/3$, the response frequency is three times the forcing frequency and every three vortices merge. This merging of three adjacent vortices into a larger structure is shown in Figure 4.9.

Figure 4.8 also shows that there is a discontinuous change in response frequency with excitation frequency that occurs at $f_f = f_m/n$, where n is an integer, as the number of vortices simultaneously merging also jumps by one integer value. One case of note occurs in the range $f_m/2 < f_f < f_m$, where the excitation frequency is close to that of the dominant instability frequency, and larger than half its first subharmonic. High forcing levels near the fundamental frequency suppress the otherwise natural growth of the fundamental, delaying vortex pairing and slowing shear layer growth.

Multiple-frequency excitation leads to additional physics. For example, one can promote or inhibit vortex pairing and shear layer growth, and cause vortex "shredding" rather than merging, by suitable adjustment of the amplitude and phase of a fundamental and subharmonic frequency [22–25]. This has important influences on the combustion instability problem, where multiple harmonics of the natural acoustic frequency are generally present simultaneously [26].

In cases where $f_f \ll f_m$, related physics leads to the coalescence of a significant number of vortices into a single larger vortex whose passage frequency equals f_f. Figure 4.10 illustrates a striking image of this phenomenon, termed "collective interaction" [27]. As illustrated by the sketch, the vortices interact so that in one region they are drawn together, further amplifying the induced flow field that causes their rotation around each other, and coalesce. In what becomes the braid region, they

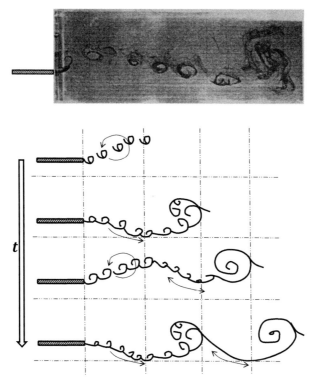

Figure 4.10 Photograph and sketch showing the "collective interaction" phenomenon where vortices form at the most amplified shear layer instability frequency, f_m, and merge to form a large structure whose passage frequency equals the forcing frequency, f_f. Adapted from Ho and Nosseir [27].

are pulled apart from each other. The large-scale vortices observed during acoustic instabilities in combustion chambers are likely the result of such a collective interaction phenomenon. A larger excitation amplitude is required for collective interaction to occur (e.g., $u'/\bar{u}_x \sim 2\%$) [27], otherwise the flow acts as an unforced shear layer. This is probably due to the fact that these subharmonic oscillations imposed by the external excitation grow very slowly. If their amplitude is too small, nonlinear growth of higher-frequency disturbances associated with the natural shear layer dynamics dominate the flow.

Finally, after vortex merging or collective interaction, the normal vortex pairing processes occur downstream, leading to vortices with passage frequencies at subharmonics of the forcing, and a shear layer growth rate that is the same as the unforced layer. However, because the subharmonic growth rate continues to decrease as the shear layer thickens, this process is slow.

4.2 Wakes and Bluff Body Flow Fields [28, 29]

This section discusses wakes, a second important set of free shear flows that occur when, for example, the flow separates over a bluff body. Bluff bodies are routinely used in combustor applications as flame holders, as shown in Figure 4.11.

The time-averaged velocity characteristics of a two-dimensional wake is illustrated in Figure 4.1(c) and Figure 4.12. The downstream evolution of the velocity is strongly influenced by the wake flow dynamics to be discussed in this section. Additionally, the velocity field only becomes self-similar far downstream at points where the wake velocity deficit is small relative to the free stream velocity. Similar to the single shear layer, the spreading rate in this self-similar region is not a universal constant, indicating that the flow "remembers" the bluff body boundary conditions [30]. A key

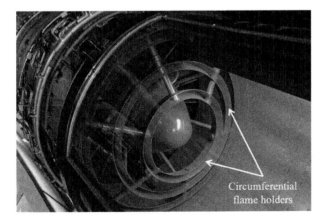

Figure 4.11 Photograph of a turbojet engine augmentor showing circumferential bluff bodies used for flame stabilization. Image courtesy of Steve Kurt.

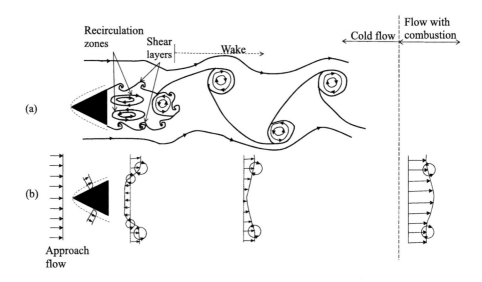

Figure 4.12 Key flow features in an isothermal bluff body flow, with instantaneous flow topology depicted on top, and time-averaged velocity profiles on the bottom. In addition, the time-averaged velocity profiles in a confined, exothermic case shown on the bottom right illustrate the flow profile transitioning from a wake to a jet. Image courtesy of S. Shanbhogue.

distinction of this flow from the free shear layer is the presence of two inflection points in the mean flow velocity profile, and the presence of two interacting shear layers; see also Section 3.6. Section 4.7 further discusses arrays of wake flows or confined wakes, which can be roughly thought of as an even larger number of interacting shear layers.

The structure of the two-dimensional bluff body flow field is shown in Figure 4.12(a) and consists of several regions, including the boundary layer along the bluff body, the separated free shear layer, and the wake [31]. The latter two of these regions can be seen in the flow visualization in Figure 4.13. The shear layer forms the boundaries of the near wake separation bubble, consisting of a recirculating flow [32] with lower pressure relative to the free stream. In a confined channel with combustion, gas expansion across the flame causes the bluff body wake profile to revert to a jet profile farther downstream, as also shown in Figure 4.12(b). This is associated with a change in sign of the mean vorticity between the wake and jet profile regions, also indicated in the figure.

The occurrence of organized flow instabilities and turbulent transition have significant influences on bluff body drag coefficient and flow properties, such as the near wake structure and entrainment rates [32]. For example, for circular cross section bluff bodies [33, 34] there are multiple distinct transitions in the drag coefficient dependence on Reynolds number (Re_D), as different instabilities appear in the flow or as the different flow regimes transition to turbulence. As Re_D is increased, the wake first transitions to turbulence. Next, the shear layer transitions, and finally the boundary layer transitions [33]. For $Re_D \lesssim 200\,000$, the circular cylinder boundary layer is

4.2 Wakes and Bluff Body Flow Fields

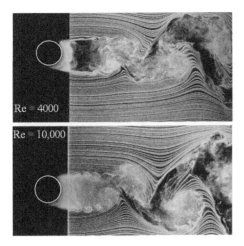

Figure 4.13 Smoke visualization of a bluff body flow field, reproduced from Prasad and Williamson [31].

laminar (referred to as the "subcritical regime") and the dynamics of the downstream flow field are largely driven by the shear layer and wake processes alone. For this Re_D range, both convective and global instabilities are present – the Kelvin–Helmholtz (KH) instability of the separated shear layer and asymmetric vortex shedding, referred to as the Bénard/von Kármán instability (BVK).

4.2.1 Parallel Flow Stability Analysis [29]

Important insights into wake stability characteristics can be obtained from a local stability analysis of a two-dimensional wake profile, similar to the analysis reported in Section 3.4.4. Consider the "top-hat" velocity and density flow profiles shown in Figure 4.14, offset from each other by a distance $\delta_\rho - \delta_u$, where ρ_a and ρ_b denote the density of the gases in the wake and free stream. This density profile resembles the case of premixed combustion. Likewise, $u_{a,0}$ and $u_{b,0}$ denote the wake (which may be negative for reverse flow) and free stream velocities, respectively. The velocity parameter and density ratios, χ and σ_ρ, are defined as

$$\chi = \frac{u_{a,0} - u_{b,0}}{u_{a,0} + u_{b,0}}, \tag{4.1}$$

$$\sigma_\rho = \frac{\rho_b}{\rho_a}. \tag{4.2}$$

Two-dimensional flows, whether jets or wakes, admit two different instability mode shapes. The "sinuous" mode is antisymmetric about the flow centerline, while the varicose mode is symmetric. For zero offset between the density and velocity

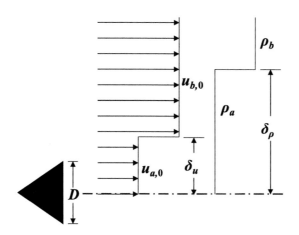

Figure 4.14 Illustration of top-hat density and velocity profiles used for wake stability analysis.

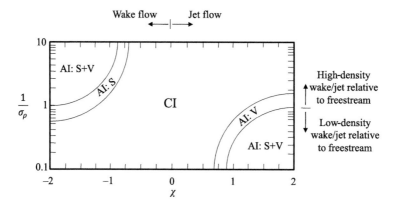

Figure 4.15 Convective/absolute stability boundaries (CI/AI) for two-dimensional, variable-density wake and jet flows, where S and V denote sinuous and varicose modes. Adapted from Yu and Monkewitz [29].

gradients, $\delta_\rho - \delta_u = 0$, the dispersion relation relating disturbance growth rate and frequency is [29]:

$$\sigma_\rho^{-1} \frac{\left[1 + \chi - \frac{1}{u_{av}}\frac{\omega}{k}\right]^2}{\left[1 - \chi - \frac{1}{u_{av}}\frac{\omega}{k}\right]^2} = -\frac{e^{\frac{kD}{2}} + se^{-\frac{kD}{2}}}{e^{\frac{kD}{2}} - se^{-\frac{kD}{2}}}, \qquad (4.3)$$

where $s = 1$ and -1 refer to the sinuous and varicose modes, respectively. This relation is derived in Exercise 4.3. Using the procedure detailed in Section 3.4.3, the regions of convective and absolute stability can be determined from this relation, as summarized in Figure 4.15 (the figure also shows jet flow stability regions, to be discussed in Section 4.3).

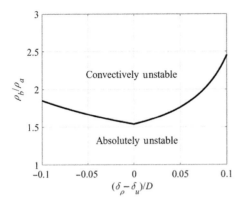

Figure 4.16 Effect of offset between density and velocity jumps, $\delta_\rho - \delta_u$ (see Figure 4.14) on absolute instability boundary for wake flow problem at $\chi = -1.67$ [35].

As can be anticipated from earlier discussions, the figure shows that local absolute instability appears at a sufficiently high velocity ratio, parameterized by χ. Note that the sinuous mode stability boundary appears before that of the varicose mode. The figure also shows that heating/cooling the center stream of the wake flow has a stabilizing/destabilizing effect on absolute instability. This occurs because the wake density change influences the manner in which the two shear layers interact. This result has important implications for combustion applications, where bluff body stabilized flames always lead to hot, low-density wakes relative to the free stream and, thus, exert a stabilizing influence on the wake flow. In addition to heating, flow stability is influenced by base bleed/blowing (stabilizing/destabilizing, respectively), insertion of porous plates in the middle of the bluff body wake, and bluff body geometry.

Offsetting the flame from the velocity shear layers promotes absolute instability. For example, Figure 4.16 plots the dependence of the σ_ρ value at which the transition to absolute instability occurs as a function of the offset, $\delta_\rho - \delta_u$, for the model profiles shown in Figure 4.14. It shows that, for a fixed shear ratio χ, increasing this offset causes the transition to absolute instability at higher density ratios. In the combustion example, this offset is, for example, a function of the flame speed which controls the flame spreading angle and position.

Next, consider axisymmetric wakes. As further expanded in the discussion around Eq. (4.7), the azimuthal distribution of disturbance modes in axisymmetric geometries can be expanded in normal modes as $e^{i(kx+m\theta)}$, where x and θ denote axial and azimuthal directions, respectively. For $k > 0$, a positive/negative m value denotes a helical disturbance winding itself in the direction of decreasing/increasing θ with increasing axial distance. (An example of such helical disturbances can be seen in Figure 4.45.) The presence of two modes with equal amplitudes and opposite-signed m values denotes a "standing" wave disturbance in the azimuthal direction, such as is illustrated in Figures 3.10 or 6.12(b). A key distinction of the axisymmetric geometry from the two-dimensional one is that, in addition to the symmetric $m = 0$ mode, there is (theoretically) an infinite number of azimuthal modes with a 2π periodicity [36]. In contrast, the two-dimensional wake admits only a single antisymmetric mode, the sinuous mode. Calculations for axisymmetric wakes, like those from spheres or streamwise-oriented cylinders, also show them to be absolutely unstable for $Re_D > O(100)$, with the first helical mode, $m = \pm 1$, having the largest growth rates [37, 38].

These local stability analyses are quite revealing for understanding the global flow dynamics. The next sections further detail the global wake and shear layer dynamics behind a bluff body.

4.2.2 Bluff Body Wake

Bluff body wake dynamics have been most extensively characterized for the circular cylinder. Many similarities in wake dynamics apparently exist for other bluff body shapes, but the Reynolds number boundaries for changes in flow features depend on the shape. As discussed in the context of Figure 3.17, the two-dimensional wake becomes *globally* unstable for $Re_D \gtrsim 50$ [39, 40], leading to the BVK instability as sketched in Figure 4.12. This sinuous, absolute flow instability can be predicted from the local analysis described in Section 4.2.1. The BVK instability is periodic with a frequency of [31, 41]:

$$f_{BVK} = St_D \cdot \frac{\bar{u}_x}{D}, \quad (4.4)$$

where St_D is the Strouhal number. For circular cylinders $St_D = 0.21$ and is independent of the Reynolds number in the post-shear-layer-transition, laminar boundary layer regime, $\sim 1000 < Re_D \lesssim 200\,000$ [42]. There are some indications that this Strouhal number value changes above $Re_D \simeq 200\,000$ as the boundary layer transitions to

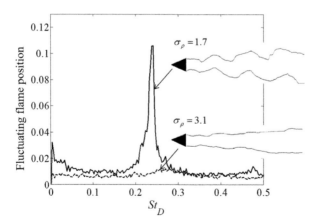

Figure 4.17 Spectra of flame position at two density ratios, showing appearance of a narrowband peak at $St_D = f\bar{u}_x/D = 0.24$ at low density ratio. Inset images show typical instantaneous flame position, showing the corresponding manifestation of sinuous flame flapping at lower flame density ratios. Images courtesy of B. Emerson [35].

turbulence [43, 44]. Measurements on spheres also demonstrate the appearance of narrowband flow oscillations at $St_D \sim 0.2$ [45].

This Strouhal number value is also a function of bluff body shape [46]. In particular, St_D is lower for "bluffer" bodies, i.e., those with higher drag and wider wakes [47]. For example, $St_D \sim 0.18$ for a 90-degree "v-gutter" and drops to 0.13 for a sharp-edged, vertical flat plate [48]. More fundamentally, St_D scales with the wake width [49] and, therefore, care must be exercised in applying Strouhal number trends from one bluff body shape to another. For example, flow separation is retarded for circular bluff bodies when the boundary layer transitions to turbulence, implying a reduction in wake width. In contrast, flow separation locations do not change with Reynolds number in bluff bodies with sharp separation edges.

Having considered basic features of a nonreacting wake, consider next the influences of combustion. As can be anticipated from Figure 4.15, the wake instability in the bluff body near-field is stabilized in the presence of combustion with sufficiently high density ratio, $\sigma_\rho = \rho_b/\rho_a$, across the flame [50–54]. This causes significant differences between the key flow features in the reacting and nonreacting flows. This transition in flow character with flame density ratio can be seen in the data shown in Figure 4.17 [55]. These data show the presence of a narrowband peak in the velocity spectrum and a sinuous flame at low density ratios that is not present at the higher density ratio.

Computations demonstrating this combustion effect are also shown in Figure 4.18, which shows instantaneous flame sheet locations at several flame density ratios [50].

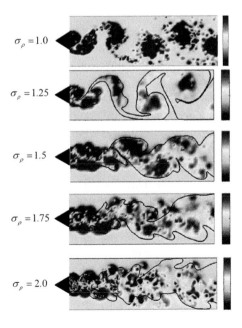

Figure 4.18 Computed vorticity contours and instantaneous flame positions at various flame density ratios σ_ρ. Image courtesy of M. Soteriou and R. Erickson [56].

These results show that the BVK instability magnitude becomes very prominent for σ_ρ less than about 2.

These images and spectra show that the high density ratio, reacting flow field is dominated by fluctuations of the shear layer. In contrast, the low density ratio or nonreacting flow field exhibits a sinuous character with strong, narrowband velocity oscillations. In turbulent flows this transition does not occur in an abrupt fashion, but rather the flow exhibits significant intermittency in the intermediate density ratio range (e.g., for $1.7 < \sigma_\rho < 2.4$ in Ref. [55]). This observation is quite significant as it shows that the dominant fluid mechanics in a facility with nonpreheated reactants, which has a higher density ratio σ_ρ, can be very different from that of a facility with highly preheated reactants, such as when the reactants have been adiabatically compressed to a high pressure ratio or partially vitiated.

4.2.3 Shear Layer Dynamics in Wake Flows

The preceding discussion shows that the BVK instability is suppressed and the unsteady flow dynamics are dominated by the separating shear layers for flames with higher σ_ρ values. As such, it is particularly important for combustion applications to give focused attention to the shear layers, as these are often the dominant source of coherent fluctuations.

We start with nonreacting flows. The separated shear layer evolves in a similar manner to a mixing layer given a sufficiently long recirculation zone length. However, a key distinction is the spatial restriction imposed on the bluff body shear layers by the merging of the two shear layers. The shear layer instability must develop and undergo sufficient amplification before the shear layer rolls up into the BVK vortices, in order to be evident and to exert influences on the flow [31]. This is clearly evident from the images in Figure 4.13 showing the difference in number of shear layer structures at $Re_D = 4000$ and $10\,000$.

Turning to the combustion problem, the flame lies nearly parallel to the flow in high-velocity flows, and thus almost directly in the bluff body shear layers. The flame is wrapped around the regions of intense vorticity, as discussed in Section 7.3.4. Moreover, there is interaction between shear-generated vorticity and flame-generated, baroclinic vorticity. Assuming that the pressure is decreasing in the flow direction, baroclinic vorticity has the opposite sign to the bluff body generated vorticity, but the same sign as the vorticity generated along the wall of the channel, if the body is confined. As such, there is a competition between shear and baroclinic vorticity sources, which can result in complete cancellation, and potentially sign reversal, of flow vorticity; i.e., the flow near and far from the bluff body is dominated by shear and baroclinically generated vorticity, respectively. Another way of thinking of this is to note that the near-field bluff body wake profile transitions into a jet profile farther downstream due to gas expansion, as shown in Figure 4.12(b). The result of this is that the time-averaged vorticity decreases with increasing axial distance and, at some point, switches sign from positive (bluff body generated vorticity) to negative (flame-generated vorticity) [57, 58].

4.2.4 Effects of Harmonic Excitation

The global instability of the nonreacting wake results in substantial differences in how this flow and free shear layers respond to external excitation, as discussed in Section 3.2. This globally unstable flow has intrinsic dynamics with narrowband oscillations in flow velocity and pressure occurring at well-defined Strouhal numbers, as illustrated by curve "1" in Figure 4.19 [46]. In the presence of low-amplitude excitation, the spectra shows the presence of at least two peaks – one corresponding to the excitation frequency, f_f, and one to the natural frequency, f_m – as shown in curve "2" in the figure. In addition, sum and difference spectral peaks may be present due to nonlinear interactions.

If the excitation amplitude is increased above some threshold value, the global instability is "entrained" or "frequency locked" into the excitation frequency. In other words, oscillations at the natural frequency, f_m, either decrease in amplitude or drift in frequency toward f_f, so that oscillations are only observed to occur at the excitation frequency, f_f, at high forcing amplitudes, as illustrated in curve "3." Frequency locking was also discussed in an example problem in Section 2.5.3.3. For globally unstable bluff bodies, this implies that the frequency of vortex shedding shifts to that of the excitation.

The excitation amplitude required to achieve frequency locking is a function of $|nf_f - f_m|$, where n is an integer [46, 59]. In other words, low-amplitude forcing is required to achieve frequency locking when f_f is close to f_m, but the lock-in amplitude monotonically rises as f_f diverges from f_m. However, if f_f is forced close to the first harmonic of f_m, then the lock-in amplitude drops again and the frequency-locked shedding occurs at the subharmonic of the excitation frequency.

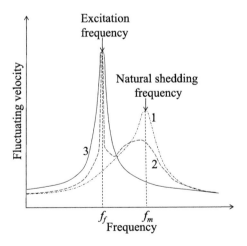

Figure 4.19 Spectra from a forced bluff body experiment [46], for cases of (1) no external forcing, showing a peak corresponding to the natural shedding frequency, (2) low-amplitude forcing, showing peaks at the natural shedding frequency and excitation frequency, and (3) high-amplitude forcing, showing a peak at only the excitation frequency.

These points have important implications on combustion instability mechanisms driven by vortex shedding, to be addressed in Sections 12.2.2 and 12.3.1. As discussed above, the wake is globally stable at high flame temperature ratios. As such, it appears likely that the flame dynamics in these cases is dominated by the harmonically excited, convectively unstable shear layer. In contrast, harmonic forcing of flames at low temperature ratios, where the wake is globally unstable, should lead to fundamentally different effects depending on whether the excitation amplitude is above or below the lock-in amplitude. This illustrates that simulating the forced response of such flames requires careful thought in the design of the experiment to ensure that the base flow's global/convective stability characteristics are preserved.

4.3 Jets

This section considers jets, another important class of free shear flows. As shown in Figure 4.20, there are a variety of free jet configurations of interest to combustors, such as the (a) basic free jet, (b) annular jet, or (c) coaxial jet [60–63]. Note how the annular/coaxial jet incorporates a wake flow field as well. Figure 4.21 illustrates actual combustor hardware with annular jets. In addition, the jet can be injected at an angle to the flow field, referred to as a "jet in cross flow" or "transverse jet" (Figure 4.20(d)).

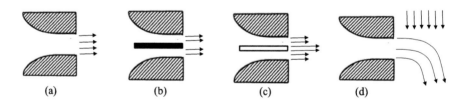

Figure 4.20 Illustration of various jet flow configurations: (a) round jet, (b) annular jet, (c) coaxial jet, (d) jet in cross flow.

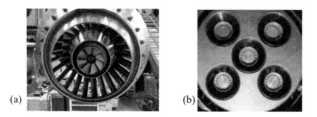

Figure 4.21 Photographs showing annular (swirling) jet configurations in two industrial burners. Images courtesy of Siemens Energy and GE Energy, respectively.

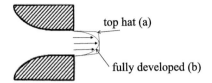

Figure 4.22 Time-averaged axial velocity profiles of a round jet with exit velocity profiles that are top hat or fully developed laminar pipe flow.

Finally, the jet itself can be nominally two dimensional, circular, or incorporate more complex shapes, such as square or elliptical jets, or round jets with canted or lobed exits. As will be briefly alluded to in this section, changing the jet shape from, for example, round to elliptical, or the exit condition from, for example, square edged to canted, has an important influence on the dynamics of the flow. Finally, the presence of swirl, i.e., streamwise rotation of the flow, in any of these configurations can have such a significant impact on the flow evolution that a separate section, Section 4.4, is dedicated to it.

Two time-averaged round jet exit velocity profiles are illustrated in Figure 4.22. Profile (a) corresponds to a top-hat exit velocity profile, as would occur in a fully developed turbulent pipe flow or in a laminar flow after a sharp area contraction. The separating free shear layers grow with downstream distance and eventually intersect at the end of the potential core. Further downstream, the flow continues spreading in the radial direction, with a corresponding decrease in axial velocity. Beyond $x/D \sim 30$, the turbulent jet flow evolves in a self-similar manner [30]. Profile (b) corresponds to a fully developed laminar pipe flow (e.g., a Poiseuille flow profile) and has no potential core, so that the downstream flow evolves from a bounded pipe flow to a free jet. The distinction between these two examples will be discussed later and will illustrate the importance of the jet exit conditions, such as boundary layer thickness and turbulence intensity, on the flow dynamics. As in wakes, a key distinction of this flow from the free shear layer is the presence of two inflection points in the mean flow velocity profile as soon as the jet exits the pipe, and the presence of two interacting shear layers.

4.3.1 Parallel Flow Stability Analysis [64]

Insights into jet stability characteristics can be obtained from a local stability analysis of inviscid, incompressible model jet profiles. In contrast to wakes, a constant-density jet exhausting into a quiescent medium is globally stable, but convectively unstable. In other words, such jets would be expected to amplify ambient disturbances and, as such, exhibit significant sensitivities to background noise characteristics, but not to display intrinsic oscillations.

Jets become globally unstable, however, with sufficient counterflow or when heated; this latter sensitivity is opposite that of wakes. For example, parallel two-dimensional jets become absolutely unstable with a counterflow velocity of $u_{counterflow} \sim 0.05 u_{jet}$ ($\chi = 1.1$), as shown in Figure 4.15 [65]. As can be seen from this low counterflow value, the absolute stability mode is weakly damped, implying that the mode behaves as a weakly damped linear oscillator, but responds very strongly when the spatiotemporal nature of the excitation is close to the natural mode oscillations [66].

Heating a top-hat jet emanating into a quiescent environment to a temperature 20% (for two dimensional, see Figure 4.15) and 30% (for axisymmetric) above ambient is enough for the flow to cross the absolute stability boundary and display intrinsic oscillations [67]. Combustor applications involve jets that are both hot and cold relative to their surroundings, depending on spatial position and application. For example, flows downstream of a jet flame will be hot relative to the ambient and incoming reactants. The injection of fuel into hot air, or injection of cooling air into hot combustion products, are two examples of cold jets relative to the ambient. Measurements in a forced, nonpremixed jet flame clearly show the suppression in spatial growth rate of shear layers [68].

We next consider the modal growth rates and propagation speeds of the spatially growing disturbances for constant-density jets. Two-dimensional jets admit two basic modes of instability – the symmetric varicose mode and the asymmetric sinuous mode. These correspond to the $s = \pm 1$ mode in the dispersion relation in Eq. (4.3). In contrast to wakes, however, it is the varicose, or symmetric, mode that is most unstable in a two-dimensional jet, as shown in Figure 4.15.

For axisymmetric jets, in addition to the symmetric mode, $m = 0$, there are also the azimuthal modes, $m = \pm 1, \pm 2, \ldots$ to consider. Calculations suggest that the symmetric $m = 0$ mode and the first helical mode, $m = \pm 1$, have the largest growth rates near the jet exit [69, 70], although higher-order modes are also unstable. However, the relative growth rates of the $m = 0$ and ± 1 modes are a sensitive function of exit velocity profile. Figure 4.23 plots the result of parallel flow calculations, showing the predicted dependence of the spatial growth rate and phase speed for these two modes. Also shown, in (c), are the corresponding radial distribution of the axial velocity used for these calculations, where a and δ denote the pipe radius and shear layer thickness, respectively. These different velocity profiles could correspond to either different jet exit profiles or axial stations, or both, as the jet evolves downstream. For example, the velocity profiles $a/\delta = 100, 25, 10, 5$, and 2.5 simulate those at different axial stations within the potential core. The velocity profile $a/\delta = 2.19$ simulates that of a fully developed jet exit profile, or that downstream of the potential core.

From these plots, we can make the following observations. First, the $m = 0$ and ± 1 modes have comparable maximum growth rates for the thinner boundary layer profiles, with the $m = 0$ mode slightly more amplified for thin boundary layers (e.g., $a/\delta = 10$), and the $m = \pm 1$ modes being slightly more amplified as the boundary layer thickens (e.g., $a/\delta = 5$). The instability growth rate is only a function of shear layer thickness, independent of burner radius for very thin boundary layers, and asymptotes to the two-dimensional shear layer result. This is

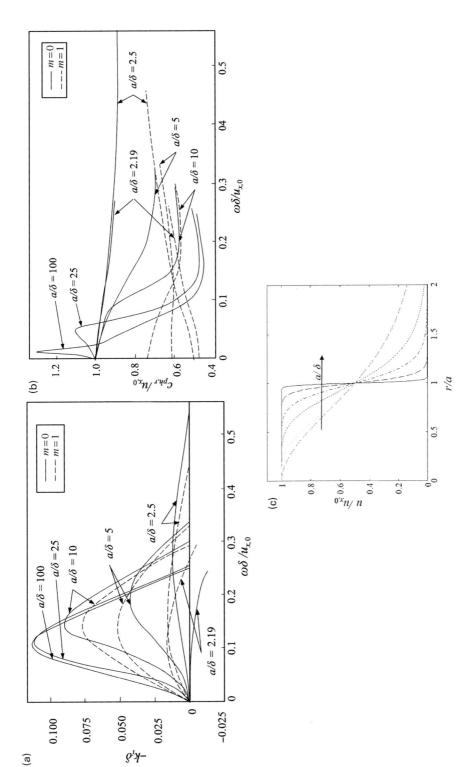

Figure 4.23 Dependence of the spatial instability (a) growth rate and (b) phase speed on the dimensionless frequency of excitation. Corresponding velocity profiles (c) used for calculations are $a/\delta = 2.19$, 2.5, 5, and 10. Adapted from Michalke [64].

evidenced by the $m = 0$ results being virtually identical for the $a/\delta = 25$ and 100 cases. As a/δ decreases, the growth rate curves start to differ somewhat by $a/\delta = 25$ and sharply by $a/\delta = 5$, indicating that in the latter case the burner radius is a better scale for the growth rate. This plot also shows that the local instability properties evolve rapidly near the jet exit. For example, it suggests that the $m = 0$ mode grows most rapidly initially in a jet with a very thin exit boundary layer. However, the $m = \pm 1$ modes become dominant as the shear layer broadens to a value of $a/\delta \gtrsim 6$ [71]. For jets in which the fully developed profile has thick exit boundary layers ($a/\delta = 2.19$), the $m = \pm 1$ modes have the largest growth rate everywhere, and the $m = 0$ and $m > 1$ modes are actually damped [70].

Turning our attention next to the phase speed plots, see Figure 4.23(b), note that both modes are dispersive and exhibit strong sensitivities to shear layer thickness for frequencies below that associated with the maximum amplification rate. At high frequencies, the phase speeds for both of these modes become less sensitive to frequency.

4.3.2 Constant-Density Jet Dynamics

It is very clear from experiments that jets, even at very high Reynolds number, exhibit large-scale structures [72]. The basic convective instability of the jet stems from the shear layer, and there are therefore a number of analogies between jet dynamics and the dynamics of two-dimensional shear layers discussed in Section 4.1. However, there are also several important differences in round jets due to induced flows from vorticity at other azimuthal locations.

Near the nozzle lip, the separating shear layer rolls up into tightly concentrated regions of vorticity, see Figure 4.24, with vortex cores and braids between them. These structures merge as they propagate downstream, causing a spreading in shear layer width and a drop in vortex passage frequency with downstream distance [73]. Multiple vortex mergings occur between the nozzle exit and the end of the potential core [74]. In other words, in the near field of jets with thin exit boundary layers, the jet is essentially an axisymmetric shear layer, destabilized by the Kelvin–Helmholtz instability, which then evolves through rollup of the vortex sheet, and then vortex pairing [75].

Depending on the experimental facility, the dominant structures near the nozzle are either symmetric or helical in nature. The parallel instability theory discussed in Section 4.3.1 suggests that this difference is controlled by the initial velocity profile exiting the jets, as well as the nature of the background noise in the facility. The significance of the jet boundary layer thickness also shows that the degree of turbulence in the internal jet flow before the exit plays an important role in the dominant modal structure in the downstream flow, due to the influence of turbulence on the jet exit velocity profile. Helical structures appear to be dominant in all cases downstream of the potential core, a result that can also be anticipated from the local stability analysis.

4.3 Jets

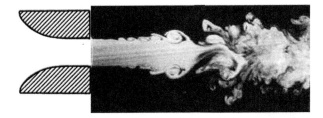

Figure 4.24 Photograph of an unstable jet illustrating shear layer rollup at the jet edges. Reproduced from Liepmann and Gharib [75].

Various secondary instabilities in the jet are dynamically significant. Vortex pairing of symmetric structures was already mentioned. In addition, the axisymmetric vortex rings can be unstable to azimuthal disturbances, as illustrated in Figure 3.10. These disturbances grow through vortex self-induction as the rotated vortex rings induce an azimuthal flow field upon other parts of the contorted ring. In addition, the flow is subject to an analogous secondary instability to that encountered in two-dimensional shear layers, where streamwise vortices develop because of the flow divergence in the braid region connecting the two vortex cores. These structures can be clearly seen in the lower Reynolds number jet visualization shown in Figure 4.25. These structures control jet entrainment downstream of the potential core, as they grow even while the shear-supported azimuthal vortex rings decay [75].

This discussion illustrates that jet flows rapidly become three dimensional downstream of the potential core. These three-dimensional effects can be due to quasi-linear processes associated with growth in helical disturbance amplitude as the mixing layer grows in thickness, or due to explicitly nonlinear dynamics associated with secondary structures. Initially, these three-dimensional structures consist of relatively orderly helical and streamwise vortices (e.g., see Figure 3.10) whose motion becomes increasingly entangled. These structures then rapidly transition to a highly disorganized flow structure [76], as can be seen from the $x/D = 10$ image shown in Figure 4.25. However, large-scale coherent structures continue to persist farther downstream, although they exhibit significant space–time random characteristics and become difficult to visualize at high Reynolds numbers.

4.3.3 Effects of Harmonic Excitation

As in shear layers, harmonic excitation has important influences on jets by amplifying and phase-locking the coherent structures in the flow. Because of the convectively unstable nature of jets, the nature of the forcing, e.g., axisymmetric or antisymmetric, strongly influences the nature of the jet response.

Axisymmetrically forced jets are well known to exhibit a maximum response at a frequency associated with the jet "preferred mode." Reported values occur at a jet Strouhal number of around $f u_x/D \sim 0.3$, and range in values from ~0.25 to 0.6 [74].

Figure 4.25 Laser-based axial cuts of a jet exit flow showing streamwise vortical structure development in the vortex core (left) and braid (right) at two downstream distances of (a, b) $x/D = 2.5$ and (c, d) 3.5. The last image shows the more disordered appearance of the jet farther downstream at (e) $x/D = 10$ ($Re_D = 5500$). Reproduced from Liepmann and Gharib [75].

This preferred mode is not, however, a global mode instability of the jet column; rather, it is the frequency of the shear layer instability as observed at the end of the potential core, where the local shear layer thickness necessarily scales with jet diameter [36].

Forcing has significant effects on vortex pairing of the separated shear layer [77], influencing both the jet spreading and entrainment. As in shear layers, the presence of multiple excitation frequencies significantly influences these pairing processes. For example, if the jet is excited with two frequencies, a fundamental and a subharmonic, the vortex pairing process can be enhanced or diminished, depending on the relative amplitudes and phases of these excitations [78]. Moreover, experiments on lower Reynolds number jets have shown that combined axial and transverse forcing leads to the jet bifurcating into two or more separate branches [79], depending on the relative amplitudes and configurations of the excitation.

4.3.4 Jets in Cross Flow

The introduction of cross flow, as illustrated in Figure 4.20(d), is an important generalization of the jet problem discussed earlier [80]. This is a critical problem for fuel/air mixers, where fuel jets are often injected into a flowing air stream. In addition, it is an important issue in rich, quick-quench, lean-burn (RQL) combustors [81–83], where air is injected into the combustion chamber downstream of the rich combustor head end, as illustrated in Figure 4.26. The next two subsections discuss nonreacting configurations first, then the effects of combustion.

4.3.4.1 Nonreacting Jets in Cross Flow

Important parameters that characterize such flows are the velocity and momentum ratios of the jet to cross flow, u_j/u_0 and $\rho_j u_j^2/\rho_0 u_0^2$. Intuitively, as these ratios approach infinity, the flow tends toward a standard jet. However, important new physics enters the problem at lower velocity ratios. One of the key macro effects of the cross flow is the mutual deflection of the jet and cross flow [84], as shown in Figure 4.27. This causes the jet to evolve from a "jet in cross flow" to a "jet in co-flow" (or to a wake flow if the axial jet velocity is lower than the cross flow). A variety of correlations for the jet trajectory, based on either jet velocity or concentration, have been proposed, typically taking the form [85]:

$$\frac{y}{D\sqrt{\frac{\rho_j u_j^2}{\rho_0 u_0^2}}} = A \left(\frac{x}{D\sqrt{\frac{\rho_j u_j^2}{\rho_0 u_0^2}}} \right)^{\mathcal{B}}. \tag{4.5}$$

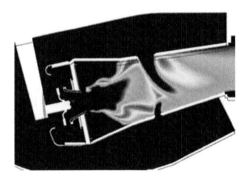

Figure 4.26 Simulation of temperature contours in an RQL combustor showing a jet in cross flow in a combustor quench section. © United Technologies Corporation. Used with permission.

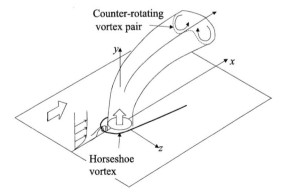

Figure 4.27 Typical time-averaged character of a jet in cross flow, adapted from Kelso et al. [87].

The coefficients vary in the ranges $1.2 < \mathcal{A} < 2.6$ and $0.28 < \mathcal{B} < 0.34$, depending on such parameters as the velocity profile of the jet exit and the Reynolds number, and whether they are based on velocity or scalar concentration [85]. The jet trajectory is largely independent of Reynolds number at high Reynolds numbers [86]. Moreover, the jet trajectory is not substantially altered in cases where a fuel jet in an oxidizing crossflow is combusting [80, 85].

In addition to bending over, the jet is distorted from an initially circular cross section to a kidney or horseshoe shape with a counter-rotating vortex pair, as shown in Figure 4.27. This vortex flow stems from a complex reorientation of the vorticity at the axisymmetric jet exit. This counter-rotating vortex pair plays an important role in jet-free stream mixing [80]; indeed, the jet velocity and concentration profiles decay faster than in free jets.

The alteration of the flow structure can also be appreciated from the two images in Figure 4.28 showing successive positions of a vortex line (a) and ring of fluid (b) originating at the jet exit. The strong stretching of the vortex line in the cross flow direction is clearly evident.

There are some similarities of this problem to that of a compliant cylinder placed in the cross flow, as the cross flow must bend around the jet, both upwards (due to its bending over) and to the side. This leads to a wake flow behind the cylinder. The adverse pressure gradient imposed on the cross flow causes the approach flow boundary layer to separate and also divide around the jet, leading to a "horseshoe vortex," also shown in Figure 4.27.

The analogy with the compliant cylinder should not be taken too far, however. In particular, the reverse flow wake behind the jet is much shorter and weaker than a physical flow blockage. To illustrate, Figure 4.29 shows a visualization of the wake behind the jet (a)–(c) and behind a rigid cylinder with the same diameter (d). The only region in which there is significant separated flow in the jet wake is very near the wall. Images (a)–(c) indicate that there is no separated flow and weak flow recirculation. This has important implications on fluidic flameholding ideas, where a jet flow is used to create a "wake" for flameholding. These images show that such a wake is very weak! Moreover, a comparison of the images for the jet wake with those of the cylinder shows that they are fundamentally different, for reasons discussed further later.

We next consider the unsteady features of the jet [90]. The first feature to note is the rollup of the jet shear layer, whose dynamics appear similar to those of the jets

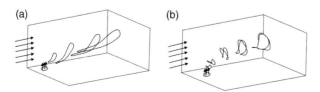

Figure 4.28 Successive images of (a) a vortex line and (b) a ring of fluid originating at the jet exit, calculated at a velocity ratio of $u_j/u_0 = 4$. Adapted from Sykes et al. [88].

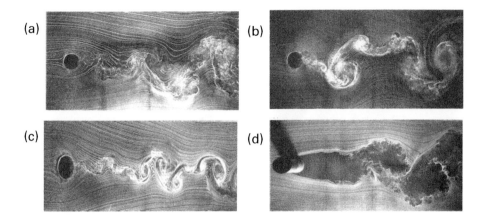

Figure 4.29 Top-down view of smoke streaklines for a jet in crossflow. (a) $Re_D = 3800$, $y/D = 0.5$, $u_j/u_0 = 8$; (b) $Re_D = 7600$, $y/D = 0.5$, $u_j/u_0 = 6$; (c) $Re_D = 11\,400$, $y/D = 1$, $u_j/u_0 = 10$; (d) circular cylinder with same diameter as jet (y denotes the height above the jet exit). Reproduced from Fric and Roshko [89].

Figure 4.30 Smoke streaklines showing shear layer rollup on the leading edge side of the jet. Reproduced from Fric and Roshko [89].

discussed in Section 4.3.2. These can be clearly seen in the smoke streaklines shown in Figure 4.30. These shear layers in constant-density jets are convectively unstable but globally stable, for values of $u_j/u_0 \gtrsim 3$. However, the flow is globally unstable at lower jet velocities, $u_j/u_0 \lesssim 3$, where this transition value is also a function of density ratio if the jet/cross flow densities are not equal [80, 91–95].

The second unsteady feature is the wake vortex structure. These wake vortices can be seen in Figures 4.29 and 4.31, showing the undulating nature of the wake region, very reminiscent of the von Kármán vortex street behind bluff bodies discussed in Section 4.2.2. Indeed, the temptation is strong to invoke the cylinder wake analogy and to view these as vortices shedding from the jet. However, this is not the case and, indeed, the differences between the jet and cylinder become sharp here. In separating flow over a bluff body, vorticity is generated at the boundaries due to the no-slip

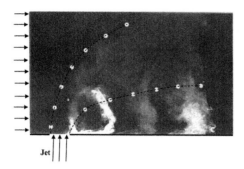

Figure 4.31 Side view of a jet in cross flow where the boundary layer has been seeded with smoke, illustrating vertical "eruptions" of smoke that lead to wake vortices ($Re_D = 3800$, $u_j/u_0 = 4$). Reproduced from Fric and Roshko [89].

condition. This vorticity is convected into the flow and rolls up in the bluff body wake. However, vorticity may not be generated at the jet–cross flow interface! As discussed in Section 1.6, there are no sources of "new" vorticity in a homogeneous flow; vorticity sources only occur at boundaries [89]. This vorticity can be stretched, bent, and diffuse, but there are no sources or sinks in the flow. In flows with density gradients, vorticity can be generated by baroclinic effects.

There are two potential sources of vorticity in the homogeneous jet in cross flow problem: that originating from the jet exit, and that in the wall boundary layer. These wake vortices derive their vorticity from the separating wall boundary layer; this can be seen from the visualizations in Figure 4.31, obtained by adding smoke to the cross flow boundary layer. The adverse pressure gradient on the downstream side of the jet induces eruptions of boundary layer fluid through "separation events," leading to the wake vortices that then convect downstream.

Finally, the instantaneous structure of the counter-rotating vortex pair is significantly more complex than illustrated in Figure 4.27. Simulations indicate that this region consists of a dense, tangled distribution of smaller structures whose maximum streamwise vorticity is substantially higher than that of the time-averaged, large-scale rollers [86].

4.3.4.2 Reacting Jets in Cross Flow

Combustion has strong influences on jets in cross flow. As might be expected from earlier discussions, the nature of these influences strongly depends on the location of the flame with respect to the shear layers and jet exit. There are several topologically different flame configurations in reacting jets in cross flow, as illustrated in Figures 4.32 and 4.33 [96–98]. Figure 4.32 shows sketches of these different conditions for (i) a fully attached flame, (ii) a partially lifted windward flame, (iii) a fully lifted windward flame, and (iv) a lifted flame. Figure 4.33 shows experimental side-view images, overlaying planar cuts of the seeded jet flow (dark) and

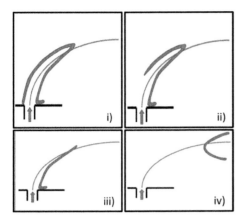

Figure 4.32 Illustration of different jet in cross flow flame stabilization configurations: (i) fully attached flame, (ii) partially lifted windward flame, (iii) fully lifted windward flame, and (iv) lifted flame. The thin and thick lines are approximations of the jet centerline and flame locations respectively [99].

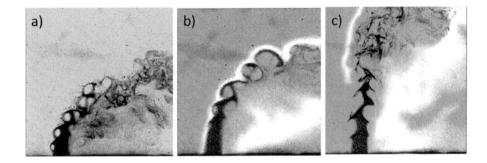

Figure 4.33 Side view of (a) nonreacting jet in cross flow, (b) reacting jet in cross flow where the flame is fully anchored, (c) reacting jet in cross flow with flame lifted on windward side of jet. Mie scattering of seeded jet scales from black (high) to gray (low). OH PLIF scales from white (high) to gray (low). Reproduced from Nair [100].

OH field (white) in (b) and (c). As these are largely nonpremixed flames, the OH field provides an indication of high-temperature regions on the oxidizer side of the flame. Figure 4.33(a) is nonreacting, (b) a fully attached flame, and (c) a partially lifted windward flame.

Several things are evident from the figures. First, Figure 4.32 shows the general sequence of flame attachment configurations as velocity/shear is increased or reaction rates decrease. In other words, only in low-velocity and/or high-reactivity mixtures does the flame fully attach. The flame will almost always lift off from the windward side first, because of the higher gradients/scalar dissipation rate relative to the leeward side. These differing gradients in the leeward and windward sides can be deduced from the thickness of the OH layers in Figure 4.33. The functional dependencies of the lift-off condition and height are a function of whether the flame is autoignition or flame-propagation initiated [98].

As alluded to several times, combustion has the general effect of reducing shear layer growth rates, as can be seen by the difference between the nonreacting case (a) and fully anchored flame (b) in Figure 4.33.

4.4 Swirling Jets and Wakes [101–107]

The presence of swirl (i.e., azimuthal rotation) in the flow fundamentally changes both its time-averaged and unsteady character. The ratio of angular to axial velocity or momentum is a key controlling parameter dividing different types of flow behavior. The typical approach to quantifying this ratio is to define a swirl number, S, as the ratio of axial flux of angular momentum to axial flux of axial momentum, S_m, or an azimuthal to axial velocity, S_V [108, 109]. The momentum flux-based definition of swirl number is given by [110]:

$$S_m = \frac{\int_0^a \rho u_x u_\theta r^2 dr}{a \int_0^a \rho u_x^2 r dr}, \qquad (4.6)$$

where a is the radius of the duct.

To illustrate the effect of swirl number on the flow structure, Figure 4.34 plots the axial velocity distribution of two jets, one with weak and one with stronger swirl. Note

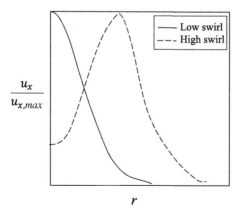

Figure 4.34 Qualitative depiction of radial variation in axial velocity profiles at low and high swirl numbers showing wake-like axial velocity profile at the jet centerline that occurs at high swirl numbers.

4.4 Swirling Jets and Wakes 145

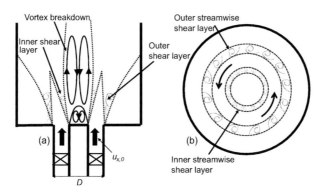

Figure 4.35 Basic swirling structure in an annular combustor arrangement viewed from the (a) side and (b) top. Image courtesy of J. O'Connor.

the presence of a "wake-like" feature in the center of the flow for the stronger swirl case, or the reverse flow shown in Figure 4.35.

The presence of swirl in a flow leads to instability and disturbance propagation mechanisms that are distinct from the shear-driven processes emphasized to this point. First, flow rotation can lead to instability, even without shear, due to centrifugal instability mechanisms, as summarized in Section 3.8. Because of this centrifugal effect, the radial variation of azimuthal velocity exerts important influences on stability. For example, thicker boundary layers on the outer wall of the jet promote centrifugal instability [111]. In contrast, for annular jets with an inner boundary layer, a thicker inner boundary layer has a stabilizing influence. This shows that viscous processes, which influence boundary layer profiles, have important influences on stability characteristics, even though centrifugal instability mechanisms are inviscid. An alternative way to state this point is that even if the key instability dynamics are inviscid, they still exhibit Reynolds number sensitivities.

Second, rotating flows support the presence of inertial waves along the axis of rotation, even in the absence of shear. The reason for this is that flow rotation introduces a restoring force, the Coriolis force, which has an analogous effect to that of the fluid having an anisotropic "fluid elasticity" (fluid elasticity and fluid inertia are jointly responsible for compressional wave propagation) [112]. Flows for which these waves can propagate upstream are referred to as "subcritical"; those for which they can only propagate downstream are referred to as "supercritical."[2] Swirling flow structure and breakdown conditions are known to exhibit significant sensitivity to downstream boundary conditions, such as nozzle contraction ratios under high swirl

---

[2] Note that there is some analogy here to global and convective instability, respectively, but this analogy should not be taken too far as globally unstable flows can be supercritical and subcritical flows can be convectively unstable [111]. In addition, the terms "subcritical" and "supercritical" should not be confused with the identical names used to describe nonlinear system bifurcations in Sections 2.5.1 and 2.5.2.

conditions [113]. This sensitivity is generally attributed to the subcritical nature of the flow.

Swirl flows exhibit a variety of behaviors depending on geometry, swirl number, Reynolds number, and many other parameters. The reader is cautioned against drawing general conclusions about flow structure from specific sources, as results are highly configuration specific. The perspective taken in this text for organizing swirl flow results is to utilize a nomenclature that differentiates the vortex breakdown/reverse flow and the shear layer regions of the flow. Although these two effects are highly coupled in swirl flows, instabilities leading to vortex breakdown can occur without shear, and shear layer instabilities can occur without swirl – thus, we can discuss how vortex breakdown features influence shear layer dynamics or vice versa, but we wish to avoid mislabeling different manifestations of these underlying physics. Indeed, it is not uncommon in the combustion literature to see references to helical shear layer modes as the "precessing vortex core," leading to a confusing nomenclature for characterizing two completely different (but coupled) sets of flow phenomena.

4.4.1 Rotating and Winding Directions of Flow Disturbances

This section introduces the nomenclature on disturbance winding and rotating directions. There are several different conventions in the literature [114–122] so the reader is encouraged to pay careful attention to these conventions when comparing results across different sources. The conventions are summarized in Figure 4.36.

Consider the following normal mode decomposition in a cylindrical coordinate system:

$$\vec{u}_1(r, \theta, x, t) = \text{Real}\left\{\hat{\vec{u}}_1(r)e^{-i(\omega t - kx - m\theta)}\right\}, \tag{4.7}$$

where the real parts of ω and k denote the temporal frequency and axial wavenumber respectively, and their imaginary parts denote the temporal and spatial growth rates. Without loss of generality, assume that the direction of swirl is in the positive θ direction. As such, spatiotemporal points of constant phase correspond to

$$\omega_r t - k_r x - m\theta = \text{constant}. \tag{4.8}$$

Thus, at a constant axial location x, Eq. (4.8) shows that $m > 0$ corresponds to modes that are co-rotating, i.e., points of constant phase advance in time in the same azimuthal direction as the swirl direction. Conversely, $m < 0$ corresponds to counter-rotating modes.

At a given time instant t, Eq. (4.8) shows that $m/k_r > 0$ corresponds to modes that are counter-winding, i.e., points of constant phase wrap helically downstream in the opposite direction to the swirling direction. Conversely, $m/k_r < 0$ corresponds to modes that are co-winding. Finally, at a given θ location (such as would be inferred from a planar velocity measurement), Eq. (4.8) shows that $k_r > 0$ corresponds to constant phase points that are convecting in the positive axial direction.

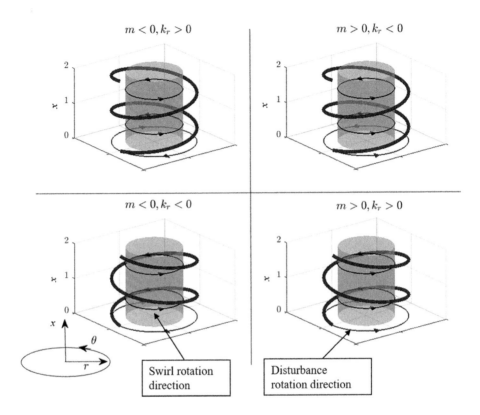

Figure 4.36 Illustration of different rotating and winding direction of surfaces of constant phase. Image courtesy of Travis Smith [123].

4.4.2 Vortex Breakdown [107, 114, 115, 117, 124]

The degree of swirl in the flow, S, has profound influences on the flow structure. A variety of helical instabilities appear as the swirl number increases, whose structure and topology also depend on the Reynolds number [118]. The particular focus of this section, however, is on the fundamental change in time-averaged flow at high swirl numbers. One of the most prominent features of high swirl number flows is the occurrence of "vortex breakdown," which is manifested as a stagnation point in the flow itself [119–121]. This stagnation point is followed by a region of reverse flow – see Figure 4.37. The outer flow not entrained into the recirculating bubble accelerates around it. In fact, to the remainder of the flow, this recirculating bubble resembles a blockage in the time-averaged flow.

Vortex breakdown can be predicted from a steady-state analysis of a rotating, axisymmetric fluid. Below some swirl number, S_A, the flow possesses a single steady-state solution manifested by a unidirectional axial flow. Above some larger swirl number, $S > S_B$, the flow again possesses a single, stable, steady solution where there is a negative flow velocity on the flow axis; this is the vortex breakdown state.

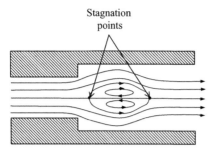

Figure 4.37 Schematic of the time-averaged flow field in a bubble-type vortex breakdown field.

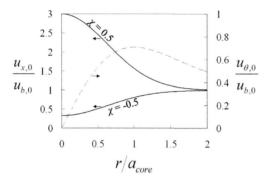

Figure 4.38 Axial and azimuthal velocity profiles used for vortex breakdown calculation, using $S_V = 0.71$ for the $u_{\theta,0}$ plot.

The flow can exist in either state in an intermediate hysteresis regime $S_A < S < S_B$, depending on the initial conditions. To illustrate, consider these boundaries for the "Q-vortex" velocity profile given by [122]:

$$\frac{u_{x,0}}{u_{b,0}} = 1 + \frac{2\chi}{1-\chi} \exp\left(-\frac{5}{4}\left(\frac{r}{a_{core}}\right)^2\right), \qquad (4.9)$$

$$\frac{u_{\theta,0}}{u_{b,0}} = \frac{S_V}{(r/a_{core})} \frac{\left(1 - \exp\left(-\frac{5}{4}\left(\frac{r}{a_{core}}\right)^2\right)\right)}{(1 - \exp(-5/4))}, \qquad (4.10)$$

where the velocity ratio is given by $\chi = \frac{u_{a,0}-u_{b,0}}{u_{a,0}+u_{b,0}}$, and $u_{a,0}$ and $u_{b,0}$ denote the centerline and large-r axial velocities, respectively. The vortex core radius is parameterized by a_{core}, which denotes the radial location of the peak angular velocity, $u_{\theta,0}$. These velocity profiles are illustrated in Figure 4.38.

The resulting calculated vortex breakdown boundaries are illustrated in Figure 4.39 [122]. These boundaries are plotted as a function of velocity based swirl number, S_V, but lines of constant momentum-based swirl number, S_m, are indicated for reference.

4.4 Swirling Jets and Wakes

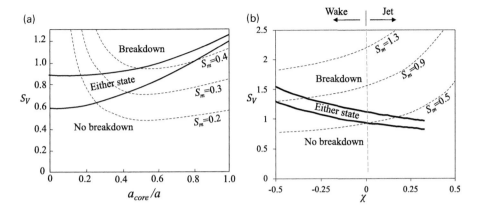

Figure 4.39 Dependence of vortex breakdown boundaries on (a) the ratio of vortex core to pipe radius, a_{core}/a, for jet flow where $\chi = 1/3$, and (b) velocity ratio, χ, at a fixed value of $a_{core}/a = 0.56$. Adapted from Rusak et al. [122].

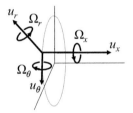

Figure 4.40 Coordinate system used in the discussion of vortex breakdown mechanisms.

The figure shows that vortex breakdown occurs at lower swirl numbers for jet flows, and flows with smaller vortex cores. Higher swirl numbers are required for wake flows to break down.

This bifurcation in time-averaged flow characteristics with increasing swirl number can be understood by considering the dynamics of the azimuthal flow vorticity [114, 124–127], as illustrated Figure 4.40. The azimuthal vorticity, Ω_θ, is of particular interest because its induced velocity is in the *negative* axial direction [126, 128]. The basic idea can be summarized as follows: In a swirling flow that necessarily possesses axial vorticity, Ω_x, the presence of a radial flow perturbation, u_r, stretches and tilts the axial vorticity and produces negative azimuthal vorticity, Ω_θ. This forces deceleration of the on-axis flow, u_x; i.e., $du_x/dx < 0$. This produces, from flow continuity considerations, a positive du_r/dr and a larger radial velocity, u_r, closing the feedback loop. This interaction may create a stagnation point and flow reversal along the vortex axis. Moreover, this discussion illustrates the strong sensitivity of swirling flows to adverse pressure gradients (or, equivalently, axial flow deceleration or radial expansions) such as occur at rapid expansions.

The conditions under which vortex breakdown occurs are a function of Reynolds number, Re, swirl number, S, upstream velocity profile, geometry, pressure gradient, and upstream and downstream boundary conditions [128]. In swirling flows, in particular, there is a strong dependence of the flow structure on downstream flow characteristics, such as nozzle contraction ratio [129]. In addition, area expansions, such as the step area change shown in Figure 4.37, strongly influence swirling flows. For example, the spreading angle of the flow passing around the breakdown bubble can exhibit discontinuous dependencies on area expansion ratio [130].

It is very common in combustors to have a centerbody in the swirling flow, primarily to avoid flashback (see Section 10.1). The centerbody essentially acts as an axisymmetric wake, which induces a recirculating flow even in the absence of swirl. In such cases, the centerbody wake and vortex breakdown bubble remain as two distinct structures at lower swirl numbers and/or with slender/aerodynamic centerbodies. At high swirl numbers and/or with large/bluff centerbodies, the vortex breakdown bubble and centerbody wake merge [131–133]. This effect has very important influences on potential flame stabilization locations in swirling flows with centerbodies, as illustrated in Figure 4.41. For example, with a single merged, central recirculation zone, multiple flame configurations are observed: (a) lifted flame, (b) flame anchored in the centerbody shear layer and spreading in a "V-flame," or (c) anchored in both outer and centerbody shear layers and spreading in an "M-flame." If the centerbody is aerodynamic and it is not possible for a flame to stabilize in its wake, potential flame shapers are (d) a modified M-flame referred to as a "BB-configuration" [133], or (a′) a lifted, aerodynamically stabilized flame [133, 134].

We next consider in more detail the dynamical character of the recirculating flow region, referred to here as the vortex breakdown bubble. As referred to above, a key feature of vortex breakdown is the presence of a flow stagnation point at the leading edge of the bubble. This stagnation point can remain relatively fixed at the flow

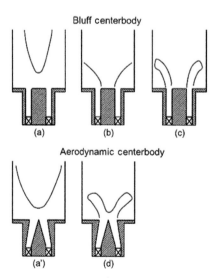

Figure 4.41 Sketch of potential flame shapes in swirl flows with centerbodies. Image courtesy of I. Chterov.

4.4 Swirling Jets and Wakes

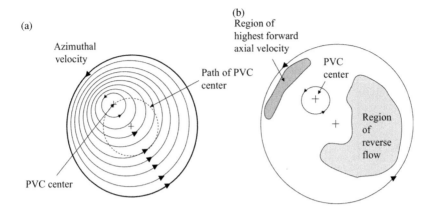

Figure 4.42 Sketch of instantaneous (a) azimuthal and (b) axial velocity inside the vortex breakdown bubble showing the precessing vortex core, following Fick et al. [136].

centerline, referred to as "axisymmetric breakdown," or can sit off-axis and rotate about the flow centerline [119, 135]. Inside the bubble, the flow is highly nonsteady and nonaxisymmetric. The fluid entrained into the bubble is entrained and ejected by an unsteady emptying and filling process that occurs in the downstream part of the bubble [119]. Moreover, the flow does not instantaneously rotate about the geometric centerline. Rather, the point of zero azimuthal velocity is located off-center and this point itself rotates about the geometric center; see Figure 4.42. This location of zero azimuthal velocity is referred to as the precessing vortex core (PVC) [136].

The frequency of rotation of the precessing vortex core increases linearly with Strouhal number based on axial flow velocity and pipe diameter [135]. The positive and negative axial flow velocity regions are not necessarily aligned opposite each other across the PVC center as depicted in the figure, however, and the relative degree of offset varies axially. This leads to a helical pattern in instantaneous axial flow velocity. Finally, as mentioned earlier, the PVC is a distinct flow feature from the other helical shear flow structures that may also be present, such as those due to shear layer instabilities.

Many of these flow features are still present, but substantially altered, in swirling flows with combustion. The vortex breakdown region changes in both shape and size, significantly influencing the spreading angle and velocity of the outer flow going around the bubble. These differences are generally attributed to the effects of gas expansion on the internal vortical structures, but also on the sub/supercritical nature of the flow. For example, because of gas expansion across the flame, the flow field downstream of the flame has higher axial velocities, with little change in azimuthal velocity. This is an effective reduction in swirl number, and pushes the flow toward a supercritical state [137].

4.4.3 Swirling Jet and Wake Dynamics

We start this section by summarizing several key results from local stability analyses of axisymmetric swirling jets and wakes [138]. A key difference from the nonswirling

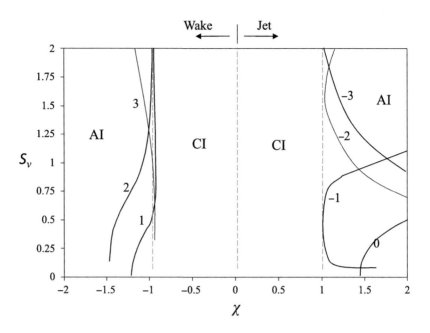

Figure 4.43 Dependence of the absolute/convective instability (AI/CI) regions on swirl parameter and amount of co-/counter-flow for a Rankine vortex model. Swirl number, S_v, is defined as $u_{\theta,max}/(u_{a,0} - u_{b,0})$. Adapted from Loiseleux et al. [138], where the direction of positive outer flow velocity for wakes is inverted.

results is that swirl breaks the $\pm m$ symmetry, implying that helical disturbances winding and/or rotating in the direction of, and counter to, the swirl have different stability characteristics.

Moreover, swirl promotes absolute instability. To illustrate, Figure 4.43 illustrates the results of a parallel flow calculation of the dependence of the absolute/convective stability boundary on the velocity ratio, χ. While nonswirling jets are only absolutely unstable at sufficiently high velocity ratios, the figure shows that jets with sufficient swirl are absolutely unstable even with no backflow at all; e.g., the $m = -2$ mode at a value of $S_v = 1.6$ [138] and the $m = -1$ mode at an even lower swirl number. The figure shows that the most unstable modes for typical swirl numbers encountered in combustors, i.e., $S_v \leq \sim 1$, are the $m = 0$, -1, and -2 modes for jets and $m = 1$ and 2 for wakes. For jets, the general trend is for a given mode to become absolutely unstable only in a certain swirl number range. For example, for $\chi = 1.25$, the $m = -1$ mode is absolutely unstable in the $0.1 < S_v < 0.9$ range, the $m = -2$ mode for $1.1 < S_v < 2.1$, the $m = -3$ mode for $1.6 < S_v < 3.1$, and so forth. The axisymmetric $m = 0$ mode instability growth rate decreases with increasing swirl number.

Experiments and simulations reveal several characteristics of swirling jet flow dynamics, distinct from the vortex breakdown bubble. First, the instability of the shear layers is strongly evident. Swirling flows exhibit shear in both the axial and

4.4 Swirling Jets and Wakes

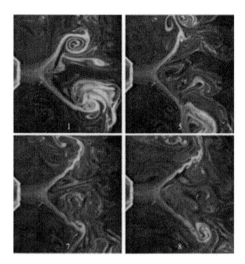

Figure 4.44 Slices through a dye-seeded swirling jet at four instances of time ($S = 1.31$, $Re_D = 626$). Reproduced from Billant et al. [119].

azimuthal directions. As such, the same Kelvin–Helmholtz mechanism already extensively discussed is also present in the azimuthal flow direction. Shear layer disturbances are not necessarily axisymmetric, but generally consist of helical disturbances. Several of these features are evident in the low Reynolds number experimental visualization in Figure 4.44. The axial cuts show that the flow, which starts as a jet, quickly divides into an annulus as it passes around the vortex breakdown region. The rollup of the outer shear layer of the jet can also be seen in the image.

At high Reynolds numbers, the flow structure becomes substantially more complex but similar bulk features are observed; this is illustrated in Figure 4.45, which plots two views of the instantaneous computed vorticity. Note that the flow exits as an annular jet, so that there are two axisymmetric shear layers. The rollup of both shear layers is clearly evident at both swirl numbers. However, the inner shear layer appears to lose its coherent nature more quickly than the outer, possibly due to strong nonlinear interactions with the vortex breakdown bubble. This is particularly evident at the higher swirl number. The outer shear layer remains coherent much farther downstream for both cases, and is particularly prevalent in the $S = 0.44$ case.

4.4.4 Effects of Harmonic Excitation

External forcing leads to modulation of the shear layer at the forcing frequency [140, 141] and also influences the vortex breakdown characteristics [142–152]. Figure 4.46 illustrates several measured axial cuts of an acoustically excited, swirl-stabilized flame, revealing the rollup of the flame by vortical structures that probably originate from collective interaction in the shear layer. The first three images were obtained at a lower forcing amplitude than the last one, hence the difference in instantaneous flame

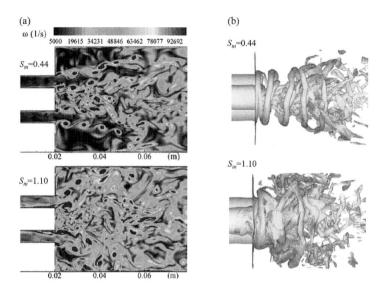

Figure 4.45 Computations of (a) vorticity field magnitude and (b) isovorticity surfaces at low and high swirl numbers. Reproduced from Huang and Yang [139].

Figure 4.46 OH PLIF images of acoustically forced, swirl-stabilized flames showing rollup of the flame [153]. Image courtesy of S. Kumar Thumuluru and B. Bellows.

length [153]. The small-scale shear layer vortices on both the inner and outer sides of the annulus are particularly evident in the last image.

The spatial structure of the forcing strongly influences which disturbance modes are excited. For example, axial forcing of the jet is axisymmetric and excites the $m = 0$ mode most strongly. Given that helical modes often dominate the natural dynamics of swirl flames, this is the reason that the dominant mode often switches from a helical disturbance under "nominal" conditions to the axisymmetric, $m = 0$, mode in the presence of an axial combustion instability [154]. In the case of transverse forcing (a common issue when a transverse acoustic duct mode is excited, such as in an annular combustor), multiple scenarios are possible. If the forcing wavelength (e.g., acoustic

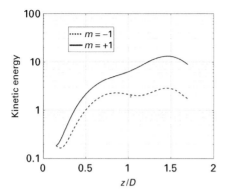

Figure 4.47 Magnitude of co- and counter-rotating modes in transversely forced swirl flame, showing equal excitation of disturbances, but differing axial growth rates. Adapted from Smith [123].

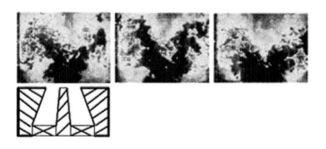

Figure 4.48 Results from OH PLIF images for a harmonically excited, lifted swirl-stabilized flame [157]. Image courtesy of S. Kumar Thumuluru and B. Bellows.

wavelength) is long relative to the jet width, and the jet is located at a pressure antinode, the forcing excites axisymmetric jet disturbances [155]. If, however, the jet is located at a pressure antinode, the $m = +1$ and -1 modes are excited with nearly equal magnitude. For example, Figure 4.47 shows the magnitude of the $m = +1$ and -1 modes as a function of downstream distance for a flow excited in this manner. It shows that they start with a nearly identical amplitude, but that the co-rotating mode has a larger axial growth rate and grows in magnitude relative to the counter-rotating mode with downstream distance.

Forcing also influences the vortex breakdown structure. The presence of geometric centerbodies is an important differentiator between different types of flow dynamics in response to forcing. For flows without centerbodies (or small centerbodies with very small wakes) a stagnation point exists downstream whose axial location is modulated by forcing [156]. Flames that are stabilized in this low-velocity region are buffeted back and forth by bulk forcing, as can be seen in Figure 4.48.

Very different dynamics are observed for flows where the centerbody wake and vortex breakdown bubble merge into a single structure. This occurs in high swirl number flows with centerbodies that are large enough to produce a nonnegligible wake. In this case, low-amplitude forcing exerts minimal influences on the vortex breakdown region as compared to the unmerged case [131]. This appears to be related to the fact that the recirculation zone is "attached" to the geometrically fixed centerbody and the forward stagnation point is on the face of its solid surface. In this case, forcing does not cause axial modulation in the location of the stagnation point.

The globally unstable nature of vortex breakdown leads to additional amplitude dependencies. Globally unstable systems execute self-excited, limit cycle motions, even in the absence of external forcing [158]. Large-amplitude forcing, however, has been shown to cause the recirculating bubble to exhibit amplitude dependence that is sometimes discontinuous [159]. In addition, large-amplitude forcing has been shown to revert the flow to a nonbreakdown state [140].

4.5 Backward-Facing Steps and Cavities

This section considers flows over backward-facing steps or cavities. Such configurations, either in two-dimensional or axisymmetric form, are routinely used in combustion systems for flame stabilization because of the forced flow separation and flow recirculation [160]. Several illustrations are shown in Figure 4.49, including (a) a nominally two-dimensional backward-facing step and (b) an axisymmetric step, or two oppositely faced two-dimensional backward-facing steps. In addition, step flows can be seen in several prior figures, such as Figures 4.45 and 4.46. A step flow was also depicted in Figure 4.1 in the free shear layer section; indeed, many dynamical

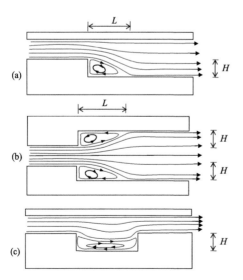

Figure 4.49 Schematic of flow configurations showing (a) a two-dimensional backward-facing step, (b) an axisymmetric or oppositely faced two-dimensional backward-facing step, and (c) a cavity.

features of these flows can be understood from Section 4.1 as long as the step height, H, is large compared to the upstream boundary layer thickness [161].

A key additional piece of physics that is present here, however, is the reattachment of the separating shear layer to the outer wall, which also leads to a recirculation zone (or multiple recirculation zones) of fluid. This reattachment length is a function of shear layer spreading rate and typically ranges between about 6 to 8 step heights, although values as low as 5 and high as 15 have been measured [162]. This reattachment length is a function of Reynolds number, increasing with Re in the laminar regime and decreasing in the turbulent regime [163]. Furthermore, for the configuration shown in Figure 4.49(b), interaction between oppositely faced shear layers can occur, leading to coupled dynamics in the opposed cavities and flow asymmetry [164]. In addition, the recirculating flow implies a shear layer with strong backflow, suggesting the possibility for global instability; see Section 3.4.4. The time-averaged recirculating flow can consist of a single rotating structure or multiple, counter-rotating flow vortices. Indeed, in the low Reynolds number limit, there are an infinite sequence of progressively smaller, counter-rotating vortices that occur as one approaches a sharp corner, referred to as "Moffatt eddies" [165]. Finally, additional separated flow regimes can occur on the opposite wall of the step for the configuration shown in Figure 4.49(a), that interact with the "primary" shear layer and recirculation zone.

Figure 4.49(c) shows a cavity flow, again exemplified by a separating shear layer and a recirculation zone. In this configuration, the shear layer can impinge on the opposite side of the cavity, leading to "cavity tones" [166, 167]. These cavity tones are narrowband, self-excited fluctuations associated with feedback between convecting vortical disturbances from the separating shear layer that impinge upon the opposite wall, generating sound that propagates upstream and further disturbs the separating shear layer (see Exercise 4.6).

4.5.1 Parallel Flow Stability Analysis

This section generalizes the shear layer analysis considered earlier to step/cavity flows. Figure 4.50 illustrates the time-averaged flow field and its key length scales

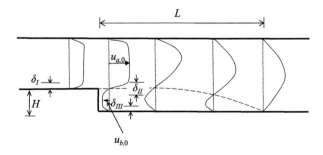

Figure 4.50 Evolution of the time-averaged velocity profile in flow over a backward-facing step. Adapted from Wee et al. [169].

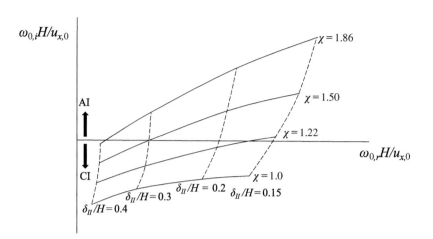

Figure 4.51 Constant-density convective/absolute (CI/AI) stability boundary dependence on velocity ratio, χ, and normalized shear layer thickness, δ_{II}/H. Adapted from Wee et al. [169].

used for this analysis [168, 169]. These include (1) the boundary layer thickness of the approach flow before separation, δ_I, (2) the shear layer thickness, δ_{II}, (3) the bottom wall boundary layer thickness, δ_{III}, (4) the step height, H, and (5) the attachment length, L. In addition, there are at least two velocity scales, the approach flow velocity, $u_{a,0}$, and the reverse flow velocity, $u_{b,0}$.

The dynamical significance of the length scales δ_I and δ_{II}, as well as the ratio of the reverse and forward flow velocity, $u_{b,0}/u_{a,0}$, on the evolution of the separating shear layer can be readily understood by revisiting the material in Section 3.4.4. In particular, the velocity ratio has important effects on convective/absolute instability boundaries.

Figure 4.51 illustrates a sample parallel stability calculation showing the absolute instability boundary dependence on velocity ratio, χ, whose definition is repeated below, and shear layer thickness, δ_{II};

$$\chi = \frac{u_{a,0} - u_{b,0}}{u_{a,0} + u_{b,0}}. \tag{4.11}$$

The plot shows that absolute instability is promoted as the reverse flow velocity $u_{b,0}$ increases and/or shear layer thickness decreases. For example, for a shear layer with thickness $\delta_{II}/H = 0.2$, the flow is absolutely unstable for $\chi > 1.28$, or equivalently for $-u_{b,0}/u_{a,0} > 0.12$.

Combustion has a stabilizing influence on absolute instability, thus shifting these lines down as the ratio of primary flow density to wake flow density increases. For example, the $\delta_{II}/H = 0.3$ line becomes neutrally stable at density ratios of $\sigma_\rho = 2.1$ and 3.0 for $\chi = 1.9$ and 2.3 (corresponding to $-u_{b,0}/u_{a,0} = 0.3$ and 0.4), respectively.

4.5.2 Unsteady Flow Structure

We next consider the flow dynamics. The separating shear layer has many similarities to free shear flows, such as vortex merging processes leading to shear layer growth. In

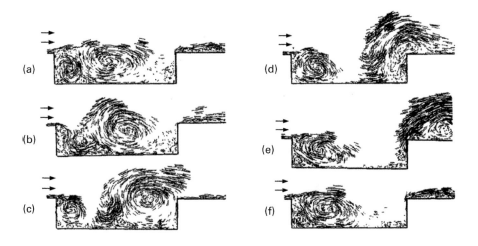

Figure 4.52 Vortex element visualization of a nonreacting $L/H = 4$ cavity flow, showing the downstream convection of the recirculating flow region. Reproduced from Najm and Ghoniem [173].

addition, streamwise structures in the flow also appear farther downstream of the separation point. After reattachment, the spanwise rollers continue to persist, but further pairing is halted [161, 170]. Harmonic excitation of cavity flows has an influence on both time-averaged flow features, such as reattachment length, L, and vortical dynamics. The effect of forcing appears analogous to that of free shear layers: by influencing the vortex merging processes discussed in Section 4.1.2.

As can be anticipated from the local stability analysis in the prior section, backward-facing steps are globally unstable in certain cases, exhibiting discrete frequency oscillations typically around the global mode frequency of $St = f\overline{u_x}/H \sim 0.1$ [158, 168, 169, 171, 172]. As discussed in Section 3.5, mean flow stability analysis requires accounting for nonlinear interactions between harmonics to predict the global mode frequencies of this flow; i.e., one cannot simply use the time-averaged flow field to predict its global mode frequencies. When globally unstable, the recirculating fluid behind the backward-facing step or cavity can periodically detach and convect downstream during the limit cycle oscillations, as indicated in Figure 4.52.

It has been found that the recirculating flow can be "locked" into cavities with certain aspect ratios, L/H. This causes the flow to be substantially less unsteady, to dramatically reduce entrainment into the cavity and to reduce cavity drag [174–176]. This observation has motivated extensive development of "trapped-vortex combustors," see Figure 4.53, which are designed with specific cavity aspect ratios that "lock" the recirculating vortex into the cavity. The recirculating flow plays a key role in stabilizing the flame, but also reduces the level of fluctuations associated with a periodically detaching recirculation zone [177]. Extensive work has been performed to develop the design rules on cavity aspect ratio to "lock" the vortex [176, 177]. The

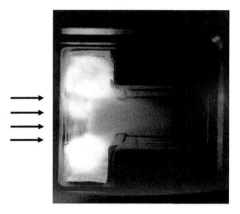

Figure 4.53 Photograph of a trapped-vortex combustor. Image courtesy of J. Zelina and D. Shouse.

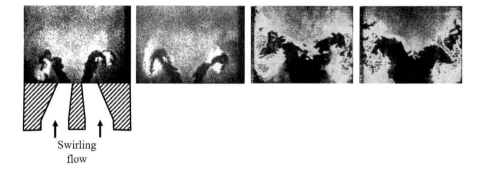

Figure 4.54 OH PLIF image of an acoustically forced swirl flame, showing flame rollup outward by large-scale vortical structures. The first two images were obtained at $\bar{u}_x = 21$ m/s and the latter two images at $\bar{u}_x = 44$ m/s. Image courtesy of S. Kumar Thumuluru and B. Bellows [153].

trapped-vortex combustor is an important case study for the combustion community as its design has been largely driven by considering the unsteady nature of the flow field, rather than just its time-averaged features. Indeed, a key design objective is to make the time-averaged and instantaneous flows of this combustor the same!

In general, however, the backward-facing step flow is highly unsteady and dominated by downstream convecting structures, particularly in the presence of acoustic excitation [178, 179]. This is clearly evident in the series of experimental images from an axisymmetric swirling combustor shown in Figure 4.54, showing the rollup of the flame by an outer wake structure. The difference in the flame stabilization point in the first two and latter two images is due to a doubling in flow velocity, causing the flame to lift off the centerbody. It is similarly evident in Figure 4.55, which illustrates images from a backward-facing step of unforced and forced flames. Note the decrease in wavelength of flame wrinkling with increased forcing frequency.

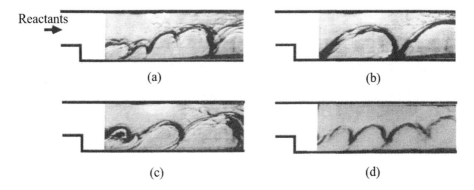

Figure 4.55 Schlieren image of (a) unforced and (b–d) forced flames. Forcing frequency is (b) 98 Hz, (c) 160 Hz, and (d) 280 Hz. Image reproduced from McManus et al. [180].

4.6 Aside: Boundary Layers [3, 181]

The structure, stability, and thickness of the boundary layer play a significant role in important combustor phenomena: the overall combustor stagnation pressure drop, wall heat transfer rates, boundary layer flashback tendencies, and the "initial conditions" for separating free shear layers at points of boundary layer separation. The discussion below summarizes several key boundary layer features.

The simple case of a laminar boundary layer on a flat plate illustrates fundamental concepts associated with boundary layer behavior. With the exception of development regions at the plate leading and trailing edges, the boundary layer is controlled by a simple balance of axial convection and transverse diffusion – momentum convects along the plate over a distance x, and diffuses normal to the plate over the boundary layer thickness, δ, occurring with time scales of x/u_0 and δ^2/ν_ν. Equating these time scales leads to the boundary layer thickness scaling, $\delta \sim (\nu_\nu x/u_0)^{1/2}$. Using the Blasius solution [3], this scaling can be converted into a quantitative formula. For example, a common definition of δ is the displacement thickness, δ_{disp}, equal to the distance that the outer flow is displaced from the plate due to the boundary layer, where $\delta_{disp} = 1.72(\nu_\nu x/u_0)^{1/2}$. Similarly, the location where the local velocity reaches $u_x = 0.99 u_0$ is $\delta_{99} = 5(\nu_\nu x/u_0)^{1/2}$.

As the flow convects along the plate, the boundary layer thickens and viscous flow instabilities develop, grow, and nonlinearly interact, eventually leading to transition to a turbulent boundary layer. As alluded to in Section 3.3, invisicidly stable flow profiles, such as the Blasius boundary layer, generally have instability onset points that are strongly sensitive to background noise and surface roughness. In extremely low noise environments, turbulence transition Reynolds numbers can be as high as $Re = u_0 x/\nu_\nu = 3 \times 10^6$, but more typical values are 5×10^5. For example, in 50 m/s air at 300 K and 1600 K, the latter value implies transition after a distance of 15 cm and 2.6 m, respectively, as a result of the change in kinematic viscosity with temperature.

Other factors that affect boundary layer stability and turbulence transition include the flow pressure gradient, wall temperature, wall blowing, and wall curvature. One important effect in combusting flows is heat addition from the walls to the flow. For example, the boundary layer in nozzle approach flows can be heated by back-conduction from hot metal surfaces, which absorb heat from recirculating hot products. Heat addition to boundary layers leads to an inflection point in the boundary layer profile and, hence, is destabilizing [3]. To illustrate, Figure 4.57 compares measured velocity profiles near the separation point of a bluff body in a reacting and nonreacting flow. In the reacting flow, the bluff body is heated by recirculating hot products, which leads to the inflection point in the velocity (see Exercise 4.2).

In addition, an adverse pressure gradient also leads to an inflection point in the mean velocity profile and, thus, is inviscidly unstable (see Exercise 4.1). In contrast, turbulent transition is delayed for flows with favorable pressure gradients. Wall blowing normal to the wall is strongly destabilizing as it thickens the boundary layer. Wall curvature in the direction of flow also has important effects. From Rayleigh's criterion for inviscid rotating flow, Eq. (3.46), it can be seen that flows that curve in on themselves (due to concave walls) are unstable, as shown in Figure 4.58. Defining a local polar coordinate system for this problem, we note that the angular velocity in this configuration decreases with increasing radius. This flow instability, a result of centrifugal mechanisms, leads to Görtler vortices [3]. In contrast, convex wall curvature flows are stabilized by centrifugal mechanisms.

Turbulent boundary layers are nominally thicker than laminar ones, as shown in Figure 4.56, and have a multizone structure. This structure features an outer region that is unaffected by viscosity, and a viscous sublayer, δ_v, which is a thin region near the wall where viscous processes dominate. Assuming $Re_x = 3.3 \times 10^6$ at a distance of $x = 1$ m along the plate and $v_v = 15$ mm^2/s (corresponding to air at 300 K), the turbulent boundary layer thickness, δ_{99}, is 8 mm and the viscous sublayer thickness, δ_v, is 0.4 mm [3]. For the same Re_x, the corresponding laminar boundary layer thickness, $\delta_{99, laminar}$, is 2.8 mm. Schlichting and Gersten [3] provide the following expression for the turbulent boundary layer thickness over a flat plate:

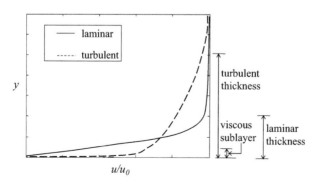

Figure 4.56 Typical laminar and turbulent velocity profiles in a flat plate boundary layer. Adapted from Schlichting and Gersten [3].

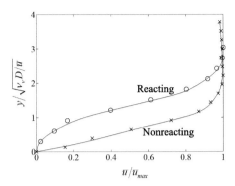

Figure 4.57 Comparison of measured velocity profiles normal to a bluff body in a nonreacting ($\bar{u} = 30$ m/s) and reacting flow ($\bar{u} = 50$ m/s). The inflection point in the reacting flow problem is due to back heating of the bluff body from recirculating products. Viscosity is calculated at the average temperature of reactants and products in the reacting case. Data courtesy of B. Emerson.

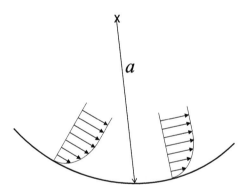

Figure 4.58 Illustration of boundary layer flow over a curved wall; the flow is destabilized through the centrifugal mechanism.

$$\frac{\delta_{99}}{x} = \frac{0.14 F(Re_x)}{\ln Re_x}, \quad (4.12)$$

where the function $F(Re_x)$ is a weak function of x with values of ~1.5 in the $10^5 < Re_x < 10^6$ range. Note that the turbulent δ_{99} grows downstream as $x/\ln x$.

4.7 Aside: Confinement Effects and Arrays of Jets/Wakes

While this chapter has focused on the stability of canonical flows in isolation, many applications exist where these canonical flow fields are aligned in multielement configurations: linearly, in two-dimensional arrays, or distributed azimuthally [182]. A multielement flow field can differ from that of a single-element flow in terms of steady and unsteady flow properties. From the point of view of stability, Section 3.6 showed examples where two point vortex arrays (a rough approximation of a jet or wake) have completely different stability characteristics than a single line of vortices. Given that a rough model of two wakes would be four sheets of vorticity with different spacing, that example shows why the stability of such a flow could differ significantly from a single wake. A related problem is that of confinement. Using the method of images, a confined, two-dimensional wake/jet is equivalent to an infinite, linear array of unconfined wakes/jets with corresponding spatial separation parameter (neglecting no-slip at the walls).

There is a large literature on the flow properties of a linear array of wakes. For example, the average flow between the cylinders vectors to one side and the average cylinder drag changes [183]. The drag force of the transversely arranged system is less than twice the drag of a single cylinder, termed "interference drag." The close interaction of the two cylinder flows introduces lift, even though the geometric arrangement is symmetric [183, 184]. The time-averaged bias in average flow intermittently flips, causing changes in the structure and dynamics of the flow field [185, 186].

Wake separation, L, strongly influences both the global mode frequency and mode shape [187]. As expected, the frequency and mode shapes are nearly the same as individual flow elements for large L/D, typically $L/D > 3.5$ [185, 188]. For smaller separations, the vortex shedding frequency of single- and two-cylinder systems are different; for example, the Strouhal number of a single cylinder is $St = 0.21$ for circular cylinders [42], but varies for a two-cylinder system from 0.41 to 0.21 for L/D ranges from 1 to 3.5 [185]. In addition, the global mode shape also varies with L/D, manifested in the symmetry of the vortex shedding from each cylinder, as well as the relative phasing between the different cylinders [188]. A variety of different symmetric and antisymmetric behaviors are possible, depending on the number of elements and the spacing, L/D [185, 189–193].

Similarly, in jet flows, the average jet velocity field is influenced [194]. For two jets, the jets may be vectored toward each other, due to lower pressure between the jets [195]. Isolated jets that are globally stable can also exhibit global instability when deployed in arrays for certain jet spacing values [196–198].

4.8 Aside: Noncircular Jets

A key feature of circular jets is that, for axisymmetric disturbance modes at least, the mutual induction of each part of the vortex sheet is azimuthally symmetric. This is not true in noncircular jets, such as rectangular, elliptical, or other jet shapes [199, 200]. Furthermore, even in circular jets with canted or lobed exits, the separating vortex sheet necessarily has both azimuthal and axial components.

Either effect, noncircularity or canted exits, leads to substantial differences in the jet dynamics and spreading rates because of vortex self-induction. For example, consider an elliptical jet. As illustrated in Figure 4.59, vorticity distributed along the

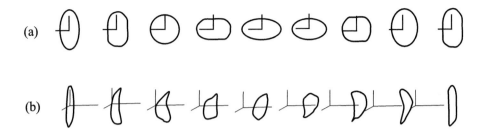

Figure 4.59 Deformation of an elliptic vortex ring viewed from (a) end on or (b) side, showing the axis switching phenomenon. Reproduced from Hussain and Husain [201].

minor axis is much closer in distance to the opposing minor axis than vorticity distributed along the major axis. Since the induced velocity scales inversely with distance from the vortex core, this causes the minor axis sections to induce a higher velocity on each other than the major axis and, therefore, they translate axially faster than the major axis. This, in turn, introduces axial vorticity and leads to switching of the major and minor axes.

Exercises

4.1. Prove that an adverse pressure gradient in a steady, constant-temperature, incompressible flow over a flat plate implies that the mean velocity profile in the boundary layer must have a point of inflection. (Hint: Write the axial momentum equation at the wall.)

4.2 Prove that boundary layer heating in a steady, zero pressure gradient, incompressible flow over a flat plate implies that the mean velocity profile in the boundary layer must have a point of inflection, such as shown in Figure 4.57. (Hint: Write the axial momentum equation at the wall.)

4.3. Derive the wake/jet dispersion relation in Eq. (4.3).

4.4. Plot the temporal varicose and sinuous instability growth rate as a function of kD at several χ values for Eq. (4.3).

4.5. Using the dispersion relation in Eq. (4.3), plot the relationship between $c_{ph,i}$ and $c_{ph,r}$ using $k\delta$ as a parameter. Overlay the inequality from Eq. (3.21) and compare the actual temporal stability characteristics to these bounds.

4.6. Work out an approximate formula for the frequency of a self-excited cavity tone in a low Mach number flow, as discussed in the context of Figure 4.49. Assume that the convection speed of the vortices is given by u_c.

References

1. Drazin P. and Reid W., *Hydrodynamic Stability*. 1981, Cambridge University Press.
2. Juniper M.P., The effect of confinement on the stability of two-dimensional shear flows. *Journal of Fluid Mechanics*, 2006, **565**: pp. 171–195.

3. Schlichting H. and Gersten K., *Boundary Layer Theory*. 8th ed. 2000, New York: Springer Press.
4. Dimotakis P., *Turbulent Free Shear Layer Mixing and Combustion*. 1991, Graduate Aeronautical Laboratories, Report # FM91–2. Pasadena, CA.
5. Soteriou M. and Ghoniem A., Effects of the free stream density ratio on free and forced spatially developing shear layers. *Physics of Fluids*, 1995, **7**: pp. 2036–2052.
6. Ho C. and Huerre P., Perturbed free shear layers. *Annual Review of Fluid Mechanics*, 1984, **16**(1): pp. 365–422.
7. Dimotakis P. and Brown G., The mixing layer at high Reynolds number: Large-structure dynamics and entrainment. *Journal of Fluid Mechanics*, 1976, **78**(03): pp. 535–560.
8. Gaster M., Kit E., and Wygnanski I., Large-scale structures in a forced turbulent mixing layer. *Journal of Fluid Mechanics*, 1985, **150**: pp. 23–39.
9. Roshko A., The plane mixing layer flow visualization results and three dimensional effects, in *The Role of Coherent Structures in Modelling Turbulence and Mixing*, J. Jimenez, ed. 1981, Berlin: Springer, pp. 208–217.
10. Lasheras J., Cho J., and Maxworthy T., On the origin and evolution of streamwise vortical structures in a plane, free shear layer. *Journal of Fluid Mechanics*, 1986, **172**: pp. 231–258.
11. Bernal L. and Roshko A., Streamwise vortex structure in plane mixing layers. *Journal of Fluid Mechanics*, 1986, **170**: pp. 499–525.
12. Ford C.L. and Winroth, P.M., On the scaling and topology of confined bluff-body flows. *Journal of Fluid Mechanics*, 2019, **876**: pp. 1018–1040.
13. Fugger C.A., Gallagher, T.P., Sykes, J.P., and Caswell, A.W., Spanwise recirculation zone structure of a bluff body stabilized flame. *Combustion and Flame*, 2020, **216**: pp. 58–61.
14. Roshko A., Structure of turbulent shear flows: A new look. *AIAA Journal*, 1976, **14**: pp. 1349–1357.
15. Hegde U. and Zinn B., Vortical mode instability of shear layers with temperature and density gradients. *AIAA Journal*, 1990, **28**: pp. 1389–1396.
16. McMurtry P., Riley J., and Metcalfe R., Effects of heat release on the large-scale structure in turbulent mixing layers. *Journal of Fluid Mechanics*, 1989, **199**: pp. 297–332.
17. Djordjevic V., Pavithran S., and Redekopp L. Stability properties of subsonic mixing layers, in *AIAA 2nd Shear Flow Conference*. 1989, Tempe, AZ: AIAA.
18. Ghoniem A. and Krishnan A. Origin and manifestation of flow-combustion interactions in a premixed shear layer, in *Proceedings of the Combustion Institute*. 1989, The Combustion Institute.
19. Weisbrot I. and Wygnanski I., On coherent structures in a highly excited mixing layer. *Journal of Fluid Mechanics*, 1988, **195**: pp. 137–159.
20. Wang S. and Yang V., Unsteady flow evolution in swirl injectors with radial entry. II. External excitations. *Physics of Fluids*, 2005, **17**: p. 045107.
21. Ho C. and Huang L., Subharmonics and vortex merging in mixing layers. *Journal of Fluid Mechanics*, 1982, **119**: pp. 443–473.
22. Patnaik P., Sherman F., and Corcos G., A numerical simulation of Kelvin–Helmholtz waves of finite amplitude. *Journal of Fluid Mechanics*, 1976, **73**(02): pp. 215–240.
23. Kelly R., On the stability of an inviscid shear layer which is periodic in space and time. *Journal of Fluid Mechanics*, 1967, **27**(04): pp. 657–689.
24. Riley J. and Metcalfe R. Direct numerical simulation of a perturbed, turbulent mixing layer, in *AIAA 18th Aerospace Sciences Meeting*. 1980, Pasadena, CA: AIAA.

25. Liu J. and Nikitopoulos D., Mode interactions in developing shear flows. *The Bulletin of the American Physical Society*, 1982, **27**: p. 1192.
26. Mastorakos E., Taylor A.M.K.P., and Whitelaw J.H., Extinction of turbulent counterflow flames with reactants diluted by hot products. *Combustion and Flame*, 1995, **102**: pp. 101–114.
27. Ho C. and Nosseir N., Dynamics of an impinging jet. Part 1. The feedback phenomenon. *Journal of Fluid Mechanics*, 1981, **105**: pp. 119–142.
28. Shanbhogue S., Husain S., and Lieuwen T., Lean blowoff of bluff body stabilized flames: Scaling and dynamics. *Progress in Energy and Combustion Science*, 2009, **35**(1): pp. 98–120.
29. Yu M. and Monkewitz P., The effect of nonuniform density on the absolute instability of two dimensional inertial jets and wakes. *Physics of Fluids*, 1990, **2**: p. 1175.
30. Pope S., *Turbulent Flows*. 2000, Cambridge University Press.
31. Prasad A. and Williamson C.H.K., The instability of the shear layer separating from a bluff body. *Journal of Fluid Mechanics*, 1997, **333**: pp. 375–402.
32. Cardell G.S., *Flow Past a Circular Cylinder with a Permeable Splitter Plate*. 1993, Pasadena CA: California Insititute of Technology.
33. Williamson C.H.K., Vortex dynamics in the cylinder wake. *Annual Reviews of Fluid Mechanics*, 1996, **28**: pp. 477–539.
34. Zdravkovich M.M., *Flow around Circular Cylinders: A Comprehensive Guide through Flow Phenomena, Experiments, Applications, Mathematical Models, and Computer Simulations*. 1997, Oxford University Press.
35. Emerson B., O'Connor, J., Juniper, M., and Lieuwen, T., Density ratio effects on reacting bluff-body flow field characteristics. *Journal of Fluid Mechanics*, 2012, **706**: pp. 219–250.
36. Wygnanski I. and Petersen R., Coherent motion in excited free shear flows. *AIAA Journal*, 1987, **25**: pp. 201–213.
37. Pier B., Local and global instabilities in the wake of a sphere. *Journal of Fluid Mechanics*, 2008, **603**: pp. 39–61.
38. Monkewitz P., A note on vortex shedding from axisymmetric bluff bodies. *Journal of Fluid Mechanics*, 1988, **192**: pp. 561–575.
39. Provansal M., Mathis C., and Boyer L., Bénard–von Kármán instability: Transient and forced regimes. *Journal of Fluid Mechanics*, 1987, **182**: pp. 1–22.
40. Plaschko P., Berger E., and Peralta-Fabi R., Periodic flow in the near wake of straight circular cylinders. *Physics of Fluids*, 1993, **5**: pp. 1718–1725.
41. Perry A.E., Chong M.S., and Lim T.T., The vortex shedding process behind two-dimensional bluff bodies. *Journal of Fluid Mechanics*, 1982, **116**: pp. 77–90.
42. Cantwell B. and Coles D., An experimental study of entrainment and transport in the turbulent near wake of a circular cylinder. *Journal of Fluid Mechanics*, 1983, **136**: pp. 321–374.
43. Bearman P., On vortex shedding from a circular cylinder in the critical Reynolds number regime. *Journal of Fluid Mechanics*, 1969, **37**: pp. 577–585.
44. Roshko A., Experiments on the flow past a cylinder at very high Reynolds numbers. *Journal of Fluid Mechanics*, 1961, **10**: pp. 345–356.
45. Achenbach E., Vortex shedding from spheres. *Journal of Fluid Mechanics*, 1974, **62**(2): pp. 209–221.
46. Blevins R.D., *Flow-Induced Vibration*. 2nd ed. 2001, Malabar, FL: Krieger Publishing Company.

47. Roshko A., On the wake and drag of bluff bodies. *Journal of the Aeronautical Sciences*, 1955, **22**: pp. 124–132.
48. Huang R.F. and Chang K.T., Oscillation frequency in wake of a vee gutter. *Journal of Propulsion and Power*, 2004, **20**(5): pp. 871–878.
49. Roshko A., *On the Drag and Shedding Frequency of Two-Dimensional Bluff Bodies.* 1954, National Advisory Committee on Aeronautics.
50. Erickson R.R., Soteriou M.C., and Mehta P.G., The influence of temperature ratio on the dynamics of bluff body stabilized flames, in *44th AIAA Aerospace Sciences Meeting and Exhibit*. 2006, Reno, NV: AIAA.
51. Hertzberg J.R., Shepherd I.G., and Talbot L., Vortex shedding behind rod stabilized flames. *Combustion and Flame*, 1991, **86**: pp. 1–11.
52. Fujii S. and Eguchi K., A comparison of cold and reacting flows around a bluff-body flame stabilizer. *Journal of Fluids Engineering*, 1981, **103**: pp. 328–334.
53. Yamaguchi S., Ohiwa N., and Hasegawa T., Structure and blow-off mechanism of rod-stabilized premixed flame. *Combustion and Flame*, 1985, **62**: pp. 31–41.
54. Anderson K.R., Hertzberg J., and Mahalingam S., Classification of absolute and convective instabilities in premixed bluff body stabilized flames. *Combustion Science and Technology*, 1996, **112**(1): pp. 257–269.
55. Emerson B., Lundrigan J., O'Connor J., Noble D., and Lieuwen T., Dependence of the bluff body wake structure on flame temperature ratio, in *49th AIAA Aerospace Sciences Meeting including the New Horizons Forum and Aerospace Exposition*. 2011, Orlando, FL: AIAA.
56. Erickson R. and Soteriou M., The influence of reactant temperature on the dynamics of bluff body stabilized premixed flames. *Combustion and Flame*, 2011, **158**(12): pp. 2441–2457.
57. Nair S., and Lieuwen, T., Near-blowoff dynamics of a bluff-body stabilized flames. *Journal of Propulsion and Power*, 2007, **23**(2): pp. 421–427.
58. Bush S.M. and Gutmark E.J., Reacting and non-reacting flow fields of a V-gutter stabilized flame. *AIAA Journal*, 2007, **45**(3): 662–672.
59. Provansal M., Mathis C., and Boyer L., Bénard–von Kármán instability: Transient and forced regimes. *Journal of Fluid Mechanics*, 1987, **182**: pp. 1–22.
60. Juniper M.P. and Candel S.M., The stability of ducted compound flows and consequences for the geometry of coaxial injectors. *Journal of Fluid Mechanics*, 2003, **482**: pp. 257–269.
61. Pal S., Moser M., Ryan H., Foust M., and Santoro R., Shear coaxial injector atomization phenomena for combusting and non-combusting conditions. *Atomization and Sprays*, 1996, **6**(2): pp. 227–244.
62. Schefer R., Wicksall D., and Agrawal A., Combustion of hydrogen-enriched methane in a lean premixed swirl-stabilized burner. *Proceedings of the Combustion Institute*, 2002, **29**(1): pp. 843–851.
63. Vandervort C., 9 ppm NOx/CO combustion system for F class industrial gas turbines. *Journal of Engineering for Gas Turbines and Power*, 2001, **123**(2): pp. 317–321.
64. Michalke A., Survey on jet instability theory. *Progress in Aerospace Sciences*, 1984, **21**: pp. 159–199.
65. Yu M. and Monkewitz P., The effect of nonuniform density on the absolute instability of two dimensional inertial jets and wakes. *Physics of Fluids A: Fluid Dynamics*, 1990, **2**: p. 1175.

66. Huerre P. and Monkewitz P., Local and global instabilities in spatially developing flows. *Annual Review of Fluid Mechanics*, 1990, **22**(1): pp. 473–537.
67. Monkewitz P. and Sohn K., Absolute instability in hot jets. *AIAA Journal*, 1988, **26**(8): pp. 911–916.
68. Furi M., Papas P., Rais R., and Monkewitz P.A., The effect of flame position on the Kelvin–Helmholtz instability in non-premixed jet flames. *Proceedings of the Combustion Institute*, 2002, **29**: pp. 1653–1661.
69. Morris P., The spatial viscous instability of axisymmetric jets. *Journal of Fluid Mechanics*, 1976, **77**(03): pp. 511–529.
70. Batchelor G. and Gill A., Analysis of the stability of axisymmetric jets. *Journal of Fluid Mechanics*, 1962, **14**(04): pp. 529–551.
71. Michalke A. and Hermann G., On the inviscid instability of a circular jet with external flow. *Journal of Fluid Mechanics*, 1982, **114**: pp. 343–359.
72. Crow S. and Champagne F., Orderly structure in jet turbulence. *Journal of Fluid Mechanics*, 1971, **48**(03): pp. 547–591.
73. Petersen R., Influence of wave dispersion on vortex pairing in a jet. *Journal of Fluid Mechanics*, 1978, **89**(03): pp. 469–495.
74. Gutmark E. and Ho C., Preferred modes and the spreading rates of jets. *Physics of Fluids*, 1983, **26**: p. 2932.
75. Liepmann D. and Gharib M., The role of streamwise vorticity in the near-field entrainment of round jets. *Journal of Fluid Mechanics*, 1992, **245**: pp. 643–668.
76. Yule A., Large-scale structure in the mixing layer of a round jet. *Journal of Fluid Mechanics*, 1978, **89**(03): pp. 413–432.
77. Zaman K. and Hussain A., Vortex pairing in a circular jet under controlled excitation. Part 1. General jet response. *Journal of Fluid Mechanics*, 1980, **101**(03): pp. 449–491.
78. Paschereit C., Wygnanski I., and Fiedler H., Experimental investigation of subharmonic resonance in an axisymmetric jet. *Journal of Fluid Mechanics*, 1995, **283**: pp. 365–407.
79. Reynolds W., Parekh D., Juvet P., and Lee M., Bifurcating and blooming jets. *Annual Review of Fluid Mechanics*, 2003, **35**(1): pp. 295–315.
80. Karagozian A.R., Transverse jets and their control. *Progress in Energy and Combustion Science*, 2010, **36**: pp. 531–553.
81. Lefebvre A.H. and Ballal D.R., *Gas Turbine Combustion: Alternative Fuels and Emissions*. 2010, Taylor and Francis.
82. Sturgess G., McKinney R., and Morford S., Modification of combustor stoichiometry distribution for reduced NOx emission from aircraft engines. *Journal of Engineering for Gas Turbines and Power-Transactions of the ASME*, 1993, **115**(3): pp. 570–580.
83. McKinney R., Sepulveda D., Sowa W., and Cheung A. The Pratt & Whitney TALON X low emissions combustor: revolutionary results with evolutionary technology, in *45th AIAA Aerospace Sciences Meeting and Exhibit*. 2007.
84. Andreopoulos J. and Rodi W., Experimental investigation of jets in a crossflow. *Journal of Fluid Mechanics*, 1984, **138**: pp. 93–127.
85. Hasselbrink Jr E.F. and Mungal M.G., Transverse jets and jet flames. Part 2. Velocity and OH field imaging. *Journal of Fluid Mechanics*, 2001, **443**: pp. 27–68.
86. Marzouk Y. and Ghoniem A., Vorticity structure and evolution in a transverse jet. *Journal of Fluid Mechanics*, 2007, **575**: pp. 267–305.
87. Kelso R., Lim T., and Perry A., An experimental study of round jets in cross-flow. *Journal of Fluid Mechanics*, 1996, **306**: pp. 111–144.

88. Sykes R., Lewellen W., and Parker S., On the vorticity dynamics of a turbulent jet in a crossflow. *Journal of Fluid Mechanics*, 1986, **168**: pp. 393–413.
89. Fric T.F. and Roshko A., Vortical structure in the wake of a transverse jet. *Journal of Fluid Mechanics*, 1994, **279**: pp. 1–47.
90. Narayanan S., Barooah P., and Cohen J., Dynamics and control of an isolated jet in crossflow. *AIAA Journal*, 2003, **41**(12): pp. 2316–2330.
91. Alves L., Kelly R., and Karagozian A., Local stability analysis of an inviscid transverse jet. *Journal of Fluid Mechanics*, 2007, **581**: pp. 401–418.
92. Alves L.S., Kelly R.E., and Karagozian, A.R., Transverse-jet shear-layer instabilities. Part 2. Linear analysis for large jet-to-crossflow velocity ratio. *Journal of Fluid Mechanics*, 2008, **602**: pp. 383–401.
93. Megerian S., Davitian J., Alves, L.S., and Karagozian, A.R., Transverse-jet shear-layer instabilities. Part 1. Experimental studies. *Journal of Fluid Mechanics*, 2007, **593**: pp. 93–129.
94. Iyer P.S. and Mahesh K., A numerical study of shear layer characteristics of low-speed transverse jets. *Journal of Fluid Mechanics*, 2016, **790**: pp. 275–307.
95. Getsinger D.R., Hendrickson C., and Karagozian A.R., Shear layer instabilities in low-density transverse jets. *Experiments in Fluids*, 2012, **53**(3): pp. 783–801.
96. Kolb M., Ahrens D., Hirsch C., and Sattelmayer T., A model for predicting the lift-off height of premixed jets in vitiated cross flow. *Journal of Engineering for Gas Turbines and Power*, 2016, **138**(8).
97. Wagner J.A., Grib S.W., Renfro M.W., and Cetegen B.M., Flowfield measurements and flame stabilization of a premixed reacting jet in vitiated crossflow. *Combustion and Flame*, 2015, **162**(10): pp. 3711–3727.
98. Sullivan R., Wilde B., Noble D.R., Seitzman J.M., and Lieuwen T.C., Time-averaged characteristics of a reacting fuel jet in vitiated cross-flow. *Combustion and Flame*, 2014, **161**(7): pp. 1792–1803.
99. Sirignano M.D., Nair V., Emerson B., Seitzman J., and Lieuwen T.C., Nitrogen oxide emissions from rich premixed reacting jets in a vitiated crossflow. *Proceedings of the Combustion Institute*, 2019, **37**(4): pp. 5393–5400.
100. Nair V., Wilde B., Emerson B., and Lieuwen T., Shear layer dynamics in a reacting jet in crossflow. *Proceedings of the Combustion Institute*, 2019, **37**(4): pp. 5173–5180.
101. Hall M.G., Vortex breakdown. *Annual Review of Fluid Mechanics*, 1972, **4**: pp. 195–217.
102. Leibovich S., The structure of vortex breakdown. *Annual Review of Fluid Mechanics*, 1978, **10**: pp. 221–246.
103. Leibovich S., Vortex stability and breakdown: Survey and extension. *AIAA Journal*, 1984, **22**: pp. 1192–1206.
104. Escudier M.P., Vortex breakdown: Observations and explanations. *Progress in Aerospace Sciences*, 1988, **25**: pp. 189–229.
105. Sarpkaya T., Turbulent vortex breakdown. *Physics of Fluids*, 1995, **7**: pp. 2301–2303.
106. Althaus W., Bruecker C., and Weimer M., Breakdown of slender vortices, in *Fluid Vortices* S.I. Green ed. 1995, Kluwer Academy, pp. 373–426.
107. Rusak Z., Axisymmetric swirling flow around a vortex breakdown point. *Journal of Fluid Mechanics*, 1996, **323**(1): pp. 79–105.
108. Tsuji H., Gupta A., Hasegawa T., Katsuki M., Kishimoto K., and Morita M., *High Temperature Air Combustion: From Energy Conservation to Pollution Reduction*. 2003, New York: CRC Press.

109. Gupta A.K., Lilley D.G., and Syred N., *Swirl Flows*. 1984, Abacus Press.
110. Beer J. and Chigier N., *Combustion Aerodynamics*. 1972, New York: John Wiley and Sons.
111. Healey J., Inviscid axisymmetric absolute instability of swirling jets. *Journal of Fluid Mechanics*, 2008, **613**: pp. 1–33.
112. Batchelor G., *An Introduction to Fluid Dynamics*. 2000, Cambridge University Press.
113. Escudier M., Nickson A., and Poole R., Influence of outlet geometry on strongly swirling turbulent flow through a circular tube. *Physics of Fluids*, 2006, **18**: pp. 1–18.
114. Rusak Z., Kapila A.K., and Choi J.J., Effect of combustion on near-critical swirling flows. *Combustion Theory and Modeling*, 2002, **6**: pp. 625–645.
115. Rusak Z. and Lee J.C., The effect of compressibility on the critical swirl of vortex flows in a pipe. *Journal of Fluid Mechanics*, 2002, **461**(1): pp. 301–319.
116. Rusak Z. and Wang S. Review of theoretical approaches to the vortex breakdown phenomenon, in *1st Theoretical Fluid Mechanics Conference, AIAA*. 1996. New Orleans, LA, paper 96–2126.
117. Wang S. and Rusak Z., On the stability of an axisymmetric rotating flow in a pipe. *Physics of Fluids*, 1996, **8**: p. 1007.
118. Faler J.H. and Leibovich S., Disrupted states of vortex flow and vortex breakdown. *Physics of Fluids*, 1977, **20**: pp. 1385–1400.
119. Billant P., Chomaz J., and Huerre P., Experimental study of vortex breakdown in swirling jets. *Journal of Fluid Mechanics*, 1998, **376**: pp. 183–219.
120. Huang Y. and Yang V., Dynamics and stability of lean-premixed swirl-stabilized combustion. *Progress in Energy and Combustion Science*, 2009, **35**(4): pp. 293–364.
121. Tangirala V., Chen R., and Driscoll J., Effect of heat release and swirl on the recirculation within swirl-stabilized flames. *Combustion Science and Technology*, 1987, **51**(1–3): pp. 75–95.
122. Rusak Z., Whiting C., and Wang S., Axisymmetric breakdown of a Q-vortex in a pipe. *AIAA Journal*, 1998, **36**(10).
123. Smith T.E., Douglas C.M., Emerson, B.L., and Lieuwen, T.C., Axial evolution of forced helical flame and flow disturbances. *Journal of Fluid Mechanics*, 2018, **844**: pp. 323–356.
124. Rusak Z. and Wang S., Review of theoretical approaches to the vortex breakdown phenomenon, in *1st Theoretical Fluid Mechanics Conference, AIAA*. 1996, New Orleans, LA, paper 96-2126.
125. Squire H.B., *Analysis of the "Vortex Breakdown" Phenomenon. Part 1*. Imperial College, Aeronautics Department, 1960, p. 102.
126. Brown G.L. and Lopez J.M., Axisymmetric vortex breakdown. Part 2. Physical mechanisms. *Journal of Fluid Mechanics*, 1990, **221**: pp. 553–576.
127. Darmofal D.L., The role of vorticity dynamics in vortex breakdown, in *24th Fluid Dynamic Conference, AIAA*. 1993, Orlando, FL. pp. 1–14.
128. Lucca-Negro O. and O'Doherty T., Vortex breakdown: a review. *Progress in Energy and Combustion Science*, 2001, **27**: pp. 431–481.
129. Escudier M. and Keller J., Recirculation in swirling flow: A manifestation of vortex breakdown. *AIAA Journal*, 1985, **23**: pp. 111–116.
130. Fu Y., Cai J., Jeng S., and Mongia H. Confinement effects on the swirling flow of a counter-rotating swirl cup in *ASME Turbo Expo: Power for Land, Sea and Air*. 2005. Reno, NV: ASME. Paper # GT2005–68622.
131. Wang S., Hsieh S., and Yang V., Unsteady flow evolution in swirl injector with radial entry. I. Stationary conditions. *Physics of Fluids*, 2005, **17**: p. 045106.

132. Sheen H.J., Chen W., and Jeng, S.Y., Recirculation zones of unconfined and confined annular swirling jets. *AIAA Journal*, 1996. **34**(3): pp. 572–579.
133. Chterev I., Foley C.W., Foti D., Kostka S., Caswell A.W., Jiang N., Lynch A., Noble D.R., Menon S., Seitzman J.M., and Lieuwen T.C., Flame and flow topologies in an annular swirling flow. *Combustion Science and Technology*, 2014, **186**(8): pp. 1041–1074.
134. Shanbhogue S.J., Sanusi Y., Taamallah, S., Habib, M.A., Mokheimer, E.M.A., and Ghoniem, A.F., Flame macrostructures, combustion instability and extinction strain scaling in swirl-stabilized premixed CH4/H2 combustion. *Combustion and Flame*, 2016, **167**: pp. 494–507.
135. Manoharan K., Frederick M., Clees S., O'Connor J., and Hemchandra S., A weakly nonlinear analysis of the precessing vortex core oscillation in a variable swirl turbulent round jet. *Journal of Fluid Mechanics*, 2019, **884**: A29.
136. Fick W., Griffiths A., and O'Doherty T., Visualisation of the precessing vortex core in an unconfined swirling flow. *Optical Diagnostics in Engineering*, 1997, **2**(1): pp. 19–31.
137. Escudier M., Vortex breakdown: observations and explanations. *Progress in Aerospace Sciences*, 1988, **25**(2): pp. 189–229.
138. Loiseleux T., Chomaz J., and Huerre P., The effect of swirl on jets and wakes: Linear instability of the Rankine vortex with axial flow. *Physics of Fluids*, 1998, **10**: p. 1120.
139. Huang Y. and Yang V., Effect of swirl on combustion dynamics in a lean-premixed swirl-stabilized combustor. *Proceedings of the Combustion Institute*, 2005, **30**(2): pp. 1775–1782.
140. Khalil S., Hourigan K., and Thompson M., Response of unconfined vortex breakdown to axial pulsing. *Physics of Fluids*, 2006, **18**: p. 038102.
141. Lu X., Wang S., Sung H., Hsieh S., and Yang V., Large-eddy simulations of turbulent swirling flows injected into a dump chamber. *Journal of Fluid Mechanics*, 2005, **527**: pp. 171–195.
142. Schildmacher K.U., Hoffmann A., Selle L., Koch R., Schulz C., Bauer H.J., Poinsot T., Krebs W., and Prade B., Unsteady flame and flow field interaction of a premixed model gas turbine burner. *Proceedings of the Combustion Institute*, 2007, **31**(2): pp. 3197–3205.
143. Schildmacher K.U., Koch R., and Bauer H.J., Experimental characterization of premixed flame instabilities of a model gas turbine burner. *Flow, Turbulence and Combustion*, 2006, **76**(2): pp. 177–197.
144. Steinberg A., Boxx I., Stoehr M., Meier W., and Carter C., Flow–flame interactions causing acoustically coupled heat release fluctuations in a thermo-acoustically unstable gas turbine model combustor. *Combustion and Flame*, 2010, **157**(12).
145. Boxx I., Stohr M., Carter C., and Meier W., Temporally resolved planar measurements of transient phenomena in a partially pre-mixed swirl flame in a gas turbine model combustor. *Combustion and Flame*, 2010, **157**(8): pp. 1510–1525.
146. Diers O., Schneider D., Voges M., Weigand P., and Hassa C. Investigation of combustion oscillations in a lean gas turbine model combustor, in *Proceedings of the ASME Turbo Expo*. 2007, ASME.
147. Giezendanner R., Weigand P., Duan X.R., Meier W., Meier U., Aigner M., and Lehmann B., Laser-based investigations of periodic combustion instabilities in a gas turbine model combustor. *Journal of Engineering for Gas Turbines and Power*, 2005, **127**: pp. 492–496.

148. Giezendanner R., Keck O., Weigand P., Meier W., Meier U., Stricker W., and Aigner M., Periodic combustion instabilities in a swirl burner studied by phase-locked planar laser-induced fluorescence. *Combustion Science and Technology*, 2003, **175**(4): pp. 721–741.
149. Fureby C., Grinstein F., Li G., and Gutmark E., An experimental and computational study of a multi-swirl gas turbine combustor. *Proceedings of the Combustion Institute*, 2007, **31**(2): pp. 3107–3114.
150. Huang Y. and Ratner A., Experimental investigation of thermoacoustic coupling for low-swirl lean premixed flames. *Journal of Propulsion and Power*, 2009, **25**(2): pp. 365–373.
151. Kang D., Culick F., and Ratner A., Combustion dynamics of a low-swirl combustor. *Combustion and Flame*, 2007, **151**(3): pp. 412–425.
152. Lee S.Y., Seo S., Broda J., Pal S., and Santoro R., An experimental estimation of mean reaction rate and flame structure during combustion instability in a lean premixed gas turbine combustor. *Proceedings of the Combustion Institute*, 2000, **28**(1): pp. 775–782.
153. Bellows B., Bobba M., Forte A., Seitzman J., and Lieuwen T., Flame transfer function saturation mechanisms in a swirl-stabilized combustor. *Proceedings of the Combustion Institute*, 2007, **31**(2): pp. 3181–3188.
154. Terhaar S., Ćosić B., Paschereit C.O., and Oberleithner, K., Suppression and excitation of the precessing vortex core by acoustic velocity fluctuations: an experimental and analytical study. *Combustion and Flame*, 2016, **172**: pp. 234–251.
155. O'Connor J., Acharya V., and Lieuwen T., Transverse combustion instabilities: acoustic, fluid mechanic, and flame processes. *Progress in Energy and Combustion Science*, 2015, **49**: pp. 1–39.
156. Khalil S., Hourigan K., and Thompson M.C., Response of unconfined vortex breakdown to axial pulsing. *Physics of Fluids*, 2006, **18**: p. 038102.
157. Thumuluru S.K. and Lieuwen T., Characterization of acoustically forced swirl flame dynamics. *Proceedings of the Combustion Institute*, 2009, **32**(2): pp. 2893–2900.
158. Hemchandra S., Shanbhogue S., Hong S., and Ghoniem A.F., Role of hydrodynamic shear layer stability in driving combustion instability in a premixed propane-air backward-facing step combustor. *Physical Review Fluids*, 2018, **3**(6).
159. O'Connor J. and Lieuwen T., Disturbance field characteristics of a transversely excited burner. *Combustion Science and Technology*, 2011, **183**(5): pp. 427–443.
160. Pitz R. and Daily J., Combustion in a turbulent mixing layer formed at a rearward-facing step. *AIAA Journal*, 1983, **21**: pp. 1565–1570.
161. Troutt T.R., Scheelke B., and Norman T.R., Organized structures in a reattaching separated flow field. *Journal of Fluid Mechanics*, 1984, **143**: pp. 413–427.
162. Eaton J. and Johnston J., A review of research on subsonic turbulent flow reattachment. *AIAA Journal*, 1981, **19**: pp. 1093–1100.
163. Armaly B., Durst F., Pereira J., and Schönung B., Experimental and theoretical investigation of backward-facing step flow. *Journal of Fluid Mechanics*, 1983, **127**: pp. 473–496.
164. Durst F., Melling A., and Whitelaw J.H., Low Reynolds number flow over a plane symmetric sudden expansion. *Journal of Fluid Mechanics*, 1974, **64**(1): pp. 111–128.
165. Moffatt H., Viscous and resistive eddies near a sharp corner. *Journal of Fluid Mechanics*, 1964, **18**(01): pp. 1–18.
166. Rockwell D. and Naudascher E., Self-sustained oscillations of impinging shear layers. *Annual Review of Fluid Mechanics*, 1979, **11**: p. 67–94.

167. Rockwell D., Oscillations of impinging shear layers. *AIAA Journal*, 1983, **21**(5): pp. 645–664.
168. Ghoniem A.F., Annaswamy A., Wee D., Yi T., and Park S., Shear flow-driven combustion instability: Evidence, simulation, and modeling. *Proceedings of the Combustion Institute*, 2002, **29**(1): pp. 53–60.
169. Wee D., Park S., Yi T., Annaswamy A.M., and Ghoniem A.F., Reduced order modeling of reacting shear flow, in *40th AIAA Aerospace Sciences Meeting and Exhibit*. 2002, Reno, NV: AIAA.
170. Bhattacharjee S., Scheelke B., and Troutt T.R., Modification of vortex interactions in a reattaching separated flow. *AIAA Journal*, 1986, **24**(4): pp. 623–629.
171. Najm H.N. and Ghoniem A.F., Coupling between vorticity and pressure oscillations in combustion instability. *Journal of Propulsion and Power*, 1994, **10**(6): pp. 769–776.
172. Driver D.M., Seegmiller H.L., and Marvin J.G., Time-dependent behavior of a reattaching shear layer. *AIAA Journal*, 1987, **25**(7): pp. 914–919.
173. Najm H. and Ghoniem A., Numerical simulation of the convective instability in a dump combustor. *AIAA Journal*, 1991, **29**: pp. 911–919.
174. Mair W.A., The effect of a rear-mounted disc on the drag of a blunt-based body of revolution. *The Aeronautical Quarterly*, 1965, **16**(4): pp. 350–359.
175. Little Jr B.H. and Whipkey R.R., Locked vortex afterbodies. *Journal of Aircraft*, 1979, **16**(5): pp. 296–302.
176. Sturgess G.J. and Hsu K.Y., Entrainment of mainstream flow in a trapped-vortex combustor, in *35th AIAA Aerospace Sciences Meeting and Exhibit*. 1997, Reno, NV: AIAA.
177. Katta V.R., Zelina J., and Roquemore W.R. Numerical studies on cavity-inside-cavity-supported flames in ultra compact combustors. in *53rd ASME International Gas Turbine and Aeroengine Congress and Exposition*. 2008, Berlin, Germany: ASME.
178. Cadou C., Karagozian A., and Smith O., Transport enhancement in acoustically excited cavity flows. Part 2. Reactive flow diagnostics. *AIAA Journal*, 1998, **36**: pp. 1568–1574.
179. Pont G., Cadou C., Karagozian A., and Smith O., Emissions reduction and pyrolysis gas destruction in an acoustically driven dump combustor. *Combustion and Flame*, 1998, **113**(1–2): pp. 249–257.
180. McManus K., Vandsburger U., and Bowman C., Combustor performance enhancement through direct shear layer excitation. *Combustion and Flame*, 1990, **82**(1): pp. 75–92.
181. White F.M., *Viscous Fluid Flow*. 3rd ed. 2006, New York: McGraw-Hill.
182. Sebastian J., Emerson B., O'Connor J., and Lieuwen T., Spatio-temporal stability analysis of linear arrays of 2D density stratified wakes and jets. *Physics of Fluids*, 2018, **30**: p. 114103.
183. Zdravkovich M., Review of flow interference between two circular cylinders in various arrangements. *Journal of Fluids Engineering*, 1977, **99**(4): pp. 618–633.
184. Hori E.-i. Experiments on flow around a pair of parallel circular cylinders, in *Proceedings of 9th Japan National Congress for Applied Mechanics*. 1959.
185. Sumner D., Wong S., Price S., and Paidoussis M., Fluid behaviour of side-by-side circular cylinders in steady cross-flow. *Journal of Fluids and Structures*, 1999, **13**(3): pp. 309–338.
186. Kim H.-J. and Durbin P., Investigation of the flow between a pair of circular cylinders in the flopping regime. *Journal of Fluid Mechanics*, 1988, **196**: pp. 431–448.
187. Spivack H.M., *Vortex Frequency and Flow Pattern in the Wake of Two Parallel Cylinders at Varied Spacing Normal to an Airstream*. 1945, Catholic University of America.

188. Bearman P. and Wadcock A., The interaction between a pair of circular cylinders normal to a stream. *Journal of Fluid Mechanics*, 1973, **61**(3): pp. 499–511.
189. Peschard I. and Le Gal P., Coupled wakes of cylinders. *Physical Review Letters*, 1996, **77**(15): p. 3122.
190. Mizushima J. and Akinaga T., Vortex shedding from a row of square bars. *Fluid Dynamics Research*, 2003, **32**(4): pp. 179–191.
191. Kang S., Characteristics of flow over two circular cylinders in a side-by-side arrangement at low Reynolds numbers. *Physics of Fluids*, 2003, **15**(9): pp. 2486–2498.
192. Meehan M., Tyagi A., and O'Connor J., Flow dynamics in a variable-spacing, three bluff-body flowfield. *Physics of Fluids*, 2018, **30**(2): p. 025105.
193. Zheng Q. and Alam M.M., Intrinsic features of flow past three square prisms in side-by-side arrangement. *Journal of Fluid Mechanics*, 2017, **826**: pp. 996–1033.
194. Smith T.E., Chterev I., Emerson B.L., Noble D.R., and Lieuwen T.C., Comparison of single- and multinozzle reacting swirl flow dynamics. *Journal of Propulsion and Power*, 2018, **34**(2): pp. 384–394.
195. Miller D.R. and Comings E.W., Force-momentum fields in a dual-jet flow. *Journal of Fluid Mechanics*, 1960, **7**(2): pp. 237–256.
196. Anderson E.A., Snyder D.O., and Christensen J., Periodic flow between low aspect ratio parallel jets. *Journal of Fluids Engineering*, 2003, **125**(2): pp. 389–392.
197. Bunderson N.E. and Smith B.L., Passive mixing control of plane parallel jets. *Experiments in Fluids*, 2005, **39**(1): pp. 66–74.
198. Subramanian H.G., Manoharan K., and Hemchandra S., Influence of nonaxisymmetric confinement on the hydrodynamic stability of multinozzle swirl flows. *Journal of Engineering for Gas Turbines and Power*, 2019, **141**(2).
199. Gutmark E. and Grinstein F., Flow control with noncircular jets. *Annual Review of Fluid Mechanics*, 1999, **31**: pp. 239–272.
200. Grinstein F. and Kailasanath K. Exothermicity and three-dimensional effects in unsteady propane square jets, *Symposium (International) on Combustion*, 1996, **26**(1): 91–96.
201. Hussain F. and Husain H., Elliptic jets. Part 1. Characteristics of unexcited and excited jets. *Journal of Fluid Mechanics*, 1989, **208**: pp. 257–320.

5 Acoustic Wave Propagation I: Basic Concepts

This chapter discusses acoustic wave propagation in combustor environments. As noted in Chapter 2, acoustic waves propagate energy and information through the medium without requiring bulk advection of the flow. For this reason, and as discussed further in this chapter, the details of the time-averaged flow have relatively minor influences on the acoustic wave field in low Mach number flows. In contrast, vortical disturbances, which propagate with the local flow field, are highly sensitive to the flow details. For these reasons, there is no analogue in the acoustic problem to the myriad different ways in which vorticity can organize and reorganize itself as in the hydrodynamic stability problem. Rather, in low Mach number flows the acoustic field is insensitive to these details and is largely controlled by the boundaries and sound speed field.

The acoustic problem has, however, its own distinctive set of rich physics. In particular, sound waves reflect off boundaries and refract around bends or other obstacles. In contrast, vortical and entropy disturbances advect out of the domain where they are excited – the only way in which they can further influence the disturbance field in the system is if they excite backward-propagating sound waves, a topic discussed in Section 6.3.2. The wave propagation nature of sound waves also implies that an acoustic disturbance in any part of the system will make itself felt in every other subsonic region of the flow. For example, an acoustic disturbance in the combustor propagates upstream and causes oscillations throughout the air flow passages, into fuel supply systems, and all other locations downstream of a sonic point.

Sound wave reflections also cause the system in which they reside to have natural acoustic modes. That is, they will exhibit natural oscillations at a multiplicity of discrete frequencies. These natural modes are intrinsically system dependent; for example, the natural modes of a combustion chamber are not only a function of the combustion chamber itself, but also of flow passages upstream of the combustion chamber. They often are the "clock" that controls the frequencies of self-excited oscillations and the frequency of excitation of vortical disturbances; their narrowband character often dominates the otherwise broadband velocity and pressure spectrum.

This is the first of two chapters on acoustic wave propagation. It introduces the concepts of standing and traveling sound waves, boundary condition effects, and natural modes of oscillation for various geometries. It is written for individuals without an acoustics background. The following chapter treats additional topics where less coverage exists in most acoustic texts, such as the effects of unsteady combustion, mean flow, inhomogeneities in temperature, and variations in cross-sectional area.

5.1 Traveling and Standing Waves

A helpful starting point for the discussion is the acoustic wave equation, presented in Eq. (2.30) for a homogeneous medium with no unsteady heat release. This is a linearized equation describing the propagation of small-amplitude disturbances – nonlinear acoustic effects are described in Section 6.6. We will first assume a one-dimensional domain and neglect mean flow, and therefore consider the wave equation

$$\frac{\partial^2 p_1}{\partial t^2} - c_0^2 \frac{\partial^2 p_1}{\partial x^2} = 0. \tag{5.1}$$

This equation possesses the following solution:

$$p_1 = F_f\left(t - \frac{x}{c_0}\right) + F_g\left(t + \frac{x}{c_0}\right), \tag{5.2}$$

$$u_{x,1} = \frac{1}{\rho_0 c_0} \left[F_f\left(t - \frac{x}{c_0}\right) - F_g\left(t + \frac{x}{c_0}\right) \right]. \tag{5.3}$$

The product $\rho_0 c_0$ is called the characteristic acoustic impedance, and relates the unsteady pressure and velocity in a traveling wave. The functions F_f and F_g are arbitrary and controlled by initial and/or boundary conditions. For example, consider the boundary value problem where $u_{x,1}(x=0,t) = 0$, and

$$p_1(x=0,t) = 2\mathcal{A}\exp(-bt). \tag{5.4}$$

The solution for the forward-propagating wave, valid at spatial locations where $x > 0$, is

$$x > 0 \quad p_1(x,t) = \mathcal{A}\exp\left(-b\left(t - \frac{x}{c_0}\right)\right), \tag{5.5}$$

$$x > 0 \quad u_{x,1}(x,t) = \frac{\mathcal{A}}{\rho_0 c_0} \exp\left(-b\left(t - \frac{x}{c_0}\right)\right). \tag{5.6}$$

The backward-propagating solution, valid at spatial locations where $x < 0$, is

$$x < 0 \quad p_1(x,t) = \mathcal{A}\exp\left(-b\left(t + \frac{x}{c_0}\right)\right), \tag{5.7}$$

$$x < 0 \quad u_{x,1}(x,t) = -\frac{\mathcal{A}}{\rho_0 c_0} \exp\left(-b\left(t + \frac{x}{c_0}\right)\right). \tag{5.8}$$

Solutions for the forward-propagating wave are plotted in Figure 5.1. Note that the disturbance created at $x = 0$ simply propagates in the positive axial direction at a speed of c_0, but with unchanged shape.

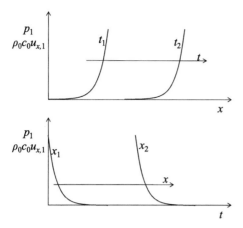

Figure 5.1 Variation of the unsteady pressure and velocity in space and time showing propagation of a disturbance at the speed of sound with unchanged shape.

The fact that the spatial and temporal profile of the disturbance stays constant is a manifestation of the nondispersive nature of acoustic waves – i.e., all frequency components travel at the same speed. This is in contrast to vortical disturbances, discussed in Chapters 3 and 4. The disturbances $F_f(t - x/c_0)$ and $F_g(t + x/c_0)$ are referred to as traveling waves – they are waves propagating in the positive and negative axial directions at the sound speed. Each wave also induces an unsteady velocity field. If the wave travels to the right and has a positive pressure, the unsteady flow is also perturbed in the positive axial direction, and vice versa. In the presence of multiple traveling waves, F_f and F_g, the unsteady pressure is the scalar superposition of the two; e.g., $p_1 = F_f + F_g$. However, the unsteady velocity is the vector sum (in a one-dimensional sense) of the induced flow fields and equals the difference between the two, $\rho_0 c_0 u_{x,1} = F_f - F_g$. This occurs because the velocity in the leftward-propagating wave is in the opposite direction to the rightward-propagating one. This can also be understood from the following general solution for a plane wave propagating in an arbitrary direction $\vec{n}$ in a three-dimensional domain:

$$p_1 = F_f\left(t - \frac{\vec{x} \cdot \vec{n}}{c_0}\right), \tag{5.9}$$

$$\vec{u}_1 = \frac{\vec{n}}{\rho_0 c_0} F_f\left(t - \frac{\vec{x} \cdot \vec{n}}{c_0}\right). \tag{5.10}$$

Because of linear superposition, a total disturbance field of N plane waves with arbitrary waveforms and propagation directions can then be written as

$$p_1 = \sum_{i=1}^{N} F_{f_i}\left(t - \frac{\vec{x} \cdot \vec{n}_i}{c_0}\right), \tag{5.11}$$

$$\vec{u}_1 = \frac{1}{\rho_0 c_0} \sum_{i=1}^{N} \vec{n}_i F_{f_i}\left(t - \frac{\vec{x} \cdot \vec{n}_i}{c_0}\right). \tag{5.12}$$

The one-dimensional solution in Eq. (5.2) can then be recognized as the special case where $N = 2$ and $\vec{n}_1 = \vec{e}_x$, $\vec{n}_2 = -\vec{e}_x$.

Most of this chapter focuses on the important case where the disturbance is harmonically oscillating. Using complex notation, we write the unsteady pressure and velocity as

$$p_1 = Real(\hat{p}_1(x, y, z) \exp(-i\omega t)), \tag{5.13}$$

$$\vec{u}_1 = Real(\hat{\vec{u}}_1(x, y, z) \exp(-i\omega t)). \tag{5.14}$$

Then, the one-dimensional acoustic field is given by

$$p_1 = Real((\mathcal{A}\exp(ikx) + \mathcal{B}\exp(-ikx))\exp(-i\omega t)), \tag{5.15}$$

$$u_{x,1} = \frac{1}{\rho_0 c_0} Real((\mathcal{A}\exp(ikx) - \mathcal{B}\exp(-ikx))\exp(-i\omega t)), \tag{5.16}$$

where $k = \omega/c_0$ is the wavenumber. Figure 5.2 plots the pressure time variation of a rightward-moving wave at several time instants, again demonstrating that the disturbance field simply consists of space–time harmonic disturbances propagating with unchanged shape at the sound speed.

An alternative way to visualize these results is to write the pressure in terms of amplitude and phase as

$$\hat{p}_1(x) = |\hat{p}_1(x)| \exp(-i\theta(x)). \tag{5.17}$$

The spatial distributions of $\hat{p}_1(x)$ and $\theta(x)$ for individual left- and rightward-moving waves are shown in Figure 5.3. If they are simultaneously present, the results look quite different, as discussed shortly. Note that the magnitude of the disturbance stays constant, but that the phase decreases/increases linearly with axial distance. The slope of these lines is $\left|\frac{d\theta}{dx}\right| = \omega/c_0$. This is an important general result that applies beyond acoustics – i.e., a harmonic disturbance propagating with a constant phase speed has a linearly varying axial phase dependence whose slope is inversely proportional to the disturbance phase speed.

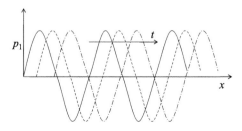

Figure 5.2 Spatial dependence of pressure in a rightward-moving wave at several time instants, showing its propagation with unchanged shape at the speed of sound.

Consider next the superposition of a left- and a rightward-traveling wave of equal amplitudes, $\mathcal{A} = \mathcal{B}$, assuming, without loss of generality, that $\mathcal{A}$ and $\mathcal{B}$ are real. Using Eq. (5.15), the disturbance field is given by

$$p_1(x,t) = 2\mathcal{A}\cos(kx)\cos(\omega t), \tag{5.18}$$

$$u_{x,1} = \frac{1}{\rho_0 c_0} 2\mathcal{A}\sin(kx)\sin(\omega t). \tag{5.19}$$

Such a disturbance field is referred to as a "standing wave." There are three fundamental differences between this solution and that of a single plane wave, which should be emphasized. These can be seen from Figure 5.4, which plots the time variation of the pressure and velocity, where the numbering is used to denote their time sequences. The corresponding spatial dependence of the amplitude and phase are indicated in the other two plots.

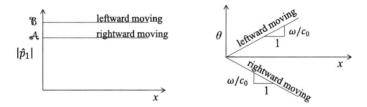

Figure 5.3 Spatial amplitude/phase variation of harmonically varying acoustic plane waves.

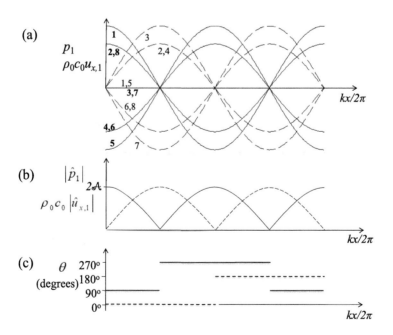

Figure 5.4 Spatial dependence of the pressure (solid) and velocity (dashed) in a standing wave. These are denoted in three ways: (a) instantaneous representations at equally spaced time intervals of 45 degrees, (b) amplitude, and (c) phase.

First, these solutions show that the amplitude of the oscillations is not spatially constant, as it was for a single traveling wave. Rather, it varies in a *spatially periodic* manner, between values that are identically zero to double that of the individual waves. Thus, the superposition of two traveling waves propagating in opposite directions leads to interference patterns in the disturbance amplitude. Second, the phase does not vary linearly with x, but has a constant phase, except across the nodes where it jumps 180 degrees. Third, the pressure and velocity have a 90 degree phase difference, as opposed to being in phase for a single plane wave. We will show next that standing wave patterns such as these are set up by wave reflections at boundaries.

5.2 Boundary Conditions: Reflection Coefficients and Impedance

We next consider acoustic boundary conditions and the reflection of traveling waves. These boundary conditions have significant influences on the overall unsteady combustor system dynamics [1, 2]. Assume that a rightward-traveling wave, given by $F_f(t - x/c_0)$, is incident on a boundary at $x = 0$ with the boundary condition $p_1(x = 0, t) = 0$. This is referred to as a "pressure release" boundary condition and causes the reflection of a wave of equal amplitude but opposite sign, so that the total disturbance field is given by

$$p_1 = F_f\left(t - \frac{x}{c_0}\right) - F_f\left(t + \frac{x}{c_0}\right) \tag{5.20}$$

or, for harmonically oscillating disturbances,

$$\begin{aligned}p_1 &= \text{Real}(\mathcal{A}(\exp(ikx) - \exp(-ikx))\exp(-i\omega t))\\ &= \text{Real}(2i\mathcal{A}\sin(kx)\exp(-i\omega t)).\end{aligned} \tag{5.21}$$

The corresponding velocity is given by

$$u_{x,1} = \frac{1}{\rho_0 c_0}\left[F_f\left(t - \frac{x}{c_0}\right) + F_f\left(t + \frac{x}{c_0}\right)\right]. \tag{5.22}$$

For a harmonically oscillating disturbance,

$$u_{x,1} = \frac{1}{\rho_0 c_0}\text{Real}(2\mathcal{A}\cos(kx)\exp(-i\omega t)). \tag{5.23}$$

The unsteady velocity at $x = 0$ is double its value in the incident wave because of the wave reflection. Now consider a "rigid wall" boundary condition, where $u_{x,1}(x = 0, t) = 0$. In this case, the disturbance field for a harmonically oscillating disturbance is given by

$$p_1 = F_f\left(t - \frac{x}{c_0}\right) + F_f\left(t + \frac{x}{c_0}\right), \tag{5.24}$$

$$u_{x,1} = \frac{1}{\rho_0 c_0}\left[F_f\left(t - \frac{x}{c_0}\right) - F_f\left(t + \frac{x}{c_0}\right)\right]. \tag{5.25}$$

Here, the unsteady pressure is doubled at $x = 0$. The solution for the harmonically oscillating case is given in Eqs. (5.18) and (5.19). A third important boundary condition is the "anechoic" one. In this case, the unsteady pressure and velocity are related at each instant of time by the characteristic impedance of the medium, i.e., $\frac{p_1(x=0,t)}{u_{x,1}(x=0,t)} = \rho_0 c_0$. In this case, the disturbance field is given by

$$p_1 = F_f\left(t - \frac{x}{c_0}\right), \tag{5.26}$$

$$u_{x,1} = \frac{1}{\rho_0 c_0} F_f\left(t - \frac{x}{c_0}\right). \tag{5.27}$$

Thus, the "boundary condition" has no influence on the incident wave – it is "anechoic."

Boundary conditions for harmonic disturbances are typically posed as a complex reflection coefficient, R, or impedance, Z. The reflection coefficient is the complex ratio of the reflected to the incident wave. The impedance is the ratio of the pressure to the velocity component *into the boundary*. For a wave traveling in the positive axial direction, they are given by

$$R = \frac{\mathcal{B}}{\mathcal{A}}, \tag{5.28}$$

$$Z = \frac{\hat{p}_1}{\vec{\hat{u}}_1 \cdot (-\vec{n})}, \tag{5.29}$$

where $\vec{n}$ points into the domain from which the wave originates. They are related by

$$R = \frac{Z/\rho_0 c_0 - 1}{Z/\rho_0 c_0 + 1}, \tag{5.30}$$

$$Z = \rho_0 c_0 \frac{1+R}{1-R}. \tag{5.31}$$

Table 5.1 summarizes the reflection coefficient and impedance of the pressure release, rigid wall, and anechoic boundary conditions.

A more general boundary condition is to consider a wave of amplitude $\mathcal{A}_{inc}$ propagating from one semi-infinite medium with impedance $\rho_{0,a} c_{0,a}$ and impinging on a second semi-infinite medium of impedance $\rho_{0,b} c_{0,b}$ at the "boundary," $x = 0$. This is an approximate model for sound propagation across an acoustically compact,

Table 5.1 Summary of reflection coefficient and impedance values of the limiting case boundary conditions.

	Reflection coefficient, R	Specific impedance, $Z/\rho_0 c_0$
Pressure release	−1	0
Rigid wall	1	∞
Anechoic	0	1

5.2 Boundary Conditions: Reflection and Impedance

premixed flame, where the gas properties jump between the reactants and the products. Assuming harmonic oscillations, the pressure and velocity in the two media are given by the general solutions:

Medium a:

$$\hat{p}_1 = \mathcal{A}_{inc}(\exp(ik_a x) + R \exp(-ik_a x)), \tag{5.32}$$

$$\hat{u}_{x,1} = \frac{\mathcal{A}_{inc}}{\rho_{0,a} c_{0,a}} (\exp(ik_a x) - R \exp(-ik_a x)); \tag{5.33}$$

Medium b:

$$\hat{p}_1 = \mathcal{A}_{inc} \mathcal{T} \exp(ik_b x), \tag{5.34}$$

$$\hat{u}_{x,1} = \frac{\mathcal{A}_{inc} \mathcal{T}}{\rho_{0,b} c_{0,b}} \exp(ik_b x), \tag{5.35}$$

where $k_a = \omega/c_{0,a}$ and $k_b = \omega/c_{0,b}$ denote the wavenumbers in mediums a and b, respectively. In analogy with the reflection coefficient, $\mathcal{T}$ is referred to as the "transmission coefficient." Solving for R leads to

$$R = \frac{\rho_{0,b} c_{0,b}/\rho_{0,a} c_{0,a} - 1}{\rho_{0,b} c_{0,b}/\rho_{0,a} c_{0,a} + 1}, \tag{5.36}$$

or, equivalently, in terms of boundary impedance Z,

$$\frac{Z}{\rho_{0,a} c_{0,a}} = \frac{\rho_{0,b} c_{0,b}}{\rho_{0,a} c_{0,a}}. \tag{5.37}$$

Thus, the incident wave is affected by changes in gas properties in a way that resembles a rigid wall, pressure release, or anechoic boundary in the limits where $\rho_{0,b} c_{0,b}/\rho_{0,a} c_{0,a}$ is $\gg 1$, $\ll 1$, or $= 1$, respectively.

We next consider the reflection coefficient and impedance in more detail in order to physically understand their meaning. The magnitude of the reflection coefficient, $|R|$, describes the net absorption or transmission of acoustic energy at the boundary, as shown in Exercise 5.5. If its value is unity, $|R| = 1$, then the incident wave is reflected back with equal amplitude. Values greater or less than unity describe amplification or damping of the incident wave by the boundary, respectively. Generally, values are less than unity except in certain very special instances, e.g., during acoustic–flame interactions due to wave amplification by the flame. These points are illustrated by Figure 5.5, which shows a polar plot of R. We can identify values of $R = 0$, $|R| < 1$, and $|R| > 1$ as anechoic, absorptive, and amplifying boundaries. We can also identify the regimes where $Real(R) = 1$ or -1 as rigid walled ($u_{x,1} = 0$) or pressure release ($p_1 = 0$) boundaries.

As shown by the polar plot, the reflection coefficient can be a complex number. This means that the phase of the reflected wave is different from the incident one. To illustrate, consider the example problem shown in Figure 5.6 where a rigid wall boundary is located at $x = L$.

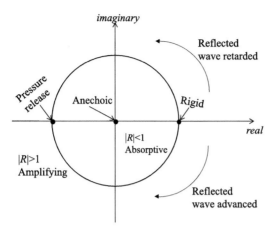

Figure 5.5 Polar plot of the reflection coefficient, R, where the solid circle denotes $|R| = 1$.

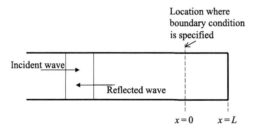

Figure 5.6 Illustration showing the physical meaning of a complex reflection coefficient, where the incident wave reflects off a rigid wall at $x = L$.

The disturbance field is

$$p_1 = Real(\mathcal{A}_{inc}(\exp(ikx) + \exp(ik(2L-x))) \exp(-i\omega t)), \quad (5.38)$$

$$u_{x,1} = \frac{1}{\rho_0 c_0} Real(\mathcal{A}_{inc}(\exp(ikx) - \exp(ik(2L-x))) \exp(-i\omega t)). \quad (5.39)$$

Suppose that the same disturbance field were to be simulated by a boundary at $x = 0$. It can be seen by inspection that this is equivalent to the $x = 0$ boundary having a reflection coefficient of

$$R = \exp(2ikL). \quad (5.40)$$

Since the magnitude, $|R|$, and phase, θ, of R are given by $R = |R| \exp(i\theta)$, this solution shows that $|R| = 1$, but R is complex. The phase angle of R equals $\theta = \omega\tau$, where $\tau = 2L/c_0$. This delay, τ, is the time required for the actual disturbance to propagate from the "pseudo-boundary" at $x = 0$ to the actual reflecting surface at $x = L$, and then to reappear at $x = 0$. As such, for real problems, the phase of R can be thought of as a time delay that accounts for the fact that the wave reflection occurs at some

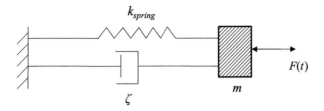

Figure 5.7 Spring–mass–damper system analogy used in acoustic boundary condition discussion.

"pseudo-boundary" either inside or outside of the domain. Returning to Figure 5.5, the angle of R describes whether the reflected wave leads or lags the incident wave in phase.

The impedance provides an alternative way to describe and physically interpret the boundary condition. To facilitate a physical interpretation of the impedance, we will model the boundary as an effective spring–mass–damper system. The sound wave interacts with this system by exerting an unsteady force on the mass because of the oscillating pressure, as illustrated in Figure 5.7. We will compare the impedance, Z, of such a system to that of the acoustic boundary defined in Eq. (5.29) which provides useful analogies between compact systems (e.g., where the duct radius is small relative to a wavelength) and these mechanical elements.

The location of the mass in this system, $x(t)$, is described by the ordinary differential equation

$$m\frac{d^2 x(t)}{dt^2} + \zeta \frac{dx(t)}{dt} + k_{spring} x(t) = F(t). \tag{5.41}$$

Assuming harmonic oscillations, the position, $x(t)$, and velocity, $v(t)$, of the mass are given by

$$x(t) = \text{Real}(\hat{x} \exp(-i\omega t)), \tag{5.42}$$

$$v(t) = \text{Real}(\hat{v} \exp(-i\omega t)), \tag{5.43}$$

where velocity and displacement are related by

$$\hat{v} = -i\omega \hat{x}. \tag{5.44}$$

Equation (5.41) can be rewritten as the following complex relation between the unsteady velocity and pressure on the surface of mass (where A denotes its cross-sectional area):

$$\left(-i\omega m + \zeta + \frac{k_{spring}}{-i\omega}\right)\hat{v} = \hat{F} = \hat{p}A. \tag{5.45}$$

Rearranging, this leads to the following expression for the complex impedance:

$$Z = \frac{\hat{p}}{\hat{v}} = \frac{1}{A}\left(-i\omega m + \zeta + \frac{k_{spring}}{-i\omega}\right). \tag{5.46}$$

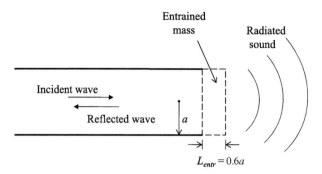

Figure 5.8 Illustration of a sound wave impinging on the end of an unflanged pipe.

This expression provides a mechanical interpretation of the acoustic impedance. The real part of the impedance represents the resistive damper element and describes acoustic absorption/transmission at the boundary. The imaginary part represents reactive spring (scaling as $1/\omega$) or mass elements (scaling as ω) which do not dissipate sound but, rather, introduce a phase shift between the pressure and velocity at the boundary. In many acoustic problems where sound is radiated into free space, the complex part of the impedance can be written as $Imag(Z) = -i\omega m_{entr}/A$, where m_{entr} is called the "entrained mass" and accounts for the kinetic energy the wave induces on the flow near the boundary.

In order to illustrate these concepts on a specific problem, consider sound incidence upon the open end of an unflanged pipe with radius a, as shown in Figure 5.8.

This boundary condition is often approximated as a pressure release, $Z = 0$, but a more accurate boundary condition that is accurate to $O(ka)^2$ is [3]:

$$\frac{Z}{\rho_0 c_0} = -0.6ika + \frac{(ka)^2}{4} + O(ka)^3. \tag{5.47}$$

First, note that the zeroth-order term for this impedance in an expansion in powers of ka is $Z = 0$. Referring to Table 5.1, this corresponds to a pressure release, $p_1 = 0$, boundary condition. The $O(ka)$ approximation of the impedance is $Z/\rho_0 c_0 = -0.6ika$. Comparing this to Eq. (5.46), we can recognize that this is equivalent to a mass element at the boundary of

$$m_{entr} = 0.6\rho_0 A a. \tag{5.48}$$

Referring to Figure 5.8, this mass is equal to that of a slug of fluid of length $L_{entr} = 0.6a$. The $O(ka)^2$ approximation of the impedance includes a resistive term,

$$\zeta = \rho_0 c_0 A \frac{(ka)^2}{4}. \tag{5.49}$$

This nonzero resistance implies that some sound is radiated out of the pipe into the domain.

The impedance in Eq. (5.47) can alternatively be written as a reflection coefficient, utilizing Eq. (5.30). To $O(ka)^2$, the reflection coefficient is given by

$$R = -\left(1 - \frac{1}{2}(ka)^2\right)e^{1.2ika} + O(ka)^3. \quad (5.50)$$

Note first that the phase of R, $\theta = 1.2ka$. Returning to the discussion of Eq. (5.40), this implies that the open end of the pipe is more accurately modeled as having an added length of $L_{entr} = 0.6a$. Adding such a length to the physical dimensions of the system is a common approach to more accurately describe its acoustic characteristics, and this example shows that this procedure is equivalent to assuming the complex impedance in Eq. (5.47). Note also that $|R| < 1$, due to the acoustic losses (more precisely, the sound radiation) at the boundary.

5.3 Natural Modes of Simple Geometries

This section analyzes the natural modes of vibrations in a constant-area duct with a homogeneous medium. The effects of density/temperature inhomogeneities and area changes are considered in Chapter 6. Start with the homogeneous wave equation with no mean flow, $\nabla^2 p_1 - \frac{1}{c_0^2}\frac{\partial^2 p_1}{\partial t^2} = 0$, and assume harmonic oscillations, $p_1(x, y, z, t) = Real(\hat{p}_1(x, y, z)\exp(-i\omega t))$. Substituting this assumed form into the wave equation leads to the Helmholtz equation for $\hat{p}_1$,

$$\nabla^2 \hat{p}_1 + \frac{\omega^2}{c_0^2}\hat{p}_1 = 0. \quad (5.51)$$

This equation can be solved using separation of variables, which is explained next. Assuming a rectangular coordinate system for now, write $\hat{p}_1(x, y, z)$ as

$$\hat{p}_1(x, y, z) = \hat{X}(x)\hat{Y}(y)\hat{Z}(z), \quad (5.52)$$

where the functions $\hat{X}$, $\hat{Y}$, and $\hat{Z}$ are only functions of x, y, and z, respectively. Substituting this into the Helmholtz equation leads to

$$\hat{Y}(y)\hat{Z}(z)\frac{d^2\hat{X}(x)}{dx^2} + \hat{X}(x)\hat{Z}(z)\frac{d^2\hat{Y}(y)}{dy^2} + \hat{X}(x)\hat{Y}(y)\frac{d^2\hat{Z}(z)}{dz^2} + k^2\hat{X}(x)\hat{Y}(y)\hat{Z}(z) = 0. \quad (5.53)$$

Note that k and ω are always related by the speed of sound, $k = \omega/c_0$, due to the nondispersive nature of acoustic wave propagation. Rearranging this equation,

$$-\frac{1}{\hat{X}(x)}\frac{d^2\hat{X}(x)}{dx^2} = \frac{1}{\hat{Y}(y)}\frac{d^2\hat{Y}(y)}{dy^2} + \frac{1}{\hat{Z}(z)}\frac{d^2\hat{Z}(z)}{dz^2} + k^2 \equiv k_x^2. \quad (5.54)$$

Because the left-hand side of this equation is only a function of x, and the right-hand side only a function of y and z, both sides must then be equal to a constant, denoted here by k_x^2. The solution for the axial component is given by

$$\hat{X}(x) = \mathcal{A}_1 \exp(ik_x x) + \mathcal{A}_2 \exp(-ik_x x). \tag{5.55}$$

The resulting equations for the y and z components can be similarly separated and equated to additional separation constants:

$$\hat{Y}(y) = \mathcal{A}_3 \exp(ik_y y) + \mathcal{A}_4 \exp(-ik_y y), \tag{5.56}$$

$$\hat{Z}(z) = \mathcal{A}_5 \exp(ik_z z) + \mathcal{A}_6 \exp(-ik_z z), \tag{5.57}$$

where the three separation constants are related by

$$k_x^2 + k_y^2 + k_z^2 = k^2 = \frac{\omega^2}{c_0^2}. \tag{5.58}$$

Determination of the values of k_x, k_y, and k_z requires specification of boundary conditions and will be discussed in Section 5.3.2. Pulling all of these results together leads to the following expression for the pressure:

$$p_1(x,y,z,t) = \text{Real}\left(\begin{array}{c} [\mathcal{A}_1 \exp(ik_x x) + \mathcal{A}_2 \exp(-ik_x x)] \\ \times [\mathcal{A}_3 \exp(ik_y y) + \mathcal{A}_4 \exp(-ik_y y)] \\ \times [\mathcal{A}_5 \exp(ik_z z) + \mathcal{A}_6 \exp(-ik_z z)] \exp(-i\omega t) \end{array}\right). \tag{5.59}$$

5.3.1 One-Dimensional Modes

We start by considering natural modes with no transverse variation. In this case, $k_x = k$, $k_y = k_z = 0$, and the unsteady pressure and velocity are then given by

$$p_1 = \text{Real}((\mathcal{A}\exp(ikx) + \mathcal{B}\exp(-ikx))\exp(-i\omega t)), \tag{5.60}$$

$$u_{x,1} = \frac{1}{\rho_0 c_0} \text{Real}((\mathcal{A}\exp(ikx) - \mathcal{B}\exp(-ikx))\exp(-i\omega t)). \tag{5.61}$$

Consider a duct of length L with rigid boundaries at both ends, $u_{x,1}(x=0,t) = u_{x,1}(x=L,t) = 0$. Applying the boundary condition at $x=0$ implies that $\mathcal{A} = \mathcal{B}$, leading to

$$u_{x,1} = \frac{1}{\rho_0 c_0} \text{Real}(2i\mathcal{A}\sin(kx)\exp(-i\omega t)). \tag{5.62}$$

The only way to satisfy the $x=L$ boundary condition for an arbitrary value of k is for $\mathcal{A} = 0$, which is a trivial solution. The alternative is for the constant $\mathcal{A}$ to be completely arbitrary and the wavenumber, k, to only equal certain discrete values:

$$k = \frac{n\pi}{L}, \quad n = 0, 1, 2, \ldots, \tag{5.63}$$

which also implies that the natural frequencies, or eigenfrequencies, of the duct are

$$f_n = \frac{\omega}{2\pi} = \frac{kc_0}{2\pi} = \frac{nc_0}{2L}. \tag{5.64}$$

5.3 Natural Modes of Simple Geometries

These natural frequencies are integer multiples of each other, i.e., $f_2 = 2f_1$, $f_3 = 3f_1$, etc. Note that the period of the first natural mode, $\mathcal{T}_1 = \frac{1}{f_1} = \frac{2L}{c_0}$, is simply the time required for a disturbance to make the "round trip" from one end of the duct to the other and back. The associated pressure standing wave mode shape, $\sin\left(\frac{n\pi x}{L}\right)$, is called an eigenfunction. The mode shape of the $n = 1$ mode is half of a sine wave and is often referred to as a "1/2-wave" mode. Assuming without loss of generality that $\mathcal{A}$ is real, the solution for the unsteady pressure and velocity is

$$p_1(x,t) = 2\mathcal{A}\cos\left(\frac{n\pi x}{L}\right)\cos(\omega t), \tag{5.65}$$

$$u_{x,1}(x,t) = \frac{2\mathcal{A}}{\rho_0 c_0}\sin\left(\frac{n\pi x}{L}\right)\sin(\omega t). \tag{5.66}$$

The spatial eigenfunctions form a complete, orthogonal set over the domain of interest. This property turns out to be generally true for solutions of the wave equation in arbitrarily complex domains under quite general conditions, as shown in Section 6.7. This property implies that an arbitrary function satisfying the boundary conditions can be built up by an appropriate superposition of these eigenfunctions, as illustrated in Section 5.4.2.

It is helpful to walk in detail through the time evolution of the unsteady pressure, velocity, and density in the duct in order to visualize their time sequence during the modal oscillation; see Figure 5.9. Start with the first natural mode, $n = 1$. At time $t = 0$ the pressure is at its maximum value and the unsteady velocity is zero; see curve 1 in Figure 5.9. The gas pressure is above and below ambient at $x = 0$ and L, respectively. The instantaneous, local gas density is correspondingly high and low, as $\rho_1 = p_1/c_0^2$; see Eq. (2.44). The spatial gradient in pressure induces a force on the gas, causing it to temporally accelerate from its zero value, as described by the linearized

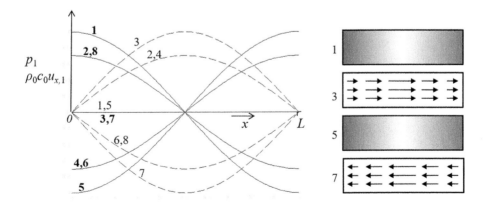

Figure 5.9 Spatial distribution of the unsteady pressure (solid) and velocity (dashed) at several time instants during natural oscillations of the $n = 1$ mode in a rigid walled duct. The shading in the right figure denotes the instantaneous pressure amplitude, and the arrows denote the direction and magnitude of the instantaneous velocity.

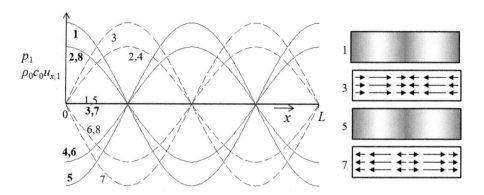

Figure 5.10 Time variation of the unsteady pressure (solid) and velocity (dashed) during natural oscillations of a rigid walled duct of the $n = 2$ mode.

momentum equation, $\rho_0 \frac{\partial u_{x,1}}{\partial t} = -\frac{\partial p_1}{\partial x}$; see curve 2. The gas motion causes the density to drop at $x = 0$ and rise at $x = L$, as described by the linearized continuity equation, $\frac{\partial \rho_1}{\partial t} = -\rho_0 \frac{\partial u_{x,1}}{\partial x}$. The unsteady pressure mirrors the density. This rightward acceleration of the flow, and drop in pressure and density, continues until the unsteady pressure drops to zero; see curve 3. Once the unsteady pressure crosses zero, the acceleration drops to zero but the velocity continues in the rightward direction. This causes the unsteady pressure to drop below zero, causing deceleration of the flow (but not yet a reversal in flow direction); see curve 4. A quarter of a period later, the velocity goes to zero, but now the unsteady pressure is at a minimum; see curve 5. This pressure gradient causes the flow direction to reverse, as it now accelerates toward the left; see curve 6.

This discussion shows the important coupling between inertia, gas compressibility, and the isentropic relationship between density and pressure in these natural mode oscillations. The motions become more complex for higher frequency modes. For example, Figure 5.10 plots the pressure and velocity variations for the $n = 2$ mode.

Note that the spatial gradients of pressure are higher, which is necessary to cause higher flow acceleration associated with the higher natural frequencies. Note also that the unsteady pressure and velocity have their antinodes and nodes at the duct center, $x = L/2$, opposite that of the $n = 1$ mode. This implies that good spatial sensor locations to measure the $n = 1$ pressure oscillations are not good locations to measure the $n = 2$ oscillations!

Consider now the mode shapes and natural frequencies of the same duct with mixed boundary conditions: $u_{x,1}(x = 0, t) = p_1(x = L, t) = 0$. The solution for the unsteady pressure and velocity is given by

$$p_1 = 2\mathcal{A} \cos\left(\frac{(2n-1)\pi x}{2L}\right) \cos(\omega t), \tag{5.67}$$

$$u_{x,1} = \frac{2\mathcal{A}}{\rho_0 c_0} \sin\left(\frac{(2n-1)\pi x}{2L}\right) \sin(\omega t), \tag{5.68}$$

$$k_n = \frac{(2n-1)\pi}{2L}, f_n = \frac{(2n-1)c_0}{4L}. \qquad (5.69)$$

The $n = 1$ mode here is called the "1/4-wave mode," as it is a quarter of a sine wave. The frequency spacing of these modes are odd multiples: $f_2 = 3f_1$ (the "3/4-wave mode"), $f_3 = 5f_1$, etc.

The last set of boundary conditions we consider are periodic: $p_1(x = 0, t) = p_1(x = L, t)$ or $u_{x,1}(x = 0, t) = u_{x,1}(x = L, t)$. This boundary condition simulates connecting the ends of a pipe to each other, which approximates azimuthal oscillations in a thin annular gap, as discussed further in Section 5.5. Solving this problem leads to

$$f_n = \frac{nc_0}{L}. \qquad (5.70)$$

For a pipe of given length L and mode number n, this is the highest frequency of the simple boundary conditions considered above. Moreover, there is a fundamental difference in mode shapes in that the values of $\mathcal{A}$ and $\mathcal{B}$ in Eq. (5.60) are *both* arbitrary. As such, the superposition of any of the following solutions will satisfy the boundary conditions (note that only two of these solutions are independent):

(1) $\hat{p} = \mathcal{A}\exp(ikx)$ – traveling wave propagating in the positive x-direction
(2) $\hat{p} = \mathcal{A}\exp(-ikx)$ – traveling wave propagating in the negative x-direction
(3) $\hat{p} = \mathcal{A}\cos(kx)$ – standing wave
(4) $\hat{p} = \mathcal{A}\sin(kx)$ – standing wave

5.3.2 Multidimensional Rectangular Duct Modes

We next consider multidimensional modes of a duct with dimensions (L, W, H) in the (x, y, z) directions. The solutions for the axial dependence of the pressure disturbance field, $\hat{X}(x)$, can be directly extracted from the previous section by replacing k_n with $k_{x,n}$. Assuming that the side walls of the duct are rigid,

$$u_{y,1}(x, y = 0 \text{ and } W, z, t) = u_{z,1}(x, y, z = 0 \text{ and } H, t) = 0,$$

allows us to solve for $\hat{Y}(y)$ and $\hat{Z}(z)$ as

$$\hat{Y}(y) = \cos\left(\frac{m\pi}{W}y\right), \quad m = 0, 1, 2, \ldots, \qquad (5.71)$$

$$\hat{Z}(z) = \cos\left(\frac{j\pi}{H}z\right), \quad j = 0, 1, 2, \ldots, \qquad (5.72)$$

where the constants are absorbed into the $\hat{X}(x)$ function. Thus, the pressure is given by

$$\hat{p}_1 = \cos\left(\frac{m\pi}{W}y\right)\cos\left(\frac{j\pi}{H}z\right)(\mathcal{A}\exp(ik_{x,n}x) + \mathcal{B}\exp(-ik_{x,n}x)) \qquad (5.73)$$

and the natural frequencies by

Table 5.2 Value of the axial eigenvalue for various boundary conditions.

Boundary condition	Axial eigenvalue, $k_{x,n}$
Rigid–Rigid (or) Pressure release–Pressure release	$k_{x,n} = \dfrac{n\pi}{L}$
Rigid–Pressure release (or) Pressure release–Rigid	$k_{x,n} = \dfrac{(2n-1)\pi}{2L}$
Periodic	$k_{x,n} = \dfrac{2n\pi}{L}$

$$f_{n,m,j} = \frac{c_0}{2\pi}\sqrt{k_{x,n}^2 + \left(\frac{m\pi}{W}\right)^2 + \left(\frac{j\pi}{H}\right)^2}, \tag{5.74}$$

where $k_{x,n}$ depends on the axial boundary conditions as summarized in Table 5.2.

The velocity can be calculated using the linearized momentum equation, $\vec{\hat{u}}_1 = \frac{1}{-i\omega\rho_0}\nabla\hat{p}_1$:

$$\hat{u}_{x,1} = -\frac{k_{x,n}}{\rho_0 c_0 k}\cos\left(\frac{m\pi}{W}y\right)\cos\left(\frac{j\pi}{H}z\right)(\mathcal{A}\exp(ik_{x,n}x) - \mathcal{B}\exp(-ik_{x,n}x)), \tag{5.75}$$

$$\hat{u}_{y,1} = \frac{1}{i\rho_0 c_0}\left(\frac{m\pi}{kW}\right)\sin\left(\frac{m\pi}{W}y\right)\cos\left(\frac{j\pi}{H}z\right)(\mathcal{A}\exp(ik_{x,n}x) + \mathcal{B}\exp(-ik_{x,n}x)), \tag{5.76}$$

$$\hat{u}_{z,1} = \frac{1}{i\rho_0 c_0}\left(\frac{j\pi}{kH}\right)\cos\left(\frac{m\pi}{W}y\right)\sin\left(\frac{j\pi}{H}z\right)(\mathcal{A}\exp(ik_{x,n}x) + \mathcal{B}\exp(-ik_{x,n}x)). \tag{5.77}$$

These mode shapes are straightforward generalizations of the one-dimensional duct modes into two or three dimensions. The indices n, m, and j describe the number of nodal lines (or 180 degree phase reversals) in a given direction. To illustrate, Figure 5.11 plots the pressure and velocity nodal lines in the transverse dimension for several duct modes. Note that $(m,j) = (0,0)$ corresponds to the one-dimensional duct modes considered in the previous section.

5.3.3 Circular Duct Modes

In this section we consider the natural modes of the various circular geometries illustrated in Figure 5.12.

A similar separation of variables technique as described earlier can be used to develop a solution for the pressure as $\hat{p}_1 = \hat{R}(r)\hat{\Theta}(\theta)\hat{X}(x)$. The general solution of the Helmholtz equation for the axial, radial, and azimuthal dependence of the disturbance field in a polar coordinate system is given by

5.3 Natural Modes of Simple Geometries

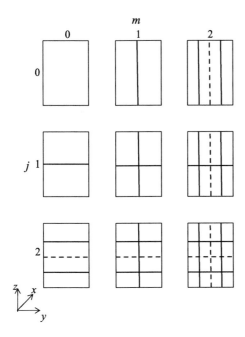

Figure 5.11 Pressure (solid) and velocity (dashed) nodal lines for several lower-order transverse modes in a rectangular duct geometry. Velocity nodes refer to zero velocity *perpendicular* to the dashed line (velocity parallel to nodal line is not necessarily zero).

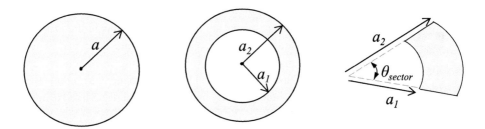

Figure 5.12 Illustration of circular, annular, and sector geometries.

$$p_1(x,r,\theta,t) = Real\left(\begin{array}{c}[\mathcal{A}_1 \exp(ik_{x,n}x) + \mathcal{A}_2 \exp(-ik_{x,n}x)][\mathcal{B}_1 J_m(k_r r) + \mathcal{B}_2 Y_m(k_r r)]\\ \times [\mathcal{C}_1 \exp(im\theta) + \mathcal{C}_2 \exp(-im\theta)]\exp(-i\omega t)\end{array}\right),$$
(5.78)

where J_m and Y_m are the Bessel functions of the first and second kind, respectively. An important practical problem is a rigid-walled, circular duct of radius a, so that $\hat{u}_{r,1}(r = a) = 0$. For this problem, $\mathcal{B}_2 = 0$ because the functions Y_m diverge at $r = 0$. The resulting solutions for the unsteady pressure, velocity, and natural frequencies/eigenvalues are [4]:

$$\hat{p}_1 = \begin{pmatrix} [\mathcal{A}_1 \exp(ik_{x,n}x) + \mathcal{A}_2 \exp(-ik_{x,n}x)] \\ \times [\mathcal{C}_1 \exp(im\theta) + \mathcal{C}_2 \exp(-im\theta)] J_m\left(\alpha_{mj}\frac{r}{a}\right) \end{pmatrix}, \quad (5.79)$$

$$\hat{u}_{x,1} = \frac{k_{x,n}}{k\rho_0 c_0} \begin{pmatrix} [\mathcal{A}_1 \exp(ik_{x,n}x) - \mathcal{A}_2 \exp(-ik_{x,n}x)] \\ \times [\mathcal{C}_1 \exp(im\theta) + \mathcal{C}_2 \exp(-im\theta)] J_m\left(\alpha_{mj}\frac{r}{a}\right) \end{pmatrix}, \quad (5.80)$$

$$\hat{u}_{r,1} = \frac{1}{ika\rho_0 c_0} \begin{pmatrix} [\mathcal{A}_1 \exp(ik_{x,n}x) + \mathcal{A}_2 \exp(-ik_{x,n}x)] \\ \times [\mathcal{C}_1 \exp(im\theta) + \mathcal{C}_2 \exp(-im\theta)] \frac{d}{d(r/a)}\left(J_m\left(\alpha_{mj}\frac{r}{a}\right)\right) \end{pmatrix}, \quad (5.81)$$

$$\hat{u}_{\theta,1} = \frac{m}{k\rho_0 c_0} \begin{pmatrix} [\mathcal{A}_1 \exp(ik_{x,n}x) + \mathcal{A}_2 \exp(-ik_{x,n}x)] \\ \times [\mathcal{C}_1 \exp(im\theta) - \mathcal{C}_2 \exp(-im\theta)] J_m\left(\alpha_{mj}\frac{r}{a}\right) \end{pmatrix}, \quad (5.82)$$

$$f_{nmj} = \frac{c_0}{2\pi}\sqrt{k_r^2 + k_{x,n}^2} = \frac{c_0}{2\pi}\sqrt{\left(\frac{\alpha_{mj}}{a}\right)^2 + k_{x,n}^2}, \quad (5.83)$$

where m is an integer and α_{mj} are the roots of

$$\frac{d}{dr}J_m(k_r a) = J'_m(\alpha_{mj}) = 0. \quad (5.84)$$

An important distinction from the rectangular coordinate problem is that these roots are not integer multiples of each other, nor can one write explicit formulae for their values. The roots of several lower-order modes are tabulated in Table 5.3.

As in Figure 5.11, we can visualize the nodal lines for the standing wave solution (i.e., where $|\mathcal{C}_1| = |\mathcal{C}_2|$) for the transverse modes (Figure 5.13). However, note that both standing and traveling waves are allowable solutions in the azimuthal direction, analogous to the periodic boundary condition solution described above Eq. (5.70). Moreover, the locations of nodal lines for the standing wave solution are completely arbitrary. In real combustors, these nodal lines may rotate in a random manner, leading to a random envelope of the unsteady pressure measured at a fixed azimuthal location, as discussed in Section 5.6.

The first four transverse mode natural frequencies are given by

$$f_{mj} = \frac{c_0}{a}\left(\frac{\alpha_{mj}}{2\pi}\right) = (0.29, 0.49, 0.61, 0.67)\frac{c_0}{a}. \quad (5.85)$$

Table 5.3 Roots of $J'_m(\alpha_{mj}) = 0$.

$\alpha_{mj}/2\pi$	$m=0$	$m=1$	$m=2$	$m=3$	$m=4$	$m=5$
$j=1$	0	0.2930	0.4861	0.6686	0.8463	1.0211
$j=2$	0.6098	0.8485	1.0673	1.2757	1.4773	1.6743
$j=3$	1.1166	1.3586	1.5867	1.8058	2.0184	2.2261
$j=4$	1.6192	1.8631	2.0961	2.3214	2.5408	2.7554
$j=5$	2.1205	2.3656	2.6018	2.8312	3.0551	3.2747

5.3 Natural Modes of Simple Geometries

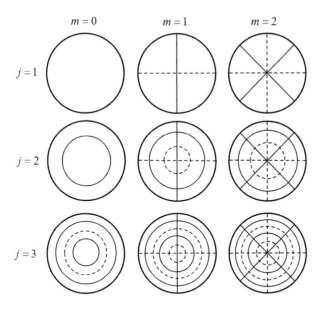

Figure 5.13 Pressure (solid) and velocity (dashed) nodal lines for several lower-order transverse modes in a circular duct. Velocity nodes refer to velocity *perpendicular* to the dashed line (velocity parallel to nodal line is not necessarily zero).

The natural frequency of the first mode can be approximated using two different approaches, using analogies to the one-dimensional duct problem with periodic and rigid boundary conditions. First, the natural frequency is equal to that of a duct with periodic boundary conditions and length $L = 2\pi a_{\it eff}$, where $a_{\it eff} = 0.55$. In other words, it is the same as the time of travel for a disturbance propagating around a circle at roughly the midpoint between the duct center and outer radius. It will be shown in Section 5.5 that this time of travel approximation becomes increasingly accurate for annular geometries as the gap width becomes smaller. Alternatively, the first natural frequency is close to the travel time for a disturbance to propagate the distance, $2a$, across the duct and back, which leads to a frequency of $f = 0.25 \, c_0/a$ (the exact expression for a rectangular duct). For reference, the corresponding first four modes of a rectangular duct with dimensions $H = W = 2a$ are (where the * denotes a repeated root):

$$f_{m,j} = \frac{\sqrt{m^2 + j^2}}{4} \frac{c_0}{a} = (0.25^*, 0.35, 0.5^*, 0.56^*) \frac{c_0}{a}. \tag{5.86}$$

Thus, the lowest natural frequency of a circular duct is roughly 16% higher than that of a square duct with the same outer dimension.

The two lowest-frequency transverse modes both have an azimuthal dependence, the $(m,j) = (1,1)$ and $(2,1)$ modes. Figure 5.14 plots instantaneous velocity vectors and pressure isocontours at one phase of the cycle for these modes. Note that although they are often referred to as "azimuthal modes," these modes have a mixed

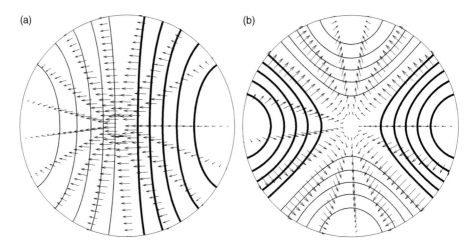

Figure 5.14 Instantaneous velocity vectors and pressure isocontours (thick and thin lines are phased 180 degrees apart from each other) for the (a) first (1,1) and (b) second (2,1) azimuthal circular duct modes as shown in Figure 5.13.

radial–azimuthal dependence. Only at radial locations near $r = a$ is the radial dependence weak and the unsteady velocity purely in the azimuthal location.

The natural frequency of the (1,1) and (2,1) "azimuthal" modes are independent of whether the azimuthal dependence is standing or, if it is traveling, on the azimuthal direction of rotation. This is analogous to the periodic boundary conditions discussed in Eq. (5.70). However, many combustors have a certain degree of swirl, i.e., azimuthal bulk rotation of the flow, which causes the azimuthal modes to actually have different frequencies depending on the direction of wave rotation.

Returning to Eq. (5.85), the third circular duct mode, $(m,j) = (0,2)$, is a purely radial mode. The associated frequency is 75% higher than the analogous rectangular duct mode, the $(m,j) = (1,1)$ mode.

The acoustics of annular and sector geometries are considered in Section 5.5 and Exercise 5.4.

5.3.4 Lumped Elements and Helmholtz Resonators

The medium through which a compressible sound wave propagates has inherent mass and stiffness. Indeed, it is the balance of this inertia and stiffness that leads to sound wave propagation in the first place. For the natural motions of simple geometries considered above, the distributed nature of mass and stiffness is key to understanding the resulting motions. For other geometries, particular regions may be idealized as "lumped elements" that are characterized by either stiffness, mass, or resistive elements. Of course, all matter, including springs, contains mass. However, for many practical problems this mass is negligible relative to that of other "mass-like" elements. Similarly, the mass-like elements have stiffness which can often be ignored.

5.3 Natural Modes of Simple Geometries

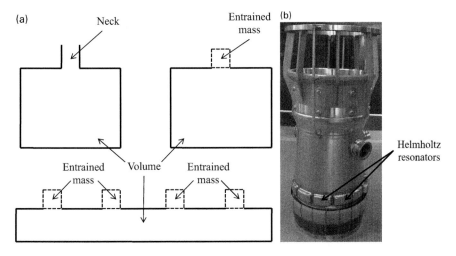

Figure 5.15 (a) Illustration of several Helmholtz resonator configurations. (b) Photograph of a commercial combustor can with Helmholtz resonators (image courtesy of Siemens Energy, Inc.).

The Helmholtz resonator is an important example of a lumped acoustic element. This device has natural modes that are much lower than might be expected based on time-of-flight analyses, where $f \sim c_0/L_{ref}$; here, L_{ref} is a characteristic axial or transverse length scale. A Helmholtz resonator consists of a rigid-walled volume coupled to a neck of smaller cross-sectional area; see Figure 5.15.

In this configuration, the neck region is essentially an incompressible mass of fluid that oscillates back and forth. This region is "mass-like" because it primarily stores fluctuating kinetic energy. The volume is "spring-like" because it primarily stores potential energy and has negligible kinetic energy. We will present a heuristic derivation of the natural frequency, noting that the natural frequency of a mechanical spring–mass system is given by

$$f = \frac{1}{2\pi}\sqrt{\frac{k_{spring}}{m}}, \tag{5.87}$$

where k_{spring} denotes the spring constant. More rigorous calculations can be performed by solving the wave equation. From Figure 5.15, it is fairly clear that the mass of the neck element is given by

$$m = \rho_{neck} A_{neck} L_{eff}, \tag{5.88}$$

where $L_{eff} = L_{neck} + L_{entr}$ and L_{entr} refers to the "entrained mass" associated with the impedance of an open-ended pipe, discussed in Section 5.2. Thus, even if $L_{neck} = 0$, as shown in Figure 5.15(a), the system still has a mass-like element. It is left as an exercise (Exercise 5.2) to show that the spring constant, k_{spring}, is given by

$$k_{spring} = \rho_{vol} c_{vol}^2 \frac{A_{neck}^2}{V_{vol}}. \tag{5.89}$$

The above expressions for k_{spring} and m are general enough to allow the gas properties in the spring and neck to differ (as may often be the case in combustion applications), which leads to the following natural frequency:

$$f = \frac{c_{neck}}{2\pi} \sqrt{\frac{A_{neck}}{L_{eff} V_{vol}}}. \tag{5.90}$$

This expression shows the geometric dependencies of f on the neck area, length, and volume size. In addition, it shows that f is independent of gas properties in the volume section, and only a function of the sound speed in the neck. In the long-wavelength limit, this natural frequency formula is independent of the geometric details of the area/volume.

Many combustors have natural Helmholtz modes, associated with, for example, small-diameter nozzles ("necks") transitioning into larger-diameter combustors ("volumes"). In addition, Helmholtz resonators are often used as reactive damping devices that act as "band gap filters" to eliminate certain natural frequencies from a combustor that may be unstable [5, 6], as illustrated in Figure 5.15(b).

5.3.5 Convective Modes

The preceding sections have shown how natural acoustic oscillations can arise that are intimately associated with acoustic wave propagation. In this section we describe additional "convective" modes that can occur. These motions arise through mixed acoustic and convective processes. For example, a convecting entropy or vortical disturbance excites an acoustic disturbance when propagating through a convergent passage (see Section 6.3.2). This sound wave then propagates upstream and excites new convecting vortical and/or entropy fluctuations. From a basic time-of-travel argument, it can be seen that this cycle repeats itself with a period equal to the travel time for the entropy/vorticity to propagate downstream, L/u_0, and the sound wave to propagate upstream, $L/(c - u_0)$, implying that

$$f = \frac{1}{T} \propto \frac{1}{(L/(c-u_0) + L/u_0)} = \frac{u_0}{L}(1 - M). \tag{5.91}$$

For low Mach number flows, this expression shows that the natural frequency of the convective mode is proportional to the convective time in the combustor. This is lower than the corresponding acoustic frequencies by a factor of the Mach number.

Low-frequency oscillations are often encountered in combustion systems under near blowoff conditions [7–10]. These are believed to be associated with convective entropy modes. Similarly, "cavity tones" [11] are a common phenomenon associated with flow over cavities and are due to convected vortical modes. These cavity tones were also discussed in the context of Figure 4.49(c) and Exercise 4.6.

5.4 Forced Oscillations

In this section we consider acoustic oscillations introduced by external forcing. Basic concepts, such as resonance, are introduced in the context of a one-dimensional duct in Section 5.4.1. Multidimensional forcing and duct cutoff modes are then treated in Section 5.4.2.

5.4.1 One-Dimensional Forcing and Resonance

A key difference between the natural mode analysis described above and forced oscillations is that the frequency of oscillations (not necessarily at the natural frequency) is imposed. Returning to the one-dimensional problem, consider a duct with forcing at $x = 0$ by a piston with velocity amplitude $\hat{V}$ and an impedance condition at $x = L$:

$$u_1(x = 0, t) = Real(\hat{V}e^{-i\omega t}), \quad (5.92)$$

$$Z(x = L) = Z_L. \quad (5.93)$$

The pressure and velocity are given by

$$\hat{p}_1(x) = \mathcal{A}e^{ikx} + \mathcal{B}e^{-ikx}, \quad (5.94)$$

$$\hat{u}_1(x) = \frac{\mathcal{A}e^{ikx} - \mathcal{B}e^{-ikx}}{\rho_0 c_0}. \quad (5.95)$$

Solving these expressions, the pressure at the piston face, $x = 0$, is given by

$$\frac{\hat{p}_1(x=0)}{\rho_0 c_0 \hat{V}} = \frac{\frac{Z_L}{\rho_0 c_0}\cos(kL) - i\sin(kL)}{\cos(kL) - \frac{Z_L}{\rho_0 c_0}i\sin(kL)}. \quad (5.96)$$

The normalized pressure amplitude at $x = 0$ is plotted as a function of dimensionless frequency in Figure 5.16 for several large, but finite, Z_L values. Note the large pressure amplitudes at the natural frequencies of the duct, $kL = n\pi$. This is referred to as "resonance."

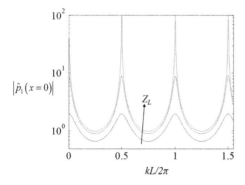

Figure 5.16 Dependence of the pressure amplitude at the piston face as a function of the dimensionless frequency for $Z_L/\rho_0 c_0 = 3, 10,$ and 100.

Systems with broadband background noise exhibit similar spectra, due to the resonant response of the system. Indeed, in some cases it has been suggested that large amplitude oscillations in combustors are not due to self-excited oscillations, but rather a manifestation of the lightly damped combustor's response at natural frequencies to background disturbances [12]. This subject is also treated in Section 2.5.3.1.

Resonance occurs at frequencies where the imaginary part of the impedance at the piston face is zero or, equivalently, where the pressure and velocity are exactly in phase.[1] In order to understand resonance, it is helpful to consider the solution in the time domain, as the frequency domain solution ignores the initial transient. Assume that the system starts with zero unsteady pressure and velocity, and that the excitation source is turned on at $t = 0$. For the time interval $0 < t < L/c_0$, the sound field consists of the waves initiated at the piston face only, and for the time interval $L/c_0 < t < 2L/c_0$ it consists of the superposition of these waves with those reflected at $x = L$. Resonance occurs when the piston motion is phased so as to reinforce the incident wave that reflects off it at $t = 2nL/c_0$, where n is an integer denoting the number of piston face reflections. The pressure amplitude in the tube then increases monotonically with time. If there is no damping, then the pressure grows without bound – this is why the harmonic solution predicts an infinite "steady-state" system response at the resonant frequency when there are no acoustic losses at the other boundary.

5.4.2 Forced Oscillations in Ducts and Cutoff Modes [4]

In a multidimensional duct, the presence of external forcing can excite motions associated with multiple natural duct modes. As shown in this section, the behavior of the system at large axial distances from the excitation location is fundamentally different depending on whether the forcing frequency is higher or lower than natural transverse duct frequencies. In order to illustrate the basic concepts, it is easiest to work through an example problem and then discuss the solution characteristics. Consider a two-dimensional duct that is excited by a source at $x = 0$ with the velocity boundary condition

$$u_{x,1}(x=0,y,t) = Real(F(y)e^{-i\omega t}). \tag{5.97}$$

The duct is of width W and has boundary conditions of zero normal velocity at the top and bottom walls:

$$u_{y,1}(x,y=0,t) = u_{y,1}(x,y=W,t) = 0. \tag{5.98}$$

In order to determine the solution, we exploit the fact that the natural modes form a complete, orthogonal set, which means that we can expand any arbitrary function

---

[1] The imaginary part of the impedance at $x = 0$ is also zero at frequencies halfway between these resonant frequencies – these points correspond to minima in power flow into the system for a given amplitude of piston velocity and a minimum in pressure at the piston face.

satisfying the wall boundary conditions as a summation over the appropriately weighted mode shapes. In other words, the unsteady pressure can be expressed as

$$p_1(x,y,t) = \text{Real}\left(\sum_{m=0}^{\infty} \left(\mathcal{A}_m \sin\left(k_{y,m}y\right) + \mathcal{B}_m \cos\left(k_{y,m}y\right)\right) e^{i(k_x x - \omega t)}\right). \tag{5.99}$$

The corresponding unsteady velocity is:

$$u_{x,1}(x,y,t) = \text{Real}\left(\frac{1}{\rho\omega}\sum_{m=0}^{\infty} k_x \left(\mathcal{A}_m \sin\left(k_{y,m}y\right) + \mathcal{B}_m \cos\left(k_{y,m}y\right)\right) e^{i(k_x x - \omega t)}\right),$$

$$u_{y,1}(x,y,t) = \text{Real}\left(\frac{1}{\rho\omega}\sum_{m=0}^{\infty} k_{y,m} \left(\mathcal{A}_m \cos\left(k_{y,m}y\right) - \mathcal{B}_m \sin\left(k_{y,m}y\right)\right) e^{i\left(k_x x - \omega t - \frac{\pi}{2}\right)}\right). \tag{5.100}$$

Applying the boundary conditions specified in Eq. (5.98) implies $\mathcal{A}_m = 0$ (from the boundary condition at $y = 0$), and

$$k_{y,m} = \frac{m\pi}{W}. \tag{5.101}$$

The coefficients $\mathcal{B}_m$ are determined from the boundary condition at $x = 0$, Eq. (5.97), by the expression

$$F(y) = \frac{1}{\rho\omega}\sum_{m=0}^{\infty} k_x \mathcal{B}_m \cos\left(\frac{m\pi y}{W}\right). \tag{5.102}$$

In general, Eq. (5.102) constitutes an infinite set of equations for an infinite number of unknown coefficients, $\mathcal{B}_m$. However, the orthogonality of the eigenfunctions over the domain of interest enables us to derive explicit expressions for $\mathcal{B}_m$.[2] Multiply both sides of Eq. (5.102) by $\cos(m'\pi y/W)$ and integrate between $y = 0$ and $y = W$ to yield

$$\int_{y=0}^{y=W} F(y) \cos\left(\frac{m'\pi y}{W}\right) dy = \frac{1}{\rho\omega}\sum_{m=0}^{\infty}\left[k_x \mathcal{B}_m \int_{y=0}^{y=W} \cos\left(\frac{m\pi y}{W}\right)\cos\left(\frac{m'\pi y}{W}\right) dy\right]. \tag{5.103}$$

Because of the orthogonality of the eigenfunctions, the integral is zero for all $m \neq m'$. As such, the original system of infinite equations for the infinite number of coefficients $\mathcal{B}_m$ can be solved exactly:

$$\mathcal{B}_m = \frac{2\rho\omega}{k_x W} \int_{y=0}^{y=W} F(y) \cos\left(\frac{m\pi y}{W}\right) dy. \tag{5.104}$$

---

[2] However, this property cannot be utilized for many important problems of interest, such as the coupling between two ducts of different diameters. For such problems, one cannot derive explicit expressions for the coefficients and the solution must be obtained numerically [13, 14].

If $F(y)$ is a constant, that is, if the excitation does not vary in the transverse direction, then only the plane wave mode is excited; i.e., $\mathcal{B}_m = 0$ for $m \neq 0$. In general, however, an axially oscillating wall piston excites all of the duct modes. However, the axial characteristics of the mth mode are fundamentally different depending on whether the forcing frequency is higher or lower than $f_{m,cutoff}$. This "cutoff frequency" exactly equals the mth natural mode of the purely transverse duct mode and is given by

$$f_{m,cutoff} = \frac{mc_0}{2W}. \tag{5.105}$$

This can be seen by noting that the axial dependence of the mth mode of the disturbance field is given by

$$\begin{aligned} f < f_{m,cutoff} &: \mathcal{B}_m e^{-\sqrt{m^2\pi^2 - (kW)^2}\frac{x}{W}} \cos\left(m\pi\frac{y}{W}\right), \\ f > f_{m,cutoff} &: \mathcal{B}_m e^{-i\sqrt{(kW)^2 - m^2\pi^2}\frac{x}{W}} \cos\left(m\pi\frac{y}{W}\right). \end{aligned} \tag{5.106}$$

This equation shows that for frequencies $f < f_{m,cutoff}$, the mth mode decays exponentially with distance. For frequencies $f > f_{m,cutoff}$, this mode propagates as a traveling wave. The exponentially decaying disturbances, corresponding to the spatial modes of higher-frequency transverse duct modes, are referred to as "evanescent waves." Although they decay axially, these multidimensional disturbances can have very strong amplitudes near the region where they are excited. The frequency $f_{1,cutoff} = c_0/2W$ is particularly significant, because below this frequency only one-dimensional plane waves propagate and all multidimensional disturbances decay exponentially. For this reason, one-dimensional analyses of the acoustic frequencies and mode shapes in complex, multidimensional geometries are often quite accurate in describing the bulk acoustic features of the system for frequencies below cutoff. As shown in Figure 5.17, although multidimensional acoustic features are excited in the vicinity of multidimensional changes in geometry (such as where nozzles transition to combustor cans) or ambient conditions (such as across the flame), the evanescent disturbances decay with distance [15].

We next consider propagating modes. Note first that their axial phase speed varies with frequency; i.e., they are dispersive:

$$c_{ph,m} = \frac{c_0}{\sqrt{1 - (f_{m,cutoff}/f)^2}}. \tag{5.107}$$

This dispersive character is due to the fact that the multidimensional disturbances propagate at a frequency-dependent angle, θ_{inc}, with respect to the axial coordinate and successively reflect off the top and bottom walls. Figure 5.18 provides a geometric picture of the acoustic field structure by depicting a number of lines of constant phase of a propagating disturbance. The solid and dashed lines denote the maximum and minimum pressure amplitude isocontours, respectively. The thick solid line cuts perpendicularly through these constant-phase surfaces and represents the propagation direction of the wave, θ_{inc}, with respect to the vertical.

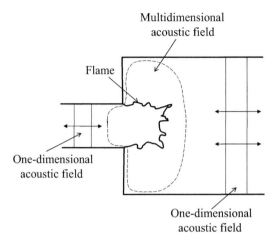

Figure 5.17 Illustration of multidimensional acoustic features near a flame and an area expansion, with a one-dimensional acoustic field in the remainder of the system.

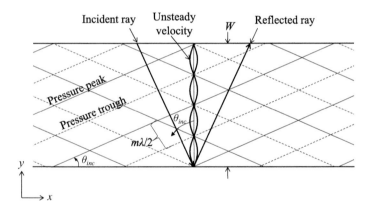

Figure 5.18 Illustration of transverse wave propagation in a duct, following Blackstock [4].

Two instantaneous representations of the transverse velocity are also shown, which must be zero at the walls because of the boundary conditions. Because the projection of the wave in the vertical direction must correspond to an integer number of half-wavelengths, we can write the following expression for θ_{inc}:

$$\cos \theta_{inc} = \frac{(m\lambda/2)}{W}. \tag{5.108}$$

Since these surfaces of constant phase propagate nondispersively normal to themselves at the speed of sound, we can also write the following expression for the corresponding "effective" axial propagation speed, or phase speed:

$$\sin \theta_{inc} = \frac{c_0}{c_{ph,m}}, \tag{5.109}$$

which, using Eq. (5.108), leads to

$$c_{ph,m} = \frac{c_0}{\sin\theta_{inc}} = \frac{c_0}{\sqrt{1-(f_{m,cutoff}/f)^2}}. \qquad (5.110)$$

Note that this expression is identical to Eq. (5.107). The excited waves propagate vertically up and down, $\theta_{inc} = 0$, at the cutoff frequency and have no axial dependence – hence the infinite axial phase speed. For frequencies below cutoff, it is not possible to satisfy the boundary conditions with the traveling wave solution sketched in the figure. As the frequency increases beyond cutoff, the phase speed drops asymptotically towards c_0, because the wave propagation angle bends progressively from 0 degrees at cutoff toward 90 degrees at frequencies much greater than cutoff.

In order to graphically illustrate the changing character of the wave field with frequency, it is helpful to work through a specific example. Assume that the wall-mounted piston has a height of h and has the following transverse axial velocity dependence:

$$F(y) = \begin{cases} v_W & \left(\frac{W}{2}-\frac{h}{2}\right) \le y \le \left(\frac{W}{2}+\frac{h}{2}\right), \\ 0 & \text{otherwise.} \end{cases} \qquad (5.111)$$

Using Eq. (5.104), the solutions for the pressure and velocity are given by

$$\frac{p_1(x,y,t)}{\rho c_0 v_W} = \text{Real}\left(\frac{h}{W}e^{i(k_x x-\omega t)} + \sum_{m=1}^{\infty}\left(\frac{4}{m\pi}\frac{kW}{k_x W}\cos\left(\frac{m\pi}{2}\right)\sin\left(\frac{m\pi h}{2W}\right)\cos\left(\frac{m\pi y}{W}\right)\right)e^{i(k_x x-\omega t)}\right), \qquad (5.112)$$

$$\frac{u_{x,1}(x,y,t)}{v_W} = \text{Real}\left(\frac{h}{W}e^{i(k_x x-\omega t)} + \sum_{m=1}^{\infty}\left(\frac{4}{m\pi}\cos\left(\frac{m\pi}{2}\right)\sin\left(\frac{m\pi h}{2W}\right)\cos\left(\frac{m\pi y}{W}\right)\right)e^{i(k_x x-\omega t)}\right), \qquad (5.113)$$

$$\frac{u_{y,1}(x,y,t)}{v_W} = \text{Real}\left(\sum_{m=1}^{\infty}\left(\frac{4}{k_x W}\cos\left(\frac{m\pi}{2}\right)\sin\left(\frac{m\pi h}{2W}\right)\sin\left(\frac{m\pi y}{W}\right)\right)e^{i(k_x x-\omega t-\frac{\pi}{2})}\right). \qquad (5.114)$$

Figure 5.19 plots the pressure magnitude and velocity vectors at two frequencies, corresponding to below and above cutoff.

Below the frequency $f_{1,cutoff}$, a multidimensional disturbance field persists in front of the piston face, but quickly evolves to the one-dimensional plane wave solution farther downstream. Above the cutoff frequency, the transverse mode becomes propagating and the multidimensional field persists indefinitely far downstream.

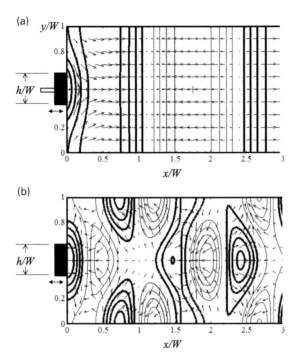

Figure 5.19 Plot of the instantaneous distribution of the pressure and velocity at frequencies (a) below cutoff, $f/f_{cutoff} = 0.8$, and (b) above cutoff, $f/f_{cutoff} = 2.55$ ($h/W = 0.3$; the thick and thin lines are phased 180 degrees apart from each other).

5.5 Aside: Annular and Sector Circular Geometries

This aside considers more complex circular geometries. Start with the annular duct with rigid walls, with boundary conditions $\hat{u}_{r,1}(r = a_1, \theta) = \hat{u}_{r,1}(r = a_2, \theta) = 0$, shown in Figure 5.12. The unsteady pressure is given by

$$\hat{p}_1 = \begin{pmatrix} [\mathcal{A}_1 \exp(ik_{x,n}x) + \mathcal{A}_2 \exp(-ik_{x,n}x)] \\ \times [\mathcal{C}_1 \exp(im\theta) + \mathcal{C}_2 \exp(-im\theta)] [\mathcal{B}_1 J_m(k_{r,j}r) + \mathcal{B}_2 Y_m(k_{r,j}r)] \end{pmatrix}, \quad (5.115)$$

where m is an integer and $k_{r,j}$ satisfies

$$J'_m(k_{r,j}a_1) Y'_m(k_{r,j}a_2) - J'_m(k_{r,j}a_2) Y'_m(k_{r,j}a_1) = 0. \quad (5.116)$$

The frequencies are given by Eq. (5.83). Thus, the solution involves Y_m in the radial dependence.

A limiting case of significant interest for annular combustors is the "thin gap" limit of $a_1 \approx a_2$ [16]. In this limit, the radial velocity can be approximated as zero for the $j = 1$ mode, and the radial dependence of the azimuthal velocity ignored. The natural azimuthal frequencies of this geometry are identical to the periodic boundary condition in Eq. (5.70), with $L = 2\pi(a_1 + a_2)/2$. In addition, the first radial mode is equal to that of a rigid–rigid duct whose length is equal to the gap width, $2(a_2 - a_1)$. These approximations turn out to be very close to exact solutions even when the gap is not thin, as shown in Exercise 5.4.

Finally, consider a sector of an annulus as shown in Figure 5.12, which is often used for subscale testing of annular combustors. The boundary conditions are

$$\hat{u}_{r,1}(r = a_1, \theta) = \hat{u}_{r,1}(r = a_2, \theta) = 0$$
$$\hat{u}_{\theta,1}(r, \theta = 0) = \hat{u}_{\theta,1}(r, \theta = \theta_{sector}) = 0. \quad (5.117)$$

The unsteady pressure is given by

$$\hat{p}_1 = \begin{pmatrix} [\mathcal{A}_1 \exp(ik_{x,n}x) + \mathcal{A}_2 \exp(-ik_{x,n}x)] \\ \times [\mathcal{C}_1 \sin(m\theta)] [\mathcal{B}_1 J_m(k_{r,j}r) + \mathcal{B}_2 Y_m(k_{r,j}r)] \end{pmatrix}, \quad (5.118)$$

where $k_{r,j}$ satisfies

$$J'_m(k_{r,j}a_1)Y'_m(k_{r,j}a_2) - J'_m(k_{r,j}a_2)Y'_m(k_{r,j}a_1) = 0 \quad (5.119)$$

and

$$m = \frac{l\pi}{\theta_{sector}}, \quad \text{where } l = 1, 2, 3, \ldots$$

The natural frequencies are given by Eq. (5.83). There are a few important differences between the acoustics of the full annulus and sector geometries. First, traveling wave solutions are not possible, and the azimuthal structure is fixed. Second, the index m is not necessarily an integer. From this follows the fact that only certain sector angles are permissible if one wishes to simulate a given standing wave in the full annulus.

5.6 Aside: Natural Modes in Annular Geometries

As noted in Section 5.3.1, natural modes could consist of standing or traveling waves for the periodic boundary condition case. This is an important case for annular geometries which are a common combustor geometry. A linear analysis is not sufficient to determine whether the acoustic mode will consist of a clockwise traveling wave, counterclockwise traveling wave, or a standing wave. Moreover, the standing waves' node/antinode locations can occur at any location. The relative dominance of standing vs. traveling waves can be quantified via the "spin ratio," defined as [17]:

$$SR = \frac{|\mathcal{C}_1| - |\mathcal{C}_2|}{|\mathcal{C}_1| + |\mathcal{C}_2|}. \quad (5.120)$$

Here, $SR = \pm 1$ indicates a purely traveling wave in the counterclockwise (+) or clockwise (−) direction, and $SR = 0$ denotes a purely standing wave. Otherwise, the wave motions are a combination of standing and traveling waves. Analysis of data from can-annular combustors shows standing, traveling, and switching between

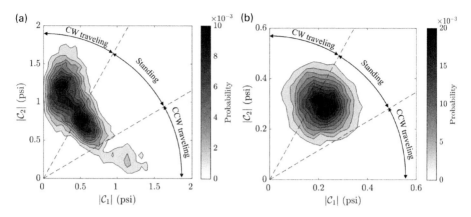

Figure 5.20 Joint probability of clockwise (CW) and counterclockwise (CCW) spinning wave amplitudes for two different operating conditions in a can combustor. Data courtesy of J. Kim.

standing and traveling modes. Figure 5.20 shows an example of these results, plotting the joint probability of the clockwise and counterclockwise traveling wave amplitudes for two different operating conditions. The bivariate PDF in Figure 5.20(a) is bimodal, indicating that the acoustic mode switches between a standing pattern and a clockwise spinning wave during the experiment. Figure 5.20(b) shows a different condition where the PDF is concentrated on the standing-wave region.

The dynamics of these traveling waves can also be visualized in the phase plane to visualize the low-dimensional attractor of limit cycling systems. While the data above was taken from a highly turbulent system, these deterministic phase dynamics can be visualized by ensemble averaging. This attractor is visualized in (SR, ϕ) phase space where $\phi = \angle \mathcal{C}_1 - \angle \mathcal{C}_2$ in Figure 5.21 for the same data shown in Figure 5.20(a). Note that the plot is periodic in the vertical direction so arrows going into the top come out of the bottom, and vice versa. Limit cycles associated with standing and traveling are represented as fixed points in these data – note that these limit cycles are all stable within their basin of attraction, as manifested by the orbits spiraling into them. It is likely that turbulent fluctuations are responsible for knocking the orbits out of each stable attractor, which is why the full phase space can be visualized.

The relative preference for standing or traveling disturbances is influenced by both linear and nonlinear factors. For example, linear unsteady heat release coupling, a topic discussed in Section 6.5 influences whether a standing or traveling wave is amplified by the unsteady heat release. This sensitivity occurs because traveling and standing waves have completely different phase relationships between pressure and velocity, being in/out of phase, or 90 degrees phased, respectively. As noted in the discussion below Eq. (6.90), the pressure–velocity phase influences how unsteady heat release amplifies or damps these modes. Nonlinear effects cause coupling of $\mathcal{C}_1(t)$ and $\mathcal{C}_2(t)$, possibly leading to competition between the two possible traveling wave solutions; see the related discussion in Section 2.5.3.4 [18].

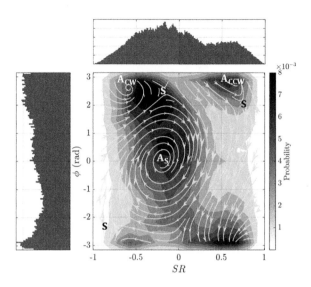

Figure 5.21 Phase space representation of the data shown in Figure 5.20(a) showing spin ratio and phase difference space, where A_S denotes the standing wave attractor, and A_{CW} and A_{CCW} denote clockwise and counterclockwise traveling wave attractors, respectively; "S" denotes saddle points. The background color denotes the joint PDF, and the plots on the top and left show PDFs of the phase angle and spin ratio, SR. Image courtesy of J. Kim.

Exercises

5.1. Plane wave initial value problem: Assume that the disturbance pressure and velocity at $t = 0$ are given by $p_1(x, t = 0) = \exp(-k|x|)$, $u_1(x, t = 0) = 0$. Develop expressions for the pressure and velocity at later times.

5.2. Derive the expression for the effective spring constant of the volume in the Helmholtz resonator problem discussed in the context of Eq. (5.89).

5.3. Derive the wavenumbers and mode shapes of the rigid-walled rectangular duct in Section 5.3.2.

5.4. Analysis of thin-gap limit for annular geometries: As detailed in Section 5.5, approximate solutions can be derived for natural acoustic modes in the thin-gap limit. Determine the dependence of the first two purely azimuthal modes ($j = 1$, $m = 1, 2$) and purely radial modes ($j = 2, 3$, $m = 0$) on the radius ratio, a_1/a_2. Normalize the natural frequencies by their thin-gap limit values so that the normalized frequencies of the first azimuthal and radial modes tend to unity as $a_1/a_2 \to 1$. Check the correspondence of the $a_1/a_2 \to 0$ results with those for a duct summarized in Table 5.3.

5.5 Reflection coefficients: Prove that acoustic energy is absorbed or transmitted by the boundary when $|R| < 1$ and that all incident energy is reflected when $|R| = 1$, as discussed in the context of Figure 5.5. This can be done by calculating the acoustic energy flux into the boundary, $I_x = p_1 u_{x,1} \Rightarrow \overline{I_x} = \frac{1}{2} Real(\hat{p}_1 \hat{u}_{x,1}^*)$.

References

1. Bothien M.R., Moeck J.P., and Paschereit C.O., Active control of the acoustic boundary conditions of combustion test rigs. *Journal of Sound and Vibration*, 2008, **318**: pp. 678–701.
2. Hield P.A. and Brear M.J., Comparison of open and choked premixed combustor exits during thermoacoustic limit cycle. *AIAA Journal*, 2008, **46**(2): pp. 517–526.
3. Levine H. and Schwinger J., On the radiation of sound from an unflanged circular pipe. *Physical Review*, 1948, **73**(4): pp. 383–406.
4. Blackstock D.T., *Fundamentals of Physical Acoustics*. 2000, John Wiley & Sons, Ltd.
5. Gysling D., Copeland G., McCormick D., and Proscia W., Combustion system damping augmentation with Helmholtz resonators. *Journal of Engineering for Gas Turbines and Power*, 2000, **122**: pp. 269–274.
6. Tang P. and Sirignano W., Theory of a generalized Helmholtz resonator. *Journal of Sound and Vibration*, 1973, **26**: pp. 247–262.
7. Sewell J.B. and Sobieski P.A., Monitoring of combustion instabilities: Calpine's experience, in *Combustion Instabilities in Gas Turbine Engines: Operational Experience, Fundamental Mechanisms and Modeling. Progress in Aeronautics and Astronautics*, T.C. Lieuwen and V. Yang, eds. 2005, Washington, DC: AIAA. pp. 147–162.
8. Abouseif G., Keklak J., and Toong T., Ramjet rumble: the low-frequency instability mechanism in coaxial dump combustors. *Combustion Science and Technology*, 1984, **36**(1): pp. 83–108.
9. Konrad W., Brehm N., Kameier F., Freeman C., and Day I., Combustion instability investigations on the BR710 jet engine. *ASME Transactions, Journal of Engineering for Gas Turbines and Power*, 1998, **120**(1): pp. 35–40.
10. Eckstein J., Freitag E., Hirsch C., and Sattelmayer T., Experimental study on the role of entropy waves in low-frequency oscillations in a RQL combustor. *Journal of Engineering for Gas Turbines and Power*, 2006, **128**: pp. 264–270.
11. Rockwell D., Oscillations of impinging shear layers. *AIAA Journal*, 1983, **21**: pp. 645–664.
12. Cohen J.M., Proscia W., and Delaat J., Characterization and control of aeroengine combustion instability: Pratt & Whitney and NASA experience, in *Combustion Instabilities in Gas Turbine Engines. Operational Experience, Fundamental Mechanisms, and Modeling*, T.C. Lieuwen and V. Yang, eds. 2005, AIAA. pp. 113–144.
13. El Sharkawy A. and Nayfeh A.H., Effect of an expansion chamber on the propagation of sound in circular ducts. *The Journal of the Acoustical Society of America*, 1978, **63**: p. 667.
14. Miles J., The analysis of plane discontinuities in cylindrical tubes. Part I. *The Journal of the Acoustical Society of America*, 1946, **17**: pp. 259–284.
15. Santosh H. and Sujith R.I., Acoustic nearfield characteristics of a wrinkled premixed flame. *Combustion Science and Technology*, 2006, **178**(7): pp. 1263–1295.
16. Stow S., Dowling A., and Hynes T., Reflection of circumferential modes in a choked nozzle. *Journal of Fluid Mechanics*, 2002, **467**: pp. 215–239.
17. Bourgouin J.-F., Durox D., Moeck J.P., Schuller T., and Candel S., Self-sustained instabilities in an annular combustor coupled by azimuthal and longitudinal acoustic modes, in *ASME Turbo Expo 2013: Turbine Technical Conference and Exposition*. 2013, American Society of Mechanical Engineers Digital Collection.
18. Schuermans B., Paschereit C., and Monkewitz P., Non-linear combustion instabilities in annular gas-turbine combustors, in *44th AIAA Aerospace Sciences Meeting and Exhibit*. 2006. Reno, NV: AIAA.

6 Acoustic Wave Propagation II: Heat Release, Complex Geometry, and Mean Flow Effects

6.1 Introduction

This chapter follows Chapter 5 by treating the additional physical processes associated with sound wave propagation through an inhomogeneous, variable-area region with bulk flow. In the rest of this section we discuss four generalizations introduced by these effects: (1) wave reflection and refraction, (2) changes in disturbance amplitude and relative amplitudes of pressure and velocity disturbances, e.g., $|\hat{p}_1|/|\hat{u}_{x,1}|$, (3) convection of disturbances, and (4) amplification/damping of disturbances. The rest of this introduction gives an overview of these four effects.

Refraction occurs in multidimensional problems where the variations in sound speed, boundaries, and/or mean flow velocity are not aligned with the wave propagation direction. This leads to a bending of the direction of wave propagation. For example, Figure 6.1 illustrates a wave propagating through a region with varying temperature, and therefore varying sound speed. Neglecting wave reflections for the moment, note that the points on the surfaces of constant phase move faster in the region of higher sound speed than in the lower, bending the wave propagation direction. Similar convective refraction effects occur for sound waves propagating through shear flows. Finally, even in uniform media, waves bend around obstacles in multidimensional geometries, as discussed in Section 6.3.3.

For example, the following multidimensional wave equation describes sound wave propagation through a region with inhomogeneities in gas properties, such as temperature, with no mean flow:

$$\frac{\partial^2 p_1}{\partial t^2} - \rho_0 c_0^2 \nabla \cdot \left(\frac{1}{\rho_0} \nabla p_1 \right) = 0. \tag{6.1}$$

Property changes also cause wave reflections, as can be understood from the worked problem in the context of Eq. (5.34). A smooth continuous property change can be thought of as the limit of a large number of very small $\rho_0 c_0$ changes, as illustrated in Figure 6.2. Furthermore, as shown in Section 6.3, area changes likewise cause wave reflections.

We can write a special case of Eq. (6.1) for quasi-one-dimensional wave propagation through a slowly varying area change (see Exercise 6.2):

6.1 Introduction

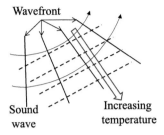

Figure 6.1 Illustration of refraction of a plane wave propagating through a temperature gradient.

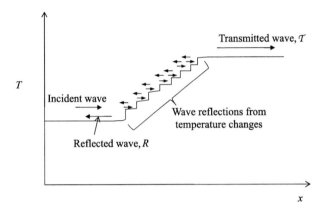

Figure 6.2 Smooth temperature variation in a duct considered as a limit of a large number of small temperature changes.

$$\frac{\partial^2 p_1}{\partial t^2} - c_0^2 \left(\frac{1}{A} \frac{dA}{dx} - \frac{1}{\rho_0} \frac{d\rho_0}{dx} \right) \frac{\partial p_1}{\partial x} - c_0^2 \frac{\partial^2 p_1}{\partial x^2} = 0 \qquad (6.2)$$

Assuming harmonic oscillations, the complex pressure is given by

$$\frac{\partial^2 \hat{p}_1}{\partial x^2} + \frac{d \ln A/\rho_0}{dx} \frac{\partial \hat{p}_1}{\partial x} + k^2 \hat{p}_1 = 0 \qquad (6.3)$$

This is a second-order differential equation with nonconstant coefficients. Approximate solutions of this equation for slowly varying coefficients are described in Section 6.8.

The second effect noted earlier is a change in the disturbance amplitude and the relative amplitudes between pressure and velocity. This effect can be understood by considering a one-dimensional wave field with no mean flow and the expression for the time-averaged acoustic energy flux, obtained from Eq. (2.55):

$$I_x = p_1 u_{x,1} \Rightarrow \overline{I_x} = \frac{1}{2} \mathrm{Real}\left(\hat{p}_1 \hat{u}_{x,1}^*\right), \qquad (6.4)$$

where the superscript * refers to complex conjugation.

As previously noted, abrupt changes of gas properties or area over an acoustic wavelength λ lead to wave reflection. However, wave reflections are negligible and the wave field consists of a traveling wave in the limits of very slow changes:

$$\frac{dA}{A} \ll \frac{dx}{\lambda}, \tag{6.5}$$

$$\frac{d\rho_0 c_0}{\rho_0 c_0} \ll \frac{dx}{\lambda}. \tag{6.6}$$

Under steady-state conditions, the time-averaged acoustic power flow, $\overline{I_x} A$, at each axial location must be a constant (if not, the acoustic energy density, and therefore the disturbance amplitude, would change in time; see Eq. (2.63)). Moreover, with no reflections, the pressure and velocity are approximately related by the plane wave relation

$$\hat{p}_1 = \rho_0 c_0 \hat{u}_{x,1}. \tag{6.7}$$

Utilizing these two facts we can write [1]:

$$\hat{p}_1 \hat{u}_{x,1}^* A = \text{constant} = \frac{A|\hat{p}_1|^2}{\rho_0 c_0} = A|\hat{u}_{x,1}|^2 \rho_0 c_0, \tag{6.8}$$

or, the unsteady pressure and velocity are related at two arbitrary axial points a and b as

$$\frac{|\hat{p}_{1,b}|}{|\hat{p}_{1,a}|} = \sqrt{\frac{A_a \rho_{0,b} c_{0,b}}{A_b \rho_{0,a} c_{0,a}}}, \tag{6.9}$$

$$\frac{|\hat{u}_{x,1b}|}{|\hat{u}_{x,1a}|} = \sqrt{\frac{A_a \rho_{0,a} c_{0,a}}{A_b \rho_{0,b} c_{0,b}}}. \tag{6.10}$$

For low Mach number flows, the time-averaged pressure is constant to $O(M^2)$, and so we can rewrite these expressions as

$$\frac{|\hat{p}_{1,b}|}{|\hat{p}_{1,a}|} = \sqrt{\frac{A_a}{A_b} \left(\frac{\gamma_b MW_b T_a}{\gamma_a MW_a T_b}\right)^{1/4}}, \tag{6.11}$$

$$\frac{|\hat{u}_{x,1b}|}{|\hat{u}_{x,1a}|} = \sqrt{\frac{A_a}{A_b} \left(\frac{\gamma_a MW_a T_b}{\gamma_b MW_b T_a}\right)^{1/4}}. \tag{6.12}$$

This shows that both the unsteady velocity and pressure amplitudes decrease as the area increases. It also shows that increases in temperature cause reductions in pressure amplitude and increases in velocity amplitude. However, area changes are a square root effect, while the temperature effect is a fourth root, suggesting that area effects are dominant, even in combustor environments with strong temperature variations.

Another way of interpreting this latter observation is to refer to the following expression for acoustic energy density from Eq. (2.64):

$$E_\Lambda = \frac{1}{2}\rho_0 u_{x,1}^2 + \frac{1}{2}\frac{p_1^2}{\rho_0 c_0^2}. \tag{6.13}$$

In a plane traveling wave, the pressure and velocity are related through the relation $p_1 = \rho_0 c_0 u_{x,1}$. Inserting this expression into Eq. (6.13), it can be seen that the potential and kinetic energy density are equal. In other words, the acoustic energy density is equipartitioned between the kinetic and potential energy term for a traveling plane wave. Changes in mean properties, such as density or γ, cause corresponding changes in the relative magnitudes of $|p_1|$ and $|u_{x,1}|$ in such a way that the kinetic and potential energy of the wave remain equal.

The third effect noted earlier is the advection of acoustic waves due to mean flow. These processes are described by the advective wave equation, Eq. (2.30), written here for the case where the mean flow velocity and gas properties are constant:

$$\frac{D_0^2 p_1}{Dt^2} - c_0^2 \nabla^2 p_1 = 0, \tag{6.14}$$

where

$$\frac{D_0}{Dt}(\ ) = \frac{\partial}{\partial t}(\ ) + \vec{u}_0 \cdot \nabla(\ ). \tag{6.15}$$

If we assume that the uniform mean flow is oriented in the axial direction, i.e., that $\vec{u}_0 = (u_{x,0}, 0, 0)$, then

$$\frac{\partial^2 p_1}{\partial t^2} + 2u_{x,0}\frac{\partial^2 p_1}{\partial x \partial t} + u_{x,0}^2 \frac{\partial^2 p_1}{\partial x^2} - c_0^2 \nabla^2 p_1 = 0. \tag{6.16}$$

Incorporation of mean flow and gas property and temperature changes simultaneously is quite complex, so the discussion in this chapter will for the most part consider each separately, with the exception of Section 6.2.3 which discusses acoustic/entropy coupling.

The fourth effect noted earlier is the amplification/damping of sound waves. We will reserve discussion of this point to Section 6.5.

6.2 Mean Flow Effects

This treatment of mean flow effects is divided into two sections: mean flow effects on wave propagation within the domain in Section. 6.2.1, and on boundary conditions in Section 6.2.2.

6.2.1 Mean Flow Effects on Wave Propagation

The effects of a uniform mean flow on wave propagation are quite straightforward and simply lead to advection of the sound wave at the flow velocity. The time domain solution of the one-dimensional form of the advected wave equation, Eq. (6.16), is given by

$$p_1 = F_f\left(t - \frac{x}{c_0 + u_{x,0}}\right) + F_g\left(t + \frac{x}{c_0 - u_{x,0}}\right), \tag{6.17}$$

$$u_{x,1} = \frac{1}{\rho_0 c_0}\left[F_f\left(t - \frac{x}{c_0 + u_{x,0}}\right) - F_g\left(t + \frac{x}{c_0 - u_{x,0}}\right)\right]. \tag{6.18}$$

Note that the propagation velocity of the right- and leftward-moving waves are now $c_0 + u_{x,0}$ (or $c_0(1 + M_0)$) and $c_0 - u_{x,0}$, respectively. This result shows that the phase speed is altered by the component of the bulk velocity in the direction of wave propagation. This result also shows that advective effects are an $O(M)$ correction to that developed without flow, and thus shows why mean flow effects on wave propagation are often neglected in low Mach number flows.

Consider next the effect of an axial mean flow, $\vec{u}_0 = (u_{x,0}, 0, 0)$, on the more general problem of multidimensional, harmonically oscillating disturbances. As before, write the unsteady pressure for a rectangular domain as

$$\hat{p}_1(x, y, z) = \hat{X}(x)\hat{Y}(y)\hat{Z}(z). \tag{6.19}$$

Substituting this expression into Eq. (6.16) leads to the same expressions for $\hat{Y}(y)$ and $\hat{Z}(z)$ as derived in Section 5.3 (assuming that the same wall boundary conditions are used) and the following equation for $\hat{X}(x)$:

$$\frac{\partial^2 \hat{X}}{\partial x^2} + \frac{2ikM}{(1 - M^2)}\frac{\partial \hat{X}}{\partial x} + \frac{(k^2 - k_x^2)}{(1 - M^2)}\hat{X} = 0, \tag{6.20}$$

whose solution is

$$\hat{X}(x) = \mathcal{A}\exp\left(ik_x^+ x\right) + \mathcal{B}\exp\left(-ik_x^- x\right) \tag{6.21}$$

where

$$k_x^\pm = k\frac{\pm M + \sqrt{1 - \left(k_y^2 + k_z^2\right)(1 - M^2)/k^2}}{(1 - M^2)}. \tag{6.22}$$

Thus, the structure of the three-dimensional disturbance field is largely the same as discussed earlier with the addition of wave advection effects. These wave advection effects can be seen explicitly by writing out the solution for the purely axial modes:

$$\hat{X}(x) = \mathcal{A}\exp\left(\frac{ikx}{1 + M}\right) + \mathcal{B}\exp\left(\frac{-ikx}{1 - M}\right). \tag{6.23}$$

Consider the problem where $\mathcal{A} = \mathcal{B}$, i.e., the left- and rightward waves have equal amplitudes. The corresponding axial and temporal dependence of the pressure is then given by

Table 6.1 Summary of advection effects on natural frequencies for different boundary conditions

Boundary condition	Axial frequency, f
Rigid–Rigid (or) Pressure release–Pressure release	$f_n = \dfrac{nc_0}{2L}(1 - M^2)$
Rigid–Pressure release (or) Pressure release–Rigid	$f_n = \dfrac{(2n-1)c_0}{4L}(1 - M^2)$
Periodic	$f_n = \dfrac{nc_0}{L}(1 \pm M)$

$$\hat{X}(x)\exp(-i\omega t) = 2\mathcal{A}\cos\left(\frac{kx}{1-M^2}\right)\exp\left(-i\omega\left(t + \frac{M}{1-M^2}\frac{x}{c_0}\right)\right). \quad (6.24)$$

Two effects can be seen in this expression. First, the axial disturbance field is not a purely standing wave, as it is when $M = 0$. The phase varies axially, as shown explicitly here by including the time component. In other words, the resulting "standing wave" is not really standing, it has an $O(M)$ axially varying phase.

Next, it can be seen by paralleling the development in Section 5.3 that the expressions for natural frequencies are altered. For purely axial disturbances with different boundary conditions, it is left as an exercise (Exercise 6.3) to prove the expressions shown in Table 6.1.

Thus, the effect of mean flow for the first two sets of boundary conditions is to decrease the natural frequency of each mode by a factor of $(1 - M^2)$. For the periodic boundary condition, the natural frequency of the forward- and backward-propagating traveling waves is different and an $O(M)$ effect.

6.2.2 Mean Flow Effects on Boundary Conditions

Having considered mean flow effects on wave propagation within the domain, which introduce $O(M)$ effects on the phase distribution and $O(M^2)$ effects on natural frequencies (for nonperiodic boundary conditions), we next discuss flow effects on boundary conditions. It is at the boundaries of the domain where mean flow generally exerts the largest influence on low Mach number flow acoustic problems. First, note that there are acoustic losses even if the reflection coefficient magnitude is unity. This can be seen by calculating the time-averaged acoustic intensity flux (see Eq. (2.68)), I_{ave}, through a boundary with reflection coefficient R. Assuming an incident disturbance with associated time-averaged intensity I_{inc}, the time-averaged energy flux through the boundary is given by

$$\frac{I_{av}}{I_{inc}} = \frac{(1+M)^2 - |R|^2(1-M)^2}{(1+M)^2}. \quad (6.25)$$

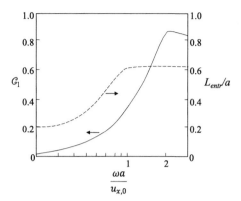

Figure 6.3 Strouhal number dependence of the magnitude and phase of the mean flow correction (defined in Eq. (6.27)) to the reflection coefficient from the end of a thin-walled pipe exit [3].

Substituting $R = 1$ leads to

$$\frac{I_{av}}{I_{inc}} = \frac{4M}{(1+M)^2}. \tag{6.26}$$

This expression shows that, for example, 20% of the incident energy on the boundary is advected out of the domain for a $M = 0.05$ flow. Thus, one study found that acoustic resonances in a pipe were almost completely suppressed for flow Mach numbers greater than about 0.4 [2].

In reality, the exit reflection coefficient, R, is altered by mean flow. An approximate expression for the magnitude of R is

$$|R| = (1 + G_1 M)(1 - G_2(ka)^2). \tag{6.27}$$

The values of these constants are a function of the shape of the pipe exit, such as wall thickness. For a thin-walled, unflanged pipe, $G_2 = 0.5$, as already discussed in the context of Eq. (5.50). As shown in Figure 6.3, the coefficient G_1 is a function of the Strouhal number, $\omega a / u_{x,0}$, and varies from $G_1 \sim 0$ ($\omega a / u_{x,0} \ll 1$) to 0.9 ($\omega a / u_{x,0} \gg 1$) [3]. Thus, for this example, advection effects are typically the leading-order pipe end loss mechanism for $M/(ka)^2 \geq 1$. The figure also shows the effects of flow on the reflection coefficient phase, as quantified by the pipe end correction, L_{entr}.

Due to modulation of the unsteady separating flow at the pipe exit, this boundary condition is also nonlinear, an effect to be discussed in Section 6.3.4.

6.2.3 Mean Flow Compressibility Effects and Acoustic/Entropy Coupling

Although the focus of this chapter is on acoustic waves, important entropy–acoustic interactions occur in high Mach number flows with varying temperature and/or Mach number, as discussed in Section 2.3.2. These points can be seen by writing the sourceless, linearized, one-dimensional momentum and energy equations in terms of pressure, velocity, and entropy perturbations:

$$\rho_0\left(\frac{\partial u_{x,1}}{\partial t} + u_{x,0}\frac{\partial u_{x,1}}{\partial x} + u_{x,1}\frac{du_{x,0}}{dx}\right) + \left(\frac{p_1}{c_0^2} - \frac{\rho_0}{c_p}s_1\right)\left(u_{x,0}\frac{du_{x,0}}{dx}\right) + \frac{\partial p_1}{\partial x} = 0, \quad (6.28)$$

$$\frac{\partial p_1}{\partial t} + u_{x,0}\frac{\partial p_1}{\partial x} + u_{x,1}\frac{dp_0}{dx} + \gamma p_1\frac{du_{x,0}}{dx} + \gamma p_0\frac{\partial u_{x,1}}{\partial x} = 0, \quad (6.29)$$

$$\frac{\partial s_1}{\partial t} + u_{x,0}\frac{\partial s_1}{\partial x} + u_{x,1}\frac{ds_0}{dx} = 0. \quad (6.30)$$

If one neglects the $u_{x,0}\frac{du_{x,0}}{dx}$ term, then Eqs. (6.28) and (6.29) can be solved for the unsteady pressure and velocity, p_1 and $u_{x,1}$, independent of the entropy. Note that the $u_{x,0}\frac{du_{x,0}}{dx}$ term is of $O(M^2)$ relative to other terms, and describes the effect of compressibility in the nominal flow. This is equivalent to saying that the acoustic field can be solved independent of the entropy field if $O(M^2)$ terms are neglected. The inverse is not true – calculating the entropy field requires a calculation of the acoustic field if there is a gradient in s_0, through the $u_{x,1}\frac{ds_0}{dx}$ term. However, if $O(M^2)$ terms are included, acoustic wave fields cannot be calculated without simultaneously solving for the entropy disturbance field simultaneously, and vice versa. For example, the acceleration of entropy fluctuations leads to the generation of sound waves. This specific example was previously discussed in the context of Eq. (2.53) and more fully in Sections 6.3.2 and 6.10.

6.2.4 Variable Temperature Effects [4–6]

The qualitative effects of gas property changes have been treated in the overview discussion at the beginning of Section 6. The intention of this section is primarily to present some explicit quantitative results for problems with constant area and no mean flow. In order to simplify the presentation, we will assume that the ratio of specific heats, γ, is constant, which implies that $\rho_0 c_0^2$ is a constant in a low Mach number flow, and focus specifically on temperature variations. Temperature variations are typically the strongest effect in combusting environments, as variations in molecular weight and specific heats ratio are substantially less significant. Neglecting mean flow, the governing equation is then given by Eq. (6.1). For harmonic oscillations,

$$\nabla^2 \hat{p}_1 + \frac{1}{T_0}\nabla T_0 \cdot \nabla \hat{p}_1 + \frac{\omega^2}{c_0^2}\hat{p}_1 = 0. \quad (6.31)$$

We will consider problems where the temperature gradient is in the axial direction only, i.e., $\nabla T_0 = (dT_0/dx, 0, 0)$. Again, expressing the pressure as $\hat{p}_1(x,y,x) = \hat{X}(x)\hat{Y}(y)\hat{Z}(z)$ and substituting leads to the following equation for $\hat{X}(x)$:

$$\frac{d^2\hat{X}}{dx^2} + \frac{1}{T_0}\frac{dT_0}{dx}\frac{d\hat{X}}{dx} + (k^2 - k_z^2 - k_y^2)\hat{X}(x) = 0, \quad (6.32)$$

where the equations for $\hat{Y}(y)$ and $\hat{Z}(z)$ are not altered by the temperature gradient since we have assumed that the temperature gradient is in the axial direction only.

This expression does not have a general solution, but fairly general solution characteristics of this type of problem are well known and discussed further in Section 6.7. We can develop approximate, quantitative solutions in the limiting cases where the length scale associated with the temperature variation is large or small relative to an acoustic wavelength. If the temperature variation occurs very gradually, we can use the approximate solutions in Section 6.8. If the temperature change occurs over a very short length scale, such as in an acoustically compact flame, we can write the following linearized jump conditions across the region, as shown in Section 6.9:

$$\text{Momentum: } [p_1] = 0, \tag{6.33}$$

$$\text{Energy: } [u_{x,1}] = 0. \tag{6.34}$$

These jump conditions show that the unsteady pressure and axial velocity remain continuous across a discontinuous temperature change.

6.2.5 Example Problem: Wave Reflection and Transmission through Variable Temperature Region

We will next work in detail through a model problem where a sound wave is incident on a temperature gradient. Here, the temperature increases linearly from $T_{0,a}$ to $T_{0,b}$ over a length L, see Figure 6.4, and we shall consider various exact and approximate solutions. The assumed temperature field is

$$T_0(x) = T_{0,a} + (T_{0,b} - T_{0,a})\frac{x}{L}. \tag{6.35}$$

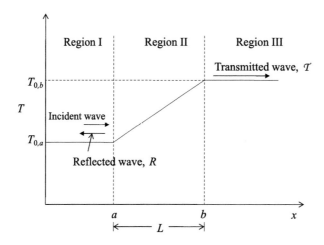

Figure 6.4 Assumed temperature distribution for model problem.

The exact solution for the pressure and velocity for this temperature profile is given by [4, 7]:

$$\hat{p}_1 = C_1 J_0 \left(\frac{2\omega L(1+\beta x/L)^{1/2}}{|\beta| c_{0,a}} \right) + C_2 Y_0 \left(\frac{2\omega L(1+\beta x/L)^{1/2}}{|\beta| c_{0,a}} \right), \quad (6.36)$$

$$\hat{u}_{x,1} = \frac{-i\beta/|\beta|}{\rho_0(x)c_0(x)} \left(C_1 J_1 \left(\frac{2\omega L(1+\beta x/L)^{1/2}}{|\beta| c_{0,a}} \right) + C_2 Y_1 \left(\frac{2\omega L(1+\beta x/L)^{1/2}}{|\beta| c_{0,a}} \right) \right), \quad (6.37)$$

where

$$\beta = \frac{(T_{0,b} - T_{0,a})}{T_{0,a}} \quad (6.38)$$

and J and Y denote Bessel functions of the first and second kind, respectively. Due to the impedance change, the incident wave is partially reflected and transmitted. Consider the dependence of the reflection and transmission coefficients as a function of the temperature jump and length scale. Start first with the exact solution. The pressure and velocity fields in Regions I, II, and III (see Figure 6.4) are given by

Region I:

$$\hat{p}_1 = \exp(ik_a x) + R \exp(-ik_a x), \quad (6.39)$$

$$\hat{u}_{x,1} = \frac{\exp(ik_a x) - R \exp(-ik_a x)}{\rho_{0,a} c_{0,a}}; \quad (6.40)$$

Region II:

$$\hat{p}_1 = C_1 J_0 \left(\frac{2k_a L(1+\beta x/L)^{1/2}}{|\beta|} \right) + C_2 Y_0 \left(\frac{2k_a L(1+\beta x/L)^{1/2}}{|\beta|} \right), \quad (6.41)$$

$$\hat{u}_{x,1} = \frac{-i\beta/|\beta|}{\rho_0(x)c_0(x)} \left(C_1 J_1 \left(\frac{2k_a L(1+\beta x/L)^{1/2}}{|\beta|} \right) + C_2 Y_1 \left(\frac{2k_a L(1+\beta x/L)^{1/2}}{|\beta|} \right) \right); \quad (6.42)$$

Region III:

$$\hat{p}_1 = T \exp(ik_b x), \quad (6.43)$$

$$\hat{u}_{x,1} = \frac{T \exp(ik_b x)}{\rho_{0,b} c_{0,b}}. \quad (6.44)$$

Matching the unsteady pressure and velocity at each interface leads to four simultaneous equations for R, T, C_1, and C_2, derived in Exercise 6.4. The exact solution of these equations can be compared with approximate solutions. If $L/\lambda \ll 1$, we can apply the jump conditions, Eqs. (6.33) and (6.34), to couple the solutions in Regions I and III to obtain

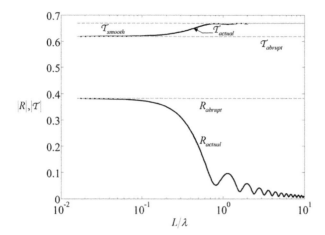

Figure 6.5 Dependence of the reflection and transmission coefficients of a plane wave incident on a temperature gradient ($\beta = 4$).

$$R_{abrupt} = \frac{(T_{0,a}/T_{0,b})^{1/2} - 1}{(T_{0,a}/T_{0,b})^{1/2} + 1} = \frac{(1+\beta)^{-1/2} - 1}{(1+\beta)^{-1/2} + 1}, \qquad (6.45)$$

$$T_{abrupt} = \frac{2(T_{0,a}/T_{0,b})^{1/2}}{(T_{0,a}/T_{0,b})^{1/2} + 1} = \frac{2(1+\beta)^{-1/2}}{(1+\beta)^{-1/2} + 1}. \qquad (6.46)$$

Similarly, we can use the $L/\lambda \gg 1$ limit derived in Eq. (6.11) to derive the asymptotic values

$$R_{smooth} = 0, \qquad (6.47)$$

$$T_{smooth} = (1+\beta)^{-1/4}. \qquad (6.48)$$

These different calculations are compared in Figure 6.5 by plotting the exact (see Exercise 6.4) and approximate solutions for R and T as a function of L/λ. Note how the reflection and transmission coefficients both asymptote toward the predicted "abrupt" and "smooth" transition limits for small and large L/λ values, respectively. For intermediate L/λ values, both R and T show undulations, due to interference patterns set up within the inhomogeneous temperature region by wave reflections.

6.2.6 Example Problem: Natural Frequencies of a Variable Temperature Region

Next, we consider exact and approximate solutions for the natural frequencies of a variable temperature region. Consider the natural frequencies of a duct of length L, with the linear temperature variation given by Eq. (6.35), and pressure release and rigid boundary conditions. Using the exact solutions for the pressure and velocity from Eqs. (6.36) and (6.37), we will consider four different sets of boundary conditions associated with pressure release and rigid-wall boundary conditions. The characteristic equations for the natural frequencies are derived in Exercise 6.5. The exact

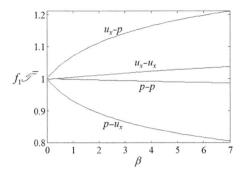

Figure 6.6 Dependence of the first natural frequency of a variable temperature duct, normalized by the time of flight, on the temperature gradient, where p–p denotes $\hat{p}_1(x=0) = \hat{p}_1(x=L) = 0$, $p - u_x$ denotes $\hat{p}_1(x=0) = \hat{u}_{x,1}(x=L) = 0$, and so forth.

solutions for the frequencies can also be compared to approximate solutions for gradually varying temperature changes, see Section 6.8. This approximate solution predicts that the fundamental natural frequency, f_1, is the inverse of the "round trip" time required for the disturbance to propagate from one end of the system to the other end and back, $\bar{\mathscr{T}}$. For example, for a duct with $\hat{p}_1(x=0) = \hat{p}_1(x=L) = 0$ boundary conditions, the period of oscillations is

$$\bar{\mathscr{T}} = 2 \int_0^L \frac{dx}{c_0(x)}, \qquad (6.49)$$

which for a constant sound speed is simply $\bar{\mathscr{T}} = 1/f_1 = 2L/c_0$.

Figure 6.6 shows the dependence of the normalized natural frequency $f_1 \bar{\mathscr{T}}$ on the temperature gradient. If the natural frequency is exclusively controlled by the sound wave propagation time, then $f_1 \bar{\mathscr{T}}$ will identically equal unity, as it does for $\beta = 0$. From this solution we can see that the natural frequency is close, but not exactly equal, to the inverse of the sound "round trip" time. Thus, it is clear that the wave travel time is a very suitable first approximation for natural frequencies.

In addition, the natural modes in the inhomogeneous temperature duct are not exactly integer multiples. We have already seen this behavior for transverse modes of circular ducts (see Table 5.3). This problem is worked in Exercise 6.6, which shows, for example, that f_2/f_1 has a value of 3.0 for the uniform temperature problem, but increases to ~3.5 for $\beta = 4.5$. This has important influences on the nonlinear acoustic processes encountered for high disturbance amplitude waves, see Section 6.6.2.1.

6.3 Variable Area and Complex Geometry Effects

This section discusses quasi-one-dimensional wave propagation through variable area regions. In certain instances, mean flow Mach numbers are quite low, and thus these effects can be analyzed quite accurately by neglecting advective effects. This analysis is considered in Section 6.3.1. However, if the area change is significant enough, the flow acceleration can lead to Mach numbers of $O(1)$ and accompanying changes to mean density, so that additional considerations are necessary. An important instance where this effect is important is in choked nozzles, and is discussed in Section 6.3.2. Finally, sharp area changes introduce the possibility of flow separation and vorticity generation, which is discussed in Section 6.3.3.

6.3.1 Baseline Results

We start by neglecting mean flow and assume a homogeneous medium. In this case, the wave equation can be deduced from Eq. (6.2). Area changes lead to wave reflections and changes in amplitude of the disturbance field (see Section 6.1). In order to illustrate these features explicitly, we will consider the coupling of two ducts with an abrupt area change, sketched in Figure 6.7 (see Exercise 6.9 for a continuous area change problem). This geometry consists of two ducts with lengths of L_a and L_b, cross-sectional areas of A_a and A_b, and rigid boundary conditions.

The solution in each duct can then be written as elementary plane waves, Eq. (5.15). The coefficients and natural frequency can be determined by applying the two boundary conditions and coupling the motions at the interface using the approximate matching conditions at the interface, see Section 6.9:

$$\text{Momentum: } [p_1] = 0, \quad (6.50)$$

$$\text{Energy: } [Au_{x,1}] = 0. \quad (6.51)$$

We will omit the details, but solving the four coupled equations leads to the following transcendental characteristic equation for the natural frequency:

$$\cos(kL_b)\sin(kL_a) + \left(\frac{A_b}{A_a}\right)\cos(kL_a)\sin(kL_b) = 0. \quad (6.52)$$

Computed solutions of this equation are plotted in several ways in Figure 6.8, where the natural frequency f_3 (i.e., the third natural mode of the duct system) is normalized by the ratios of the individual and total duct lengths, L_a/λ, L_b/λ, and $(L_a + L_b)/\lambda$. Values of 0.25 and 0.5 correspond to quarter- and half-wave modes, for instance. In the trivial case where $A_b/A_a = 1$, the natural modes of this system are integer multiples of 0.5, i.e., $(L_a + L_b)/\lambda = n/2$. The plots at an area ratio of $A_b/A_a = 2$ show related

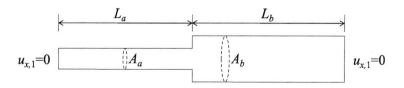

Figure 6.7 Illustration of the example problem with two coupled ducts with an abrupt area change.

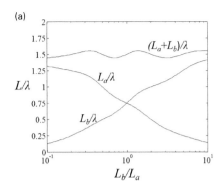
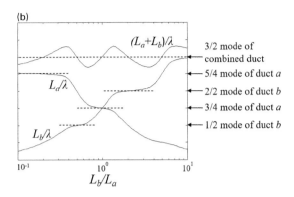

Figure 6.8 Dependence of the third natural mode of a connected duct problem on L_b/L_a for a duct with an area ratio of (a) $A_b/A_a = 2$ and (b) $A_b/A_a = 10$.

behavior, with only slight deviation from $(L_a + L_b)/\lambda = 3/2$. Thus, this nonnegligible area change has only a moderate influence on the natural frequencies, and as such could be approximated quite closely by simple time-of-flight arguments. This same conclusion was reached for temperature variation effects in Section 6.2.6. Moving to a larger area ratio, $A_b/A_a = 10$, requires a more involved discussion, as it shows multiple ways to interpret the natural frequency. The natural modes continue to remain reasonably close to $(L_a + L_b)/\lambda = 3/2$, although the deviations are larger.

Another interpretation of these results can be achieved by relating these natural frequencies to the natural modes of the *individual* pipe sections. This can be understood by considering, for example, modal oscillations in the large-diameter duct. If the area of duct a is very small, this junction resembles a rigid termination and natural frequencies exist which are essentially independent of the length of the smaller duct. In this case, the natural modes occur at frequencies of $L_b/\lambda \sim n/2$, i.e., half-waves of the smaller duct. In contrast, the rapid area expansion resembles a pressure release boundary condition to the smaller duct, so natural modes in this system are close to 1/4, 3/4, or higher-order modes. With this in mind, note from Figure 6.8 that the natural frequencies in some ranges of L_b/L_a are closely related to multiples of the quarter-wave frequencies of the smaller duct, $L_a/\lambda = 3/4$ and 5/4. In contrast, values associated with half-wave modes, $L_b/\lambda \sim 1/2$, 2/2, 3/2, of the larger-area duct are also seen for other L_b/L_a ranges. As such, for large enough area ratios, although the natural frequencies are system modes, they can be reasonably approximated as natural modes of individual pipe sections. The reader is referred to Exercises 6.7 and 6.8 for further results from this and a related problem. Exercise 6.8 works through a problem with mixed pressure release/rigid boundary conditions where time of flight approximations are not valid at the lowest natural frequency, as this geometry leads to a Helmholtz mode (see Section 5.3.4).

6.3.2 Nozzles and Diffusers

Consider the different problems sketched in Figure 6.9. For the nozzle flow, we denote the flow Mach number entering the nozzle as M_a, and exiting the nozzle as M_b. For a

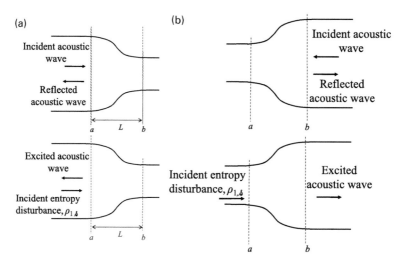

Figure 6.9 Illustration of (a) nozzle and (b) diffuser geometries used in discussion.

choked nozzle, $M_b = 1$. For the diffuser, M_a and M_b denote the Mach number entering and exiting it, respectively.

The conditions relating the fluctuating properties at stations a and b can be obtained by integrating the governing equations through the property change. Explicit results can be obtained in cases where the nozzle/diffuser element is quasi-one-dimensional and compact – i.e., when the flux of fluctuating mass, momentum, or energy at stations a and b are equal at each instant of time because of negligible volume accumulation. Approximate nozzle compactness criteria for incident acoustic and entropy waves are $kL(1 - M_a) \ll 1$ and $kL/M_a \ll 1$, respectively.

We start with a choked, compact, isentropic nozzle. Assuming an isentropic base flow, the mass and energy equations can be combined to show that the following relation holds at all axial positions between stations a and b [8, 9]:

$$\frac{u_{x,1}}{c_0} - \frac{M p_1}{2 p_0} + \frac{M \rho_1}{2 \rho_0} = 0. \tag{6.53}$$

This expression can be used as a boundary condition for the disturbance field in the region upstream of station a. For example, if $\rho_1 = 0$ in the approach flow and $M_b = 1$, this expression can be rearranged into the following impedance condition for sound waves propagating in the upstream regions (see Exercise 6.11):

$$\left.\frac{Z}{\rho_0 c_0}\right|_a = \frac{2}{(\gamma - 1) M_a}. \tag{6.54}$$

In a low Mach number flow, the leading-order expression for the reflection coefficient is then

$$R = 1 - (\gamma - 1) M_a + O(M_a^2). \tag{6.55}$$

This impedance is purely real because of the compactness assumption – this implies that the phase of the reflected wave is the same as the incident disturbance, but reduced in magnitude. If the finite length of the nozzle is accounted for, imaginary terms are present which describe phase delays in the reflected wave [10]. This equation also shows that for low approach flow Mach numbers, $M_a \ll 1$, the choked nozzle resembles an $R = 1$ rigid-wall boundary. As such, using a rigid-wall boundary condition at the nozzle entrance, station a, is a good first approximation.

The resistive nature of the nozzle impedance implies some sound wave transmission through the nozzle. In addition, acoustic energy is advected through the nozzle. Both imply "dissipation" of the sound field within the domain. The magnitude of energy losses can be calculated by determining the net acoustic energy flux through station a. For example, consider an acoustic disturbance incident on the nozzle from upstream. Combining Eqs. (6.25) and (6.54), the time-averaged energy flux (see Eq. (2.68)) out of the nozzle, normalized by the intensity of the incident wave, is

$$\frac{I_{ave}}{I_{inc}} = 1 - \left(\frac{2-(\gamma-1)M_a}{2+(\gamma-1)M_a}\frac{1-M_a}{1+M_a}\right)^2. \tag{6.56}$$

For $M_a \ll 1$, the leading-order term is

$$\frac{I_{ave}}{I_{inc}} = 2(\gamma+1)M_a + O(M_a^2). \tag{6.57}$$

Utilizing this expression shows that, for example, if $\gamma = 1.4$ and $M_a = 0.05$, then 24% of the acoustic energy incident on the nozzle is radiated or advected out of the system.

Returning to the compact nozzle equation, Eq. (6.53), consider the sound generation by an advected entropy fluctuation through a choked nozzle. Assuming an incident entropy fluctuation $s_1/c_p = -\rho_{1,s}/\rho_0$, the backward-propagating acoustic wave can be calculated from Eq. (6.53) as

$$\frac{\rho_{1,\Lambda}}{\rho_{1,s}} = \frac{p_{1,\Lambda}/c_0^2}{\rho_{1,s}} = \frac{M_a}{2+(\gamma-1)M_a}. \tag{6.58}$$

The compactness requirement for this equation to be valid, i.e., $kL/M_a \ll 1$, is quite severe and will generally not be satisfied. In the more general case, a fairly significant level of phase cancellation can occur through the nozzle due to the rapidly varying phase of the advecting entropy disturbance [9]. Figure 6.10 presents data illustrating sound transmission through the nozzle by a convected entropy disturbance, illustrating the increased sound production with Mach number as suggested by the theory. Also plotted on the graph are the predictions of the quasi-one-dimensional, compact nozzle theory.

While the above results were derived assuming a choked nozzle, Section 6.10 presents more general expressions for compact, unchoked nozzles ($M_b > M_a$) or diffusers ($M_b < M_a$).

6.3.3 Wave Refraction and Injector Coupling [12, 13]

As noted in Section 2.2.3, acoustic oscillations bend around obstacles, propagate into side branches, and spread out when propagating into a larger-area section, all

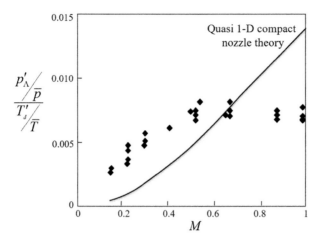

Figure 6.10 Measured ratio of the amplitude of a sound wave transmitted through a nozzle, $p'_A/\bar{p}$, to that of the incident entropy disturbance, $T'_s/\bar{T}$, as a function of the nozzle Mach number. Adapted from Bake et al. [11].

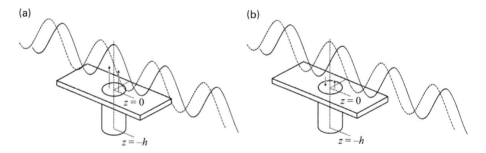

Figure 6.11 Schematic of transverse pressure fluctuations as experienced by nozzle for pressure antinode (a) and pressure node (b) forcing cases. Solid and dashed lines denote fluctuating pressure and transverse velocity, respectively.

examples of refraction. Combustion systems have a variety of multidimensional features where wave refraction occurs, such as waves propagating from a smaller duct into a larger one, or propagating sideways into injectors or mixing passages (e.g., see Figure 2.2) during transverse oscillations. This problem is illustrated in Figure 6.11, showing a nearly one-dimensional wave aligned in one direction (referred to as the transverse direction in this section), that induces acoustic flow normal to it (referred to as axial direction). This figure shows two configurations for the acoustic wave inducing the motions, where a nozzle is situated near the velocity node and another near the pressure node.

These transverse oscillations lead to axial acoustic flow oscillations in the nozzle, referred to as "injector coupling," where acoustic motions induce flow rate oscillations in the injection/premixing sections. They also excite the hydrodynamic instabilities of the nozzle jet flow, as discussed in Section 4.4.4. An important limiting case is where the nozzle cross-sectional area is small relative to that of the main passage. In this limit, the pressure and axial velocity at the nozzle outlet are controlled by the disturbance waveform in the absence of the nozzle, i.e., the acoustic field at the nozzle location if the nozzle were not actually there. More generally, one can solve for the wave motions in the main and side passages simultaneously using the "T-junction" relations derived in Section 6.9.

We can consider three limiting cases: "pressure antinode" (Figure 6.11(a)), "pressure node" (Figure 6.11(b)), and "traveling wave" scenarios. The pressure antinode case is accompanied by zero transverse velocity at the nozzle outlet centerline, but large pressure fluctuations, symmetrically distributed across the nozzle centerline. These pressure fluctuations lead to nearly axisymmetric, axial velocity disturbances. As discussed further in Section 12.3.1.3, these axial flow pulsations dominate the heat release response and so have a special significance for the combustor's stability. The transverse flow oscillations serve as the "clock" that controls the natural frequency of the wave motions and the structure of the acoustic field, but it is the induced axial velocity fluctuations that actually excite the heat release oscillations that, in turn, excite the transverse modes. An approximate expression for the amplitude of these axial velocity oscillations is derived in Exercise 6.10.

The pressure node case exhibits large transverse velocity fluctuations in the center of the combustor. Because of the centerline pressure node, the pressure fluctuations have a 180 degree phase difference on the two sides of the nozzle centerline. As shown in Figure 6.11(b), this leads to axial velocity fluctuations that are phased approximately 180 degrees apart on the left and right sides of the nozzle (but smaller in amplitude as the pressure fluctuations are small if the nozzle is acoustically compact). As noted in Section 4.4.4, this excites odd-order hydrodynamic modes, with a dominant $|m| = 1$. The third limiting case, the traveling wave, can be understood in the linear limit from the pressure node and antinode cases because a traveling wave is the complex superposition of two standing waves, $\exp(ikx) = \cos(kx) + i\sin(kx)$.

Figure 6.12 shows example data from both cases. The left images show the pressure-node case where the nozzle experiences significant transverse flow oscillations, and a relatively weak unsteady pressure field. Closer to the nozzle, however, an axial asymmetric breathing in and out of the nozzle cavity in the axial direction can be seen. This asymmetric forcing leads to the excitation of strong helical shear layer modes [13]. The right image shows the symmetric forcing case, where the nozzle is at a pressure antinode and acoustic velocity node. Near the nozzle, significant axisymmetric vortical velocity fluctuations are excited by bulk axial velocity fluctuations in and out of the nozzle, because of the large pressure oscillations. This leads to the strong excitation of axisymmetric shear layer modes.

In a combustor undergoing transverse oscillations, such as a can or annular system, different nozzles may be located at different points in the standing wave. Example data for an annular combustor demonstrating this point are shown in Figure 6.13, which shows a view looking towards the nozzle from the combustor end. These images show chemiluminescence fluctuations, clearly indicating axisymmetric flame

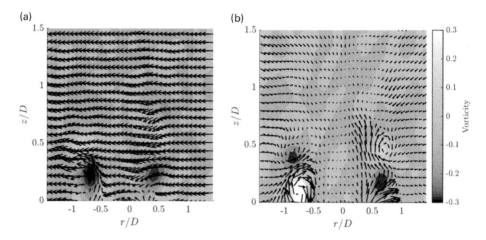

Figure 6.12 Experimental results of PIV measurements of velocity (vectors) and normalized vorticity (color) fluctuation for an annular nozzle located at a (a) pressure node (velocity antinode) and (b) pressure antinode. Shading denotes vorticity. Image courtesy of J. O'Connor [12].

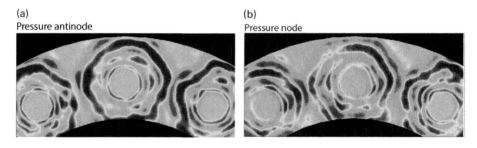

Figure 6.13 Examples of axisymmetric and azimuthally periodic flame oscillations in an annular combustor, viewed end on [14].

luminosity for a nozzle located at the pressure antinode case, and an azimuthally periodic disturbance for a nozzle located at a pressure node.

6.3.4 Unsteady Vorticity Generation and Acoustic Damping

Flow separation is another important effect during area changes. This leads to acoustic–vortical coupling and the damping of sound waves [15–17]. Moreover, it leads to significant sensitivity of the acoustic loss term to geometric details (e.g., fillet radii), flow velocities, and disturbance amplitudes, as discussed in this section. The basic flow processes are shown in Figure 6.14, which illustrates flow through an area change.

The basic ideas can be understood from steady flows, so we start the discussion there. Even in steady flows, the configuration shown in Figure 6.14(b) is known to

6.3 Variable Area and Complex Geometry Effects

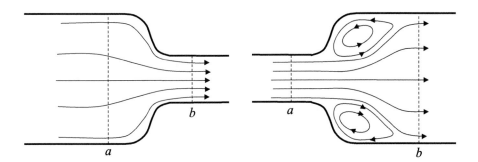

Figure 6.14 Illustration of (a) attached and (b) separating flow through an area change.

cause losses as the kinetic energy of the flow in the smaller-area channel is not completely converted into pressure in the larger-area section where the flow velocity is subsequently lower. For example, if there are no losses, such as might be approached in the flow configuration in (a), the stagnation pressure at stations b and a remains the same:

$$p_{T,b} = p_{T,a}. \tag{6.59}$$

For a low Mach number flow, this leads to

$$p_b + \frac{\rho_b u_{x,b}^2}{2} = p_a + \frac{\rho_a u_{x,a}^2}{2} \tag{6.60}$$

In contrast, losses occur when the flow separates. This situation can be analyzed by assuming that the flow exits as a jet, oriented purely in the axial direction, and that the transverse pressure is constant along the separation plane. With these assumptions, the inviscid momentum equation can be integrated to yield

$$(p_a - p_b)A_b + \rho_a u_{x,a}^2 A_a - \rho_b u_{x,b}^2 A_b = 0, \tag{6.61}$$

The dynamic pressure loss for a low Mach number flow is then given by

$$\frac{[p_b + \rho_b u_{x,b}^2/2] - [p_a + \rho_a u_{x,a}^2/2]}{\rho_a u_{x,a}^2/2} = 1 - 2\frac{A_a}{A_b} + \frac{\rho_a A_a^2}{\rho_b A_b^2}. \tag{6.62}$$

If the density remains constant, the right-hand side simplifies to $(1 - A_a/A_b)^2$. Thus, significant stagnation pressure losses occur when $A_b \gg A_a$.

These expressions can readily be generalized to a quasi-steady flow by repeating the procedures outlined in Section 6.9 for compact transitions. The pressure recovery jump condition, obtained by linearizing the instantaneous analogue of Eq. (6.60), yields, to $O(M)$,

$$\text{No loss:} \quad p_{1b} + \rho_{0b} u_{x,0b} u_{x,1b} = p_{1,a} + \rho_{0a} u_{x,0a} u_{x,1a}. \tag{6.63}$$

Similarly, the linearized instantaneous analogue of Eq. (6.61), to $O(M)$ is given by

$$\text{Separating flow:} \quad 2\rho_{0a} u_{x,0a} A_a (u_{x,1b} - u_{x,1a}) + (p_{1b} - p_{1a})A_b = 0. \tag{6.64}$$

Note that the inclusion of an $O(M^2)$ term raises the potential for acoustic–entropy coupling across the jump through terms such as $\rho_1 u_{x,0}^2$. These results show that the momentum jump condition is geometry specific, as opposed to the mass and energy conditions derived in Section 6.9.

There are two important generalizations of these results – non-quasi-steadiness and nonlinearity. Non-quasi-steadiness comes from two sources. First, accumulation terms lead to a phase lag between the fluctuations at the left and right faces of the control volume. This leads to an additional complex term in the jump conditions, leading to an "end correction," L_{eff}, also discussed in Section 6.9. Second, the separating flow dynamics themselves are non-quasi-steady. Assuming a fluid mechanic time scale of $D/u_{x,0}$ shows that this effect should scale with a Strouhal number, $St = fD/u_{x,0}$, and thus is more important at lower frequencies than the acoustic accumulation term.

Consider next nonlinear effects, which arise from the interaction between pressure, which is linear in disturbance amplitude, and momentum flux, which is nonlinear. For example, returning to Eq. (6.61) and retaining terms that are second order in disturbance amplitude introduces additional terms of the form $\rho_0 u_{x,1}^2$. In separating flows, these terms lead to losses that grow nonlinearly with amplitude. Moreover, additional nonlinear processes appear if flow reversal occurs. Flow reversal across the interface must always occur when there is no mean flow. In the presence of mean flow, reversal occurs if $u_{x,1} > u_{x,0}$. In this case, the flow has different types of behavior depending on whether it is flowing from the small to the large, or from the large to the small sections, as shown in Figure 6.15. For example, a common assumption in analyses of this problem is to use a separating flow condition during outflow, and an isentropic flow condition during inflow [2, 18].

In either case, this leads to a resistive term that is nonlinear in amplitude. With no mean flow, the problem is intrinsically nonlinear. To illustrate, Figure 6.16 plots the unsteady pressure drop across an area contraction [19], showing the monotonically increasing unsteady pressure difference (which equates to acoustic energy loss) with the amplitude of the fluctuation velocity.

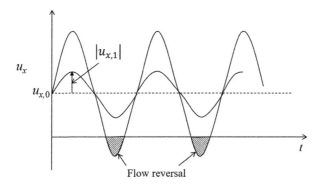

Figure 6.15 Illustration of instantaneous axial velocity for two velocity oscillation amplitudes, showing flow reversal at large-amplitude oscillations.

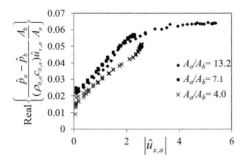

Figure 6.16 Data illustrating velocity amplitude dependence of the unsteady pressure difference across a rapid area discontinuity [19]. Subscripts a and b denote large and small pipe cross sections, respectively. Data courtesy of D. Scarborough.

Situations with and without mean flow are both of significant interest to combustion applications. Problems with mean flow have common application in flow from smaller-diameter nozzles or inlet sections into the larger-diameter combustors. Problems without mean flow have common application in the development of acoustic absorbers, such as Helmholtz resonators. In these applications, there is generally a cross flow velocity, which also influences the resistivity [20].

These effects are generally important sources of nonlinearity in combustors, as they become important when $u_{x,1} \sim O(u_{x,0})$. In contrast, acoustic or gas dynamic nonlinearities arising in the interior of the fluid domain become important when $u_{x,1} \sim O(c_0)$ [21], as also discussed in Section 6.6.2.1. Thus, these unsteady flow separation induced nonlinearities become important at much lower amplitudes than gas dynamic ones in low Mach number flows.

6.4 Acoustic Damping Processes

Although the topic of acoustic losses has been mentioned in several prior sections, it is useful to have a focused discussion of the topic. First, since energy is ultimately conserved, we must first define what is meant by "acoustic damping." From the perspective of a resonant sound field in a confined domain, damping is any process that decreases the acoustic energy at the *frequency band of interest* within the *domain of interest*. This can occur because of (1) transfer of acoustic energy to vortical or entropic disturbances, (2) transfer of acoustic energy by convection or radiation of acoustic energy out of the domain, or from (3) transfer of acoustic energy out of the frequency of interest within the domain. These different processes are illustrated in Figure 6.17.

Boundary layer losses are fundamentally associated with the transfer of acoustic energy into vortical and entropic fluctuations at walls due to the no-slip and thermal boundary conditions, as discussed in Section 2.3.1. Thus, a sound wave decays in the direction of propagation as $\exp(-\varsigma_{BL} x)$. An approximate expression for the decay coefficient ς_{BL} is [22]:

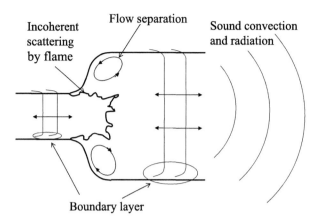

Figure 6.17 Schematic of acoustic damping processes.

$$\varsigma_{BL} = \frac{1}{c_0}\sqrt{\frac{\pi f v_v}{a^2}}\left(1 + \frac{\gamma - 1}{\sqrt{Pr}}\right). \tag{6.65}$$

This is valid for a no-slip, isothermal wall and the case where the laminar acoustic boundary layer thickness, $\delta_\Lambda \sim (f/v_v)^{1/2}$, is small relative to the duct radius, a. Note the square root dependence of this term on frequency, $\sqrt{f}$. Turbulent boundary layer results are more complex and depend on the relative thickness of the acoustic boundary layer to the viscous sublayer and log-layer [23].

Acoustic losses also occur due to vibrational relaxation, viscosity, and thermal conductivity outside of the boundary layer. Because the length scale associated with these relaxation processes is the acoustic wavelength, λ_Λ, rather than the acoustic boundary layer thickness, δ_Λ, this loss term is almost always negligible unless one is considering long-range or ultra-high-frequency sound propagation (e.g., 10s of kHz). This dissipation term has an f^2 scaling.

In addition, sound radiates out of the domain, is convected out of the domain, and is converted into vortical energy at points of flow separation, which can occur both within the domain of interest and at the boundaries. These effects are detailed in Sections 6.3.2 and 6.3.4. These types of losses, i.e., radiation, convection, and flow separation losses, are typically dominant over boundary layer losses in combustor-like geometries. For example, one study found that boundary layer losses become significant relative to these other losses only in ducts with length to diameter ratios greater than about 100 [2].

Last, we consider the "scattering" of acoustic energy out of the frequency of interest. Mechanisms that do not directly dissipate acoustic energy can transfer energy from the excited modes to other modes; e.g., as shown in Sections 2.1 and 6.6, nonlinear processes transfer energy from a certain frequency f_0 to higher harmonics ($2f_0, 3f_0, \ldots$) or subharmonics ($f_0/2, f_0/3, \ldots$). Particularly at these higher frequencies, acoustic energy is more readily dissipated by the previously discussed viscous and radiation mechanisms. Acoustic energy is also transferred from narrowband, coherent oscillations to spectrally distributed, incoherent oscillations by random modulation processes. For example, such spectral broadening occurs during reflection and scattering of a sound wave from a randomly flapping flame front [24]. The energy transfer mechanism is caused by a random Doppler shift of the reflected and transmitted waves. This reduces the amplitude of coherent acoustic disturbances reflected at the flame by a factor of $1 - 2(k\delta_T)^2$, where δ_T denotes the turbulent flame brush thickness [25]. The relative significance of these frequency scattering effects relative to other dissipation processes has not yet been assessed for combustion environments.

6.5 Unsteady Heat Release Effects

Oscillations in heat release generate acoustic waves. For unconfined flames, this is manifested as broadband noise emitted by turbulent flames [26–28]. For confined flames, these oscillations generally manifest themselves as discrete tones at the natural acoustic modes of the system [29–32]. The fundamental mechanism for sound generation is the unsteady gas expansion as the mixture reacts.

To illustrate, consider the wave equation with unsteady heat release by returning to the derivation of the wave equation in Chapter 2 and retaining the unsteady reaction terms in the energy equation, Eq. (1.48):

$$\frac{\partial^2 p_1}{\partial t^2} - c_0^2 \nabla^2 p_1 = (\gamma - 1) \frac{\partial \dot{q}_1}{\partial t}. \tag{6.66}$$

This equation neglects temperature gradients and mean flow. The acoustic source term describes sound wave production by heat release induced unsteady gas expansion. To illustrate the sound generation process, consider a combustion process in a free-field environment – i.e., an environment with no reflecting boundaries (this topic is also discussed in Section 12.4.3). The solution of Eq. (6.66) is [33]:

$$p_1(\vec{x},t) = \frac{\gamma-1}{4\pi c_0^2} \int_{\vec{y}} \frac{1}{|\vec{y}-\vec{x}|} \frac{\partial \dot{q}_1(\vec{y}, t - |\vec{y}-\vec{x}|/c_0)}{\partial t} d\vec{y}. \tag{6.67}$$

This expression shows that the unsteady heat release at each point, $\vec{y}$, induces sound proportional to the first derivative of the heat release rate, $\frac{\partial \dot{q}_1}{\partial t}$. Moreover, the acoustic pressure at the measurement point, $\vec{x}$, is related to the integral of the unsteady heat release over the combustion region, $\vec{y}$, at a retarded time $t - |\vec{y} - \vec{x}|/c_0$ earlier. Depending on the size of the combustion region relative to the acoustic wavelength, these different disturbances can arrive at nearly the same time (if the combustion region is small relative to a wavelength), or at significantly different times. In the latter case, significant phase cancellation/reinforcement can occur. If the combustion region is much smaller than an acoustic wavelength (i.e., it is acoustically compact), then disturbances originating from different points in the flame, $\vec{y}$, arrive at the observer point with negligible phase shift. In this case, Eq. (6.67) becomes

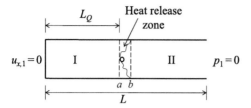

Figure 6.18 Illustration of the combustor used for the stability analysis model problem.

$$p_1(\vec{x}, t) = \frac{\gamma - 1}{4\pi c_0^2 a} \frac{\partial}{\partial t} \dot{Q}_1(t - a/c_0), \qquad (6.68)$$

where

$$\dot{Q}_1(t) = \int_{\vec{y}} \dot{q}_1(\vec{y}, t) d\vec{y} \qquad (6.69)$$

and a is some average distance between the combustion region and the observer. Equation (6.68) shows that in the compact flame case, the distribution of the heat release is unimportant; what matters is the total, spatially integrated value.

The effect of unsteady heat release in a confined environment is related, but has some key differences. Referring to Figure 6.18, a simple expression can be derived in the case where the flame region is acoustically compact. As shown in Section 6.9, unsteady heat release induces the following jump conditions:[1]

$$u_{x,1b} - u_{x,1a} = \frac{1}{A} \frac{\gamma - 1}{\gamma p_0} \dot{Q}_1, \qquad (6.70)$$

$$p_{1b} = p_{1a}. \qquad (6.71)$$

This expression neglects $O(M)$ effects and assumes that the duct cross-sectional area at interfaces a and b is the same, A. It shows that unsteady heat release causes a jump in acoustic velocity across the flame. This is due to the unsteady gas expansion associated with the heat release disturbance. The unsteady heat release exerts no influence on the pressure to leading order in flame compactness ratio [34].

Unsteady heat release causes both amplification/damping and shifts in phase of sound waves traversing the flame zone. The relative significance of these two effects depends on the relative phase of the unsteady pressure and heat release [35, 36]. As shown in Section 2.4, heat release disturbances in phase with the unsteady pressure cause amplification of the acoustic field. Heat release disturbances that are phased at 90 or 270 degrees with respect to the pressure shift the phase of sound waves traversing the heat release region. The first effect is typically the most important and

---

[1] Although not the primary focus of this chapter, unsteady heat release also generates vortical and entropic disturbances. Approximate jump conditions similar to those given above can also be derived, but require the much more restrictive assumption of a convectively compact flame zone.

can cause systems with unsteady heat release to exhibit self-excited oscillations. The second effect causes shifts in the natural frequencies of the system.

6.5.1 Thermoacoustic Stability Model Problem

In order to illustrate the effects of unsteady heat release explicitly, this section works through an example problem in detail, closely following McManus et al. [37]. Consider the combustor shown in Figure 6.18, consisting of two regions separated by a flame. We will assume rigid and pressure release boundary conditions at $x = 0$ and $x = L$, respectively, and that the flame is acoustically compact and located at $x = L_Q$.

Neglecting mean flow effects, the unsteady pressure and velocity in the two regions are given by (where, without loss of generality, we set each traveling wave's phase reference to the point at $x = L_Q$):

Region I:

$$p_1(x,t) = \left(\mathcal{A}_I e^{ik_I(x-L_Q)} + \mathcal{B}_I e^{-ik_I(x-L_Q)}\right)e^{-i\omega t},$$
$$u_{x,1}(x,t) = \frac{1}{\rho_{0,I} c_{0,I}}\left(\mathcal{A}_I e^{ik_I(x-L_Q)} - \mathcal{B}_I e^{-ik_I(x-L_Q)}\right)e^{-i\omega t}; \tag{6.72}$$

Region II:

$$p_1(x,t) = \left(\mathcal{A}_{II} e^{ik_{II}(x-L_Q)} + \mathcal{B}_{II} e^{-ik_{II}(x-L_Q)}\right)e^{-i\omega t},$$
$$u_{x,1}(x,t) = \frac{1}{\rho_{0,II} c_{0,II}}\left(\mathcal{A}_{II} e^{ik_{II}(x-L_Q)} - \mathcal{B}_{II} e^{-ik_{II}(x-L_Q)}\right)e^{-i\omega t}. \tag{6.73}$$

Applying the boundary/matching conditions leads to the following three algebraic equations:

$$\mathcal{A}_I e^{-ik_I L_Q} - \mathcal{B}_I e^{ik_I L_Q} = 0, \tag{6.74}$$

$$\mathcal{A}_{II} e^{ik_{II}(L-L_Q)} + \mathcal{B}_{II} e^{-ik_{II}(L-L_Q)} = 0, \tag{6.75}$$

$$\mathcal{A}_I + \mathcal{B}_I = \mathcal{A}_{II} + \mathcal{B}_{II}. \tag{6.76}$$

To obtain the fourth equation, we make use of the velocity jump condition across the flame as described in Eq. (6.70):

$$u_{II}(L_Q^+,t) - u_I(L_Q^-,t) = (\gamma - 1)\frac{\dot{Q}_1}{\rho_{0,I} c_{0,I}^2}, \tag{6.77}$$

where $\dot{Q}_1$ is the oscillatory heat release rate of the flame, defined in Eq. (6.69). Solving this problem requires specifying the form of the unsteady heat release. This problem is a key focus of Chapter 12, where it is shown that a good model (although not complete or perfect) for the heat release perturbations is that they are proportional to the velocity perturbation at the flame holder, with a time delay:

$$\dot{Q}_1 = A \frac{\rho_{0,I} c_{0,I}^2}{\gamma - 1} n u_{x,1}(x = L_Q, t - \tau). \tag{6.78}$$

This is a "velocity-coupled" flame response model; i.e., it assumes that the unsteady heat release is proportional to the unsteady flow velocity, multiplied by the gain factor, n, and delayed in time by τ. This time delay could originate from, for example, the convection time associated with a vortex that is excited by the sound waves. Using this model, Eq. (6.77) becomes

$$\frac{\rho_{0,I} c_{0,I}}{\rho_{0,II} c_{0,II}} (\mathcal{A}_{II} - \mathcal{B}_{II}) - (\mathcal{A}_I - \mathcal{B}_I)(1 + n e^{i\omega\tau}) = 0. \tag{6.79}$$

Solving the system of equations in Eqs. (6.74)–(6.76) and (6.79) leads to the characteristic equation

$$\frac{\rho_{0,I} c_{0,I}}{\rho_{0,II} c_{0,II}} \cos(k_I L_Q) \cos(k_{II}(L - L_Q)) - (1 + n e^{i\omega\tau}) \sin(k_I L_Q) \sin(k_{II}(L - L_Q)) = 0, \tag{6.80}$$

where $k_I = \omega/c_{0,I}$ and $k_{II} = \omega/c_{0,II}$.

This equation can be numerically solved for arbitrary parameter values. In order to obtain analytic solutions, we will next assume that the flame is located at the midpoint of the duct, i.e., $L = 2L_Q$, and that the temperature jump across the flame is negligible, so that $c_{0,I} = c_{0,II}$ and $\rho_{0,I} = \rho_{0,II}$. Also, denote $k_I = k_{II} = k$. Using this in Eq. (6.80) yields the characteristic equation

$$\cos kL = n e^{i\omega\tau} \sin^2(kL/2). \tag{6.81}$$

If $n = 0$, then there is no unsteady heat release, and the eigenvalues of the system are given by

$$k_{n=0}L = \frac{(2n-1)\pi}{2}. \tag{6.82}$$

Equation (6.81) is a transcendental expression and cannot be solved analytically in the general case. If $n \ll 1$ it can be solved approximately, by expanding the solution around the $n = 0$ result in a Taylor series. To first order in n, this leads to

$$k = k_{n=0}\left(1 - \frac{n \exp(i\omega_{n=0}\tau)}{2k_{n=0}L \sin k_{n=0}L} + O(n^2)\right). \tag{6.83}$$

Note that the wavenumber and frequency have an imaginary component, as opposed to prior problems where they were completely real. Considering the time component and expanding $\omega = \omega_r + i\omega_i$,

$$\exp(-i\omega t) = \exp(-i\omega_r t)\exp(\omega_i t), \tag{6.84}$$

showing that the imaginary component corresponds to exponential growth or decay in time and space. Thus, $\omega_i > 0$ corresponds to a linearly unstable situation, referred to as combustion instability. Conversely, $\omega_i < 0$ corresponds to a stable situation where unsteady heat release actually damps oscillations.

6.5 Unsteady Heat Release Effects

We can write out the real and imaginary parts of the frequency in Eq. (6.83) as

$$\omega_r = \omega_{n=0}\left[1 + (-1)^n \frac{n\cos(\omega_{n=0}\tau)}{(2n-1)\pi}\right], \qquad (6.85)$$

$$\omega_i = (-1)^n \frac{n c_0 \sin(\omega_{n=0}\tau)}{2L}. \qquad (6.86)$$

Starting with the real part, ω_r, Eq. (6.85) shows that unsteady heat release causes a frequency shift, but that the effect is a small correction to the $n = 0$ value, for $n \ll 1$. Moreover, the frequency shift oscillates as $\cos(\omega_{n=0}\tau)$ and, for a given τ value, switches sign for successive mode index values n (not to be confused with heat release gain, n). For example, if $\tau = 0$, then the frequency shifts down for the $n = 1$ (1/4-wave mode), up for the $n = 2$ (3/4-wave mode), and so forth. As discussed later, both of these trends are due to the phase between unsteady pressure and heat release.

The nonzero imaginary part is more fundamentally significant, as it shows that self-excited oscillations exist for certain parameter values. We will analyze these conditions in detail, starting with the $n = 1$, or 1/4-wave, mode. Assuming $n > 0$, self-excited oscillations occur when $\sin(\omega_{n=0}\tau) < 0$. Defining $\mathcal{T}_{1/4}$ as the $n = 0$ quarter-wave mode period and a new index, $m = 0, 1, 2, \ldots$, self-excited oscillations occur when

Unstable 1/4-wave mode ($n = 1$):

$$m - 1/2 < \frac{\tau}{\mathcal{T}_{1/4}} < m. \qquad (6.87)$$

This expression shows that self-excited oscillations occur for bands of $\tau/\mathcal{T}_{1/4}$ values, such as $0.5 < \tau/\mathcal{T}_{1/4} < 1$, $1.5 < \tau/\mathcal{T}_{1/4} < 2$, and so forth, as illustrated in Figure 6.19.

The reasons for self-excited oscillations in this $\tau/\mathcal{T}$ range can be understood by referring back to Rayleigh's criterion, Eq. (2.66), and the conditions under which unsteady heat release adds energy to the acoustic field. The discussion around this equation showed that energy is added to the acoustic field when the product of the unsteady pressure and heat release is greater than zero. This is equivalent to stating that the magnitude of the phase between the unsteady pressure and heat release is less than 90 degrees in the frequency domain. We will next show that Eq. (6.87) is an equivalent statement of Rayleigh's criterion for the $n = 1$ mode. Using the equations for the unsteady pressure and velocity, Eqs. (6.72) and (6.78), the product of the unsteady pressure and heat release at $x = L/2$ is given by

$$p_1(x = L/2, t)\dot{Q}_1(t) \propto \frac{n}{2}\sin\left(\frac{(2n-1)\pi}{2}\right)\cos(\omega_{n=0}t)\sin(\omega_{n=0}(t-\tau)). \qquad (6.88)$$

Thus, for the $n = 1$ mode, it can be seen that in order for $\overline{p_1(x = L/2, t)\dot{Q}_1(t)} > 0$,

$$\overline{p_1(x = L/2, t)\dot{Q}_1(t)} > 0 \Rightarrow m - 1/2 < \frac{\tau}{\mathcal{T}_{1/4}} < m. \qquad (6.89)$$

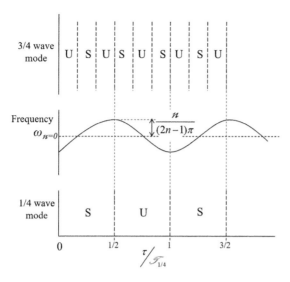

Figure 6.19 Dependence of the stable/unstable (S/U) regions for the model problem on the ratio of heat release time delay, τ, to the quarter-wave period, $\mathcal{T}_{1/4}$. The frequency shift for the quarter-wave mode is also plotted.

This is the exact condition derived in Eq. (6.87). Thus, the conditions under which self-excited oscillations occur are exactly those where $p_1 \dot{Q}_1 > 0$.

Next, consider the $n = 2$ mode, which is the 3/4-wave mode. This mode is unstable when $\sin(\omega_{n=0}\tau) > 0$. Since the $n = 0$ period of this mode $\mathcal{T}_{3/4} = \mathcal{T}_{1/4}/3$, we can write the conditions for self-excited oscillations using the same index, m, as

Unstable 3/4-wave mode ($n = 2$):

$$\frac{m}{3} < \frac{\tau}{\mathcal{T}_{1/4}} < \frac{m+1/2}{3}. \tag{6.90}$$

These are also plotted in Figure 6.19. The stability inequality for τ is also normalized by $\mathcal{T}_{1/4}$ in order to plot the stability conditions for these two modes on the same graph (note that $\mathcal{T}_{1/4}$ is essentially a combustor parameter that is a function of its length and sound speed). Referring to Figure 6.19, it can be seen that instability modes overlap. In other words, from an overall combustor stability point of view, there are only a few narrow "$\tau/\mathcal{T}_{1/4}$ windows" where both modes are simultaneously stable. Including additional modes further decreases this stability window. Of course, this analysis does not include combustor damping processes which narrow the width of the instability regions and may cause higher modes to become completely stable. However, this problem does illustrate the challenges in having a combustor where all modes are simultaneously stable.

A few other points are worthy of mention. First, note that the condition for instability for the $n = 1$ mode is $\sin(\omega_{n=0}\tau) < 0$, while it is $\sin(\omega_{n=0}\tau) > 0$ for the $n = 2$ mode. This is due to the $(-1)^n$ factor in Eq. (6.86). The reason for this can be understood from the $\sin((2n-1)\pi/2)$ term in the $p_1 \dot{Q}_1$ product expression,

Eq. (6.88). The sign of this product flips with successive n values – this is due to the phase of the unsteady pressure and velocity at $x = L/2$ alternating with each successive mode. Moreover, this illustrates an important point in combustion instability problems. Fundamentally, instabilities occur when the pressure and heat release are correctly phased. However, the unsteady heat release is often driven by unsteady velocity oscillations (so-called "pressure coupling," the direct response of the unsteady heat release to pressure, is often weak, as discussed in Section 12.1). As such, overall system stability is then controlled by the phase between unsteady pressure and velocity, and between unsteady velocity and heat release.

A second point is the frequency shift induced by unsteady heat release. Heat release oscillations at 90 or 270 degrees with respect to the pressure cause phase shifts, and therefore natural frequency shifts, of sound waves traversing the flame. The dependence of the frequency shift for the 1/4-wave mode on τ is plotted in Figure 6.19. Note that the largest frequency shifts occur at the τ values where oscillations are not amplified, and that the center of instability bands coincide with points of no frequency shift.

An interesting generalization of this problem which we do not discuss here is active control of self-excited oscillations by adding a secondary acoustic energy source that is appropriately phased to damp oscillations, also analyzed in McManus et al. [37]. Indeed, active control has emerged as a viable technique to eliminate instabilities and has been demonstrated in a variety of laboratory and field settings [38–40]. Typical implementations utilize feedback from a pressure sensor which is filtered and used to drive a secondary fuel injector, which excites heat release oscillations [39, 40].

Finally, while this section has focused on self-excited oscillations, significant oscillation levels can also occur under lightly damped conditions, where background noise levels can be amplified significantly, as also discussed in Section 2.5.3.1. In fact, several studies have concluded that large-amplitude oscillations were noise driven [41]. The basic analysis approach described in this section can also be used, by analyzing a forced oscillation problem instead of an eigenvalue problem. For example, this can be done by replacing the heat release source term in Eq. (6.78) with a forcing term. Solutions of the resulting equations for the pressure or velocity amplitude show very high values for $n - \tau$ values near the boundaries of regions marked stable ("S") in Figure 6.19.

6.5.2 Further Discussion of Thermoacoustic Instability Trends

In this section we present several experimental results illustrating combustor stability trends and show how they can be interpreted with the aid of the model problem described earlier and general considerations on Rayleigh's criterion. The model problem in Section 6.5.1 emphasized the importance of two parameters, the heat release time delay, τ, and acoustic period, $\mathcal{T}$, in controlling instability conditions. This time delay is controlled by fluid mechanics, flame location, and geometry, as illustrated in Figure 6.20, and is considered in detail in Chapter 12 and its exercises.

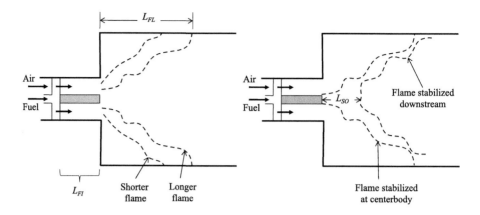

Figure 6.20 Illustration showing the significance of the fuel injector location, L_{FI}, flame length, L_{FL}, and flame standoff distance, L_{SO}, in controlling the time delay, τ.

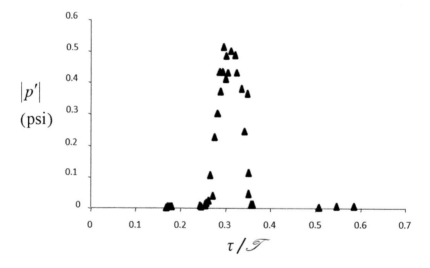

Figure 6.21 Data illustrating variation of instability amplitude with normalized time delay [42]. Image courtesy of D. Santavicca.

For example, it can equal (1) the time required for fuel to convect from injection point to flame, or (2) the time for a vortex to convect from a flow separation point to the flame. As such, both physical combustor design parameters, such as fuel injector location in a premixing passage (L_{FI}), and flame parameters, such as flame length (L_{FL}) or standoff distance (L_{SO}), have important influences on thermoacoustic stability boundaries.

Figure 6.21 illustrates data from a variable-length combustor that clearly illustrate these points [42]. The x-axis plots the measured natural frequency multiplied by an estimated time delay ($\propto L_{FL}/\bar{u}_x$). These data clearly illustrate the nonmonotonic variation of instability amplitude with the parameter $\tau/\mathscr{T}$, as would be expected

6.5 Unsteady Heat Release Effects

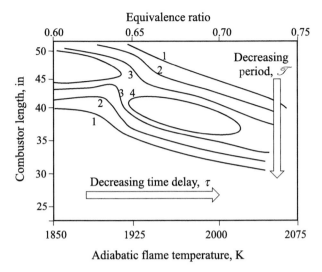

Figure 6.22 Measured pressure amplitude isocontours (in psi) as a function of the fuel/air ratio and combustor length [42]. Data courtesy of D. Santavicca.

from a criterion such as Eq. (6.86), which predicts that $\omega_i \sim \sin(\omega \tau)$. These data also show that instability occurs in a "band" of width $\Delta(\tau/\mathcal{T}) \sim 0.1$. As shown in Figure 2.4, the point where the instability amplitude peaked was where the unsteady pressure and heat release were exactly in phase.

Such data can be thought of as a "slice" obtained from a broader parametric mapping, such as shown in Figure 6.22, where both combustor length (i.e., acoustic period, $\mathcal{T}$) and fuel/air ratio are varied.

Increasing fuel/air ratio causes a decrease in the flame length, L_{FL}, and therefore the time delay. This plot shows that such a shift at a fixed combustor length causes the instability amplitude to exhibit a nonmonotonic variation with respect to fuel/air ratio similar to that for combustor length. Moreover, the figure shows that if one follows the center of the instability band, increases in fuel/air ratio occur at shorter combustor lengths. This occurs because the resulting decreases in time delay (flame length) cause instabilities to occur at a shorter combustor length, with a shorter period, $\mathcal{T}$. In other words, the instability island shifts in such a way that the ratio $\tau/\mathcal{T}$ stays roughly constant.

A similar point can be reached in an alternate way using Figure 6.23. This figure plots the dependence of the combustor instability amplitude on the axial location of the fuel injection point. This figure shows the nonmonotonic variation of instability amplitude with axial injector location, L_{FI}, due to the more fundamental variation of fuel convection time delay, τ. Also shown in the figure is the corresponding variation of instability frequency with injector location. The biggest change in frequency is observed near the stability boundary, as also predicted by the model problem in Section 6.5.1.

The switching of dominant combustor instability frequencies/modes with time delay suggested in Figure 6.19 is also evident from both the spectra and instability

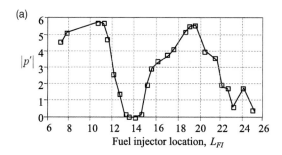
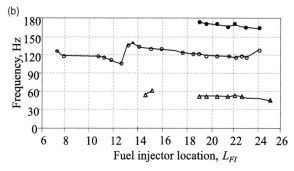

Figure 6.23 Measured dependence of the (a) instability amplitude and (b) frequency on the axial location of the fuel injector. Data obtained from Lovett and Uznanski [43].

amplitudes in Figure 6.24. These data were obtained by varying the nozzle velocity in the combustor, and so the indicated velocity is inversely proportional to time delay. The figures show the dependence of instability amplitude and the switching of dominant frequency on mean velocity.

An important takeway from these results is the intrinsically nonmonotonic variation of instability amplitude with the underlying parameters. For example, a variation in fuel/air ratio, combustor length, or flow velocity can cause the instability amplitude to increase, decrease, or exert no effect. While caution should be exercised in applying linear results to limit cycling systems, the analysis in Section 6.5.1 provides a convenient framework with which to interpret these results – namely, that the effect of some parameter on instability growth rate (i.e., to cause an increase, decrease, or little change) depends on where in the $\tau/\mathcal{T}$ space the system is operating.

6.6 Nonlinear Effects and Limit Cycles

The prior section showed that unsteady heat release could lead to self-excited oscillations. Within the linearized approximations of that analysis, the acoustic perturbation amplitudes grow exponentially in time, see Eq. (6.84). However, as the amplitudes grow, nonlinear effects grow in significance and the system is attracted to a new orbit in phase space, such as a limit cycle; see Figure 6.25 and Section 2.6. This limit cycle

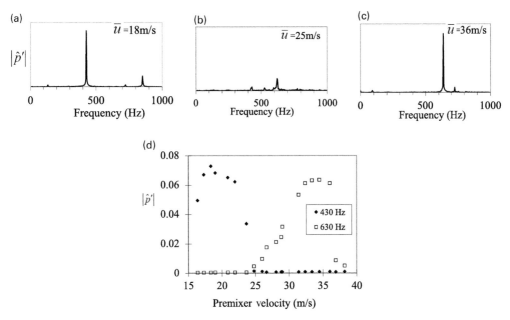

Figure 6.24 (a–c) Measured spectra of combustor pressure at three velocities. (d) Dependence of the excited instability mode amplitude on the mean velocity in the combustor inlet. Obtained from measurements by the author [44].

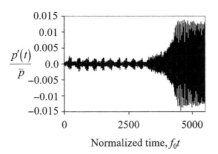

Figure 6.25 Data showing growth and saturation of instability amplitude. Data obtained by the author.

oscillation can consist of relatively simple oscillations at some nearly constant amplitude, but in real combustors the amplitude more commonly "breathes" up and down in a somewhat random or quasi-periodic fashion.

This section describes the aforementioned nonlinear processes. Since the governing wave equation is now nonlinear, approximate methods are needed to solve the problem. Section 6.6.1 first provides an overview of a procedure for developing approximate systems of modal and amplitude equations for these nonlinear oscillations. Section 6.6.2 then gives an overview of the different physical processes leading to nonlinearities.

6.6.1 Formulation of Modal and Amplitude Equations [45–49]

6.6.1.1 Modal Equations

Acoustic oscillations in combustors are manifested as narrowband oscillations near the natural acoustic frequencies of the system, superposed on a broadband noise background. This property can be exploited using a Galerkin approach to derive a set of equations for the nonlinear space–time dynamics of these natural modes [50, 51]. This approach takes advantage of the fact that natural acoustic modes form a complete orthogonal set, see Sections 6.7 and 5.4.2, and any disturbance field can be written as an appropriate sum of these modes. While formally rigorous, in order for the summation to be truncated to an analytically manageable number of modes, the nonlinear effects must introduce only "small" distortions of the wave field from its linear value.

The basic ideas can be illustrated by considering the one-dimensional example

$$\frac{\partial^2 p_1}{\partial t^2} - c_0^2 \frac{\partial^2 p_1}{\partial x^2} = F_h(x,t) \qquad (6.91)$$

subject to a rigid and pressure release boundary condition at the left-side and right-side boundary of the combustor respectively:

$$u_{x,1}(x=0,t) = 0, \qquad (6.92)$$

$$p_1(x=L,t) = 0, \qquad (6.93)$$

where $F_h(x,t)$ denotes effects not included in the wave operator, such as nonlinearities, mean flow effects, unsteady heat release, and so forth. The disturbance pressure field is expanded in terms of the natural acoustic modes of the homogeneous version of Eq. (6.91) as

$$p_1(x,t) = p_0 \sum_{n=1}^{\infty} \eta_n(t)\psi_n(x). \qquad (6.94)$$

The corresponding velocity can be obtained from the linearized momentum equation

$$u_1(x,t) = -\frac{p_0}{\rho_0} \sum_{n=1}^{\infty} \int [\eta_n(t')dt'] \frac{d\psi_n(x)}{dx}. \qquad (6.95)$$

Typically, the integral term is eliminated by using the linearized energy equation in the absence of heat release to derive the relation

$$u_1(x,t) = \sum_{n=1}^{\infty} \frac{1}{\gamma k_n^2} \frac{d\eta_n(t)}{dt} \frac{d\psi_n(x)}{dx}. \qquad (6.96)$$

The mode shapes, ψ_n, are described by the Helmholtz equation:

$$\frac{d^2\psi_n}{dx^2} + k_n^2\psi_n = 0, \qquad (6.97)$$

where the mode shapes are chosen to satisfy the boundary conditions of Eqs. (6.92) and (6.93):

6.6 Nonlinear Effects and Limit Cycles

$$\psi_n(x) = \cos(k_n x), \tag{6.98}$$

$$k_n = \frac{(2n-1)\pi}{2L}, \quad n = 1, 2, 3, \ldots \tag{6.99}$$

Now, multiply Eq. (6.97) by p_1 and Eq. (6.91) by ψ_m/c_0^2, and add them to obtain

$$\frac{\psi_m}{c_0^2}\frac{\partial^2 p_1}{\partial t^2} + k_m^2 p_1 \psi_m = \frac{\psi_m}{c_0^2} F_h(x,t) - \left[p_1 \frac{d^2 \psi_m}{dx^2} - \psi_m \frac{\partial^2 p_1}{\partial x^2}\right]. \tag{6.100}$$

Next, substitute Eq. (6.94) for p_1 in Eq. (6.100) and integrate over the domain. Utilizing the orthogonality property of acoustic modes yields

$$\frac{d^2 \eta_m}{dt^2} + \omega_m^2 \eta_m = \frac{1}{\rho_0 E_m}\int_0^L F_h(x,t)\psi_m(x)dx - \frac{c_0^2}{\rho_0 E_m}\int_0^L \left[p_1 \frac{d^2 \psi_m}{dx^2} - \psi_m \frac{\partial^2 p_1}{\partial x^2}\right]dx, \tag{6.101}$$

where $E_m = \int_0^L \psi_m^2(x)dx = \frac{L}{2}$.

Thus, we have converted the original partial differential equation, Eq. (6.91), into an infinite set of oscillator equations for the time dynamics, $\eta_m(t)$, of each mode. There are two terms on the right-hand side of Eq. (6.101), reflecting the effects of sources, F_h, and boundary conditions on these modal dynamics. In order to see that the second term is a boundary condition term, note that

$$\iiint_V [p_1 \nabla^2 \psi_m - \psi_m \nabla^2 p_1]dV = \iint_A \underbrace{[p_1 \nabla \psi_m - \psi_m \nabla p_1]}_{\vec{F}_b(x,t)} \cdot \vec{n}\, dA, \tag{6.102}$$

where $\vec{n}$ is the unit, outward-pointing vector from the surface and we have written a more general, three-dimensional form to illustrate the surface integral form explicitly. Equation (6.101) can be rewritten as

$$\frac{d^2 \eta_m}{dt^2} + \omega_m^2 \eta_m = \frac{1}{\rho_0 E_m}\int_0^L F_h(x,t)\cos(k_m x)dx - \frac{c_0^2}{\rho_0 E_m A}\iint_A \vec{F}_b(x,t) \cdot \vec{n}\, dA. \tag{6.103}$$

6.6.1.2 Derivation of Modal Amplitude Equations

This section describes additional simplifications that can be made in order to analyze Eq. (6.103). We can exploit the fact that the oscillations are essentially periodic, and utilize the van der Pol decomposition to write the temporal dynamics, $\eta(t)$, as [52]:

$$\eta_m(t) = \mathcal{A}_m(t)\cos(\omega_m t) + \mathcal{B}_m(t)\sin(\omega_m t). \tag{6.104}$$

Because we have introduced two unknowns, $\mathcal{A}_m(t)$ and $\mathcal{B}_m(t)$, in place of $\eta_m(t)$, we can prescribe one arbitrary relationship for these quantities, which we take as

$$\frac{d\mathcal{A}_m(t)}{dt}\cos(\omega_m t) + \frac{d\mathcal{B}_m(t)}{dt}\sin(\omega_m t) = 0. \tag{6.105}$$

The free constraint is chosen such that the first derivative of Eq. (6.104) is not a function of derivatives in $\mathcal{A}_m(t)$ and $\mathcal{B}_m(t)$. This anticipates the approximation we will make later that changes in $\mathcal{A}_m(t)$ and $\mathcal{B}_m(t)$ in time are slow relative to the period of oscillations. Substituting Eq. (6.104) into Eq. (6.103), we obtain

$$-\omega_m \frac{d\mathcal{A}_m(t)}{dt} \sin(\omega_m t) + \omega_m \frac{d\mathcal{B}_m(t)}{dt} \cos(\omega_m t)$$

$$= \frac{1}{p_0 E_m} \int_0^L F_h(x,t) \cos(k_m x) dx - \frac{c_0^2}{p_0 E_m A} \iint_A \vec{F}_b(x,t) \cdot \vec{n}\, dA. \tag{6.106}$$

Separate equations for $d\mathcal{A}_m(t)/dt$ and $d\mathcal{B}_m(t)/dt$ can be derived by substituting Eq. (6.105) into Eq. (6.106):

$$\frac{d\mathcal{A}_m(t)}{dt} = -\frac{\sin(\omega_m t)}{\omega_m p_0 E_m} \int_0^L F_h(x,t) \cos(k_m x) dx + \frac{c_0^2 \sin(\omega_m t)}{\omega_m p_0 E_m A} \iint_A \vec{F}_b(x,t) \cdot \vec{n}\, dA,$$

$$\tag{6.107}$$

$$\frac{d\mathcal{B}_m(t)}{dt} = \frac{\cos(\omega_m t)}{\omega_m p_0 E_m} \int_0^L F_h(x,t) \cos(k_m x) dx - \frac{c_0^2 \cos(\omega_m t)}{\omega_m p_0 E_m A} \iint_A \vec{F}_b(x,t) \cdot \vec{n}\, dA.$$

$$\tag{6.108}$$

Often, it is more convenient to look at the dynamics of the total oscillation amplitude, $\mathcal{R}_m(t) = [\mathcal{A}_m^2(t) + \mathcal{B}_m^2(t)]^{1/2}$, and phase, called a Krylov–Bogolyubov decomposition [53]. This naturally leads to the following alternative polar decomposition for amplitude, $\mathcal{R}_m(t)$, and phase, $\theta_m(t)$:

$$\eta_m(t) = \mathcal{R}_m(t) \sin(\omega_m t + \theta_m(t)), \tag{6.109}$$

where we utilize the free constraint

$$\frac{d\mathcal{R}_m(t)}{dt} \sin(\omega_m t + \theta_m(t)) + \mathcal{R}_m(t) \frac{d\theta_m(t)}{dt} \cos(\omega_m t + \theta_m(t)) = 0. \tag{6.110}$$

The equations for $\mathcal{R}_m(t)$ and $\theta_m(t)$ are

$$\frac{d\mathcal{R}_m(t)}{dt} = \frac{\cos(\omega_m t + \theta_m(t))}{\omega_m p_0 E_m} \int_0^L F_h(x,t) \cos(k_m x) dx - \frac{c_0^2 \cos(\omega_m t + \theta_m(t))}{\omega_m p_0 E_m A} \iint_A \vec{F}_b(x,t) \cdot \vec{n}\, dA,$$

$$\tag{6.111}$$

$$\frac{d\theta_m(t)}{dt} = -\frac{\sin(\omega_m t + \theta_m(t))}{\omega_m p_0 E_m \mathcal{R}_m(t)} \int_0^L F_h(x,t) \cos(k_m x) dx + \frac{c_0^2 \sin(\omega_m t + \theta_m(t))}{\omega_m p_0 E_m A \mathcal{R}_m(t)} \iint_A \vec{F}_b(x,t) \cdot \vec{n}\, dA.$$

$$\tag{6.112}$$

Note that although the polar decomposition is more physically intuitive, the resulting equations are more complex to actually solve.

6.6.1.3 Example Application of van der Pol Decomposition and Method of Averaging

To illustrate the application of this decomposition approach, consider the dynamics of a single mode (we will drop the m subscript from the previous section), with the "source terms" chosen to coincide with the nonlinear oscillator equation previously analyzed in Section 2.5.3.2:

$$\frac{d^2\eta}{dt^2} + \omega^2 \eta = -2\zeta\omega(\beta\eta^2 - 1)\frac{d\eta}{dt}. \tag{6.113}$$

Substituting Eq. (6.113) into Eqs. (6.107) and (6.108),

$$\frac{d\mathcal{A}(t)}{dt} = 2\zeta\omega \sin(\omega t)(\mathcal{B}(t)\cos(\omega t) - \mathcal{A}(t)\sin(\omega t))$$
$$\times \begin{pmatrix} \beta\mathcal{A}^2(t)\cos^2(\omega t) + \beta\mathcal{B}^2(t)\sin^2(\omega t) \\ +2\beta\mathcal{A}(t)\mathcal{B}(t)\sin(\omega t)\cos(\omega t) - 1 \end{pmatrix} \tag{6.114}$$

$$\frac{d\mathcal{B}(t)}{dt} = -2\zeta\omega \cos(\omega t)(\mathcal{B}(t)\cos(\omega t) - \mathcal{A}(t)\sin(\omega t))$$
$$\times \begin{pmatrix} \beta\mathcal{A}^2(t)\cos^2(\omega t) + \beta\mathcal{B}^2(t)\sin^2(\omega t) \\ +2\beta\mathcal{A}(t)\mathcal{B}(t)\sin(\omega t)\cos(\omega t) - 1 \end{pmatrix}. \tag{6.115}$$

These equations are quite difficult to solve analytically, so we spend the rest of this section illustrating the powerful approximate Krylov–Bogolyubov method, or method of time averaging [53]. This exploits the fact that the forcing terms on the right-hand sides of these amplitude equations consist of terms that are constant in time and those that oscillate about zero at integer multiples of the frequency, ω. For example,

$$\sin^2(\omega t) = 1/2 - \cos(2\omega t)/2. \tag{6.116}$$

The key idea is that the problem has both fast and slow time scales, associated with the frequency of oscillations and the variation of the amplitude. The rapidly oscillating forcing terms in Eqs. (6.114) and (6.115) have a negligible effect on $\mathcal{A}(t)$, since they largely cancel over the slow time scale at which $\mathcal{A}(t)$ varies. Instead, the dynamics of $\mathcal{A}(t)$ are largely controlled by the constant or very slowly varying terms. The essence of the method of averaging is then to neglect these rapidly oscillatory forcing terms. Applying this method to the amplitude equations leads to

$$\frac{d\mathcal{A}(t)}{dt} = -\zeta\omega\left[\frac{\beta}{4}\mathcal{A}^3(t) + \frac{\beta}{4}\mathcal{A}(t)\mathcal{B}^2(t) - \mathcal{A}(t)\right], \tag{6.117}$$

$$\frac{d\mathcal{B}(t)}{dt} = -\zeta\omega\left[\frac{\beta}{4}\mathcal{B}^3(t) + \frac{\beta}{4}\mathcal{A}^2(t)\mathcal{B}(t) - \mathcal{B}(t)\right]. \tag{6.118}$$

These equations are substantially simpler than the unaveraged ones. We will seek limit cycle solutions where $d\mathcal{A}(t)/dt = d\mathcal{B}(t)/dt = 0$:

$$0 = \mathcal{A}(t)\left[\beta(\mathcal{B}^2(t) + \mathcal{A}^2(t)) - 4\right], \tag{6.119}$$

$$0 = \mathcal{B}(t)\left[\beta\left(\mathcal{A}^2(t) + \mathcal{B}^2(t)\right) - 4\right]. \tag{6.120}$$

These two expressions can be solved simultaneously for $\mathcal{A}$ and $\mathcal{B}$, which lead to the following solution for the limit cycle amplitude, $\mathcal{R}$, in Eq. (6.109):

$$\mathcal{R} = \sqrt{\mathcal{A}^2 + \mathcal{B}^2} = \frac{2}{\sqrt{\beta}}. \tag{6.121}$$

This approach has been used extensively by Culick and Yang to work out amplitude equations for several nonlinear combustion dynamics problems; the reader is referred to the references for example applications [45–48].

6.6.2 Sources of Nonlinearities

Section 6.6.1 described approximate analysis methods to incorporate nonlinear effects into acoustic analyses. In this section we consider the specific physical process which introduces these nonlinearities. These nonlinearities may be due to gas dynamic processes, which describe nonlinear effects on wave propagation, considered in Section 6.6.2.1. They can also be due to the combustion process, discussed in Section 6.6.2.2, which causes acoustic source terms to exhibit a nonlinear amplitude dependence. Finally, they can be due to nonlinear effects at combustor boundaries, considered in Section 6.6.2.3.

6.6.2.1 Gas Dynamic Nonlinearities [54–56]

Gas dynamic nonlinearities within the combustor volume are introduced by processes described by nonlinear terms in the Navier–Stokes and energy equations, e.g., advective terms such as $\rho \vec{u} \cdot \nabla \vec{u}$ in the Navier–Stokes equation or the nonlinear pressure–density relationship. Such terms generally become significant when the amplitudes of the fluctuating pressure, density, or velocity become of the order of the mean pressure, density, or speed of sound, respectively. Consequently, these terms are generally not important when the relative amplitudes of the acoustic disturbances, $p_{1,\Lambda}/p_0$, $u_{1,\Lambda}/c_0$, or $\rho_{1,\Lambda}/\rho_0$, are small. On the other hand, when very large amplitude oscillations are encountered, these nonlinear processes strongly affect the characteristics of the oscillations [47, 57–59]. These nonlinear terms transfer energy to higher-order acoustic modes with an amplitude-dependent energy transfer rate. Extensive analyses of these nonlinearities have been performed in the context of rocket instabilities [46, 48, 54, 59–64].

To illustrate, we reproduce Awad and Culick's [54] analysis of nonlinear effects on the longitudinal modes of a duct with rigid–rigid boundary conditions. The amplitude equations below were derived using the methods detailed in Sections 6.6.1.1 and 6.6.1.2, where the source term F_h is replaced by the relevant nonlinear gas-dynamic terms (truncated to second order in perturbation amplitude) and the unsteady heat release is assumed to exhibit a linear dependence on disturbance amplitude:

$$\frac{d\mathcal{A}_n}{dt} = a_n\mathcal{A}_n + \theta_n\mathcal{B}_n + n\frac{\beta}{2}\sum_{i=1}^{\infty}[\mathcal{A}_i(\mathcal{A}_{n-i} - \mathcal{A}_{i-n} - \mathcal{A}_{n+i}) - \mathcal{B}_i(\mathcal{B}_{n-i} + \mathcal{B}_{i-n} + \mathcal{B}_{n+i})],$$

(6.122)

$$\frac{d\mathcal{B}_n}{dt} = a_n\mathcal{B}_n - \theta_n\mathcal{A}_n + n\frac{\beta}{2}\sum_{i=1}^{\infty}[\mathcal{A}_i(\mathcal{B}_{n-i} + \mathcal{A}_{i-n} - \mathcal{B}_{n+i}) + \mathcal{B}_i(\mathcal{A}_{n-i} - \mathcal{A}_{i-n} + \mathcal{A}_{n+i})],$$

(6.123)

where $\beta = (\gamma + 1/8\gamma)\omega_1$. The parameters a_n and θ_n represent the amplitude growth rate and frequency shift induced by the unsteady heat release and boundary conditions. For example, for the problem considered in Section 6.5.1, these are equivalent to $a_n = 2\omega_{i,n}$ in Eq. (6.86) and $\theta_n = (n\omega_{n=0})^2 - \omega_{r,n}^2$ in Eq. (6.85). Thus, the linearized solution for the pressure field is given by

$$p_1(x,t) = \sum_{i=1}^{\infty} \mathcal{C}_i e^{a_i t} \sin((\omega_i - \theta_i)t)\psi_i(x),$$

(6.124)

where we write the solution for the pressure, instead of the pressure amplitudes, in order to explicitly illustrate the frequency shift effect, θ_n. An explicit analytic solution to the full nonlinear equations can be obtained from an analytical solution obtained by truncating to two modes. This assumption implies that the first mode, $n = 1$, is unstable and the second is stable, as the energy added to the $n = 1$ mode is then passed at an amplitude-dependent rate to the $n = 2$ mode where it is damped (if these modes are both unstable, then more modes must be included). Seeking a limit cycle solution of the form[2] $d\mathcal{A}_n/dt = d\mathcal{B}_n/dt = 0$, leads to the following limit cycle amplitudes for the first two modes:

$$\mathcal{A}_1 = \frac{1}{\beta}\left[-a_1 a_2\left(1 + \frac{\theta_1^2}{a_1^2}\right)\right]^{1/2},$$

(6.125)

$$\mathcal{B}_1 = 0,$$

(6.126)

$$\mathcal{A}_2 = \frac{a_1}{\beta},$$

(6.127)

$$\mathcal{B}_2 = \frac{\theta_1}{\beta}.$$

(6.128)

Since the phase is arbitrary, $\mathcal{B}_1$ has been set to zero. Equation (6.125) shows that a physically meaningful limit cycle is only possible if $a_1 > 0$ and $a_2 < 0$, given the two-mode approximation. This equation shows that the limit cycle amplitudes of $\mathcal{A}_1$ and $\mathcal{A}_2$ are both proportional to the unstable mode growth rate, a_1. This result shows that limit cycle amplitudes are proportional to the rate at which the unsteady heat release is pumping energy into the unstable first mode. Interestingly, it also shows that the frequency shift, θ_1, influences the limit cycle amplitude. This dependence of limit

---

[2] In addition, more complex system orbits are possible, such as where the limit cycle amplitude is modulated at some low-frequency, quasi-periodic orbits, or chaotic orbits [54, 65].

cycle amplitude on frequency shift is apparently due to the efficiency with which energy is passed from one mode to another through nonlinear processes. For example, in an acoustic system with rigid–rigid boundaries, the natural acoustic modes of the system without heat release, f_n, are all integer multiples of the frequency of the first mode, $f_n = nf_1$. Furthermore, nonlinear processes cause oscillations at f_1 to excite other frequencies, nf_1. Analysis shows that the rate of energy transfer from lower- to higher-frequency acoustic modes by nonlinear effects is inversely proportional to this difference, $nf_1 - f_n$ [66]. Indeed, in one application where large-amplitude, forced oscillations were desirable, nonconstant-area ducts (whose natural modes are not integer multiples of each other) were specifically developed in order to achieve the largest possible oscillations without shock wave formation [67].

6.6.2.2 Combustion Process Nonlinearities

Combustion process nonlinearities are introduced by the nonlinear dependence of the heat release rate on the acoustic disturbance amplitude. As detailed in Chapter 12, there are a variety of factors that can cause this nonlinear response. For example, the fact that premixed flames propagate normal to themselves is an inherently nonlinear phenomenon. Larger-amplitude wrinkles are smoothed out at faster relative rates than smaller wrinkles. Several examples of heat release nonlinearities are presented in Section 12.3.3. For example, referring to the experimental data in Figure 12.13, note the linear dependence of the heat release amplitude on velocity for velocity disturbance amplitudes of up to about 80% of the mean velocity. Above velocity amplitudes of 80%, the unsteady heat release saturates. In addition, the phase exhibits amplitude dependence at much smaller velocity amplitudes. An important point from these and other results is that nonlinearities in heat release often occur when the ratio $u'/\bar{u} \sim O(1)$. Consequently, in these cases the relevant velocity scale that determines when nonlinearities are important is the mean velocity and not the sound speed (as in the case of gas-dynamic nonlinearities).

6.6.2.3 Boundary-Induced Nonlinearities

The nonlinearities in processes that occur at or near the combustor boundaries also affect the combustor dynamics as they are introduced into the analysis of the problem through nonlinear boundary conditions. Such nonlinearities are caused by, e.g., flow separation at sharp edges or rapid expansions, which cause stagnation pressure losses and a corresponding transfer of acoustic energy into vorticity. These nonlinearities were discussed in Section 6.3.4 and, similar to the combustion nonlinearities previously discussed, become significant when $u'/\bar{u} \sim O(1)$. Also, wave reflection and transmission processes through choked and unchoked nozzles become amplitude dependent at large amplitudes [68].

6.7 Aside: Sturm–Liouville Eigenvalue Problems [69]

The ordinary differential equations describing harmonic acoustic motions can often be cast in the following general form, referred to as a Sturm–Liouville eigenvalue problem:

$$\frac{d}{dx}\left(F_p(x)\frac{d\hat{p}(x)}{dx}\right) - F_q(x)\hat{p}(x) + k^2 F_w(x)\hat{p}(x) = 0, \qquad (6.129)$$

where k^2 denotes the eigenvalues of the equation for which a nontrivial solution that satisfies the boundary condition exists at $x = 0$ and L. For example, for the radial dependence of acoustic oscillations in a round duct, $\hat{R}(r)$, we can identify $F_p = r$, $F_q = m^2/r^2$, and $F_w = r^2$. Similarly, $F_p = F_w = A(x)$ and $F_q = 0$ for the axial dependence of wave motions in a variable-area duct (see Eq. (6.3)). The variable-temperature problem can also be cast in this form. There are certain general properties that can be developed for the solutions of this equation that are the focus of this section.

Problems where (1) F_p, F_q, F_w are real, (2) the boundary impedance Z is imaginary (see below), and (3) the functions F_w, $F_p > 0$ in the $(0, L)$ domain, are referred to as "regular" Sturm–Liouville problems. Fairly general conclusions can be worked out for the characteristics of the eigenvalues and eigenfunctions in this case. The set of eigenvalues, k^2, are referred to as the "spectrum" of the problem. The spectrum of a regular Sturm–Liouville problem consists of an infinite number of discrete eigenvalues.

6.7.1 Orthogonality of Eigenfunctions

An important property of this problem is the orthogonality of the eigenfunctions. This can be shown by identifying k_n^2 and k_m^2 as two different eigenvalues and $\hat{p}_n(x)$ and $\hat{p}_m(x)$ as the associated eigenfunctions satisfying Eq. (6.129). As shown in Exercise 6.12, these expressions can be manipulated to yield

$$(k_n^2 - k_m^2)\int_0^L F_w(x)\hat{p}_n(x)\hat{p}_m(x)dx + \left(F_p(x)\left[\frac{d\hat{p}_n}{dx}(x)\hat{p}_m(x) - \hat{p}_n(x)\frac{d\hat{p}_m}{dx}(x)\right]\right)_{x=0}^{x=L} = 0.$$

(6.130)

The second term vanishes for a variety of homogeneous, purely imaginary impedance type or periodic boundary conditions. Recall that the impedance is defined as $Z = \hat{p}_1/(\vec{u}_1 \cdot (-\vec{n}))$, where $\vec{n}$ points into the domain, which for a one-dimensional domain implies the following definition:

$$x = 0 : Z = -i\omega\rho_0 \frac{\hat{p}_1}{d\hat{p}_1/dx}, \qquad (6.131)$$

$$x = L : Z = i\omega\rho_0 \frac{\hat{p}_1}{d\hat{p}_1/dx}. \qquad (6.132)$$

Note the inversion of signs for the definition of Z at 0 and L. The second term in Eq. (6.130) then vanishes for any combination of boundary conditions at $x = 0$ and L that satisfy a pressure release ($\hat{p}_1 = 0$), rigid ($d\hat{p}_1/dx = 0$), or purely imaginary impedance[3] (Z) boundary condition. In addition, this term vanishes for periodic boundary conditions, such as $\hat{p}_1(x=0) = \hat{p}_1(x=L)$.

If the surface term vanishes, then

$$\left(k_n^2 - k_m^2\right) \int_0^L F_w(x)\hat{p}_n(x)\hat{p}_m(x)dx = 0. \tag{6.133}$$

This shows that the eigenfunctions are orthogonal with respect to the weight function, F_w, in the interval $(0, L)$. Although not proved here, we can also show that the eigenfunctions form a complete orthogonal system. This implies that any continuous function, $F(x)$, that satisfies the boundary conditions can be expanded in a series of the form

$$F(x) = \sum_{n=1}^{\infty} A_n \hat{p}_n(x),$$

$$A_n = \frac{\int_0^L F_w(x)F(x)\hat{p}_n(x)dx}{\int_0^L F_w(x)\hat{p}_n^2(x)dx}. \tag{6.134}$$

6.7.2 Real/Imaginary Characteristics of Eigenvalues

A second important property to investigate is the conditions under which the eigenvalues are real. Note that a real eigenvalue implies harmonic oscillations in time that do not grow or decay. In contrast, an imaginary part of the eigenvalue implies that the oscillations grow or decay in time, implying that energy is being added either from or to the boundary or within the domain itself. As shown in Exercise 6.13, the eigenvalues of Eq. (6.129) subject to the boundary conditions described above are real.

6.7.3 Asymptotic Representation of Eigenfunctions

General expressions can be developed for the high-frequency (i.e., $n \gg 1$) eigenvalues and associated eigenfunctions. The reader is referred to Section 6.8 for a related treatment of this topic.

6.8 Aside: Wave Propagation through Regions with Slowly Varying Properties

In the limit of slowly varying property changes relative to a wavelength, asymptotic solutions of the wave equation with nonconstant coefficients can be developed. These

---

[3] However, note that the impedances of each mode, Z_n and Z_m, must be the same, implying a frequency-independent impedance.

are closely related to the high-frequency solutions of Sturm–Liouville problems described in Section 6.7. To illustrate, return to Eq. (6.2),

$$\frac{d^2\hat{p}_1}{dx^2} + \frac{d\ln(A(x)/\rho_0(x))}{dx}\frac{d\hat{p}_1}{dx} + \frac{\omega^2}{c_0^2(x)}\hat{p}_1 = 0. \tag{6.135}$$

In the limit of slowly varying area, density, and sound speed, the leading-order solution of this equation is given by (corrections to this expression have terms of $O(d^2()/dx^2)$ or $O((d()/dx)^2)$ of the area, density, and/or sound speed) [69, 70]:

$$\hat{p}_1(x) = \sqrt{\frac{\rho_0(x)c_0(x)}{A(x)}} \left[\mathcal{C}_1 \exp\left(i\omega \int \frac{dx'}{c_0(x')}\right) + \mathcal{C}_2 \exp\left(-i\omega \int \frac{dx'}{c_0(x')}\right) \right]. \tag{6.136}$$

First, note that the amplitude is rescaled at every axial position by $\sqrt{\frac{\rho_0(x)c_0(x)}{A(x)}}$, which is identical to the scaling derived in Eq. (6.9), derived under more restrictive conditions using energy flux considerations. Second, the phase simply accounts for time of flight in a medium with varying sound speed. In other words, if the sound speed is constant, the time taken for a disturbance to propagate through a distance L is given by L/c_0. If the sound speed varies spatially, this time of flight generalizes to $\int_0^L \frac{dx}{c_0(x)}$.

6.9 Aside: Approximate Methods for Linearized Jump Conditions across Compact Zones

This section describes methods for analyzing the relationship between unsteady flow properties across the zone of rapid variation in temperature, area, etc. In the vicinity of such transitions, the acoustic field is generally multidimensional. However, assuming that the acoustic mode under consideration is below the cutoff frequency of the duct, then these multidimensional disturbances rapidly decay, as described in Section 5.4.2. Thus, we will assume one-dimensional oscillations at sectional interfaces; see Figure 6.26.

The relationship between flow variables at the interfaces a and b can be obtained by integrating the conservation equations over the indicated control volume. We will start with the continuity equation, Eq. (1.8). Integrating this equation over the control volume shown in Figure 6.26, linearizing, and assuming one-dimensional flow at the up- and downstream interface leads to

$$\frac{\partial}{\partial t}\int_{CV} \rho_1 dV + [(\rho_0 u_{x,1} + \rho_1 u_{x,0})A]_a^b = 0, \tag{6.137}$$

where the notation $[\,]_a^b$ means $[\,]_b - [\,]_a$. It is assumed that the normal component of the velocity is zero at the walls, so that the surface integral is only nonzero at interfaces a and b. This equation can be readily generalized to include elements with more than a single inlet or outlet, such as a T-junction where the $[\,]_a^b$ convention generalizes to $\sum_{outlets} - \sum_{inlets}$. The volume integral terms describe accumulation within the control volume. Assuming harmonic oscillations, rewrite Eq. (6.137) as

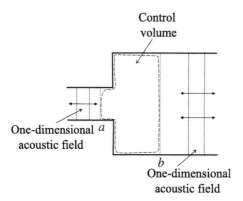

Figure 6.26 Sketch showing the procedure for matching one-dimensional oscillations at interfaces up- and downstream of an area or temperature variation.

$$-i\omega \rho_{1,a} A_a L_{\mathit{eff}} + [(\rho_0 u_{x,1} + \rho_1 u_{x,0})A]_a^b = 0, \quad (6.138)$$

where the "effective length" (analogous to the end corrections described in Section 5.2) is

$$L_{\mathit{eff}} = \frac{1}{-i\omega \rho_{1,a} A_a} \frac{\partial}{\partial t} \int_{CV} \rho_1 dV. \quad (6.139)$$

It can be shown that the unsteady term is of $O(kL_{\mathit{eff}})$ relative to the flux terms, and thus is negligible for acoustically compact transition regions. Neglecting the accumulation terms causes the flux terms on the upstream and downstream sides to equal each other at every instant in time. Retaining the unsteady term (describing inertial and/or compressibility effects) incorporates a phase delay between the inlet and outlet because of accumulation in the control volume. Multipole expansion techniques can be used to derive these jump conditions to higher order in kL_{eff} [34].

The energy jump condition can be similarly derived by using the following form of the energy equation, which assumes constant γ:

$$\frac{\partial}{\partial t}\left(\frac{p}{\gamma-1} + \frac{1}{2}\rho|\vec{u}|^2\right) + \nabla \cdot \left(\left(\frac{\gamma p}{\gamma-1} + \frac{1}{2}\rho|\vec{u}|^2\right)\vec{u}\right) = \dot{q}. \quad (6.140)$$

Integrating this equation over the volume, linearizing, and assuming one-dimensional flow at the upstream and downstream interface leads to

$$\frac{\partial}{\partial t}\int_{CV}\left(\frac{p_1}{\gamma-1} + 2\rho_0 \vec{u}_0 \cdot \vec{u}_1\right) dV + \left[\frac{\gamma}{\gamma-1}(p_1 u_{x,0} + p_0 u_{x,1})A\right]_a^b = \dot{Q}_1. \quad (6.141)$$

Neglecting the accumulation term leads to

$$\left[\frac{\gamma}{\gamma-1}(p_1 u_{x,0} + p_0 u_{x,1})A\right]_a^b = \dot{Q}_1. \quad (6.142)$$

Equation (6.142) shows that the heat release oscillations, $\dot{Q}_1$, lead to a jump in acoustic velocity across the flame. The analogous quasi-steady momentum jump conditions are presented in Eqs. (6.63) and (6.64).

6.10 Aside: Wave Interactions with Compact Nozzles or Diffusers [8]

This section presents general expressions for wave interactions with compact nozzles or diffusers, see Figure 6.9. The reflection coefficient or excited acoustic wave amplitude is a function of both M_a and M_b. The reflection coefficient for an unchoked nozzle is given by the following more general form of Eq. (6.54):

$$R = \left(\frac{M_b - M_a}{1 - M_a}\right)\left(\frac{1 + M_a}{M_b + M_a}\right)\left[\frac{1 - \frac{1}{2}(\gamma - 1)M_a M_b}{1 + \frac{1}{2}(\gamma - 1)M_a M_b}\right]. \quad (6.143)$$

Similarly, the amplitude of the backward-propagating acoustic wave excited by an entropy disturbance advecting through the nozzle is given by

$$\frac{p_{1,\Lambda}}{p_{1,s}} = \frac{p_{1,\Lambda}/c_0^2}{\rho_{1,s}} = \left(\frac{M_b - M_a}{1 - M_a}\right)\left[\frac{M_a}{2 + (\gamma - 1)M_a M_b}\right]. \quad (6.144)$$

The reflection coefficient for an upstream-propagating sound wave into the diffuser interface is similarly given by

$$R = -\left(\frac{M_b - M_a}{M_b + M_a}\right)\left(\frac{1 - M_b}{1 + M_b}\right)\left[\frac{2 - (\gamma - 1)M_a M_b}{2 + (\gamma - 1)M_a M_b}\right]. \quad (6.145)$$

Finally, the amplitude of the forward-propagating sound wave excited by an entropy disturbance propagating through the diffuser (originating from upstream) is

$$\frac{p_{1,\Lambda}}{p_{1,s}} = \frac{p_{1,\Lambda}/c_0^2}{\rho_{1,s}} = -\left(\frac{M_b - M_a}{1 + M_b}\right)\left[\frac{M_b}{2 + (\gamma - 1)M_a M_b}\right]. \quad (6.146)$$

Exercises

6.1. Derive the inhomogeneous wave equation in Eq. (6.1).
6.2. Derive the quasi-one-dimensional variable area and gas property equation in Eq. (6.2). You can do this by utilizing the quasi-one-dimensional momentum and energy equations derived in Chapter 1, Exercise 1.7.
6.3. Derive the natural frequencies and mode shapes of a one-dimensional rigid–rigid duct with an axial mean flow with Mach number M. Plot the time variation of the unsteady pressure and discuss the effect of mean flow on the standing wave structure.
6.4. In this problem we consider the reflection and transmission of a wave incident on a linearly varying temperature field (see Section 6.2.5). By matching the unsteady pressure and velocity at each interface, derive the four simultaneous equations for R, $\mathscr{T}$, $\mathscr{C}_1$, and $\mathscr{C}_2$.

6.5. Consider the natural frequencies of a duct of length L with the linear temperature variation given by Eq. (6.35), and with pressure release and rigid-wall boundary conditions. Using the exact solutions for the pressure and velocity from Eqs. (6.36) and (6.37), derive the characteristic equations for the natural frequency for the four different sets of boundary conditions $(p-p, v-p, p-v, v-v,$ where $p-p$ refers to pressure release boundary conditions at $x=0$ and L, etc).

6.6. As discussed in Section 6.2.6, the ratios of natural longitudinal frequencies of variable temperature ducts are not integer multiples. Calculate and plot the ratio of the frequencies of the first two modes, f_2/f_1, as a function of temperature ratio, β, for the four different boundary conditions analyzed in that section.

6.7. Natural modes in a two-duct area system: Return to the abrupt area change problem, Eq. (6.52). Demonstrate that, for certain duct length ratios L_b/L_a, the natural frequencies are independent of duct area ratio.

6.8 Consider a variant of the two-duct problem shown in Figure 6.7, where the $x=0$ boundary condition is pressure release, rather than rigid as considered in the text. Derive the characteristic equation for the eigenfrequency. Discuss the solution characteristics.

6.9. Consider the acoustics of a duct with an exponentially varying area given by $A(x) = A_0 \exp(mx)$. Derive expressions for the pressure and velocity and compare these solutions to the approximate solution developed in Section 6.8.

6.10. As discussed in Section 6.3.3, acoustic waves propagate into side branches, leading to induced flow oscillations. Assume that the acoustic wavelength is large relative to the side branch diameter, and that the nozzle impedance is given by Z_{-h} at the lower end of the nozzle section, $z=-h$ (see Figure 6.11). For a given pressure oscillation, $\hat{p}_1(z=0, \omega)$, derive an expression for the induced nozzle flow oscillations, $\hat{u}_{z,1}(z=0, \omega)$, at the side branch outlet, $z=0$.

6.11. Develop an expression for the acoustic impedance of a compact choked nozzle, Eq. (6.54), using the relation Eq. (6.53).

6.12. Prove that the solutions of the Sturm–Liouville problem are orthogonal with respect to a weight function $F_w(x)$, Eq. (6.133).

6.13. As discussed in Section 6.7, the conditions under which eigenvalues are real or imaginary is important in determining whether natural oscillations are amplified or damped. Prove that the eigenvalues of the regular Sturm–Liouville problem subject to the boundary conditions described below Eq. (6.130) are real.

6.14. Derive the advective wave equation in Eq. (6.20). Assume that gas properties and the mean velocity are constant.

References

1. Subrahmanyam P.B., Sujith R.I., and Lieuwen T.C., A family of exact transient solutions for acoustic wave propagation in inhomogeneous, non-uniform area ducts. *Journal of Sound and Vibration*, 2001, **240**(4): pp. 705–715.
2. Ingard U. and Singhal V., Effect of flow on the acoustic resonances of an open-ended duct. *The Journal of the Acoustical Society of America*, 1975, **58**: pp. 788–793.

3. Peters M., Hirschberg A., Reijnen A., and Wijnands A., Damping and reflection coefficient measurements for an open pipe at low Mach and low Helmholtz numbers. *Journal of Fluid Mechanics*, 1993, **256**: pp. 499–534.
4. Kumar B. and Sujith R., Exact solution for one-dimensional acoustic fields in ducts with polynomial mean temperature profiles. *Journal of Vibration and Acoustics*, 1998, **120**.
5. Cummings A., Ducts with axial temperature gradients: An approximate solution for sound transmission and generation. *Journal of Sound and Vibration*, 1977, **51**(1): pp. 55–67.
6. Cummings A., Sound generation and transmission in flow ducts with axial temperature gradients. *Journal of Sound and Vibration*, 1978, **57**(2): pp. 261–279.
7. Sujith R.I., Waldherr G.A., and Zinn B.T., An exact solution for one-dimensional acoustic fields in ducts with an axial temperature gradient. *Journal of Sound and Vibration*, 1995, **184**(3): pp. 389–402.
8. Marble F. and Candel S., Acoustic disturbance from gas non-uniformities convected through a nozzle. *Journal of Sound and Vibration*, 1977, **55**(2): pp. 225–243.
9. Williams J. and Howe M., The generation of sound by density inhomogeneities in low Mach number nozzle flows. *Journal of Fluid Mechanics*, 1975, **70**(03): pp. 605–622.
10. Goh C.S. and Morgans A.S., Phase prediction of the response of choked nozzles to entropy and acoustic disturbances. *Journal of Sound and Vibration*, 2011, **330**(21): pp. 5184–5198.
11. Bake F., Kings N., and Roehle I., Fundamental mechanism of entropy noise in aero-engines: Experimental investigation. *Journal of Engineering for Gas Turbines and Power*, 2008, **130**: p. 011202.
12. O'Connor J., Acharya V., and Lieuwen T., Transverse combustion instabilities: Acoustic, fluid mechanic, and flame processes. *Progress in Energy and Combustion Science*, 2015, **49**: pp. 1–39.
13. Blimbaum J., Zanchetta M., Akin T., Acharya V., O'Connor J., Noble D., and Lieuwen T., Transverse to longitudinal acoustic coupling processes in annular combustion chambers. *International Journal of Spray and Combustion Dynamics*, 2012, **4**(4): pp. 275–298.
14. Dawson J.R. and Worth N.A., The effect of baffles on self-excited azimuthal modes in an annular combustor. *Proceedings of the Combustion Institute*, 2015, **35**(3): pp. 3283–3290.
15. Bechert D., Sound absorption caused by vorticity shedding, demonstrated with a jet flow. *Journal of Sound and Vibration*, 1980, **70**(3): pp. 389–405.
16. Dowling A. and Hughes I., Sound absorption by a screen with a regular array of slits. *Journal of Sound and Vibration*, 1992, **156**(3): pp. 387–405.
17. Howe M.S., *Acoustics of Fluid–Structure Interactions*. 1st ed. Cambridge Monographs on Mechanics. 1998, Cambridge: Cambridge University Press, p. 572.
18. Zinn B., A theoretical study of non-linear damping by Helmholtz resonators. *Journal of Sound and Vibration*, 1970, **13**(3): pp. 347–356.
19. Scarborough D.E., *An Experimental and Theoretical Investigation of a Fuel System Tuner for the Suppression of Combustion Driven Oscillations*. 2010, Georgia Institute of Technology.
20. Peat K., Ih J., and Lee S., The acoustic impedance of a circular orifice in grazing mean flow: comparison with theory. *Acoustical Society of America Journal*, 2003, **114**: pp. 3076–3086.
21. Culick F.E.C., *Unsteady Motions in Combustion Chambers for Propulsion Systems*. 2006, RTO/NATO. Document # AG-AVT-039.
22. Temkin S., *Elements of Acoustics*. 2001, Acoustical Society of America.

23. Ronneberger D. and Ahrens C., Wall shear stress caused by small amplitude perturbations of turbulent boundary-layer flow: An experimental investigation. *Journal of Fluid Mechanics*, 1977, **83**(03): pp. 433–464.
24. Lieuwen T., Analysis of acoustic wave interactions with turbulent premixed flames. *Proceedings of the Combustion Institute*, 2002, **29**(2): pp. 1817–1824.
25. Lieuwen T., Theory of high frequency acoustic wave scattering by turbulent flames. *Combustion and Flame*, 2001, **126**(1–2): pp. 1489–1505.
26. Strahle W.C., On combustion generated noise. *Journal of Fluid Mechanics*, 1971, **49**(02): pp. 399–414.
27. Putnam A., Combustion roar of seven industrial burners. *Journal of the Institute of Fuel*, 1976, **49**(400): pp. 135–138.
28. Rajaram R. and Lieuwen T., Acoustic radiation from turbulent premixed flames. *Journal of Fluid Mechanics*, 2009, **637**: pp. 357–385.
29. McManus K., Han F., Dunstan W., Barbu C., and Shah M. Modeling and control of combustion dynamics in industrial gas turbines, in *ASME Turbo Expo 2004: Power for Land, Sea, and Air*, 2004, Vienna, Austria: ASME.
30. Stone C. and Menon S., Swirl control of combustion instabilities in a gas turbine combustor. *Proceedings of the Combustion Institute*, 2002, **29**(1): pp. 155–160.
31. Menon S. and Jou W.H., Large-eddy simulations of combustion instability in an axisymmetric ramjet combustor. *Combustion Science and Technology*, 1991, **75**(1–3): pp. 53–72.
32. Bhatia R. and Sirignano W., One-dimensional analysis of liquid-fueled combustion instability. *Journal of Propulsion and Power*, 1991, **7**: pp. 953–961.
33. Dowling A., Thermoacoustic sources and instabilities, in *Modern Methods in Analytical Acoustics: Lecture Notes*, D.G. Crighton, ed. 1992, Springer.
34. Lieuwen T. and Zinn B.T., Application of multipole expansions to sound generation from ducted unsteady combustion processes. *Journal of Sound and Vibration*, 2000, **235**(3): pp. 405–414.
35. Tang Y., Waldherr G., Jagoda J., and Zinn B., Heat release timing in a nonpremixed Helmholtz pulse combustor. *Combustion and Flame*, 1995, **100**(1–2): pp. 251–261.
36. Menon S., Secondary fuel injection control of combustion instability in a ramjet. *Combustion Science and Technology*, 1994, **100**(1–6): pp. 385–393.
37. McManus K., Poinsot T., and Candel S., A review of active control of combustion instabilities. *Progress in Energy and Combustion Science*, 1993, **19**(1): pp. 1–29.
38. Lieuwen T.C. and Yang V., *Combustion Instabilities in Gas Turbine Engines: Operational Experience, Fundamental Mechanisms, and Modeling*, T.C. Lieuwen and V. Yang, eds. 2005, AIAA.
39. Candel S.M., Combustion instabilities coupled by pressure waves and their active control. *Symposium (International) on Combustion*, 1992, **24**(1): pp. 1277–1296.
40. Sattinger S.S., Neumeier Y., Nabi A., Zinn B.T., Amos D.J., and Darling D.D., Sub-scale demonstration of the active feedback control of gas-turbine combustion instabilities. *Journal of Engineering for Gas Turbines and Power*, 2000, **122**(2): pp. 262–268.
41. Cohen J.M., Proscia W., and Delaat J., Characterization and Control of Aeroengine Combustion Instability: Pratt & Whitney and NASA Experience, in *Combustion Instabilities in Gas Turbine Engines: Operational Experience, Fundamental Mechanisms, and Modeling*, T.C. Lieuwen and V. Yang, eds. 2005, AIAA. pp. 113–144.

42. Gonzalez-Juez E.D., Lee J.G., and Santavicca D., A study of combustion instabilities driven by flame–vortex interactions, in *41st AIAA/ASME/SAE/ASEE Joint Propulsion Conference & Exhibit*, 2005, pp. 1–10.
43. Lovett J.A. and Uznanski K.T. Prediction of combustion dynamics in a staged premixed combustor, in *ASME Turbo Expo 2002: Power for Land, Sea, and Air*. 2002. Amsterdam, Netherlands: ASME.
44. Lieuwen T.C., Experimental investigation of limit-cycle oscillations in an unstable gas turbine combustor. *Journal of Propulsion and Power*, 2002, **18**(1): pp. 61–67.
45. Culick F. and Yang V., Prediction of the stability of nonsteady motions in solidpropellant rocket motors, in *Nonsteady Burning and Combustion Stability of Solid-Propellant*, L. De Luca, E.W. Price, and M. Summerfield, eds. 1992. pp. 719–780.
46. Culick F., Nonlinear behavior of acoustic waves in combustion chambers. *Acta Astronautica*, 1976, **3**(9–10): pp. 715–734.
47. Wicker J.M., Greene W.D., Kim S.I., and Yang V., Triggering of longitudinal combustion instabilities in rocket motors: nonlinear combustion response. *Journal of Propulsion and Power*, 1996, **12**(6): pp. 1148–1158.
48. Yang V., Kim S., and Culick F., Triggering of longitudinal pressure oscillations in combustion chambers. I. Nonlinear gas dynamics. *Combustion Science and Technology*, 1990, **72**(4–6): pp. 183–214.
49. Zinn B. and Powell E. Nonlinear combustion instability in liquid-propellant rocket engines, *Symposium (International) on Combustion*, 1971, 13(1): 491–503.
50. Portillo J.E., Sisco J.C., Yu Y., Anderson W.E., and Sankaran V., Application of a generalized instability model to a longitudinal mode combustion instability, in *43rd AIAA/ASME/SAE/ASEE Joint Propulsion Conference & Exhibit*. 2007, Cincinnati, OH.
51. Krediet H.J., Krebs W., Portillo J.E., and Kok J., Prediction of thermoacoustic limit cycles during premixed combustion using the modified Galerkin approach, in *46th AIAA/ASME/SAE/ASEE Joint Propulsion Conference & Exhibit*. 2010, Nashville, TN.
52. Nayfeh A.H. and Mook D.T., *Nonlinear Oscillations*. 1995, Wiley.
53. Krylov N. and Bogolyubov N., *Introduction to Non-Linear Mechanics*. 1949, Princeton University Press.
54. Awad E. and Culick F., On the existence and stability of limit cycles. for longitudinal acoustic modes in a combustion chamber. *Combustion Science and Technology*, 1986, **46**(3): pp. 195–222.
55. Paparizos L.G. and Culick F., The two-mode approximation to nonlinear acoustics in combustion chambers. I. Exact solution for second order acoustics. *Combustion Science and Technology*, 1989, **65**(1): pp. 39–65.
56. Mitchell C., Crocco L., and Sirignano W., Nonlinear longitudinal instability in rocket motors with concentrated combustion. *Combustion Science and Technology*, 1969, **1**(1): pp. 35–64.
57. Zinn B. and Powell E., Nonlinear combustion instability in liquid-propellant rocket engines. *Symposium (International) on Combustion*, 1970, **13**(1): pp. 491–503.
58. Culick F., Burnley V., and Swenson G., Pulsed instabilities in solid-propellant rockets. *Journal of Propulsion and Power*, 1995, **11**(4): pp. 657–665.
59. Culick F., Non-linear growth and limiting amplitude of acoustic oscillations in combustion chambers. *Combustion Science and Technology*, 1971, **3**(1): pp. 1–16.
60. Zinn B., *A Theoretical Study of Nonlinear Transverse Combustion Instability in Liquid Propellant Rocket Motors*. 1966, Princeton University.

61. Zinn B., A theoretical study of nonlinear combustion instability in liquid-propellant rocket engines. *AIAA Journal*, 1968, **6**: pp. 1966–1972.
62. Culick F., Nonlinear behavior of acoustic waves in combustion chambers II. *Acta Astronautica*, 1976, **3**(9–10): pp. 735–757.
63. Kim S., *Nonlinear Combustion Instabilities in Combustion Chambers*. 1989, University Park, PA (USA): Pennsylvania State University.
64. Yang V. and Culick F.E.C., On the existence and stability of limit cycles for transverse acoustic oscillations in a cylindrical combustion chamber. 1: Standing modes. *Combustion Science and Technology*, 1990, **72**(1): pp. 37–65.
65. Kabiraj L., Sujith R.I., and Wahi P., Bifurcations of self-excited ducted laminar premixed flames. *Journal of Engineering for Gas Turbines and Power*, 2012, **134**(3): 031502.
66. Burnley V.S. and Culick F.E.C., On the energy transfer between transverse acoustic modes in a cylindrical combustion chamber. *Combustion Science and Technology*, 1999, **144**(1): pp. 1–19.
67. Lawrenson C.C., Lipkens B., Lucas T.S., Perkins D.K., and Van Doren T.W., Measurements of macrosonic standing waves in oscillating closed cavities. *The Journal of the Acoustical Society of America*, 1998, **104**: pp. 623–636.
68. Zinn B. and Crocco L., Periodic finite-amplitude oscillations in slowly converging nozzles. *Acta Astronautica*, 1968, **13**(5–6): pp. 481–488.
69. Courant R. and Hilbert D., *Methods of Mathematical Physics*. 1989, John Wiley & Sons, Inc.
70. Nayfeh A.H., *Perturbation Methods*. 1973, Wiley.

7 Flame Sheet and Flow Interactions

Chapters 2–6 focused on disturbances in combustor environments and how they evolve in space and time. This chapter initiates the second section of this book, Chapters 7–9, which focus on reactive processes and their interactions with the flow. The flame acts as a volume/energy source that leads to rapid changes in flow properties or their derivatives, such as velocity, vorticity, or entropy. Wrinkling on the flame also leads to modification of the approach flow velocity field.

This chapter also analyzes flame surface dynamics. While much of the analysis is quite general for arbitrary surface dynamics in reacting flows, such as the dynamics of an isotherm or isoconcentration surface, it is most useful in the fast-chemistry limit where the flame is very thin relative to other length scales. Kinetically controlled phenomena are then treated in Chapter 8, which discusses ignition processes, and Chapter 9, which discusses premixed and nonpremixed flames.

Section 7.1 provides a general overview of different types of surfaces – passive, propagating, and constant property. The latter two surfaces are good models for premixed and nonpremixed flames respectively, whose dynamics are discussed in further detail in Section 7.2. Then, Section 7.3 works out the jump conditions across flamelets. A general observation is that the treatment of nonpremixed flames in each section of this chapter is almost always shorter than that of premixed flames. The reasons for this can be understood from Section 7.1 – namely that one cannot derive general expressions for the nonpremixed flame based on the local conditions, such as velocity. Rather, its evolution is governed by the entire mixture fraction field, which is geometry dependent.

Section 7.4 then introduces the topic of stretching of surfaces. This topic will be considered in detail in Section 9.3 in the context of premixed flame stretch, but is introduced from a kinematic viewpoint here. It also works through a specific example in detail, showing how vortices lead to such stretching. Section 7.5 treats the modification of the approach flow by a wrinkled premixed flame. This approach flow modification plays a key role in several topics analyzed in subsequent chapters, such as flame instabilities, flame flashback, and flame stabilization in shear layers.

7.1 Surface Dynamics [1, 2]

While combustion always occurs over a distributed volumetric region, in many applications of interest its thickness is much smaller than other hydrodynamic and acoustic length scales. In this case, the external flow and internal chemical kinetics can be decoupled. Such "flamelet" approaches are the most common for analytical treatment of laminar and turbulent flames. For premixed systems, the flamelet surface separates unburnt reactants from burnt products, while for nonpremixed systems it separates fuel and oxidizer.

This section discusses the dynamics and evolution of fields and isosurfaces, from both a general mathematical standpoint and specifically for combustion-related systems. Defining a general scalar field as $\Psi(\vec{x},t)$, a surface can be defined through an equation of the form $\Psi(X(\vec{x},t),t) = \Psi_o$, where $X(\vec{x},t)$ denotes the locus of points on the surface.

Passive surfaces follow the flow, but a general surface may move with respect to the flow with a velocity of s_d. This relative velocity, s_d, is defined by

$$D\Psi/Dt = \partial\Psi/\partial t + \vec{u} \cdot \nabla\Psi = s_d |\nabla\Psi|, \tag{7.1}$$

so that s_d is given by

$$s_d\Big|_{\Psi_o} = \frac{D\Psi/Dt}{|\nabla\Psi|}\Big|_{\Psi_o}. \tag{7.2}$$

Consider three types of surfaces: *material* surfaces, *propagating* surfaces, and *constant-property* surfaces. A material surface (e.g., a surface of passive tracer particles) is a passive interface between two fluids that is advected by the flow, $\vec{u}$. There is no relative motion between it and the flow, so that $s_d = 0$. A propagating surface is a surface which propagates normal to itself at a given velocity relative to the fluid. Premixed flames are propagating surfaces whose relative propagation velocity is referred to as the displacement speed, s_d. The displacement speed is a property of the mixture, ambient conditions, flame stretch rate, and so forth, as detailed in Section 9.1.2. Note that there is another laminar flame speed, the consumption speed, s_c, which is defined in Section 9.3.4; all premixed flame speeds in this chapter refer to the displacement speed.

A constant-property surface is a surface where Ψ is constant. Nonpremixed flame sheets are constant-property surfaces, as they occur along the constant-property surface where the mixture fraction, Z, has its stoichiometric value, $Z = Z_{st}$. Constant-property surfaces do not propagate, yet they are also not passive scalars where $s_d = 0$; rather, their relative speed is given by Eq. (7.2). An explicit formula for the flow speed with respect to a given mixture fraction isosurface for a nonpremixed flame can be determined within the approximations of the mixture fraction equation derived in Eq. (1.26). For example, the velocity of the $Z = Z_{st}$ surface with respect to the flow is given by

$$s_d\Big|_{Z=Z_{st}} = \frac{DZ/Dt}{|\nabla Z|}\Big|_{Z=Z_{st}} = \frac{\nabla \cdot (\rho \mathscr{D} \nabla Z)}{\rho|\nabla Z|}\Big|_{Z=Z_{st}}. \tag{7.3}$$

One can similarly derive an equation for surface velocity under more general equations by casting the species or energy equation in the form of Eq. (7.2) and evaluating it at a given mass fraction value, $Y_{i,ref}$, or temperature, T. In that case, there would be a reaction rate term in addition to the diffusive term on the right-hand side of the expression in Eq. (7.3),

$$s_d\big|_{Y_{i,ref}} = \left[\frac{\dot{w}_i + \nabla\cdot(\rho\mathscr{D}_i\nabla Y_i)}{\rho|\nabla Y_i|}\right]_{Y_{i,ref}}. \tag{7.4}$$

For both a material surface and a propagating surface with constant s_d, the evolution of any surface element is fully described for a given velocity field and initial condition, and thus the evolution of each surface element can be solved independently of others. However, the temporal evolution of any surface element of a constant-property surface is controlled by the entire property field. This has important implications for the solution methods, since rather than considering evolution equations for a constant-property surface, one must solve for the Ψ field and extract the surface properties from the property field. For example, consider a propagating and constant-property surface flame embedded in a velocity field given by $\vec{u}(x,y,z,t)$, where the velocity field at the flame sheet is given by $\vec{u}(X(\vec{x},t),t)$. The premixed flame dynamics are only a function of $\vec{u}(X(\vec{x},t),t)$; this implies that, for a given $\vec{u}(X(\vec{x},t),t)$ along the flame surface, its space–time dynamics are the same for a variety of different velocity fields. In contrast, the space–time dynamics of the constant-property surface are a function of the entire velocity field, $\vec{u}(x,y,z,t)$, not just its value at the reaction sheet.

Consider a surface that is initially regular, defined as having finite curvature everywhere with no self-intersections or critical points. A material and constant-property surface remains regular during its evolution, while a propagating surface can develop singularities, i.e., values of infinite local curvature, and self-intersections. For premixed flames, this manifests itself as the formation of cusps, discussed extensively in Section 11.2.3, which are discontinuities in flame slope for constant-s_d flames.

7.2 Field Equations for Premixed and Nonpremixed Flames

Following this general introduction to surfaces, this section discusses field equations more specifically for premixed and nonpremixed flames. The objective is to present the equations and discuss their general properties. Solutions of the equations for different problems of interest are presented in Chapter 11.

7.2.1 Premixed Flames [3]

Consider a field variable, $G(\vec{x},t)$, in which the flame lies on the surface $G(X(\vec{x},t),t) = 0$. Away from the flame, where $G(\vec{x},t) \neq 0$, the value of G has no physical meaning. The utility of defining the larger field, however, lies in the ability to naturally handle multiconnected and multivalued surfaces. In a flame-fixed (Lagrangian) coordinate system, the fact that $G(X(\vec{x},t),t) = 0$ on the flame implies that

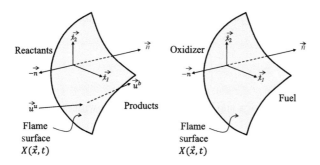

Figure 7.1 Instantaneous snapshot of premixed (left) and nonpremixed (right) flame sheets.

$$\frac{D}{Dt}G(\vec{x},t)\Big|_{\text{at the flame front}} = 0. \tag{7.5}$$

This expression can be converted into an Eulerian form,

$$\frac{\partial G}{\partial t} + \vec{v}_F \cdot \nabla G = 0, \tag{7.6}$$

where $\vec{v}_F$ denotes the flame front velocity, which can be decomposed into the local flow velocity vector and the scalar flame propagation speed with respect to the reactants, s_d^u, as

$$\vec{v}_F = \vec{u}^u - s_d^u \, \vec{n}, \tag{7.7}$$

where $\vec{n}$ is the normal vector at the flame front pointing to the products and $\vec{u}^u$ is the reactant velocity at the flame front. Because of discontinuities in flow velocity (see Section 7.3.1), note the necessity to define superscripts that differentiate between variables on the unburned (u) and burned (b) sides of the premixed flame.

Defining G as negative in the reactants and positive in the products, the flame normal is

$$\vec{n} = \nabla G / |\nabla G|. \tag{7.8}$$

Combining Eqs. (7.6), (7.7), and (7.8),

$$\frac{\partial G}{\partial t} + \vec{u}^u \cdot \nabla G = s_d^u \, |\nabla G|. \tag{7.9}$$

Equation (7.9) will be referred to as the "G-equation" [4]. It is quite general and can describe flames with complex, multiconnected, multivalued surfaces, such as the flame shown in Figure 7.2.

7.2.2 Nonpremixed Flames

Most analytical work on nonpremixed flame surface dynamics utilizes a mixture fraction formulation, due to its analytical tractability in the absence of source terms.

Figure 7.2 Instantaneous edge of a highly contorted premixed flame front.

As discussed in Section 1.4, the mixture fraction is defined as the mass ratio of material at each spatial location having its origin in the fuel stream, i.e., $Z = Z(\vec{x},t)$. Thus, Z takes values of zero and unity in the pure oxidizer and pure fuel streams, respectively, and is given for a simplified three-species system (consisting of fuel and oxidizer reacting to form a single product) by

$$Z = Y_F + \frac{Y_{Pr}}{(\varphi_{ox}+1)}, \qquad (7.10)$$

where φ_{ox} is the stoichiometric mass ratio of oxidizer to fuel. More general expressions can also be developed for systems with a larger set of intermediate species [5].

The mixture fraction has no source term if all species have equal diffusivity coefficients (see Exercise 1.8) $\mathscr{D}$, and is given by

$$\rho \frac{DZ}{Dt} - \nabla\cdot(\rho\mathscr{D}\,\nabla Z) = 0. \qquad (7.11)$$

To solve this equation, the density must be related to the mixture fraction. The simplest approach is to assume an isodensity field, although this is certainly an oversimplification for combustion applications. However, rather than assuming constant density, the less restrictive assumption of constant $\rho\mathscr{D}$ can be utilized (although it actually varies as $T^{1/2}$ for perfect gases). The modified form of the mixture fraction equation, to be referred to throughout this text as the Z-equation, is given by

$$\frac{\partial Z}{\partial t} + \vec{u}\cdot\nabla Z = \mathscr{D}\,\nabla^2 Z. \qquad (7.12)$$

The flame lies on the locus of points defining the flame sheet, given by the parametric equation $Z(X(\vec{x},t),t) = Z_{st}$.

Finally, note that for premixed combustion the mixture fraction is uniform everywhere, assuming equidiffusive species, and thus the mixture fraction based conservation equation provides no new information.

7.2.3 Comparison of Premixed and Nonpremixed Flame Evolution Equations

It is helpful to compare the dynamics and governing features of the Z-equation, given by Eq. (7.12), for nonpremixed flames with the G-equation, given by Eq. (7.9), for premixed flames. The two expressions have the same convection operator on the left-hand side, which shows that flow perturbations in the direction normal to the flame sheet move the flame sheet. However, the right-hand sides of these two expressions are different; the premixed flame expression has the normal flame propagation operator, $s_d|\nabla G|$, while the nonpremixed flame expression has a diffusion operator, $\mathcal{D}\nabla^2 Z$. This difference is significant and reflects, among other things, the fact that nonpremixed flames do not propagate. Moreover, this propagation term makes the premixed flame dynamics equation nonlinear, while the nonpremixed flame dynamics equation is linear (assuming $\vec{u}$ and $\mathcal{D}$ are not functions of Z).

Another significant difference is that the G-equation is physically meaningful and valid *only* at the flame itself, where $G(\vec{x},t) = 0$. Although it can be solved away from the flame, the resulting G values have no physical significance. In contrast, the Z-equation describes the physical values of the mixture fraction field *everywhere*. This point was also alluded to earlier for constant-property surfaces, whose dynamics are determined by the instantaneous property field, rather than simply the velocity at the surface. Thus, the entire mixture fraction field must be solved in the nonpremixed problem and the $Z(\vec{x},t) = Z_{st}$ surface (which generally cannot be expressed explicitly) extracted from the resulting solution field. The practical implication of this result is that developing explicit solutions for the nonpremixed flame problem is not as straightforward as for the premixed problem.

These examples can be made explicit in cases where the flame position is a single-valued function of some coordinate, as shown in Figure 7.3. For reasons explained later, we define the flame position as a function of y and x for the premixed and

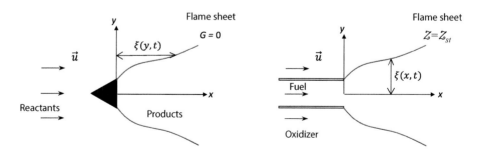

Figure 7.3 Examples showing definition of flame position for premixed and nonpremixed flame configurations.

nonpremixed cases, respectively. For example, in the two-dimensional case, the instantaneous position of the reaction sheet is given by $x = \xi(y,t)$, and $G(x,y,t) \equiv x - \xi(y,t)$. Substituting this expression into Eq. (7.9) leads to the explicit expression for flame position, Eq. (11.19), that we will use extensively in Chapter 11 to solve a number of problems. Note, however, that this substitution for G arbitrarily assigns values to the G field away from the flame itself, namely that G varies linearly with distance from the flame. Since the G field is completely arbitrary away from the flame this is allowable. However, we cannot make an analogous substitution for the nonpremixed system, such as $Z(\vec{x},t) \text{-} Z_{st} = y - \xi(x,t)$, as this assigns values to the Z field away from $Z(\vec{x},t) = Z_{st}$. Rather, the entire mixture fraction field must be solved for, and its position, $\xi(x,t)$, extracted from this larger solution. Examples of this procedure are shown in Section 11.3.4.

7.3 Jump Conditions [3, 6–10]

Having considered the equations describing front dynamics, this section considers the conditions relating flow properties on both sides of the flame, i.e., "jump conditions."

7.3.1 Premixed Jump Conditions

This section develops the conditions relating properties just up- and downstream of a premixed flame sheet. For simplicity of notation we will not always show the arguments, but note that all quantities in this section are evaluated at the flame surface. For example, although the velocity is a field quantity and defined throughout the domain $\vec{u}(x,y,z,t)$ where it is used, in this section it is evaluated at the flame, $\vec{u}(X(\vec{x},t),t)$. A schematic of the flame and flow is shown in Figure 7.1. The superscripts u and b denote values on the unburned and burned side of the flame, respectively. The local normal to the flame front, $\vec{n}$, points from the unburned to the burned side. We also define $\vec{t}_1$ and $\vec{t}_2$ as vectors tangential to the flame sheet.

Starting with mass conservation, note that the mass flux entering the flame sheet is given by

$$\dot{m}''_{in} = \rho^u (\vec{u}^u - \vec{v}_F) \cdot \vec{n} . \tag{7.13}$$

The quantity $(\vec{u}^u - \vec{v}_F) \cdot \vec{n}$ is equal to the laminar displacement speed, s_d^u; as all flame speeds in this section are displacement speeds (as opposed to consumption speeds), we will suppress the d subscript for notational simplicity:

$$s^u = (\vec{u}^u - \vec{v}_F) \cdot \vec{n} . \tag{7.14}$$

Similarly, the burning velocity of the front with respect to the burned gases is given by

$$s^b = (\vec{u}^b - \vec{v}_F) \cdot \vec{n} . \tag{7.15}$$

When integrated over a control volume, the mass conservation equation in Eq. (1.8) states that the difference between the mass flux through the front and back faces of the control volume is equal to the unsteady mass accumulation and mass flux through the sides of the control volume. In the limit of an infinitely thin flame, this implies that the mass flux into and out of the flame are equal, i.e., $\rho^u s^u = \rho^b s^b$. More generally, one can develop the following jump condition for mass flux across the flame:

$$[\rho s] = \Phi_m, \tag{7.16}$$

where the square bracket denotes the difference between the quantity evaluated just upstream (i.e., the unburned side) and downstream (i.e., the burned side) of the flame:

$$[X] = X^b - X^u. \tag{7.17}$$

The quantity Φ_m denotes sources of mass flux and can be anticipated to be small in the limit of thin flames. This and other source terms are discussed in Section 7.6.

We next consider the momentum equation. Since this is a vector equation, we will consider the momentum flux normal and tangential to the flame separately. The normal momentum flux into the flame is given by $\rho^u s^u \vec{u}^u \cdot \vec{n}$. The momentum equation states that the difference between the normal momentum flux into and out of the control volume equals the sum of the net force on the control volume and any unsteady momentum accumulation within the control volume. The net normal force on the control volume equals the sum of the pressure difference across the flame and contributions from normal viscous stresses and body forces. Neglecting viscous terms and body forces, this leads to [7]:

$$[\rho s \, \vec{u} \cdot \vec{n} + p] = \Phi_{M,n}, \tag{7.18}$$

where $\Phi_{M,n}$ denotes sources of momentum flux.

The tangential momentum equation takes the form

$$[\rho s (\vec{n} \times \vec{u} \times \vec{n})] = \vec{\Phi}_{M,t}. \tag{7.19}$$

Note that $\vec{u}_t = \vec{n} \times \vec{u} \times \vec{n}$ denotes the velocity component tangential to the flame front. Neglecting the source term, $\vec{\Phi}_{M,t}$, leads to the approximation that the tangential velocity components are continuous across the flame front.

Last, we consider the energy equation using the stagnation enthalpy formulation, Eq. (1.29). This equation states that the difference between the stagnation enthalpy flow rate into and out of the control volume equals the sum of the net heat flux into the control volume and any work terms associated with viscous stresses and body forces. Assuming that the interfaces of the control volume are placed before and after any regions of spatial gradients in temperature, or other variables causing heat flux, see Eq. (1.28), this heat flux term is zero. Neglecting work done by viscous and body forces, this leads to

$$[\rho s h_T] = \Phi_E. \tag{7.20}$$

7.3.2 Nonpremixed Jump Conditions

Jump conditions based on one-step kinetics stipulate that the diffusive fluxes of fuel and oxidizer into the reaction sheet occur in stoichiometric proportions (see Section 9.5 and Eq. (9.30) specifically), and that the jump in sensible enthalpy gradient on the fuel and oxidizer side is directly proportional to the fuel/oxidizer diffusive flux (i.e., the mass burning rate). The reader is referred to section 5.5.1 in Law [11] or section 3.1.5 in Williams [12] for these nonpremixed flame derivations.

Note that all the jump conditions described in the previous section can be equally applied to determining the relationships for pressure, velocity, vorticity, and so forth for a nonpremixed flame, by utilizing the expression for s_d from Eq. (7.3). However, there is no density jump across a nonpremixed flame, only a jump in density gradient. Because of the continuity of density, Eq. (7.16) immediately implies that the velocity of the front with respect to the gases, s_d, is the same on both sides of the nonpremixed flame. Similarly, working through the other relations in Sections 7.3.1 and 7.3.4 shows that the normal velocity and vorticity, tangential velocity and vorticity, and pressure must similarly be continuous.

To summarize, a key difference between the premixed and nonpremixed jump conditions lies in the jump in scalars and velocity across the premixed flame, while these quantities are continuous across nonpremixed flames. Rather, it is gradients in properties that may be discontinuous for nonpremixed flames.

7.3.3 Velocity and Pressure Relations [13]

The flame has important influences on the velocity and pressure fields because of gas expansion. This section works out these relations by applying the expressions presented in the previous section with the source terms, Φ, set to zero. A schematic showing the coordinate system and velocity fields just upstream and downstream of the flame is shown in Figure 7.4.

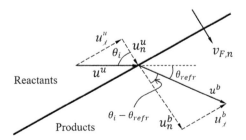

Figure 7.4 Schematic showing velocity components and angles at the premixed flame.

As shown in the figure, the incoming flow, u^u, approaches the flame horizontally, which is oriented at an angle of θ_i with respect to the normal. The approach flow is decomposed into a component tangential, u_t^u, and normal, u_n^u, to the flame. The burned gas flow is accelerated to the value u^b, and refracted by the angle θ_{refr} across the flame. The flame itself is moving at the velocity $\vec{v}_F$, with the velocity magnitude in the direction normal to itself given by $v_{F,n}$.

Following Eqs. (7.14) and (7.15), we can write the following relationships between the normal velocity, burning velocity, and velocity of the flame sheet:

$$u_n^u = s^u + v_{F,n}, \quad (7.21)$$

$$u_n^b = s^b + v_{F,n}. \quad (7.22)$$

From the mass conservation in Eq. (7.16), we can write the following relations between the burning velocities relative to the reactants and products:

$$s^b = \frac{\rho^u}{\rho^b} s^u. \quad (7.23)$$

Define the gas expansion ratio across the flame as

$$\sigma_\rho = \frac{\rho^u}{\rho^b}. \quad (7.24)$$

The tangential velocity components remain unchanged across the flame, as shown by Eq. (7.19):

$$u_t^u = u_t^b \equiv u_t. \quad (7.25)$$

We can then write the following relationships for the burned gas velocity:

$$u^b = \sqrt{(u_t^b)^2 + (u_n^b)^2} = \sqrt{(u^u \sin \theta_i)^2 + (\sigma_\rho u^u \cos \theta_i - (\sigma_\rho - 1)v_{F,n})^2}. \quad (7.26)$$

This equation quantifies the acceleration of the flow across the flame due to gas expansion. Referring to Figure 7.4, the deflection angle of the burned gas as it crosses the flame, θ_{refr}, can be readily calculated using trigonometric relations:

$$\tan(\theta_i - \theta_{refr}) = \frac{u_t^b}{u_n^b} = \frac{u^u \sin \theta_i}{\sigma_\rho u^u \cos \theta_i - (\sigma_\rho - 1)v_{F,n}}. \quad (7.27)$$

An image illustrating the deflection or refraction of the flow across the flame is shown in Figure 7.5.

While these expressions are valid instantaneously, they are more amenable to insight if we assume steady flow, so that the flame is stationary, $v_{F,n} = 0$ (or, more generally, if we revert to a flame-fixed coordinate system). Note the following trigonometric relations:

$$\frac{s^u}{u^u} = \cos \theta_i, \quad (7.28)$$

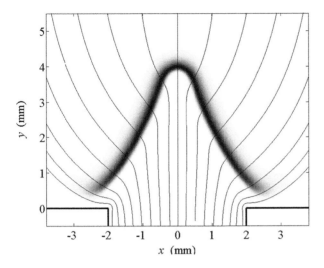

Figure 7.5 Computed streamlines showing flow refraction as it traverses the flame. The shaded region denotes reaction rate. Image courtesy of S. Hemchandra.

$$\frac{u^u_t}{u^u} = \sin \theta_i. \tag{7.29}$$

We can then write the following expression for the burned gas velocity:

$$u^b = u^u \sqrt{1 + \left(\sigma_p{}^2 - 1\right)\cos^2\theta_i}. \tag{7.30}$$

This equation quantifies the acceleration of the flow across the flame by gas expansion. The corresponding deflection of the flow across the flame is given by

$$\tan\left(\theta_i - \theta_{refr}\right) = \frac{\tan \theta_i}{\sigma_p}. \tag{7.31}$$

This can be simplified to

$$\theta_{refr} = \theta_i - \tan^{-1}\left(\frac{\tan \theta_i}{\sigma_p}\right). \tag{7.32}$$

This relationship between refraction angle and incidence angle is plotted in Figure 7.6 for several gas expansion ratios.

Note that the refraction angle is identically zero if there is no gas expansion, $\sigma_p = 1$, or in the limits of normal or tangential incidence to the flame. The flow refraction angle, θ_{refr}, monotonically increases with gas expansion ratio, σ_p, at all approach flow angles, θ_i. However, θ_{refr} depends nonmonotonically on the incident flow angle, θ_i. For low approach flow angles where $\theta_i < \theta_i|_{\theta_{refr,\,max}}$, the refraction angle monotonically rises with approach flow angle. The maximum refraction angle of the burned gas, $\theta_{refr,\,max}$, is a function of gas expansion ratio:

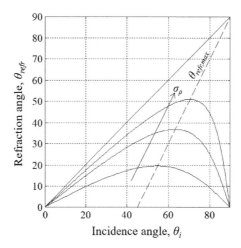

Figure 7.6 Dependence of flame refraction angle, θ_{refr}, on flow incidence angle, θ_i, at gas expansion ratios of $\sigma_\rho = 2, 4, 8$, and ∞.

$$\tan \theta_{refr,\,max} = \frac{\sigma_\rho - 1}{2\sqrt{\sigma_\rho}}. \tag{7.33}$$

The corresponding incidence angle at which the flow refraction angle is largest is given by

$$\tan\left(\theta_i|_{\theta_{refr},\,max}\right) = \sqrt{\sigma_\rho}. \tag{7.34}$$

Finally, note that for very large approach flow angles, $\theta_i = 90 - \varepsilon$, such as occurs when $s^u \ll u^u$, the flow passes through the flame with negligible velocity change, i.e.,

$$u^b = u^u + O(\varepsilon^2). \tag{7.35}$$

There is an $O(\varepsilon)$ influence on the flow angle:

$$\theta_{refr} = (\sigma_\rho - 1)\varepsilon + O(\varepsilon^3). \tag{7.36}$$

Consider next the pressure drop across the flame. Using the normal momentum jump condition, Eq. (7.18), we can write

$$p^u + \rho^u(s^u)^2 = p^b + \rho^b(s^b)^2. \tag{7.37}$$

Note that this expression is valid at each instant of time and does not assume steady flow. The flame speed with respect to the burned gas, s^b, can be written in terms of s^u to yield

$$p^b - p^u = -(\sigma_\rho - 1)\rho^u(s^u)^2. \tag{7.38}$$

This quantifies the pressure drop across the flame, reflecting the force required to accelerate the fluid to the burned gas velocity. This flow acceleration is a direct result of gas expansion. This expression is independent of the relative angle between the flame and the flow, since the tangential velocity remains unchanged across the flame. For low Mach number flows, the fractional pressure change relative to the static pressure is negligible, and is of $O((M^u)^2)$:

$$\frac{p^b - p^u}{p^u} = \frac{-(\sigma_\rho - 1)(s^u)^2}{\mathscr{R}T^u} = -\gamma(\sigma_\rho - 1)\left(\frac{s^u}{c^u}\right)^2 = O((M^u)^2). \tag{7.39}$$

However, the fractional pressure change relative to the dynamic head in the flow, $\rho^u(u^u)^2$, is of $O(1)$. In the flame-fixed coordinate system $(v_{F,n} = 0)$,

$$\frac{p^b - p^u}{\rho^u(u^u)^2} = -(\sigma_\rho - 1)\cos^2\theta_i. \tag{7.40}$$

We next consider the stagnation pressure change across the flame. Recall that for low Mach number flows, the stagnation pressure, p_T, equals the dynamic head, $p + (\rho u^2)/2$, to $O(M^4)$. This can be seen from the following series expansion:

$$p_T = p\left(1 + \frac{(\gamma - 1)}{2}M^2\right)^{\gamma/(\gamma-1)} \approx p\left(1 + \frac{(\gamma - 1)}{2}\frac{\gamma}{(\gamma - 1)}M^2 + O(M^4)\right), \tag{7.41}$$

$$p_T \approx p + \frac{\gamma p \, u^2}{2 \, c^2} = p + \frac{\rho u^2}{2}. \tag{7.42}$$

The leading-order stagnation pressure drop across the flame in a steady flow is then given by

$$\frac{p_T^b - p_T^u}{\rho^u(u^u)^2} = -\frac{(\sigma_\rho - 1)}{2\sigma_\rho}\left((\sigma_\rho - 1)\cos^2\theta_i + 1\right). \tag{7.43}$$

These fractional changes in static and stagnation pressure are plotted in Figure 7.7. To illustrate typical results, there is a static pressure drop of $1.5\rho^u(u^u)^2$ at a gas expansion ratio of $\sigma_\rho = 4$ and $\theta_i = 45$. Not all of this pressure is irreversibly lost, as it partially goes into flow acceleration. However, some of it is lost, as the figure also shows that there is a dynamic pressure loss at the same condition of $\sim 0.9\rho^u(u^u)^2$.

The loss of stagnation pressure across the flame is related to the corresponding entropy rise. The entropy change can be deduced from the $T - ds$ relation, Eq. (1.2). In order to write the simplest expression, we will assume constant specific heat and molecular weight, so that the entropy change across the flame is

$$\frac{s^b - s^u}{\mathscr{R}} = \frac{c_p}{\mathscr{R}}\ln\frac{T^b}{T^u} - \ln\frac{p^b}{p^u} + \sum_{i=1}^{N}\frac{1}{\mathscr{R}MW_i}\int_u^b \frac{\mu_i}{T}dY_i. \tag{7.44}$$

The pressure ratio in this equation is equal to unity, with a correction of $O\left((s^u/c^u)^2\right)$. Similarly, the static temperature ratio is also equal to the density ratio to $O\left((s^u/c^u)^2\right)$; i.e.,

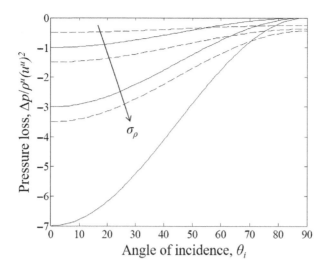

Figure 7.7 Dependence of static (solid) and stagnation (dashed) pressure changes across the flame on flame incidence angle at density ratios of $\sigma_\rho = 2, 4$, and 8.

$$\frac{T^b}{T^u} = \sigma_\rho + O\left(\left(\frac{s^u}{c^u}\right)^2\right). \tag{7.45}$$

This leads to the following leading-order expression for entropy jump across the flame, which is directly proportional to the sum of the logarithm of the gas expansion ratio, σ_ρ, and the chemical reaction contribution:

$$\frac{s^b - s^u}{\mathcal{R}} = \frac{\gamma}{\gamma - 1} \ln \sigma_\rho + \sum_{i=1}^{N} \frac{1}{MW_i} \int_u^b \frac{\mu_i}{T} dY_i. \tag{7.46}$$

7.3.4 Vorticity Relations and Vortex–Flame Interactions

7.3.4.1 General Considerations for a Prescribed Flame Position [10, 13]

In this section we consider the effect of the flame on fluid vorticity for inviscid flows with the source terms, Φ, described in Section 7.3.1, set to zero. It is helpful to begin with an overview of the different terms in the vorticity transport equation for inviscid flows, as shown below. The reader is referred to Section 1.6 for a more detailed discussion of this equation:

$$\frac{D\vec{\Omega}}{Dt} = (\vec{\Omega} \cdot \nabla)\vec{u} - \vec{\Omega}(\nabla \cdot \vec{u}) - \frac{\nabla p \times \nabla \rho}{\rho^2}. \tag{7.47}$$

The first term on the right-hand side of this equation, $(\vec{\Omega} \cdot \nabla)\vec{u}$, is the vortex stretching/bending term. Because of the refraction and acceleration of the flow by

7.3 Jump Conditions

the flame, this term has significant influences on the vorticity incident on the flame. Refraction bends the inclination of vortex tubes, potentially generating vorticity in different components than the incident tube, while flow acceleration stretches them, increasing their strength. The dilatation term, $\vec{\Omega}\left(\nabla\cdot\vec{u}\right)$, acts to decrease the vortex tube strength. Note that both of these terms modify vorticity that already exists – they equal zero if the vorticity is zero. The third term, the baroclinic source $\frac{\nabla p\times\nabla\rho}{\rho^2}$, generates new vorticity that did not exist in the approach flow.

In this section we prescribe the flame position in order to obtain analytical solutions. For actual quantitative calculations, this assumption is relevant to situations with very weak vortices, $u_\theta/s^u \ll 1$, so that the flame position is barely perturbed. In general, the flame position must be solved for simultaneously with flow; this problem is discussed in Section 7.3.4.2.

The velocity jump conditions across the flame discussed in Section 7.3.3 suggest that there will be corresponding changes in vorticity. However, this conclusion is actually too simplistic, as the processes leading to jumps in velocity and vorticity are different. For example, in the special case of a parallel, irrotational flow entering the flame at some angle θ_i and leaving at the angle θ_{refr}, such as shown in Figure 7.4, no vorticity is produced at the flame. Although the flow is refracted across the flame, there are no shear or velocity gradients in the post-flame gases. Rather, as we will show in this section, vorticity production is intimately linked to streamline curvature of the approach flow gases and variations in flow properties along the flame front. Thus, the mere fact that the flow velocity changes across the flame does not imply vorticity production. This point can also be understood from analysis of Crocco's equation in Exercises 1.4–1.6 of Chapter 1.

This point suggests that there is a significant distinction between the jump conditions for vorticity considered next, and those developed earlier for pressure and velocity. Vorticity jump conditions depend not only on the local values of variables, but also on their gradients. In contrast, the pressure, velocity, and entropy jump conditions depend only on the local value of the variables upstream and downstream of the flame. For these reasons, the development in this section is more involved than those in the earlier section.

Start by decomposing the local velocity and vorticity vectors into their components normal and tangential to the flame,

$$\vec{u} = u_n\vec{n} + \vec{u}_t, \tag{7.48}$$

where the subscript and unit vector n is along the local normal and t is the vector component along the flame surface; see Figure 7.8.

The vorticity vector is given by

$$\vec{\Omega} = \nabla\times\vec{u} = \Omega_n\vec{n} + \vec{\Omega}_t, \tag{7.49}$$

where

$$\begin{aligned}\Omega_n &= (\nabla\times\vec{u}_t)\cdot\vec{n} \\ \vec{\Omega}_t &= \vec{\Omega} - \Omega_n\vec{n}.\end{aligned} \tag{7.50}$$

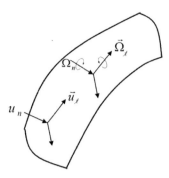

Figure 7.8 Coordinate system illustrating velocity and vorticity components at the flame.

Using the orthogonality of $\vec{n}$ and $\vec{u}_t$, i.e., $\vec{n} \cdot \vec{u}_t = 0$, the tangential component of vorticity can be written as (see Exercise 7.1):

$$\vec{\Omega}_t = \vec{n} \times \left(\frac{\partial \vec{u}_t}{\partial n} + (\vec{u}_t \cdot \nabla) \vec{n} - \nabla_t u_n \right). \quad (7.51)$$

The normal component, Ω_n, depends only on the tangential derivatives of the tangential components of the velocity. Because there is no jump in tangential velocity across the flame, see Eq. (7.19), this immediately implies that the tangential derivatives are continuous across the flame (normal derivatives are discontinuous). Hence, the normal vorticity can be seen from inspection to remain unchanged across the flame:

$$\Omega_n^u = \Omega_n^b. \quad (7.52)$$

There is a jump in tangential vorticity, which for a steady flow and flame position is given by

$$\vec{\Omega}_t^b - \vec{\Omega}_t^u = \vec{n} \times \left(\left(\frac{1}{\rho^b} - \frac{1}{\rho^u} \right) \nabla_t (\rho u_n) - \frac{1}{\rho u_n} (\rho^b - \rho^u)(\vec{u}_t \cdot \nabla_t)\vec{u}_t \right). \quad (7.53)$$

Derivation of this expression is left as a series of exercises (Exercises 7.1, 7.2, and 7.3). Note that the ρu_n product is constant through the flame by virtue of the assumption of steady flame/flow (combine Eqs. (7.16), (7.21), and (7.22)), so either the burned or unburned value can be used. Both terms in this expression are proportional to the density jump across the flame. The first term is also proportional to the tangential derivatives of the normal mass flow into the flame. The second term is proportional to tangential velocity variations along the flame. These points can be seen more clearly for a two-dimensional flow field, where only the out-of-plane vorticity (along t_2) is nonzero:

$$\text{Two-dimensional case: } \Omega_{t_2}^b - \Omega_{t_2}^u = \left(\left(\frac{1}{\rho^b} - \frac{1}{\rho^u} \right) \frac{\partial}{\partial t_1} (\rho u_n) - \frac{1}{\rho u_n} (\rho^b - \rho^u) \left(u_{t_1} \frac{\partial u_{t_1}}{\partial t_1} \right) \right). \quad (7.54)$$

An alternative representation for this two-dimensional jump condition is [13]:

$$\Omega_{f_2}^b - \Omega_{f_2}^u = -\left(1 - \frac{1}{\sigma_\rho}\right) \frac{s^u \, d\tan^2\theta_i}{2 \quad dt_1}, \tag{7.55}$$

where θ_i is the local angle between the flame surface normal and the flow vector. This expression shows that the vorticity jump is related to the density jump across the flame and the rate of change of the angle between the flow and the flame along the flame surface. Another illustrative variant for two-dimensional flows that explicitly shows the effect of gas expansion on the approach flow vorticity, $\Omega_{f_2}^u$, and curvature of the approach flow streamlines is [13]:

$$\Omega_{f_2}^b = \frac{\Omega_{f_2}^u}{\sigma_\rho} - \left(1 - \frac{1}{\sigma_\rho}\right)\left(\frac{u^u}{a_\psi} + \frac{\partial u^u}{\partial s_\psi}\right), \tag{7.56}$$

where a_ψ is the curvature of streamlines in the unburned gas and s_ψ is the coordinate along a streamline. This expression shows that the vorticity in the burned gas is equal to that in the unburned fluid decreased by the gas expansion ratio, σ_ρ, because each fluid element expands by this factor in passing through the flame. It also shows a vorticity source term associated with the curvature of the streamlines in the unburned fluid and the rate of expansion of the stream tubes. Further discussion of these vorticity jump conditions is presented in Section 7.7.

7.3.4.2 Considerations for a Coupled Flame and Flow Field

The above expressions for vorticity relations across the flame are formally correct and provide some insight into the mechanisms through which vorticity is produced or modified by the flame. However, these expressions do not present the entire story, as solving them requires specifying a flame position. In reality, the flame position itself is generally unknown and is a function of the disturbance field. As such, the above results are only quantitatively useful for analyzing flame interactions with very weak vortices, where the flame position is hardly altered. Otherwise, the flame position and flame–flow angles are themselves unknowns and must be solved for simultaneously. In this section we consider two-way interactions associated with strong vortex–flame interactions in order to show the additional prevalent physics.

This problem is not amenable to analytical solutions and therefore we revert to discussions from computational and experimental studies to elucidate its various aspects. Consider first the situation where the vortex axis is tangential to the flame front [14–17]. Vortices cause the flame to wrap around them, and "scramble" the local alignment of the density and pressure gradients, causing different signs of baroclinic torque along the flame [16]. For example, Figure 7.9 illustrates a counter-rotating vortex pair, with a rotational velocity that is of the order of the flame speed, $u_\theta/s^u \sim O(1)$, impinging on a flame [14–16]. The pressure gradient is in the direction of the flow. The resulting flame wrinkling causes significant flame-generated vorticity through the baroclinic mechanism. This baroclinic vorticity has the opposite sign to the incident vortex – this can be seen by evaluating the sign of the baroclinic term $\nabla p \times \nabla \rho$ using the ∇p and $\nabla \rho$ vectors illustrated in Figure 7.9. The combined effects

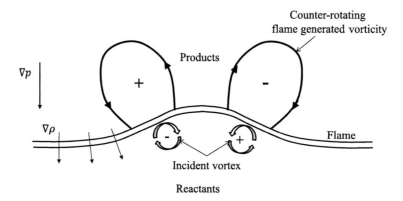

Figure 7.9 Illustration of a counter-rotating vortex pair generated by an incident vortex pair of strength $u_\theta/s^u \sim O(1)$ in a flow with a favorable pressure gradient. Incident vortices are completely attenuated and baroclinic torque generates counter-rotating vortices that may be stronger than the incident vortices.

of dilatation and this flame-generated vorticity causes the incident vortex to completely disappear as it passes through the flame!

The counter-rotating vortex pair that appears on the product side has opposite directions of rotation and a strength that exceeds that of the incident vortex. Note also that this oppositely signed vorticity tends to damp the flame wrinkling imposed by the incident vortex, moderating the wrinkling effect. Increasing vortex strength, or an adverse pressure gradient, causes some of the incident vortex to survive, leading to pockets of co-rotating and counter-rotating vorticity in the products side of the flame. As such, the flame wrinkling and vorticity characteristics are a highly nonlinear function of u_θ/s^u, and qualitatively different phenomena occur at different vortex strengths. Moreover, these results are strongly dependent on the direction and magnitude of pressure gradients in the flow, such as if the bulk flow is accelerating or decelerating. For example, in a flow with no nominal pressure gradient, flame wrinkling controls the direction of both the local pressure gradient and density gradient. In contrast, in a flow with a strong pressure gradient, the direction of the pressure gradient will be less sensitive to flame wrinkling. Finally, sufficiently strong vortices can cause local flame extinction, a phenomenon discussed in Section 9.10, leading to little change in the incident vortex as it passes through the flame.

This effect is also illustrated by flames stabilized in acoustically perturbed shear layers, as shown in Figure 7.11. In these situations, the strength of the unsteady vorticity associated with the separating shear layer is modulated. At low disturbance amplitudes, Figure 7.10(a), the flame is only weakly wrinkled by the fluctuating vortical velocity, and the sign of the baroclinic torque remains constant along the flame. As the vortex strength increases, the amplitude of perturbations of the flame position and flame angle grow. Above a certain amplitude, the flame angle passes through the vertical, wraps around the vortex core, and causes the direction of the density gradient to change sign along the flame; see Figure 7.10(b).

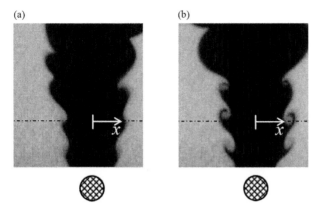

Figure 7.10 Mie scattering images of harmonically excited flames showing flame rollup by vorticity concentrations in the separating shear layers at (a) low and (b) high disturbance amplitudes [18]. The vorticity profile along the dashed line is plotted in Figure 7.11. Images courtesy of S. Shanbhogue.

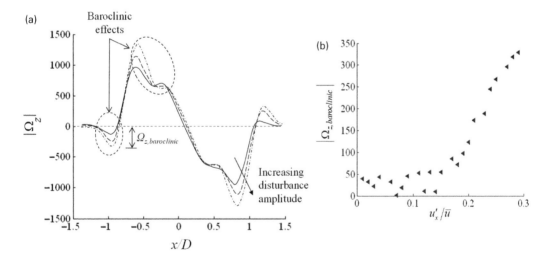

Figure 7.11 (a) Measured distribution of vorticity across the transverse cut indicated in Figure 7.10. (b) Amplitude dependence of vorticity generated through the baroclinic mechanism [18].

This wrapping of the flame around the vortex leads to baroclinic vorticity production that in some regions is of the same sign as the shear-generated vorticity, leading to a vorticity magnitude "overshoot." This can be seen in Figure 7.11(a), which plots the instantaneous vorticity magnitude along a transverse cut of the flame [18]. Note the increase in vorticity relative to the unforced case in the shear layer section closest to the centerline. Farther out from the centerline, the density gradient is of opposite sign

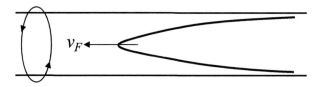

Figure 7.12 Illustration of a flame propagating toward the left inside a rotating tube [19].

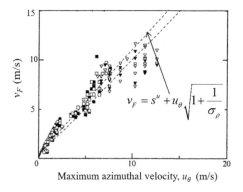

Figure 7.13 Dependence of flame velocity, v_F, on rotational velocity, u_θ. Adapted from Ishizuka [19].

and, for large disturbance magnitudes, leads to generation of countersigned vorticity through baroclinic effects. The nonlinear dependence of this baroclinic effect on disturbance amplitude is shown in Figure 7.11(b), which plots the magnitude of the vorticity peak of this countersigned vorticity.

The above discussion considered the situation where the axis of the vortex tube was tangential to the flame front. We next consider the situation where the axis of the vortex tube is normal to the flame front. Because such a vortex directly induces a velocity field that is only tangential to the flame front, it is not immediately clear that any interaction occurs, since Eq. (7.52) shows that tangential velocity and normal vorticity remain unchanged in passing through the flame. However, the velocity induced by baroclinic vorticity substantially increases the complexity and richness of this problem, as discussed next.

An interesting prototypical problem to illustrate these effects is that of a flame propagating through a rotating tube filled with premixed gases, as illustrated in Figure 7.12 [19]. The center of the flame propagates through the reactants with a speed, v_F, that monotonically increases with the rotation velocity of the gases in the tube, u_θ.

Typical correlations for v_F are [19]:

$$v_F = s^u + u_\theta F(\sigma_\rho), \qquad (7.57)$$

where σ_ρ is the density ratio across the flame. Data showing the increase of v_F with rotational velocity u_θ is shown in Figure 7.13.

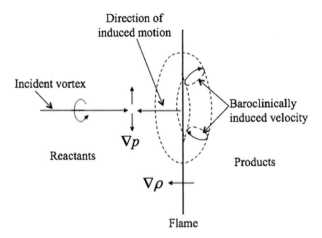

Figure 7.14 Schematic of a vortex normally incident on the flame front, showing the mechanism responsible for baroclinic vorticity production.

The precise mechanisms responsible for flame acceleration are not completely understood, but it seems clear that it is at least initially associated with baroclinic vorticity production [20], as shown in Figure 7.14.

Reverting to a cylindrical coordinate system, the incident vortex induces a radial pressure gradient. For a steady, inviscid flow with no radial component, it is given by

$$\frac{\partial p}{\partial r} = \frac{\rho u_\theta^2}{r}. \tag{7.58}$$

Note that this pressure gradient for a normally impinging vortex is orthogonal to the density gradient, producing azimuthal vorticity. This azimuthal vorticity, in turn, induces an axial flow field in the direction of the reactants, as shown in the figure. This allows the centerline of the flame to propagate into the resulting negative velocity region. Moreover, the curved front modifies the flow ahead of the flame in a way that also leads to flame front acceleration, as described in Section 7.5.

7.3.4.3 Flame Influences on Turbulent Flow Fluctuations

In this section we briefly consider the turbulent combustion problem, where the flame interacts with a random collection of eddies with a distribution of strengths, length scales and time scales. The nonlinear nature of this problem, emphasized in prior sections for even a single vortex, is further complicated by interactions of the flame with multiple vortices. It is this mutual interaction between the flame and the flow field that makes understanding the turbulent combustion problem so challenging.

However, certain general comments can be made that elucidate some key underlying physics of weak turbulence interactions with the flame [21, 22]. As discussed in Section 7.3.3, the flow normal to the flame accelerates because of gas expansion, while the tangential component remains constant. As a result, incoming isotropic turbulent flow is anisotropic downstream of the flame [23]. Pressure effects cause the flow to eventually revert to isotropy further downstream. The flame also alters the spectra of the turbulent fluctuations. For weak turbulence, the flame increases turbulent fluctuations at low frequencies and decreases them at higher turbulence intensities, with an overall enhancement of the average turbulent kinetic energy of the flow per unit mass [23].

Another important process is the increase in viscosity in the product gases due to higher temperatures. For example, the kinematic viscosity, v_ν, can increase by a factor of almost 20 (e.g., from 16×10^{-6} m^2/s at 300 K to 256×10^{-6} m^2/s at 1600 K), a very significant amount! The increase in the viscosity leads to a reduction in the product-side Reynolds number, which could result in relaminarization [5, 24] and/or damping of high-frequency turbulent fluctuations.

7.4 Stretching of Material and Flame Surfaces

Flame–vortex interactions provide an important introduction to the related topic of flame stretch, by which we mean the increase or decrease in length of material fluid elements. Flame stretch leads to an alteration in the flame structure and can even lead to flame extinction. We will consider stretching of surfaces from a kinematic perspective in this section, and consider the influences on flame structure and properties in Section 9.3.

7.4.1 Stretching of Material Surfaces

To start the discussion, consider the evolution of a material line in a flow, see Figure 7.15, consisting of a nondiffusive passive scalar, such as a line of dye. This material line changes in length due to tangential velocity gradients along the line, or the motion of curved portions of the line, as shown in the figure. We refer to this process as "stretch," which is due to both hydrodynamic straining and curved front motion. Note, however, the distinction between general hydrodynamic "flow strain" or deformation in shape of a material fluid volume, given by the tensor $\underline{\underline{S}} = \frac{1}{2}\left[\nabla \vec{u} + \left(\nabla \vec{u}\right)^T\right]$, and "stretch" of the material line. While the two are related, they are not the same.

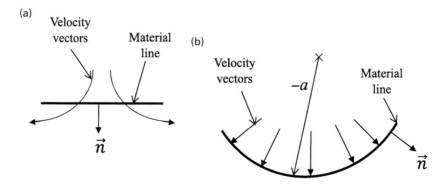

Figure 7.15 Stretching of a material line by (a) tangential velocity gradients along the line, and (b) motion of a curved line.

The fractional rate of change of length of a material line is given by [11, 25]:

$$\text{Material line:} \quad \frac{1}{L}\frac{dL}{dt} = \frac{\partial u_t}{\partial t} + (\vec{u}\cdot\vec{n})\nabla\cdot\vec{n}, \tag{7.59}$$

where $-\nabla\cdot\vec{n}$ equals the inverse of the radius of curvature, $1/a$, for a cylinder and $2/a$ for a sphere. The first term, $\frac{\partial u_t}{\partial t}$, shows that changes in tangential velocity along the line cause stretch, as illustrated in Figure 7.15(a). The second term, $(\vec{u}\cdot\vec{n})\nabla\cdot\vec{n}$, shows that the velocity normal to the line causes stretch if the line is curved, as shown in Figure 7.15(b). Generalizing this expression to three-dimensional flows, the fractional rate of change of area of a material surface is given by (see Figure 7.1 for nomenclature):

$$\text{Material surface:} \quad \frac{1}{A}\frac{dA}{dt} = \frac{\partial u_{t_1}}{\partial t_1} + \frac{\partial u_{t_2}}{\partial t_2} + (\vec{u}\cdot\vec{n})\nabla\cdot\vec{n}. \tag{7.60}$$

An alternative expression of Eq. (7.60) for an arbitrary coordinate system is

$$\frac{1}{A}\frac{dA}{dt} = -\vec{n}\cdot\nabla\times(\vec{u}\times\vec{n}) + (\vec{u}\cdot\vec{n})\nabla\cdot\vec{n}, \tag{7.61}$$

where the operator $-\vec{n}\cdot\nabla\times(\vec{u}\times\vec{n})$ is the tangential gradient operator that describes the variation of the tangential flow velocity in the tangential direction, also denoted as $\nabla_t\cdot u_t$.

7.4.2 Premixed Flame Stretch

Having considered passive scalar stretching, we next consider flame stretch. Analogous processes to those described above occur with flames, where flame surface area is increased by tangential flow gradients along the flame, and motion of the curved flame. The key difference is that the premixed flame propagates normal to itself – it is not a passive interface – so that the surface motion term in Eq. (7.60), $\vec{u}\cdot\vec{n}$, is replaced by flame motion $\vec{v}_F\cdot\vec{n} = \vec{u}\cdot\vec{n} - s^u$. We will denote the flame stretch rate by κ:

$$\kappa = \frac{1}{A}\frac{dA}{dt} = -\vec{n}\cdot\nabla\times(\vec{u}\times\vec{n}) + (\vec{v}_F\cdot\vec{n})\nabla\cdot\vec{n}. \tag{7.62}$$

Note that expanding and contracting flames have negative and positive radii of curvature, a, respectively. Also, κ has units of inverse time, showing that it can be used to define a stretching time scale, $\tau_\kappa = 1/\kappa$. The ratio of this time scale to a characteristic chemical time, τ_κ/τ_{chem}, is often referred to as the Karlovitz number, Ka, which can also be related to the inverse of a Damköhler number, Da, with τ_κ being replaced by the flow time scale.

7.4.3 Example Problem: Stretching of Material Line by a Vortex

This section considers an example problem to explicitly illustrate the distortion and stretching of material lines by a vortex. Consider a material line in a polar coordinate

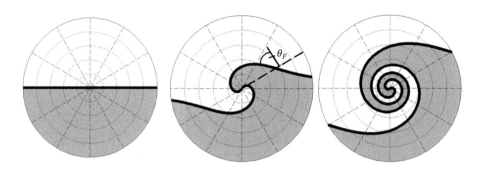

Figure 7.16 Winding of a material line by a vortex at increasing time instants from left to right ($\mathcal{G}/2\pi a_{core} = 0, 4,$ and 13).

system whose location is given by the function $\theta_L(r,t)$ that is initially aligned with the x-axis at $t = 0$, as illustrated in Figure 7.16(a). A two-dimensional vortex is centered at the origin with zero radial velocity, $u_r = 0$, and azimuthal velocity given by $u_\theta(r,t)$. Each element on the line is transported through a differential angle equal to the time integral of the azimuthal velocity:

$$\text{initially right of center: } \theta_L(r,t) = \int_0^t \frac{u_\theta(r,t)}{r} dt, \tag{7.63}$$

$$\text{initially left of center: } \theta_L(r,t) = \pi + \int_0^t \frac{u_\theta(r,t)}{r} dt. \tag{7.64}$$

Note that solid body rotation, where $u_\theta(r,t)$ increases linearly with r, causes the entire line to rotate with no distortion. To illustrate a result explicitly, consider the following profile which transitions from solid body rotation for $r \ll a_{core}$ to a potential vortex for $r \gg a_{core}$:

$$u_\theta(r,t) = \frac{\mathcal{G}}{2\pi(r/a_{core})} (1 - \exp(-(r/a_{core})^2)). \tag{7.65}$$

In general, a_{core} varies with time due to diffusion, but it is fixed as a constant here in order to explicitly integrate Eq. (7.63). Thus, the material line points initially right of center evolve as

$$\theta_L(r,t) = \frac{\mathcal{G}t}{2\pi(r^2/a_{core})} (1 - \exp(-(r/a_{core})^2)). \tag{7.66}$$

Figure 7.16 plots the location of the material line at several time instants, showing it winding up in the solid body vortex core, with two long trailing arms that asymptote out to the undisturbed portions of the line at large radial locations.

It is useful to change coordinate systems in order to see the velocity field with respect to the line. At each instant t, define a local coordinate system (n, t) at the material line element $\theta_L(r,t)$. Following Marble [26], the velocity field in this frame is given by

$$u_t(n,t) = r\frac{\partial(u_\theta/r)}{\partial r}[\sin(2\theta_F)t/2 + \cos(2\theta_F)n], \quad (7.67)$$

$$u_n(n,t) = -r\frac{\partial(u_\theta/r)}{\partial r}\sin(2\theta_F)t/2, \quad (7.68)$$

where $90 - \theta_F$ denotes the local instantaneous angle of the flame with respect to a ray emanating from the origin at the given θ_L, see Figure 7.16(b), given by

$$r\frac{\partial \theta_L(r,t)}{\partial r} = \cot\theta_F. \quad (7.69)$$

Then, using Eq. (7.59), the local stretch rate is given by

$$\frac{1}{L}\frac{dL}{dt} = \frac{\partial u_t(n=0,t=0)}{\partial t} + u_n(n=0,t=0)\nabla\cdot\vec{n} = r\frac{\partial(u_\theta/r)}{\partial r}\sin(2\theta_F)/2. \quad (7.70)$$

Exercise 7.4 provides an example calculation of the stretch rate showing that it is zero at $r=0$, increases to a maximum at some r value, and asymptotes to zero as $r\to\infty$. These zero stretch rates at the limiting r values occur because there is no stretch in the very center, which is rotating as a solid body, and because of the vanishing velocity gradients at large r.

7.5 Influence of Premixed Flames on the Approach Flow

This section analyzes how the approach flow is modified by the presence of premixed flames. The discussion in Section 7.3.3 on refraction of the flow across the flame, and the explicit expressions relating θ_i and θ_{refr}, are deceptively simplistic. The reason for this is that the angle between the flame and flow, θ_i, is not generally known because it is influenced by the flame, and therefore must be determined as part of the solution [27, 28]. Thus, there is a difference between the "nominal" angle between the flame and flow, θ_{nom}, and the actual angle, θ_i. Moreover, this relationship cannot be localized – i.e., we cannot generate a simple expression relating θ_{nom} and θ_i based on, for example, the density ratio, σ_ρ, and local features of the flame. Rather, this relationship is influenced by the flow, flame, and boundary conditions everywhere. While a general treatment of this issue is complicated by its intrinsic dependence on the geometric specifics of the problem, we present the results of a simplified problem in the rest of this section in order to illustrate a few general ideas. The main point we wish to show in the rest of this section is that flames bulging into the reactants (i.e., a convex wrinkle) cause the approach flow streamlines to diverge, and the flow ahead of the flame to decelerate. This is accompanied by a corresponding adverse pressure gradient ahead of the flame. The opposite happens when the flame is concave with respect to the reactants.

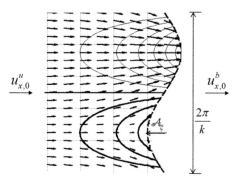

Figure 7.17 Illustration of the upstream velocity vectors and perturbed pressure field isobars induced by a wrinkle on a premixed flame. Thick and thin isobars denote positive and negative perturbation values.

A linearized solution capturing flame–flow coupling can be obtained by considering the modification of a nominally uniform approach flow that is induced by a small flame wrinkle, such as illustrated in Figure 7.17. We will assume the following nominal and perturbed flame positions:

$$\begin{aligned}\xi_0 &= 0\\ \xi_1 &= A_\xi e^{iky} e^{-i\omega t}.\end{aligned} \qquad (7.71)$$

The key objective of this analysis is to determine how the perturbation in flame position, A_ξ, influences the approach flow pressure and velocity.

We must be more precise in describing where this flame wrinkle originates, and will consider two variants of this problem. First, we will consider the forced response of the flame to a disturbance with a given frequency, $f = \omega/2\pi$, and disturbance length scale $L_{wrinkle} = 2\pi/k$, i.e., we will assume that a harmonic perturbation is carried toward the flow from upstream with axial velocity amplitude A_2^u, see Eq. (7.90). We will then relate the distortion that the resultant flame wrinkle imposes on the approach flow. We will also consider the unforced problem and analyze whether the steady-state solution is stable to small disturbances. As discussed in Section 9.8, flames may be unstable, meaning that wrinkles on the flame will grow. For this problem the frequency and wavenumber are related through a dispersion relation and cannot be independently specified. For the unforced problem, it is important to realize that while the results apply at every instant of time, the actual flame wrinkle, A_ξ, is growing exponentially in time.

The flame wrinkle perturbs the nominal velocity and pressure in both the upstream and downstream regions of the flame. The characteristics of these perturbed fields, such as their wavenumbers and propagation speeds, can be directly related to those of the wrinkle characteristics by means of the conservation equations and the dynamic flame front evolution equation. The velocity components and pressure can be decomposed into their base value and perturbation as

7.5 Influence of Premixed Flames on Approach Flow

$$u_x = u_{x,0} + u_{x,1},$$
$$u_y = u_{y,0} + u_{y,1} = u_{y,1}, \qquad (7.72)$$
$$p = p_0 + p_1.$$

The solutions for these flow perturbations are detailed in Section 7.8, where it is shown that they are functions of the parameters $St = \frac{\omega}{ku^u_{x,0}} = \frac{f L_{wrinkle}}{u^u_{x,0}}$ and $\sigma_\rho = \frac{\rho^u}{\rho^b} = \frac{u^b_{x,0}}{u^u_{x,0}}$. Figure 7.17 plots a representative result obtained from this solution showing the velocity and pressure fields induced by the flame wrinkle. A very important result from this plot is the effect of the flame bulge into the reactants – note that this causes a deceleration of the approach flow velocity and a corresponding adverse pressure gradient in the approach flow. This is important because the point of the flame that penetrates farthest into the reactant streams must, by geometric necessity, be oriented in this way. This result shows that the leading edge of a premixed flame always causes streamline divergence and flow deceleration. This result is very significant and will be returned to at multiple points in this text – it leads to inherent flame instabilities (Section 9.8), and plays an important role in flame flashback (Section 10.1.2.2) and flame stabilization in shear layers (Section 9.7.3).

These induced velocity and pressure fields are illustrated in Figure 7.18 by plotting the pressure and velocity distribution along the convex flame bulge. The figure clearly shows the flow deceleration and adverse pressure gradient imposed by a convex wrinkle. Although not shown, the opposite occurs for a concave wrinkle.

The magnitude of the adverse pressure gradient and flow deceleration at the "leading edge" of the flame front is quantified in Figure 7.19 as a function of density ratio across the flame, σ_ρ. As expected, these terms are identically zero in the limit of no density change, $\sigma_\rho = 1$, but grow monotonically with σ_ρ. For example, the curve shows that a wrinkle on the flame of dimensionless magnitude $k \mathcal{A}_\xi = \varepsilon$ induces a flow deceleration and adverse pressure gradient of 1.5ε and 1.1ε, respectively, at $\sigma_\rho = 4$

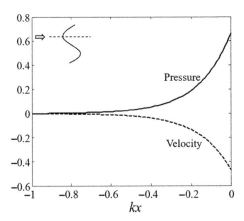

Figure 7.18 Axial distribution of the normalized pressure perturbation, $(1/k\mathcal{A}_\xi)(\mathcal{A}^u_1/u_{x,0})(1-iSt)$, and velocity perturbation, $(1/k\mathcal{A}_\xi)(\mathcal{A}^u_1/u_{x,0})$, for $St = 1$ and $\sigma_\rho = 3$, along the line sketched in the top left illustration, where the flame bulges into the reactants.

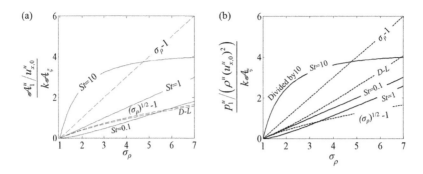

Figure 7.19 Dependence of the (a) normalized velocity and (b) pressure perturbations across a wrinkled flame on the flame density ratio and Strouhal number.

and $St = 1$. Also shown on the figure is the unforced result, calculated from the Darrius–Landau dispersion relation between ω and k that is presented in Section 9.8. Finally, the figure also shows curves of $\sqrt{\sigma_\rho} - 1$ and $\sigma_\rho - 1$ for reference.

For similar reasons, the flame also changes the characteristics of the turbulent fluctuations exciting it [29].

7.6 Aside: Finite Flame Thickness Effects on Flame Jump Conditions

This section presents the jump condition source terms derived in Section 7.3.1 [7]. First, the source terms scale as $O(\delta_F/L)$, where δ_F denotes the flame thickness and L the flow length scale. Thus, these terms are quite small except in instances where the flame curvature or vortex size is of the order of the flame thickness. In these cases, the inner flame structure and flame speed themselves are altered and, in fact, definitions of the burning velocity become ambiguous (see Section 9.3.4). These cases are discussed further in Chapter 9. A second important point is that the specific expressions for the jump conditions depend on the location within the flame region at which they are applied. In other words, changing the location within the flame zone at which the jump conditions are applied by some fraction of δ_F changes the jump condition source term by $O(\delta_F/L)$. Utilizing a jump condition location that leads to no mass sources in the flame, i.e., $\Phi_m = 0$, leads to the following momentum flux source terms [7]:

$$\Phi_{M,n} = -\frac{2\delta_F \rho^u (s^u)^2 I_\sigma}{a}, \tag{7.73}$$

$$\vec{\Phi}_{M,t} = -\delta_F s^u \vec{\nabla}_t (\rho^u s^u I_\sigma), \tag{7.74}$$

where $1/a$ denotes the mean curvature of the flame and

$$I_\sigma = \frac{4}{3}(Pr + 1)((T^b/T^u)^{3/2} - 1) - 2(T^b/T^u - 1). \tag{7.75}$$

The operator $\vec{\nabla}_t( ) = \vec{n} \times \nabla( ) \times \vec{n}$ denotes derivatives in the direction tangential to the flame front. The normal momentum flux source term at the interface, Eq. (7.73), is analogous to the jump condition across a droplet interface due to surface tension. In other words, the jump in normal stress across the curved interface equals a "surface tension coefficient" (which is negative) multiplied by the radius of curvature of the interface [7]. This negative "surface tension" term is a function of the gas expansion across the flame. Equation (7.74) also shows that there is a jump in tangential velocity across the flame that is related to the tangential derivative of the "surface tension" coefficient. This term is analogous to Marangoni forces [30] which arise when surface tension is nonuniform along the interface.

7.7 Aside: Further Analysis of Vorticity Jump Conditions across Premixed Flames

This section provides further analysis of the vorticity jump across a premixed flame. An alternative way to analyze the vorticity jump conditions in Section 7.3.4 is to directly integrate the vorticity equation, Eq. (7.47). The steady-state, scalar equation for a specific component of vorticity, Ω_i, is

$$(\vec{u} \cdot \nabla)\Omega_i = (\vec{\Omega} \cdot \nabla) u_i - \Omega_i (\nabla \cdot \vec{u}) - \left(\frac{\nabla p \times \nabla \rho}{\rho^2} \right)\bigg|_i. \tag{7.76}$$

This can be rearranged as

$$\nabla \cdot (\Omega_i \vec{u} - \vec{\Omega} u_i) = -\left(\frac{\nabla p \times \nabla \rho}{\rho^2} \right)\bigg|_i. \tag{7.77}$$

Consider first the normal component of the equation. Substituting the normal component for the index i and integrating this expression along a streamline yields

$$\underbrace{(u_n^b \Omega_n^b - u_n^u \Omega_n^u)}_{\text{I}} - \underbrace{(u_n^b \Omega_n^b - u_n^u \Omega_n^u)}_{\text{II}} = -\int_u^b \frac{\nabla p \times \nabla \rho}{\rho^2}\bigg|_n dA. \tag{7.78}$$

Notice that there is cancellation between the first bracketed term (I), the sum of vortex convection and dilatation terms, and the second bracketed term (II), the vortex stretching/bending term. Consequently, the jump in the normal component of the baroclinic term is identically zero:

$$\int_u^b \frac{\nabla p \times \nabla \rho}{\rho^2}\bigg|_n dA = 0. \tag{7.79}$$

Consider next the tangential vorticity component from Eq. (7.77). Integrating this expression along a streamline (note that this expression is really two equations, one for each tangential direction),

Flame Sheet and Flow Interactions

$$\underbrace{\left(u_n^b \Omega_t^b - u_n^u \Omega_t^u\right)}_{\text{I}} - \underbrace{\left(u_t^b \Omega_n^b - u_t^u \Omega_n^u\right)}_{\text{II}} = -\int_u^b \left.\frac{\nabla p \times \nabla \rho}{\rho^2}\right|_t dA. \tag{7.80}$$

Because the tangential velocity and normal vorticity are continuous across the flame, the terms labeled II exactly cancel, implying that the vortex stretching/bending term does not influence the evolution of vorticity tangential to the flame. Rearranging the resulting expression yields

$$\Omega_t^b = \frac{\Omega_t^u}{\sigma_\rho} - \frac{1}{u_n^b} \int_u^b \left.\frac{\nabla p \times \nabla \rho}{\rho^2}\right|_t dA. \tag{7.81}$$

This expression shows results similar to Eq. (7.56), where the burned vorticity component equals the unburned component reduced by the gas expansion ratio. The baroclinic term contains the streamline curvature and expansion properties similar to that in the second terms of Eq. (7.56).

Notice also the different way in which Ω_t^u appears in Eqs. (7.54) and (7.56), i.e., by itself or scaled by $1/\sigma_\rho$. This difference reflects the fact that approach flow vorticity terms are implicitly hidden in other terms as well. If we assume that the approach flow is irrotational, we can derive the following general expression [10]:

$$\vec{\Omega}_t^b = -\frac{(1 - \sigma_\rho)^2}{\sigma_\rho} (\vec{n} \times \vec{u}_t \cdot \nabla) \vec{n}. \tag{7.82}$$

This expression shows that vorticity generation can only occur if the reactant velocity has a nonzero component tangential to the flame, and the flame normal varies in the tangential flame direction.

The above discussion implicitly assumed steady state through its neglect of flame front motion, so we next discuss the additional effects of unsteadiness. The normal velocity component relative to the moving flame front, u_{rel}, is given by

$$u_{rel} = u_n - v_{F,n}, \tag{7.83}$$

where $v_{F,n}$ is the velocity of flame motion along the local normal direction. The normal jump conditions across the flame are valid for u_{rel}. The jump condition for the normal component of vorticity remains the same as in the stationary flame case, given by Eq. (7.52). The key aspect of flame motion that influences the vorticity jump is the time-varying local normal of the flame with respect to the flow. The flame sheet appears to be rotating to an observer through either flame front rotation or streamline curvature. Using this relation and unsteady forms of the equations used in the steady case, and following the same procedure, we can derive the following jump condition [10, 13]:

$$\vec{\Omega}_t^b - \vec{\Omega}_t^u = \vec{n} \times \left(\left(\frac{1}{\rho^b} - \frac{1}{\rho^u}\right) \nabla_t (\rho u_{rel}) - \frac{1}{\rho u_{rel}} (\rho^b - \rho^u) \left(D_t \vec{u}_t + v_F D_t \hat{n}\right)\right), \tag{7.84}$$

where D_t is the tangential total time derivative with respect to an observer moving along the flame front with a total velocity $v_{F,n} \vec{n} + \vec{u}_t$. Note that this jump condition is similar to that in the steady case, and the additional terms are introduced due to flame front motion.

The presence of unsteadiness in the form of acoustic waves can also cause nonzero baroclinic terms, even for a uniform flow incident on a flat flame. In this case, the unsteady pressure gradient is independent of the mean pressure gradient and an obliquely incident wave will lead to a misaligned fluctuating pressure and mean density gradient. This scenario was also discussed in Section 2.3.2 as an example of the excitation of vorticity by acoustic disturbances in inhomogeneous flows.

7.8 Aside: Linearized Analysis of Flow Field Modification by the Flame

This section details the analysis whose results are presented in Section 7.5, showing that a wrinkled flame modifies the approach flow field. Assuming an incompressible flow upstream and downstream of the flame, the linearized perturbation equations are

$$\text{Mass:} \quad \frac{\partial u_{x,1}}{\partial x} + \frac{\partial u_{y,1}}{\partial y} = 0, \tag{7.85}$$

$$\text{Axial momentum:} \quad \rho_0 \left(\frac{\partial u_{x,1}}{\partial t} + u_{x,0} \frac{\partial u_{x,1}}{\partial x} \right) = -\frac{\partial p_1}{\partial x}, \tag{7.86}$$

$$\text{Transverse momentum:} \quad \rho_0 \left(\frac{\partial u_{y,1}}{\partial t} + u_{x,0} \frac{\partial u_{y,1}}{\partial x} \right) = -\frac{\partial p_1}{\partial y}, \tag{7.87}$$

$$\text{Flame position:} \quad \frac{\partial \xi_1}{\partial t} - u_{x,1} = 0. \tag{7.88}$$

The assumed form of the wrinkle on the flame, Eq. (7.71), implies the following form for all perturbation quantities:

$$F_1(x, y, t) = \hat{F}(x) e^{iky} e^{-i\omega t}, \tag{7.89}$$

where F represents u_x, u_y, or p.

Substituting this form into Eqs. (7.85)–(7.87) leads to a coupled system of ordinary differential equations, whose solutions in an unbounded domain (so that exponentially growing solutions are set to zero) are (see Exercise 7.5):

$$\text{Unburned:} \quad \begin{aligned} \frac{u_{x,1}^u}{u_{x,0}^u} &= \left(\frac{\mathcal{A}_1^u}{u_{x,0}^u} e^{kx} + \frac{\mathcal{A}_2^u}{u_{x,0}^u} e^{i(St)kx} \right) e^{iky} e^{-i\omega t}, \\ \frac{u_{y,1}^u}{u_{x,0}^u} &= \left(i \frac{\mathcal{A}_1^u}{u_{x,0}^u} e^{kx} - St \frac{\mathcal{A}_2^u}{u_{x,0}^u} e^{i(St)kx} \right) e^{iky} e^{-i\omega t}, \\ \frac{p_1^u}{\rho_0^u \left(u_{x,0}^u \right)^2} &= -\frac{\mathcal{A}_1^u}{u_{x,0}^u} (1 - iSt) e^{kx} e^{iky} e^{-i\omega t}; \end{aligned} \tag{7.90}$$

Flame Sheet and Flow Interactions

Burned:
$$\frac{u^b_{x,1}}{u^b_{x,0}} = \left(\frac{\mathcal{A}^b_1}{u^b_{x,0}}e^{-kx} + \frac{\mathcal{A}^b_2}{u^b_{x,0}}e^{\frac{i(St)kx}{\sigma_\rho}}\right)e^{iky}e^{-i\omega t},$$

$$\frac{u^b_{y,1}}{u^b_{x,0}} = \left(-i\frac{\mathcal{A}^b_1}{u^b_{x,0}}e^{-kx} - \frac{St\,\mathcal{A}^b_2}{\sigma_\rho u^b_{x,0}}e^{\frac{i(St)kx}{\sigma_\rho}}\right)e^{iky}e^{-i\omega t}, \qquad (7.91)$$

$$\frac{p^b_1}{\rho^b_0\left(u^b_{x,0}\right)^2} = -\frac{\mathcal{A}^b_1}{u^b_{x,0}}\left(1+\frac{iSt}{\sigma_\rho}\right)e^{-kx}e^{iky}e^{-i\omega t}.$$

Here, the controlling parameters are $St = \frac{\omega}{ku^u_{x,0}} = \frac{f_{L,wrinkle}}{u^u_{x,0}}$ and $\sigma_\rho = \frac{\rho^u}{\rho^b} = \frac{u^b_{x,0}}{u^u_{x,0}}$. The coefficients $\mathcal{A}^u_1$, $\mathcal{A}^u_2$, $\mathcal{A}^b_1$, and $\mathcal{A}^b_2$ must be related through flame matching conditions. The linearized forms of the mass, normal momentum, and tangential momentum jump conditions at the flame are (see Exercise 7.6):

$$\begin{aligned}
p^u_1 &= p^b_1 \\
u^u_{x,1} &= u^b_{x,1} = \frac{\partial \xi_1}{\partial t} \\
u^u_{y,1} + u^u_{x,0}\frac{\partial \xi_1}{\partial y} &= u^b_{y,1} + u^b_{x,0}\frac{\partial \xi_1}{\partial y}.
\end{aligned} \qquad (7.92)$$

Applying these matching conditions at the nominal flame position ($x = 0$) to the solutions in Eqs. (7.90)–(7.91), leads to

$$\begin{aligned}
\frac{1}{kA_\xi}\frac{\mathcal{A}^u_1}{u^u_{x,0}} &= \frac{1}{2}\left(\frac{\sigma_\rho - 1}{\sigma_\rho}\right)\left(\frac{St^2 - \sigma_\rho}{iSt - 1}\right) \\
\frac{1}{kA_\xi}\frac{\mathcal{A}^u_2}{u^u_{x,0}} &= -\left[iSt + \frac{1}{2}\left(\frac{\sigma_\rho - 1}{\sigma_\rho}\right)\left(\frac{St^2 - \sigma_\rho}{iSt - 1}\right)\right] \\
\frac{1}{kA_\xi}\frac{\mathcal{A}^b_1}{u^b_{x,0}} &= -\frac{1}{2}(\sigma_\rho - 1)\left(\frac{St^2 - \sigma_\rho}{\sigma_\rho + iSt}\right) \\
\frac{1}{kA_\xi}\frac{\mathcal{A}^b_2}{u^b_{x,0}} &= -iSt\left[\frac{1}{\sigma_\rho} + \frac{1}{2}(\sigma_\rho - 1)\left(\frac{St^2 - \sigma_\rho}{\sigma_\rho + iSt}\right)\right].
\end{aligned} \qquad (7.93)$$

Plots of these solutions are presented in Section 7.5.

Exercises

7.1. The presence of density jump at a flame interface leads to several influences on the flow field, one of which is to cause a vorticity jump. The effects on the vorticity component normal and tangential to the flame are different. Using the definition of vorticity in Eq. (7.49), derive the expression for the vorticity component tangential to the flame in Eq. (7.51).

7.2. Consider the steady-state momentum equation written in a coordinate system that is tangential to the sheet:

$$-\nabla_{\!t} p = \rho\big((\vec{u}\cdot\nabla)\vec{u}\big)_{\!t}. \qquad (7.99)$$

Expand the right-hand side of this equation to obtain the following relation for pressure gradient tangential to the flame:

$$-\nabla_{\!t} p = \rho u_n\left(\frac{\partial \vec{u}_t}{\partial n} + (\vec{u}_t\cdot\nabla)\,\vec{n}\right) + \rho(\vec{u}_t\cdot\nabla)\vec{u}_t. \qquad (7.100)$$

7.3. The results from the previous problems can be used to derive the steady-state tangential vorticity jump condition. Apply the jump conditions to both Eqs. (7.51) and (7.100) to obtain the jump conditions in Eq. (7.53).

7.4. Derive an expression for the stretch rate of a two-dimensional material line, $\frac{1}{L}\frac{dL}{dt}$, subjected to the vortex-induced velocity field in Eq. (7.65). Plot the results.

7.5. Derive the linearized solutions in Eqs. (7.90) and (7.91) for harmonically propagating disturbances in an incompressible flow.

7.6. The unknown coefficients in Eqs. (7.90) and (7.91) ($\mathcal{A}_1^u$, $\mathcal{A}_2^u$, $\mathcal{A}_1^b$, and $\mathcal{A}_2^b$) are determined using the linearized form of the flame matching/jump conditions, Eqs. (7.16), (7.18), and (7.19). Linearize these jump conditions to obtain Eq. (7.92).

7.7. Derive the expression for $\theta_{refr,\,max}$, $\tan\theta_{refr,\,max} = \frac{\sigma_\rho - 1}{2\sqrt{\sigma_\rho}}$, presented in Section 7.3.3.

7.8. Derive the expression for $\theta_i|_{\theta_{refr},\,max}$, $\tan\left(\theta_i|_{\theta_{refr},\,max}\right) = \sqrt{\sigma_\rho}$, presented in Section 7.3.3.

7.9. Derive the limiting expressions for burned gas velocity and flame refraction angle for large approach flow angles presented in Section 7.3.3, $u^b = u^u + O(\varepsilon^2)$ and $\theta_{refr} = (\sigma_\rho - 1)\varepsilon + O(\varepsilon^3)$.

References

1. Magina N., Acharya V., and Lieuwen T., Forced response of laminar non-premixed jet flames. *Progress in Energy and Combustion Science*, 2019, **70**: pp. 89–118.
2. Pope S., The evolution of surfaces in turbulence. *International Journal of Engineering Science*, 1988, **26**(5): pp. 445–469.
3. Markstein G., *Nonsteady Flame Propagation*. 1964. Published for and on behalf of Advisory Group for Aeronautical Research and Development, North Atlantic Treaty Organization by Pergamon Press.
4. Williams F., Turbulent combustion, in *The Mathematics of Combustion*, J.D. Buckmaster, ed. 1985, SIAM, pp. 97–131.
5. Poinsot T. and Veynante D., *Theoretical and Numerical Combustion*. 2005, RT Edwards, Inc.
6. Matalon M. and Matkowsky B.J., Flames as gas-dynamic discontinuities. *Journal of Fluid Mechanics*, 1982, **124**(Nov): pp. 239–259.

7. Class A.G., Matkowsky B.J., and Klimenko A.Y., A unified model of flames as gas-dynamic discontinuities. *Journal of Fluid Mechanics*, 2003, **491**: pp. 11–49.
8. Maxworthy T., Discontinuity properties of laminar flames. *Physics of Fluids*, 1961, **4**(5): pp. 558–564.
9. Liberman M.A., Bychkov V.V., Golberg S.M., and Book D.L., Stability of a planar flame front in the slow-combustion regime. *Physical Review E*, 1994, **49**(1): pp. 445–453.
10. Hayes W., The vorticity jump across a gasdynamic discontinuity. *Journal of Fluid Mechanics*, 1957, **2**(06): pp. 595–600.
11. Law C.K., *Combustion Physics*. 1st ed. 2006, Cambridge University Press.
12. Williams F.A., *Combustion Theory: The Fundamental Theory of Chemically Reacting Flow Systems*. 1994, Perseus Books.
13. Emmons H.W., Flow discontinuities associated with combustion, in *Fundamentals of Gas Dynamics*, H.W. Emmons, ed. 1958, Princeton University Press.
14. Mueller C.J., Driscoll J.F., Reuss D.L., and Drake M.C. Effects of unsteady stretch on the strength of a freely-propagating flame wrinkled by a vortex *Symposium on Combustion*, 1996, **26**(1): 347–355.
15. Mueller C.J., Driscoll J.F., Reuss D.L., Drake M.C., and Rosalik M.E., Vorticity generation and attenuation as vortices convect through a premixed flame. *Combustion and Flame*, 1998, **112**(3): pp. 342–358.
16. Louch D.S. and Bray K.N.C., Vorticity in unsteady premixed flames: Vortex pair-premixed flame interactions under imposed body forces and various degrees of heat release and laminar flame thickness. *Combustion and Flame*, 2001, **125**(4): pp. 1279–1309.
17. Katta V.R., Roquemore W.M., and Gord J., Examination of laminar-flamelet concept using vortex/flame interactions. *Proceedings of the Combustion Institute*, 2009, **32**(1): pp. 1019–1026.
18. Shanbhogue S.J., Plaks D.V., Nowicki G., Preetham and Lieuwen T.C., Response of rod stabilized flames to harmonic excitation: shear layer rollup and flame kinematics, in *42nd AIAA/ASME/SAE/ASEE Joint Propulsion Conference & Exhibit*. 2006, Sacramento, CA.
19. Ishizuka S., Flame propagation along a vortex axis. *Progress in Energy and Combustion Science*, 2002, **28**(6): pp. 477–542.
20. Ashurst W.T., Flame propagation along a vortex: The baroclinic push. *Combustion Science and Technology*, 1996, **112**(1): pp. 175–185.
21. Aldredge R. and Williams F., Influence of wrinkled premixed-flame dynamics on large-scale, low-intensity turbulent flow. *Journal of Fluid Mechanics*, 2006, **228**: pp. 487–511.
22. Clavin P. and Williams F., Theory of premixed-flame propagation in large-scale turbulence. *Journal of Fluid Mechanics*, 1979, **90**(03): pp. 589–604.
23. Furukawa J., Noguchi Y., Hirano T., and Williams F., Anisotropic enhancement of turbulence in large-scale, low-intensity turbulent premixed propane-air flames. *Journal of Fluid Mechanics*, 2002, **462**: pp. 209–243.
24. Bill R.G.J. and Tarabanis K., The effect of premixed combustion on the recirculation zone of circular cylinders. *Combustion Science and Technology*, 1986, **47**: pp. 39–53.
25. Aris R., *Vectors, Tensors, and the Basic Equations of Fluid Mechanics*. 1990, Dover Publications.
26. Marble F.E., Growth of a diffusion flame in the field of a vortex, in *Recent Advances in the Aerospace Sciences*, C. Casci and C. Bruno eds. 1985, Springer, pp. 395–413.

27. Fabri J., Siestrunck R., and Foure L. On the aerodynamic field of stabilized flames. *Symposium (International) on Combustion*, 1953, **4**(1): 443–450.
28. Tsien H.S., Influence of flame front on the flow field. *Journal of Applied Mechanics*, 1951, **18**(2): pp. 188–194.
29. Searby G. and Clavin P., Weakly turbulent, wrinkled flames in premixed gases. *Combustion Science and Technology*, 1986, **46**(3): pp. 167–193.
30. Scriven L.E. and Sternling C.V., Marangoni effects. *Nature*, 1960, **187**(4733): pp. 186–188.

8 Ignition

8.1 Overview

This chapter describes the processes associated with spontaneous (or "autoignition") and forced ignition. The forced ignition problem is of significant interest in most combustors, as an external ignition source is almost always needed to initiate reaction. Two examples where the autoignition problem is relevant for flowing systems are illustrated in Figure 8.1 [1–10]. Figure 8.1(a) depicts the autoignition of high-temperature premixed reactants in a premixing duct. This is generally undesirable and an important design consideration in premixer design. Figure 8.1(b) depicts the ignition of a jet of premixed reactants by recirculating hot products. In this case, autoignition plays an important role in flame stabilization and the operational space over which combustion can be sustained. Although not shown, autoignition can also occur during the injection of a fuel, air, or premixed reactants jet into a stream of hot fuel, air, or products. For example, a vitiated H_2/CO stream reacts with a cross-flow air jet in RQL combustors [11].

Figure 8.2 shows several canonical configurations used to study ignition that will be referred to in this chapter. These are (a) the ignition of premixed reactants by hot gases, (b) the ignition of a nonpremixed flame by either a hot fuel or air stream, or an external spark, and (c) stagnating flow of fuel or premixed reactants into a hot gas stream [12].

The rest of this chapter is broken into two main sections, dealing with autoignition (Section 8.2) and forced ignition (Section 8.3). Both sections start with treatment of ignition in homogeneous mixtures and subsequently add the layers needed to understand real systems with inhomogeneities and losses, involving the interplay of thermal, kinetic, and molecular and convective transport processes. The many subtle and fascinating aspects associated with the chemical kinetics of this problem are well covered in several excellent texts [13–15] and so are discussed only briefly here.

In closing this introduction, it is helpful to discuss what we mean by the term "ignition." From a practical point of view, ignition (or at least forced ignition) in a steady flowing combustor is deemed to have occurred if, for example, a spark leads to the initiation and stabilization of a self-sustaining flame. However, we must differentiate the following three things that must occur for a spark event to lead to a self-sustaining flame: (1) initiation of an ignition kernel, (2) spread of a self-sustaining

8.1 Overview

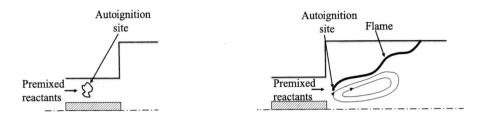

Figure 8.1 Illustration of two autoignition problems.

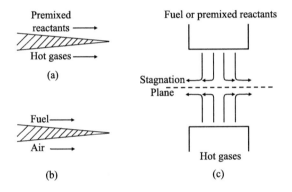

Figure 8.2 Illustration of several canonical ignition problems discussed in the text.

reaction front, and (3) stabilization of the flame [16]. A computed image showing all three processes is shown in Figure 8.3.

These three processes involve different physical mechanisms, and some of these may occur without the others; i.e., (1) does not imply (2), or (2) does not imply (3). For example, an ignition kernel may not become self-sustaining, as discussed in Section 9.7.5. Similarly, a self-sustaining flame may be initiated that simply blows out of the system if the ignitor location is not optimized. This occurs if a spark is situated downstream of the recirculating gas flow in the wake of a bluff body, leading to events (1) and (2), but not (3). For example, a set of images showing the successful initiation of a propagating flame that subsequently blows out of a swirl burner is depicted in Figure 8.4.

Processes (1) and (2) are coupled, so we will include discussion of both of them in this chapter. Both processes are equally significant for understanding autoignition, while process (2) is the focus of forced ignition analyses. For example, a reacting kernel is almost always initiated by a forced ignition event. However, this reaction kernel can subsequently become quenched due to rapid heat losses if the kernel is too small, discussed in Section 8.3, or if the propagating reaction wave has "edges," discussed in Section 9.7. Finally, process (3), which is associated with flame stabilization, is discussed in Section 10.2 and not considered further in this chapter.

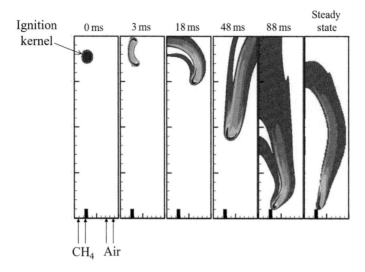

Figure 8.3 Computed images showing ignition, flame propagation, and stabilization of a nonpremixed methane/air flame. Adapted from Briones et al. [17].

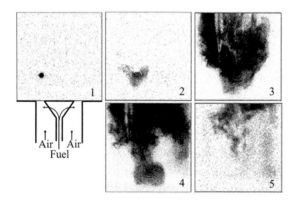

Figure 8.4 Illustration of the successful initiation of a self-sustaining reaction front that subsequently blows out of the combustor. Reproduced from Ahmed et al. [18].

8.2 Autoignition

8.2.1 Ignition of Homogeneous, Premixed Reactants

In this section, we start discussion of autoignition by considering a homogeneous, premixed mixture. In general, ignition is associated with both thermal and radical chain-branching processes, although one of these may dominate in certain instances. Thermal ignition refers to the feedback between exothermic reactions and the strong temperature sensitivity of the reaction rate. Release of a small amount of heat causes a temperature increase which leads to an exponential increase in heat release rate,

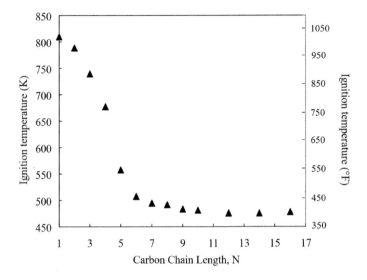

Figure 8.5 Dependence of autoignition temperature on carbon chain length for normal alkanes, C_NH_{2N+2}, at atmospheric pressure [24].

accelerating the reaction rate. Spontaneous ignition can also occur through radical chain-branching processes which may be entirely nonthermal [14, 19]. Radical concentrations grow geometrically with time when the rate of chain branching exceeds the rate of chain destruction.

The processes required to initiate chain-branched ignition are quite different from those that control the initiation steps of flames once they are self-propagating. For the ignition problem, the mixture starts with only reactant species and no radicals. In contrast, self-sustaining flames have a radical pool that is formed in the highly reactive portion of the flame that diffuses into unburned reactants. For example, the dominant radical chain initiation step for hydrogen ignition at low temperatures is $H_2 + O_2 \rightarrow HO_2 + H$, with additional contributions from the $H_2 + M \rightarrow H + H + M$ at higher temperatures. In contrast, in established H_2 flames, $H + O_2 \rightarrow OH + O$ is the dominant initiation reaction [14, 20, 21].

The conditions under which a mixture autoignites is a function of temperature, pressure, and fuel type, as well as other parameters. For example, increasing the temperature of a homogeneous fuel/air mixture beyond some "ignition temperature," T_{ign}, will lead to ignition. Typical results for T_{ign} of atmospheric-pressure n-alkane mixtures are shown in Figure 8.5,[1] showing that $500 \text{ K} < T_{ign} < \sim 800 \text{ K}$. Methane ($CH_4$) has the highest autoignition temperature, followed by ethane (C_2H_6) and then propane (C_3H_8). This result follows from the fact that the larger molecules decompose into reactive fragments more easily than the smaller ones [13]. Note also how

---

[1] It is important to recognize, however, that such results reflect not only fundamental properties of the mixture, but also the device in which the measurements were obtained.

the ignition temperatures of the larger n-alkanes are quite similar to each other [19, 22, 23].

The ignition demarcation boundary is often a nonmonotonic function of pressure and temperature. This nonmonotonicity is due to the multiple competing chemical pathways for reaction intermediates, since the balance between growth and destruction paths changes with pressure and temperature. For example, consider H_2/air autoignition tendencies at a temperature of 800 K. This mixture autoignites for pressure ranges between 0.001 atm (referred to as the "first explosion limit") $< p <$ 0.1 atm ("second explosion limit") and $p > \sim 2$ atm ("third explosion limit"), but not for $p < 0.001$ atm or 0.1 atm $< p < 2$ atm. The second explosion limit is controlled by isothermal, chain-branching processes, while the third limit involves an interplay between thermal and kinetic processes [14]. More discussion of these specific limits is included later in the context of Figure 8.10.

If a given mixture is at conditions where it will autoignite, the next important question is the autoignition time, τ_{ign}, i.e., the time required for autoignition at some given pressure and temperature. In real systems, τ_{ign} is a function of the fuel/air mixing time history, and for liquid systems, atomization and evaporation as well. If these other processes are fast relative to τ_{ign}, then the autoignition time becomes purely kinetics limited. In contrast, if these processes are slow relative to τ_{ign}, then kinetic rates are unimportant for predicting ignition times. Typical data illustrating autoignition times are plotted in Figure 8.6 for n-alkane mixtures, showing the exponential decrease in ignition times with increase in temperature for the methane and propane blends. The larger n-alkane fuels exhibit a nonmonotonic variation of τ_{ign} in the 700 K–900 K negative-temperature coefficient (NTC) region. These data also show

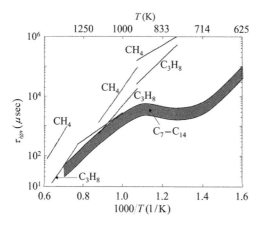

Figure 8.6 Compilation of ignition time delay data for n-alkane/air mixtures ranging from methane to n-tetradecane obtained from references [23, 25–29]. "C_7–C_{14}" denotes n-alkanes ranging from C_7H_{16} to $C_{14}H_{30}$. The data show values nominally at 12 atm pressure and $\phi = 1$, with two exceptions: the Beerer et al. [25] data set (lowest temperature CH_4 and C_3H_8) taken at $\phi = 0.4 - 0.6$, and the Holton et al. [29] data set (intermediate temperature CH_4 and C_3H_8) taken at 1 atm and extrapolated to 12 atm by dividing the ignition time by $12^{1.385}$.

the much shorter ignition times of propane than methane, as well as the even shorter ignition times of the larger n-alkanes, a result that could be anticipated from the earlier discussion. Note that these larger n-alkanes have nearly identical ignition time characteristics, similar to the points made in regards to Figure 8.5. Finally, data generally show that τ_{ign} varies with pressure as p^{-1} for these n-alkane fuels.

The dependence of τ_{ign} upon stoichiometry can be quite different from properties of a propagating flame [3, 25], such as flame thickness or burning velocity. For example, while CH_4–air flames generally exhibit flame speeds that peak near $\phi \sim 1$, their ignition time minima occur very fuel-lean, e.g., at values of $\phi \sim 0.1$. This is due to the self-inhibiting effect of CH_4 on induction chemistry. To illustrate, consider the ignition times calculated from a plug flow reactor for methane/air using GRI Mech 3.0. At 1050 K and 1 atm, the ignition time minimum is ~150 ms and occurs very fuel-lean, at $\phi \sim 0.05$. The ignition time at $\phi = 1$ is 450 ms, a factor of three difference. At 1300 K, the ignition time minimum is 5 ms and occurs at $\phi \sim 0.07$; at $\phi = 1$ it is 11 ms. We will return to these points in Section 8.2.2.3 for nonpremixed flames.

8.2.2 Effects of Losses and Flow Inhomogeneity

Any lossless system described by a single-step, exothermic reaction must spontaneously ignite through a thermal mechanism given sufficient time, since the reaction rate at ambient conditions, though small, is nonzero. However, losses modify this result as the small rate of heat release can be dissipated. To illustrate, consider Figure 8.7, which plots notional dependencies of heat release and heat loss rates on temperature [30, 31].

As shown in the heat release rate curve, the rate of reaction is an exponential function of temperature. The heat loss rate is drawn as a linear function of temperature difference from ambient, $T - T_0$, simulating convective heat transfer, with curves I and II denoting lower and higher loss coefficients.

For heat loss curve I, the heat generation term always exceeds the heat loss term, implying that no steady state is possible and the mixture will thermally self-ignite. For curve II, two possible steady states are indicated, T_a and T_b. The low-temperature

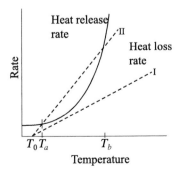

Figure 8.7 Dependence of heat loss and heat generation rates on temperature.

steady-state point, T_a, is stable to inherent temperature perturbations and will not spontaneously ignite; e.g., an increase in temperature will cause the heat loss rate to exceed the heat release rate and a subsequent temperature decrease. This example shows that conditions are possible where heat loss processes inhibit self-ignition. The higher-temperature steady-state point, T_b, is not stable and all temperatures exceeding this value will autoignite. The figure also illustrates how this critical temperature is a function of the heat loss characteristics of the device in which the measurement is being obtained.

8.2.2.1 Model Problem Illustrating Convective Loss Effects [32]

Similarly, other processes effectively constitute loss mechanisms which may influence ignition boundaries. To illustrate, we will next consider a well-stirred reactor model problem in detail. A well-stirred reactor is a prototypical reactive system in which there are no gradients in mixture properties. In this problem, the loss process is convection, which sweeps intermediate radicals or higher-temperature gas out of the system.

Starting with mass conservation, the differential form of the continuity equation, Eq. (1.8), can be integrated over a control volume to obtain

$$\iiint_{CV} \frac{\partial \rho}{\partial t} dV + \iint_{CS} \rho(\vec{u} \cdot \vec{n}) dA = 0. \tag{8.1}$$

Integrating this equation over the well-stirred reactor volume, V, and recognizing that there are no spatial gradients (by definition), yields

$$V \frac{d\rho}{dt} = \dot{m}_{in} - \dot{m}_{out}. \tag{8.2}$$

Applying the same sequence of steps to the species equation, Eq. (1.21), yields

$$V \left(Y_i \frac{d\rho}{dt} + \rho \frac{dY_i}{dt} \right) + \dot{m}_{out} Y_{i,out} - \dot{m}_{in} Y_{i,in} = \dot{w}_i V. \tag{8.3}$$

Assuming an adiabatic reactor, neglecting shear and body forces and kinetic energy, the energy equation, Eq. (1.29) becomes

$$V \left(h \frac{d\rho}{dt} + \rho \frac{dh}{dt} - \frac{dp}{dt} \right) + \dot{m}_{out} h_{out} - \dot{m}_{in} h_{in} = 0. \tag{8.4}$$

We will next assume steady-state operation, leading to the simplified mass conservation equation

$$\dot{m}_{in} = \dot{m}_{out} = \dot{m}. \tag{8.5}$$

Since the conditions at the outlet are the same as the conditions within the reactor (one of the peculiarities of a well-stirred reactor), the *out* subscript will subsequently be removed from all variables. The species equation becomes

$$\dot{w}_i V + \dot{m}(Y_{i,in} - Y_i) = 0. \tag{8.6}$$

The energy equation reduces to constant enthalpy:

$$h = h_{in}. \tag{8.7}$$

Assuming that the mixture-average specific heat is independent of temperature, we can write the enthalpy as

$$h = \sum_{i=1}^{N} Y_i h_{f,i}^0 + c_p(T - T_{ref}). \tag{8.8}$$

Using Eq. (8.7) yields

$$\sum_{i=1}^{N} (Y_i - Y_{i,in}) h_{f,i}^0 + c_p(T - T_{in}) = 0. \tag{8.9}$$

An alternative form of the energy equation can be developed by multiplying Eq. (8.6) by $h_{f,i}^0$ and summing over all species:

$$V \sum_{i=1}^{N} \dot{w}_i h_{f,i}^0 + \dot{m} \sum_{i=1}^{N} (Y_{i,in} - Y_i) h_{f,i}^0 = 0. \tag{8.10}$$

This expression can then be substituted into Eq. (8.9) to obtain

$$\dot{m} c_p (T - T_{in}) + V \sum_{i=1}^{N} \dot{w}_i h_{f,i}^0 = 0. \tag{8.11}$$

This equation states that the net flux of thermal energy out of the reactor is equal to the rate of heat release by reactions.

Assume a one-step, irreversible reaction to prescribe the reaction rate term, $\dot{w}_i$:

$$a(Fuel) + b(Ox) \rightarrow e(Prod), \tag{8.12}$$

where *Fuel*, *Ox*, and *Prod* are the fuel, oxidizer, and product, respectively, and a, b, and e are the stoichiometric coefficients. Next, assume that the oxidizer is present in large excess and can be assumed constant in composition. We will use the following expression for the fuel reaction rate:

$$\frac{\dot{w}_F}{MW_F} = -A \frac{\rho Y_F}{MW_F} e^{-E_a / \mathcal{R}_u T}. \tag{8.13}$$

Molar concentrations are related to mass fractions by

$$[X_i] = \frac{\rho Y_i}{MW_i}. \tag{8.14}$$

From stoichiometry, the following relationships exist:

$$\frac{\dot{w}_{Fuel}/MW_{Fuel}}{-a} = \frac{\dot{w}_{Ox}/MW_{Ox}}{-b} = \frac{\dot{w}_{Prod}/MW_{Prod}}{e}. \tag{8.15}$$

Using these relationships, the heat release term of the energy equation, Eq. (8.11), can be expressed as

$$\sum_{i=1}^{N} \dot{w}_i h_{f,i}^0 = \frac{\dot{w}_{Fuel}}{MW_{Fuel}} \left(MW_{Fuel} h_{f,Fuel}^0 + \frac{b}{a} MW_{Ox} h_{f,Ox}^0 - \frac{c}{a} MW_{Prod} h_{f,Prod}^0 \right) = \dot{w}_{Fuel} Q, \tag{8.16}$$

where Q is the heat release per kilogram of fuel reacted. Substituting Eqs. (8.13) and (8.16) into Eq. (8.11) yields

$$\dot{m} c_p (T - T_{in}) - V \mathcal{A} \rho Y_{Fuel} e^{-E_a / \mathcal{R}_u T} Q = 0. \tag{8.17}$$

In order to nondimensionalize this expression it can be divided through by $\dot{m} c_p T_{in}$:

$$\frac{T}{T_{in}} - 1 - \frac{\rho V}{\dot{m}} \frac{Y_{Fuel,in} Q}{c_p T_{in}} \frac{Y_{Fuel}}{Y_{Fuel,in}} \mathcal{A} e^{-E_a / (\mathcal{R}_u (T/T_{in}) T_{in})} = 0. \tag{8.18}$$

Two time scales can be defined from Eq. (8.18):

$$\begin{aligned} \tau_{res} &= \frac{\rho V}{\dot{m}} : \text{residence time,} \\ \tau_{chem} &= \frac{1}{\mathcal{A}} : \text{chemical time.} \end{aligned} \tag{8.19}$$

Defining the following dimensionless quantities, $\tilde{T} = \frac{T}{T_{in}}$, $\tilde{Q} = (Y_{Fuel,in} Q)/c_p T_{in}$, $\tilde{Y}_{Fuel} = Y_{Fuel}/Y_{Fuel,in}$, $\tilde{E} = \frac{E_a}{\mathcal{R}_u T_{in}}$, and the Damköhler number, $Da = \frac{\tau_{res}}{\tau_{chem}}$, leads to the following form of the energy equation:

$$\tilde{T} - 1 - Da \tilde{Q} \tilde{Y} e^{-\tilde{E}/\tilde{T}} = 0. \tag{8.20}$$

A similar expression for the fuel species equation can be obtained by dividing Eq. (8.6) by $\dot{m} Y_{Fuel,in}$:

$$1 - \tilde{Y} - Da \tilde{Y} e^{-\tilde{E}/\tilde{T}} = 0. \tag{8.21}$$

Solving for $\tilde{Y}$ yields

$$\tilde{Y} = \frac{\tilde{Q} + 1 - \tilde{T}}{\tilde{Q}}. \tag{8.22}$$

Finally, substituting this expression into Eq. (8.20) gives

$$1 - \tilde{T} + (1 - \tilde{T} + \tilde{Q}) Da e^{-\tilde{E}/\tilde{T}} = 0, \tag{8.23}$$

which represents a transcendental equation for a single unknown $\tilde{T}$.

Having completed the problem formulation, we next discuss the solution characteristics, which are a function of the three parameters, Da, $\tilde{Q}$, and $\tilde{E}$. Note that $Da \to 0$ and ∞ correspond to no reaction ($\tilde{T} = 1$, $\tilde{Y} = 1$) and complete reaction ($\tilde{T} = 1 + \tilde{Q}$, $\tilde{Y} = 0$), respectively. The behavior for intermediate Da values exhibits an interesting dependence on $\tilde{E}$ and $\tilde{Q}$. To illustrate, Figure 8.8 presents the solution to Eq. (8.23) for various values of the nondimensional activation energy, $\tilde{E}$. For the lowest activation energy shown, $\tilde{E} = 2$, $\tilde{T}$ is a single-valued function of Da. The reactor outlet temperature gradually rises from $\tilde{T} = 1$ to $\tilde{T} = 1 + \tilde{Q}$ with increases in Da. When $\tilde{E} = 8$, $\tilde{T}$ exhibits a much sharper dependence on Da, and has a vertical tangent point

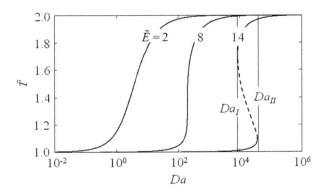

Figure 8.8 Calculated dependence of the nondimensional temperature on the Damköhler number at three different activation energies for $\tilde{Q}=1$.

(the shift of this curve to the right is a consequence of not rescaling A as $\tilde{E}$ increases, so that the reaction rate drops exponentially with increasing $\tilde{E}$).

For $\tilde{E}$ values greater than 8, a fundamentally new behavior occurs: $\tilde{T}$ is a multivalued function of Da and contains two vertical tangent points, as demonstrated by the curve for $\tilde{E} = 14$. This curve has three solution branches. At low Damköhler numbers, the low-temperature, nearly frozen state is the only steady solution. However, this branch disappears at the fold point (Da_{II}) with increasing Da, signifying a discontinuous jump in steady-state solution to the upper, nearly equilibrium, branch. The location of this jump corresponds to conditions where the mixture spontaneously autoignites. Similarly, once on this upper branch, the solution remains there while the Damköhler number is subsequently reduced until the second fold point, Da_I; here, there is a discontinuous jump from the high-temperature region to a low-temperature region. This jump represents extinction.

This example shows that for $\tilde{E} \gg 1$, only low-temperature solutions, $\tilde{T} \simeq 1$, exist at low Da values, $Da < Da_I$, and only near-equilibrium solutions, $\tilde{T} \simeq 1 + \tilde{Q}$, exist at high Da values, $Da > Da_{II}$. There is also an intermediate range of Damköhler values where hysteresis occurs and two steady-state solutions are possible (the middle dashed branch is unstable and not a physically realizable steady-state solution, as shown in Exercise 8.3). In this region, the steady-state characteristics of the system are not uniquely specified by Da, but also depend on the initial conditions. Along the lower section of the curve the temperature is low, and thus a larger Damköhler number is required before significant reaction occurs. However, along the upper branch, the Damköhler number must be decreased beyond Da_I before extinction occurs. As shown in Exercise 8.2, multivalued solutions exist for $\tilde{E} > 4(1 + \tilde{Q})/\tilde{Q}$.

One of the most physically significant aspects of high activation energy chemistry, which typifies combustion reactions, is the discontinuous relationship between steady-state reaction progress (i.e., what fraction of the fuel is consumed) and control parameters like residence time. Moreover, it also shows that steady-state solutions for a large range of fuel consumption fractions are not possible at high-$\tilde{E}$ conditions.

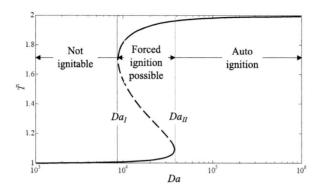

Figure 8.9 Calculated dependence of the nondimensional temperature on the Damköhler number for $\tilde{Q} = 1$ and $\tilde{E} = 14$, showing three ignition regimes in Damköhler number space.

Rather, for $\tilde{E} \gg 1$, fuel consumption essentially does not occur at all or completely occurs.

Having considered this problem in general, we now discuss it specifically for cases where $\tilde{E} \gg 1$ from the point of view of ignition. Figure 8.9 shows the three ignition regimes in Damköhler number space:

High Damköhler numbers, $Da > Da_{II}$: no steady-state, nonreacting solution exists – this mixture will always autoignite. In this regime, a key quantity of interest is the autoignition time, τ_{ign}.

Intermediate Damköhler numbers, $Da_I < Da < Da_{II}$: both steady-state, nonreacting (frozen) and "mostly reacted" solutions exist. For the purposes of this ignition discussion, we will assume that we are on the lower branch. This mixture will not spontaneously self-ignite – it can persist indefinitely in the unreacted state. However, forced ignition is possible. If the mixture temperature is forced above the middle, unstable, branch, the solution will proceed to the upper branch. The important question here is the requirement to force ignition, such as ignition energy. We will show in Section 9.7.2 that this intermediate range consists of two additional Da ranges where ignition waves with edges will and will not propagate successfully.

Low Damköhler numbers, $Da < Da_I$: The mixture will neither ignite spontaneously or with forced ignition. All ignition attempts will fail.

8.2.2.2 Diffusive Loss Effects on Ignition

Diffusion serves as another important loss process that influences ignition boundaries by transporting radicals or heat away from the ignition kernel in regions of high gradients. These high gradients occur in locations of high strain or scalar dissipation rate, topics to be discussed in Sections 9.3 and 9.6. It is generally found that autoignition temperatures increase with strain level. To illustrate, Figure 8.10 plots the calculated dependence of the ignition boundaries on strain rate for nonpremixed

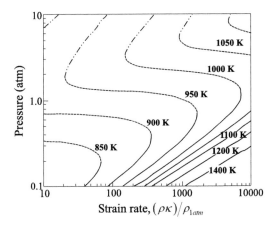

Figure 8.10 Calculated dependence of the nonpremixed ignition boundary as a function of strain rate. Mixtures spontaneously ignite to the left of each curve (—, first ignition limit; - -, second ignition limit; – ·· –, third ignition limit). The fuel stream consists of a 60% hydrogen in nitrogen mixture and the oxidizer stream consists of heated air. Figure obtained from Kreutz and Law [33].

hydrogen/air [33]. In this calculation, hot air is stagnated against a counterflowing hydrogen/nitrogen blend at ambient temperature, as illustrated in Figure 8.2(c). The ignition temperature increases significantly with strain; at 5 atm, it increases from roughly 750 K for the unstrained flame to 1100 K at a density-weighted strain rate of $(\rho\kappa)/\rho_{(1\,atm)} = 10^4 \, s^{-1}$.

The sensitivity of the ignition boundary to strain rate is a function of the relative time scales associated with the ignition induction chemistry and the effective "residence time" imposed by strain. For example, note the significant sensitivity of the first and third ignition limits to strain, in contrast to the relative insensitivity of the second ignition limit. Similar points regarding higher required ignition temperatures with increasing strain can be seen for the more complex n-decane fuel in Figure 8.11.

In unsteady/turbulent flows, the flame strain rate varies spatially and temporally. In these cases, autoignition can occur even if the time-averaged strain rate or scalar dissipation exceeds the steady-state ignition boundary, if the unsteady excursions pass into an ignitable parameter space for sufficient time [16].

8.2.2.3 Ignition Times in Inhomogeneous Mixtures

In this section we treat the effect of inhomogeneity in composition and temperature on autoignition times [35]. In such flows, the ignition kernel will originate at the point with the shortest ignition time. This minimum ignition time condition does not generally coincide with the condition where the reaction rate for self-sustaining flame is highest; e.g., τ_{ign} has a minimum at very lean stoichiometries for CH_4–air flames. Analogously, nonpremixed systems have ignition time minima at the "most reacting" mixture fraction value, Z_{MR}, that differs from the stoichiometric mixture fraction, Z_{st}. This effect can be purely kinetic, as the optimal parameter values most conducive to

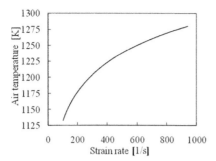

Figure 8.11 Measured strain rate dependence of the air stream temperature needed for the ignition of prevaporized n-decane (at 408 K) in a counterflow geometry at atmospheric pressure. The fuel stream is 15% n-decane by volume of fuel/nitrogen blend [34].

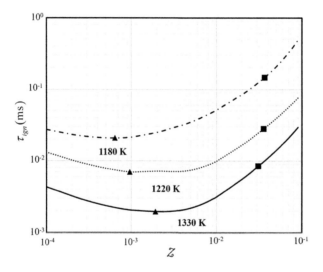

Figure 8.12 Dependence of computed autoignition times on mixture fraction, defining the most reactive mixture fraction, Z_{MR}, for a natural gas/air mixture. The square (■) and triangle (▲) denote Z_{st} and Z_{MR}, respectively. Calculations courtesy of N. Noiray [36].

autoignition chemistry may not be the same as those once the flame is propagating. In addition, thermal effects are important in cases where one of the gases is at a substantially higher temperature than the other. For example, increasing air temperature will progressively bias Z_{MR} toward lower values, and vice versa for problems where the fuel is heated (e.g., in RQL combustors).

An example set of calculations is shown in Figure 8.12, plotting the calculated ignition time as a function of mixture fraction. These curves were calculated for a mixture of vitiated combustion products from a primary burner, a secondary set of dilution air at ambient temperature, and ambient-temperature fuel. The three curves correspond to three different levels of combustion product/secondary air mixing ratios. As such, higher temperatures correspond to larger fractions of vitiated air

Figure 8.13 Schematic of an isomixture fraction field for a nonpremixed jet flame [16].

and, consequently, lower O_2 levels. Several points are evident. First, note that the minimum in ignition time (denoted by ▲) occurs at mixture fractions that are at least 10 times leaner than stoichiometric (denoted by ■) in all cases, emphasizing how autoignition sites will be preferentially located in very fuel-lean regions. Second, they show how this minimum ignition time shifts with temperature. If the composition at all three temperatures were identical, the value of Z_{st} would be identical, and Z_{MR} would shift to leaner values with increases in temperature. However, because the O_2 fraction is simultaneously varying, the Z_{MR} curve actually shifts in the opposite direction.

To further consider these ideas, consider the schematic in Figure 8.13 of the instantaneous mixture fraction field of a cold fuel jet issuing into a co-flowing hot air jet [16].

The figure shows several instantaneous isomixture fraction lines, including the stoichiometric line, Z_{st}, a representative fuel-lean line, Z_{lean}, and fuel-rich line, Z_{rich}. Also shown on the graph is a line indicated by Z_{MR}, which can be greater or less than Z_{st}. This Z_{MR} isoline corresponds to the mixture fraction value at which the autoignition time is minimum and the ignition kernel originates – in other words, autoignition does not preferentially occur at Z_{st}, the line along which the propagating flame exists, but at the minimum ignition time line, Z_{MR}.

These results suggest that autoignition sites in nonpremixed turbulent flames should be preferentially located along the Z_{MR} line at locations where diffusive losses due to turbulent straining (i.e., the scalar dissipation rate) are low. To illustrate, Figure 8.14 overlays computed contours of heat release rate, showing that they are centered on the Z_{MR} isoline [37]. Note the spatially "spotty" distribution of the heat release rate, whose local maxima coincide with regions of low scalar dissipation rate, χ. This is indicated by the dashed vertical lines. These locations should exactly coincide in the quasi-steady limit, but in reality, finite ignition times imply that the temporal mixture fraction history will also play a role in the ignition site location. [16]. One implication of this result is that turbulent nonpremixed flames have shorter autoignition times than laminar ones, due to faster mixing out of high scalar dissipation rate regions and increased probability of finding spatial regions with low local scalar dissipation values [37].

Similarly, in premixed flame ignition problems with gradients in mixture composition, autoignition sites do not coincide with spatial locations where the stoichiometry corresponds to the fastest propagating flame characteristics, but with stoichiometries associated with the shortest ignition times. In the same way, ignition sites in premixed

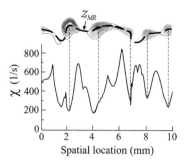

Figure 8.14 Computed overlay of the instantaneous heat release rate and most reactive mixture fraction contour, Z_{MR}, (top) and scalar dissipation rate, χ (bottom) [37]. Image courtesy of D. Thévenin.

flows with temperature gradients are biased towards the hot products, which can lead to results that differ from one's expectations derived from self-sustaining flames. For example, in a counterflow burner problem where premixed reactants are stagnated into hot products (see Figure 8.2), the ignition kernel is often located on the products side of the stagnation line [38]. This implies that fuel or air reaches the ignition kernel through diffusion. If the fuel and air have significantly different molecular weights, the stoichiometry of the ignition kernel can differ markedly from that of the bulk premixed flow. For example, a study of premixed hydrogen/air stagnating against hot nitrogen [38] showed that even though the overall mixture was fuel lean, the stoichiometry of the ignition kernel was very rich, because of the substantially higher diffusion rates of hydrogen relative to oxygen. This result shows the sensitivity of the premixed ignition problem to both the relative diffusion rates of the fuel and oxidizer and the concentrations of fuel/oxidizer in the hot co-flow. For example, using the same type of reasoning, consider the ignition of a rich premixed propane/air blend. If the hot igniting gas is completely inert, then the ignition kernel would likely consist of a lean propane/air kernel located in the hot gas side, because of the higher diffusivity of oxygen relative to propane. In contrast, suppose that the hot igniting gas consists of recirculated products from the rich flame, which would then consist of H_2, CO, CO_2, and H_2O. In this case, the ignition kernel would consist of an H_2/air or H_2/CO/air mixture.

8.2.2.4 Ignition Waves in Inhomogeneous Mixtures

In a perfectly homogeneous system under conditions where autoignition occurs, the mixture reacts uniformly after some ignition time scale, τ_{ign}. However, any real system contains spatial gradients in τ_{ign}, due to local variations in temperature or stoichiometry, and so different points in the flow will autoignite at different time instances. Ignition will occur first at the point with minimum ignition time, followed by the points with slightly longer ignition time, and so forth. These points are illustrated in Figure 8.15, which shows an example of a region with τ_{ign} gradients (denoted by the isocontours), as well as "streamlines" that are normal to these isocontours. The

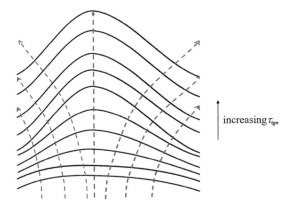

Figure 8.15 Notional example illustrating autoignition wave direction and velocity in a system with spatially varying τ_{ign}, given by $\tau_{ign}(x, y, z)$. Solid lines represent contours of constant τ_{ign}, and gray dashed lines show the direction ignition waves propagate.

subsequent ignition of different points in the flow, starting with those points with the lowest τ_{ign} and progressing toward regions with higher τ_{ign}, will locally appear as a reaction wave (but not a "flame," a term we will reserve for a coupled reaction–diffusion wave), propagating in the direction of the gradient of the τ_{ign} field at a speed of $v_{ign} = 1/|\nabla \tau_{ign}|$. In other words, for a homogeneous system with no τ_{ign} gradients, v_{ign} is infinite. It is the local gradients in τ_{ign} that lead to propagating ignition waves. We will return to this ignition wave velocity in Section 9.2, where we will compare its relative value to the laminar flame speed.

The fact that such ignition time gradients can lead to waves with very high speeds has been emphasized as a mechanism for deflagration to detonation transition [39, 40] when v_{ign} reaches values between the speed of sound and the detonation speed.

8.3 Forced Ignition

In many instances, combustible mixtures are in the hysteresis zone indicated in Figure 8.9, where either the frozen or nearly fully reacted state is possible. These systems require external forcing to initiate ignition, in order to move the system to the upper, reacting branch. We discuss issues associated with forced ignition in this section.

A pilot flame, or highly heated surface, is one potential approach to initiate reaction. For piloted flames, the reaction is already initiated by the pilot and the problem is not so much one of ignition, but of how (or whether) the flame subsequently propagates outward. This distinction between ignition kernel initiation and flame propagation is emphasized in Section 8.1. As such, this topic is deferred to edge flame propagation in Section 9.7.

Sparks are a common transient source used to deposit energy into a localized region of the gas. This is a classical topic [13–15], so only the key concepts are summarized here. Questions that must be addressed include (1) the relationship between the probability of successful ignition and the energy deposition to the gas for a given ignition system, (2) the effect of ignition system characteristics such as distance between electrodes or spark duration on this relationship, and (3) the effect of fuel and ambient conditions, such as fuel composition, stoichiometry, pressure, and turbulence intensity.

Many of these issues can be reasonably scaled using simple concepts. An important effect of the spark is to initiate a small region of high-temperature gas, typically a toroidal gas volume, around the spark gap. Vigorous reaction and heat release is initiated in this high-temperature region. The rate of heat release is proportional to the volume of this high-temperature region. However, the gradients in temperature and species also lead to thermal and radical losses to the gases outside of this volume through diffusion; these losses are proportional to the surface area of the ignition kernel. This leads to a balance equation for energy/radical generation and loss. If the spark kernel is too small, the high spatial gradients will cause heat/radical losses to exceed the chemical energy generation rate and the ignition attempt will fail. In contrast, if the spark kernel is large enough, the heat release rate exceeds the losses and the ignition attempt will be successful. These two scenarios are illustrated schematically in Figure 8.16.

Figure 8.17 shows several instantaneous shadowgraph images after a spark discharge for both air and methane/air. These images accentuate the high-temperature gradient front, as illustrated in Figure 8.16. Note that the images for both the nonreacting and reacting gases look similar for times up to about 1 ms after spark discharge, as the heat due to chemical reactions is negligible and the expanding temperature front dynamics are dominated by the spark's energy deposition. After about 2 ms, however, the difference is quite clear, as the temperature gradient is clearly weakening in the nonreacting case, while an expanding flame is evident in the reacting case.

This physical reasoning and the associated energy balance leads to the prediction that sufficient spark energy, E_{min}, must be deposited to raise a volume of combustion gases with a length scale δ_k to the adiabatic flame temperature; for a spherical volume,

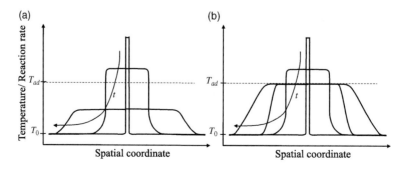

Figure 8.16 Notional representation of spatial profiles for temperature/reaction rate during (a) failed and (b) successful ignition attempts, following Ronney [41].

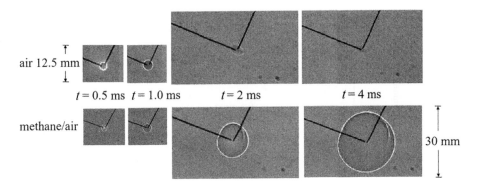

Figure 8.17 Shadowgraph of air (top) and methane/air (bottom) after spark energy deposition, both initially at room temperature and pressure. Images reproduced from Eisazadeh-Far et al. [42].

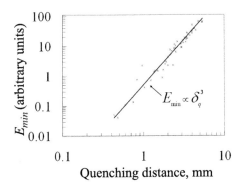

Figure 8.18 Measured dependence of minimum ignition energy on quenching distance for methane and propane fuels for a range of diluents, pressures, and turbulence intensities [44].

$E_{min} \sim \delta_k^3$ [43]. This critical ignition kernel size, δ_k, is proportional to the laminar flame thickness, δ_F, or quenching distance, δ_q (i.e., the minimum gap a combustible mixture can propagate through without extinction), as shown in Exercise 8.4. This basic idea is well borne out by experiments, such as shown in Figure 8.18. This figure plots the measured minimum ignition energy as a function of quenching distance for a variety of fuel/air blends. The fit line through the data is given by δ_q^3.

This direct relationship between minimum spark energy and flame thickness/quenching distance implies that E_{min} is a strong function of fuel composition and ambient conditions. Moreover, forced ignition and autoignition sensitivities can be quite different, reflecting the different kinetics associated with low-temperature, initiating reactions and high-temperature reactions. For example, near-stoichiometric mixtures require lower ignition energies than very lean or very rich mixtures (although E_{min} is usually located slightly rich or lean). Similarly, H_2/air flames ignite easier than CH_4/air flames – both examples are consistent with premixed flame thickness trends. Similarly, as shown in Figure 8.19, E_{min} decreases with increasing pressure, which is

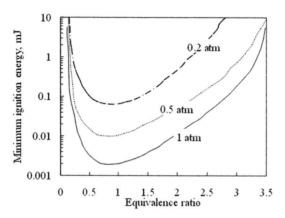

Figure 8.19 Dependence of the minimum ignition energy of a C_3H_8/O_2 mixture on stoichiometry and pressure. Obtained from figure 170 in Lewis and von Elbe [15].

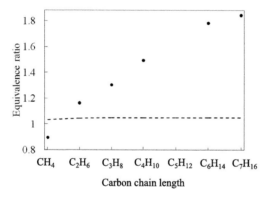

Figure 8.20 Dependence of the equivalence ratio on fuel type at which ignition energy is minimized (data obtained in air at atmospheric pressure). Also indicated is the calculated equivalence ratio at which the adiabatic flame temperature peaks (dashed line) Ignition data obtained from figure 171 in Lewis and von Elbe [15].

also consistent with the inverse dependence of flame thickness/quenching distance on pressure [45].

Moreover, the stoichiometry of the reactive layer in the expanding reaction front may differ from the average value of the premixed mixture because of differences in diffusivity of the fuel and oxidizer. This topic is directly related to that of flame stretch, which is discussed further in Section 9.3. In an outwardly propagating ignition kernel, these differential diffusion effects shift the stoichiometry value of peak reaction rate toward leaner values for fuels that are lighter than oxygen and toward rich values for fuels that are heavier than oxygen. This effect is illustrated in Figure 8.20, which plots the dependence of the equivalence ratio at which E_{min} is minimized for several normal-alkane fuels. Note that E_{min} for methane occurs at slightly lean conditions, while for heavier normal alkanes E_{min} occurs at richer stoichiometries.[2] This shift is not correlated with equilibrium flame temperature, as illustrated by the dashed line that indicates the corresponding equivalence ratio at which the adiabatic flame temperature peaks.

The dependencies of minimum ignition energy on fuel/air ratio, pressure, and fuel composition more fundamentally reflect the characteristics of the underlying volumetric heat release source required for a successful ignition. A variety of other parameters related to loss terms also influence ignition energy. A convective flow velocity through the ignition region can sharply increase ignition energy or, alternatively, decrease the probability of ignition for a given spark energy. Apparently, this occurs by stretching out the ignition kernel and increasing gradients and, consequently, heat and radical losses [45]. Similarly, minimum ignition energy generally rises with increasing turbulence intensity, also presumably due to the straining of the spark kernel interface which increases loss rates. Finally, the size of the spark gap and the time interval over which the energy is deposited also influence the required ignition energy [41, 46].

In turbulent flows where there are substantial fluctuations in conditions at the spark initiation location, the E_{min} concept must be replaced by statistical ideas of the probability of successful ignition for a given energy input. In particular, ignition of nonpremixed flames becomes particularly stochastic, since the local mixture fraction at the igniter may fluctuate appreciably. Clearly, ignition will be more difficult at time instants where nearly pure air or fuel are located at the igniter, as opposed to time instants when $Z = Z_{MR}$. Moreover, if the igniter is located in spatial regions where the time-averaged stoichiometry/mixture fraction is not located near a minimum in τ_{ign}, increases in turbulence intensity may actually increase ignition probability by intermittently introducing faster-igniting mixtures to the igniter site.

In addition to sparks, a variety of other approaches can be used to ignite flows, including plasmas, lasers, hot walls, or hot gas streams. We will discuss the hot wall ignition problem briefly next. This is an interesting boundary layer problem, with interacting momentum, energy, and species boundary layers, as illustrated in Figure 8.21 [14, 47]. In addition, the wall can have both direct thermal effects and kinetic effects through deactivation of radicals or catalytic promotion of reactions [48].

This problem can be posed in several ways. For example, we may be interested in the wall temperature, $T_{w,0}$, required for reaction to occur in a flow of a given velocity u_x over an axial distance of L [49]. Equivalently, the problem can be posed as the axial distance, L, required to ignite a given flowing mixture at a given wall temperature. It seems clear that ignition is promoted with high wall temperatures or long "residence times," $\tau = L/u_x$. For example, a thermal ignition analysis using one-step kinetics with a reaction rate of the form $\dot{\omega} \propto \exp(-E_a/\mathscr{R}_u T)$ predicts an ignition location of [50]:

$$\frac{L}{u_{x,0}} \propto \exp(E_a/\mathscr{R}_u T_{w,0}). \tag{8.24}$$

Typical data for this configuration are plotted in Figure 8.22, showing this exponential temperature sensitivity.

---

[2] While the C_3H_8/O_2 ignition energy in Figure 8.19 does minimize lean, the corresponding C_3H_8/air data in Lewis and von Elbe [15] also peak rich.

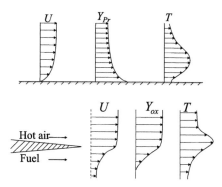

Figure 8.21 Thermal, velocity, and species boundary layers over a hot wall and in a free shear layer.

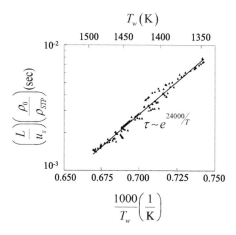

Figure 8.22 Dependence of ignition conditions on wall surface temperature (stoichiometric ethanol/air mixture at 350 K). Data obtained from figure 8.3-4 in Toong [47].

Similarly, a co-flowing stream of hot gases, such as illustrated in Figure 8.2, can lead to forced ignition of a flow [51–54]. Purely thermal boundary layer analyses predict an ignition location expression similar to Eq. (8.24), with $T_{w,0}$ replaced by the hot gas stream temperature [55]. Recall the prior discussion of this problem in Section 8.2.2.3, however, where it was noted that diffusive effects introduce additional kinetic complexities, such as causing the ignition site stoichiometry to differ from that of the premixed mixture average.

Exercises

8.1 Well-stirred reactor problem. Write an explicit result for the dimensionless temperature and fuel mass fraction exiting a well-stirred reactor in the zero activation energy limit $(\tilde{E} \to 0)$. Contrast this solution with the high activation energy results.

8.2 Determine the conditions for which the well-stirred reactor model analyzed in Section 8.2.2.1 exhibits multiple steady-state solutions.

8.3 Stability of the intermediate temperature solution for the well-stirred reactor problem. As noted in Section 8.2.2.1, the intermediate dashed solution in Figure 8.9 is unstable and thus will not be observed in the steady state. Prove this formally and provide a physical argument explaining this result.

8.4 Forced ignition. Derive the ignition energy scaling discussed in Section 8.3, namely that $E_{min} \sim \delta_F^3$, where δ_F is the laminar flame thickness. This can be done by deriving an energy equation for a spherical control volume, balancing energy released by reaction and lost through conduction. The size of the control volume required for these two processes to balance leads to this scaling.

References

1. He X., Walton S., Zigler B., Wooldridge M., and Atreya A., Experimental investigation of the intermediates of isooctane during ignition. *International Journal of Chemical Kinetics*, 2007, **39**(9): pp. 498–517.
2. Meier W., Boxx I., Arndt C., Gamba M., and Clemens N., Investigation of auto-ignition of a pulsed methane jet in vitiated air using high-speed imaging techniques. *Journal of Engineering for Gas Turbines and Power*, 2011, **133**(February): pp. 021504(1)–021504(6).
3. Spadaccini L. and Colket III M., Ignition delay characteristics of methane fuels. *Progress in Energy and Combustion Science*, 1994, **20**(5): pp. 431–460.
4. Tan Y., Dagaut P., Cathonnet M., and Boettner J.C., Oxidation and ignition of methane-propane and methane-ethane-propane mixtures: Experiments and modeling. *Combustion Science and Technology*, 1994, **103**(1): pp. 133–151.
5. Dooley S., Won S.H., Chaos M., Heyne J., Ju Y., Dryer F.L., Kumar K., Sung C.J., Wang H., and Oehlschlaeger M.A., A jet fuel surrogate formulated by real fuel properties. *Combustion and Flame*, 2010, **157**(12), pp. 2333–2339.
6. Ihme M. and See Y.C., Prediction of autoignition in a lifted methane/air flame using an unsteady flamelet/progress variable model. *Combustion and Flame*, 2010, **157**(10): pp. 1850–1862.
7. Gokulakrishnan P., Klassen M., and Roby R., Development of detailed kinetic mechanism to study low temperature ignition phenomenon of kerosene, in *ASME Turbo Expo 2005: Power for Land, Sea, and Air*. 2005, Reno, NV: ASME, Paper #GT2005–68268.
8. Sung C.J. and Law C., Fundamental combustion properties of H2/CO mixtures: Ignition and flame propagation at elevated pressures. *Combustion Science and Technology*, 2008, **180**(6): pp. 1097–1116.
9. He X., Donovan M., Zigler B., Palmer T., Walton S., Wooldridge M., and Atreya A., An experimental and modeling study of iso-octane ignition delay times under homogeneous charge compression ignition conditions. *Combustion and Flame*, 2005, **142**(3): pp. 266–275.
10. Walton S., He X., Zigler B., and Wooldridge M., An experimental investigation of the ignition properties of hydrogen and carbon monoxide mixtures for syngas turbine applications. *Proceedings of the Combustion Institute*, 2007, **31**(2): pp. 3147–3154.
11. Lefebvre A.H. and Ballal D.R., *Gas Turbine Combustion: Alternative Fuels and Emissions*. 2010, Taylor and Francis.
12. Fotache C., Tan Y., Sung C., and Law C., Ignition of CO/H2/N2 versus heated air in counterflow: Experimental and modeling results. *Combustion and Flame*, 2000, **120**(4): pp. 417–426.

13. Glassman I. and Yetter R.A., *Combustion*. 2008, Academic Press.
14. Law C.K., *Combustion Physics*. 2006, Cambridge University Press.
15. Lewis B. and von Elbe G., *Combustion, Flames and Explosions of Gases*. 3rd ed. 1987, Academic Press.
16. Mastorakos E., Ignition of turbulent non-premixed flames. *Progress in Energy and Combustion Science*, 2009, **35**(1): pp. 57–97.
17. Briones A.M., Aggarwal S.K., and Katta V.R., Effects of H2 enrichment on the propagation characteristics of CH4-air triple flames. *Combustion and Flame*, 2008, **153**(3): pp. 367–383.
18. Ahmed S., Balachandran R., Marchione T., and Mastorakos E., Spark ignition of turbulent nonpremixed bluff-body flames. *Combustion and Flame*, 2007, **151**(1–2): pp. 366–385.
19. Westbrook C.K., Chemical kinetics of hydrocarbon ignition in practical combustion systems. *Proceedings of the Combustion Institute*, 2000, **28**(2): pp. 1563–1577.
20. Mueller M.A., Yetter R.A., and Dryer F.L., Measurement of the rate constant for H + O2 + M → HO2 + M (M = N2, Ar) using kinetic modeling of the high-pressure H2/O2/NOx reaction. *Symposium (International) on Combustion*, 1998, **27**(1): pp. 177–184.
21. Zambon A. and Chelliah H., Explicit reduced reaction models for ignition, flame propagation, and extinction of C2H4/CH4/H2 and air systems. *Combustion and Flame*, 2007, **150**(1–2): pp. 71–91.
22. Shen H.P.S. and Oehlschlaeger M.A., The autoignition of C_8H_{10} aromatics at moderate temperatures and elevated pressures. *Combustion and Flame*, 2009, **156**(5): pp. 1053–1062.
23. Shen H.P.S., Steinberg J., Vanderover J., and Oehlschlaeger M.A., A shock tube study of the ignition of *n*-heptane, *n*-decane, *n*-dodecane, and *n*-tetradecane at elevated pressures. *Energy and Fuels*, 2009, **23**(5): pp. 2482–2489.
24. Zabetakis M.G., *Flammability Characteristics of Combustible Gases and Vapors*. 1965, Washington DC: Bureau of Mines.
25. Beerer D., McDonell V., Samuelsen S., and Angello L., An experimental ignition delay study of alkane mixtures in turbulent flows at elevated pressures and intermediate temperatures. *Journal of Engineering for Gas Turbines and Power*, 2011, **133**: p. 011502.
26. Penyazkov O., Ragotner K., Dean A., and Varatharajan B., Autoignition of propane-air mixtures behind reflected shock waves. *Proceedings of the Combustion Institute*, 2005, **30**(2): pp. 1941–1947.
27. Petersen E.L., Rohrig M., Davidson D.F., Hanson R.K., and Bowman C.T., High-pressure methane oxidation behind reflected shock waves. *Symposium (International) on Combustion*, 1996, **26**(1): pp. 799–806.
28. Petersen E.L., Hall J.M., Smith S.D., de Vries J., Amadio A.R., and Crofton M.W., Ignition of lean methane-based fuel blends at gas turbine pressures. *Journal of Engineering for Gas Turbines and Power*, 2007, **129**: pp. 937–944.
29. Holton M.M., Gokulakrishnan P., Klassen M.S., Roby R.J., and Jackson G.S., Autoignition delay time measurements of methane, ethane, and propane pure fuels and methane-based fuel blends. *Journal of Engineering for Gas Turbines and Power*, 2010, **132**: pp. 1–9.
30. Semenov N.N., *Some Problems in Chemical Kinetics and Reactivity*. Vol. I. 1959, Princeton University Press.
31. Semenov N.N., *Some Problems in Chemical Kinetics and Reactivity*. Vol. II. 1959, Princeton University Press.
32. Peters N., *Fifteen Lectures on Laminar and Turbulent Combustion*. ERCOFTAC Summer School, Aachen, Germany, 1992.

33. Kreutz T. and Law C., Ignition in nonpremixed counterflowing hydrogen versus heated air: Computational study with detailed chemistry. *Combustion and Flame*, 1996, **104**(1–2): pp. 157–175.
34. Seiser R., Seshadri K., Piskernik E., and Linan A., Ignition in the viscous layer between counterflowing streams: Asymptotic theory with comparison to experiments. *Combustion and Flame*, 2000, **122**(3): pp. 339–349.
35. Mukhopadhyay S. and Abraham J., Influence of compositional stratification on autoignition in *n*-heptane/air mixtures. *Combustion and Flame*, 2010, **158**(6): pp. 1064–1075.
36. Ebi D., Doll U., Schulz O., Xiong Y., and Noiray N., Ignition of a sequential combustor: Evidence of flame propagation in the autoignitable mixture. *Proceedings of the Combustion Institute*, 2019, **37**(4): pp. 5013–5020.
37. Hilbert R. and Thévenin D., Autoignition of turbulent non-premixed flames investigated using direct numerical simulations. *Combustion and Flame*, 2002, **128**(1–2): pp. 22–37.
38. Zheng X., Blouch J., Zhu D., Kreutz T., and Law C., Ignition of premixed hydrogen/air by heated counterflow. *Proceedings of the Combustion Institute*, 2002, **29**(2): pp. 1637–1643.
39. Oran E.S. and Gamezo V.N., Origins of the deflagration-to-detonation transition in gas-phase combustion. *Combustion and Flame*, 2007, **148**(1): pp. 4–47.
40. Zel'dovich Y.B., Librovich V.B., Makhviladze G.M., and Sivashinsky G.I., On the development of detonation in a non-uniformly preheated gas. *Astronautica Acta*, 1970, **15**(5–6): pp. 313–321.
41. Ronney P.D. Laser versus conventional ignition of flames. *Optical Engineering*, 1994, **33**: pp. 510–521.
42. Eisazadeh-Far K., Parsinejad F., Metghalchi H., and Keck J.C., On flame kernel formation and propagation in premixed gases. *Combustion and Flame*, 2010, **157**(12): pp. 2211–2221.
43. Chen Z., Burke M.P., and Ju Y., On the critical flame radius and minimum ignition energy for spherical flame initiation. *Proceedings of the Combustion Institute*, 2011, **33**(1): pp. 1219–1226.
44. Ballal D. and Lefebvre A., Ignition and flame quenching in flowing gaseous mixtures. *Proceedings of the Royal Society of London. A. Mathematical and Physical Sciences*, 1977, **357**(1689): p. 163.
45. Ballal D.R. and Lefebvre A.H. The influence of flow parameters on minimum ignition energy and quenching distance. *Symposium (International) on Combustion*, 1975, **15**: pp. 1473–1481.
46. Ballal D.R. and Lefebvre A. A general model of spark ignition for gaseous and liquid fuel-air mixtures. *Symposium (International) on Combustion*, 1981, **18**(1): pp. 1737–1746.
47. Toong T.Y., *Combustion Dynamics: The Dynamics of Chemically Reacting Fluids*. 1983, McGraw-Hill.
48. Vlachos D., The interplay of transport, kinetics, and thermal interactions in the stability of premixed hydrogen/air flames near surfaces. *Combustion and Flame*, 1995, **103**(1–2): pp. 59–75.
49. Turcotte D.L., *An Experimental Investigation of Flame Stabilization in a Heated Turbulent Boundary Layer*. 1958, California Institute of Technology.
50. Law C. and Law H., Thermal-ignition analysis in boundary-layer flows. *Journal of Fluid Mechanics*, 1979, **92**(01): pp. 97–108.
51. Moreau P., Experimental determination of probability density functions within a turbulent high velocity premixed flame. *Symposium (International) on Combustion*, 1981, **18**(1): pp. 993–1000.

52. Markides C. and Mastorakos E., An experimental study of hydrogen autoignition in a turbulent co-flow of heated air. *Proceedings of the Combustion Institute*, 2005, **30**(1): pp. 883–891.
53. Marble F.E. and Adamson Jr T.C., Ignition and combustion in a laminar mixing zone. *Jet Propulsion*, 1954, **24**(2): pp. 85–94.
54. Dunn M., Bilger R., and Masri A., Experimental characterisation of a vitiated coflow piloted premixed burner, in *Proceedings of Fifth Asia-Pacific Conference on Combustion*, 2005, pp. 417–420.
55. Law C. and Law H., A theoretical study of ignition in the laminar mixing layer. *Journal of Heat Transfer*, 1982, **104**: p. 329.

9 Internal Flame Processes

Chapter 8 considered ignition, and the processes associated with autoignition and forced ignition of a nonreactive mixture. In this chapter we focus on premixed and nonpremixed flames and the key physics controlling burning rates and extinction processes. Section 9.1 summarizes basic issues associated with the structure and burning rate of steady, premixed flames in homogeneous, one-dimensional flow fields. This includes discussions of the effects of pressure, temperature, and stoichiometry on burning rates. Section 9.2 then discusses how these results are modified by inhomogeneities in mixture composition, and the competition between autoignition waves and deflagration waves. Section 9.3 discusses how these one-dimensional characteristics are altered by inhomogeneities in the flow field relative to the flame, referred to as flame stretch. We then discuss how these lead to changes in burning rate and, for large enough levels of stretch, cause the flame to extinguish. Section 9.4 treats the effects of unsteadiness in pressure, fuel/air ratio, and stretch rate. Specifically, we discuss how the flame acts as a low-pass filter to disturbances in most cases, and that its sensitivity to disturbances diminishes with increasing frequency. These results have important implications for many combustion instability phenomena, where the flame is perturbed by time-varying flow and composition variations.

We then move to nonpremixed flames. Section 9.5 reviews nonpremixed flames in the fast chemistry limit, while Section 9.6 discusses finite-rate kinetic effects. This section shows that large gradients in fuel and oxidizer concentrations can lead to flame extinction.

Most real flames have "edges." For example, flame edges occur in a flame stabilized in a shear layer or an expanding reaction front after forced ignition of a nonpremixed flame [1]. Edge flames are discussed in Section 9.7, where we show important generalizations that cannot be understood from the continuous reaction front analysis in earlier sections.

Finally, Section 9.8 considers intrinsic flame instabilities and shows that, in many instances, flames are inherently unstable and develop wrinkles through internal and hydrodynamic feedback processes. These "flame instabilities" are to be distinguished from the "combustor instabilities" discussed in Section 6.5, which involve a feedback process between the flame and the combustor system itself.

9.1 Premixed Flame Overview [2–4]

This section gives an overview of key characteristics and the internal structure of unstretched, premixed deflagrations – that is, scenarios where a spatially uniform reactant approach flow is combusted in a flat flame. While this section is included for completeness, readers are referred to standard combustion texts for detailed treatments of laminar flames. We briefly review premixed flame structure in Section 9.1.1, and then present typical dependencies of key flame properties on stoichiometry, pressure, and temperature in Section 9.1.2.

9.1.1 Premixed Flame Structure

Consider first the structure of premixed flames. In order to provide a framework with which to interpret this structure for real flames, we start by considering the structure of a flame described by single-step, high activation energy kinetics. In this case, the flame consists of the two zones shown in Figure 9.1: (1) a nonreacting convective–diffusive zone where incoming reactants are preheated by diffusive flux from the (2) reactive–diffusive zone that serves as a chemical reactor. This chemical reactor is highly nonadiabatic as all of the chemical energy released is used to preheat the incoming reactants through diffusion.

The synergistic relation between these two zones – the preheating of the cold reactants in the preheat zone to a temperature where high activation energy kinetic rates become appreciable, and the chemical reactor that supplies all of the generated heat into this preheating zone – leads to a self-perpetuating, propagating reaction front. This front consumes fresh mixture with a mass burning rate per unit area of

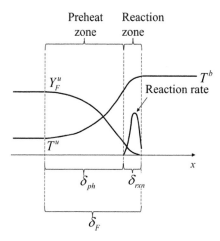

Figure 9.1 Schematic of the premixed flame structure for single-step, high activation energy kinetics.

$\dot{m}_F'' = \rho^u s^{u,0}$ and moves into fresh reactants with a velocity of $s^{u,0}$. The superscript 0 is used here to denote the unstretched value. Burned (b) and unburned (u) quantities are indicated with a superscript.

This two-zone flame structure, as opposed to, say, a single region of continuous reaction that progressively consumes all of the reactant, is a direct result of the high activation energy of the reaction rate. In high activation energy kinetics, the rate of reaction is exponentially sensitive to temperature, and the mixture is essentially chemically frozen until it is brought to high temperatures. To illustrate, increasing T^u from 300 K to 1500 K increases the reaction rate by a factor of e^{53}, or $\sim 10^{23}$, with an activation temperature of $E_a/\mathscr{R}_u = 20\,000$ K. Once the mixture temperature is sufficiently increased, reaction happens very quickly in the reactive–diffusive layer.

In addition to characterizing velocities and burning rates, s^u and $\dot{m}_F'' = \rho^u s^{u,0}$, we are also interested in the thickness of the premixed flame, δ_F^0, and the relative thickness of the reaction and preheat zones, $\delta_{rxn}^0/\delta_{ph}^0$. As shown in Exercise 9.1, the relative thickness of the reaction, δ_{rxn}^0, and preheat zone, δ_{ph}^0, is

$$\frac{\delta_{rxn}^0}{\delta_{ph}^0} \sim \frac{(T^{b,0})^2}{(T^{b,0} - T^u)(E_a/\mathscr{R}_u)} \equiv \frac{1}{Ze}, \qquad (9.1)$$

where we have defined a dimensionless activation energy, the Zeldovich number, Ze. This equation explicitly shows that $\delta_{rxn}^0/\delta_{ph}^0$ decreases monotonically with increases in activation temperature, $E_a/\mathscr{R}_u$. In other words, for low unburned reactant temperatures T^u and high activation energies, the reaction zone is much thinner than the preheat zone, $\delta_{rxn}^0/\delta_{ph}^0 \ll 1$, and therefore the overall flame thickness is essentially the same as the preheat zone thickness, i.e., $\delta_F^0 \sim \delta_{ph}^0$.

The Zeldovich number is not only a function of the activation energy, $E_a/\mathscr{R}_u$, but also the dimensionless temperature rise across the flame. Thus, Eq. (9.1) also shows that an increase in reactant temperature, T^u, decreases Ze, and therefore the ratio $\delta_{rxn}^0/\delta_{ph}^0$ increases. Indeed, for sufficiently high reactant temperature, this two-zone model is a not a good description of the reaction structure as nonnegligible reaction rates occur at the unburned temperature and the problem tends toward that of autoignition, see Section 8.2.2.4.

We will next summarize some other basic flame scalings. Assuming that the reaction rate, $\dot{w}$, scales with pressure and temperature as

$$\dot{w} \propto p^n e^{-(E_a/\mathscr{R}_u)/T}, \qquad (9.2)$$

then the flame thickness and burning rates are given by (see Exercise 9.2):

$$\dot{m}_F'' \propto \left[(\lambda/c_p)\dot{w}(T^{b,0})\right]^{1/2} \propto \left[(\lambda/c_p)p^n e^{-(E_a/\mathscr{R}_u)/T^{b,0}}\right]^{1/2}, \qquad (9.3)$$

$$s^{u,0} \propto p^{(\frac{n}{2}-1)}\left[(\lambda/c_p)e^{-(E_a/\mathscr{R}_u)/T^{b,0}}\right]^{1/2}, \qquad (9.4)$$

$$\delta_F^o \sim \delta_{ph}^o \sim (\lambda/c_p)/\dot{m}_F''. \qquad (9.5)$$

These expressions show that increases in pressure cause increases in mass burning rate and decreases in flame thickness for reaction order $n > 0$. Increasing pressure, however, causes a decrease in burning velocity, $s^{u,0}$, for $n < 2$. Finally, a more general discussion of wave speeds of reaction–diffusion equations is included in Section 9.11, demonstrating how this scaling is controlled by the functional form of the reaction rate.

With these basic ideas as a backdrop, consider more realistic flame structures. This can be done by computationally integrating the full conservation equations using detailed chemical kinetics mechanisms. For example, Figures 9.2–9.4 plot profiles through a methane/air flame. Start with Figure 9.2, which plots the value of the convection, diffusion, and heat release terms from the energy equation where, at any location, the convection term equals the sum of the diffusion and heat release terms. Note how the convection and diffusion terms balance each other up to a temperature of about 1000 K, and the heat release term is negligible. However, only at temperatures of about 1600–1800 K do the diffusion and heat release terms balance. At other temperatures, all three terms are of similar magnitude or the heat release and convection terms balance. As such, while the single-step, high activation energy picture in Figure 9.1 provides some guidance in understanding, there are significant points of difference from the real, multistep kinetics. These are driven by a significantly broader zone of heat release, both at lower temperatures and at high temperatures in the drawn out recombination zone.

In the major species plotted in Figure 9.3, note the significant levels of the intermediates H_2 and CO which are derived from fuel decomposition, as well as the different spatial regions in which the major products H_2O and CO_2 are formed due to the oxidation of these intermediates. A key point to keep in mind is that reaction and heat release rates are one-to-one related in single-step kinetics. This is not the case in multistep systems and, indeed, a significant number of initiation reactions commence upstream of the heat release zone because of upstream diffusion of radicals, such as H,

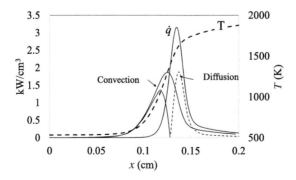

Figure 9.2 Computed convection, diffusion, and heat release rate term profiles from the energy equation through a CH_4/air flame ($\phi = 0.7$, $T^u = 533$ K, $p = 1$ atm, GRI 3.0 mechanism), along with temperature. Diffusion is positive for the solid line, negative for the dashed lines. Image courtesy of H. Johnson II.

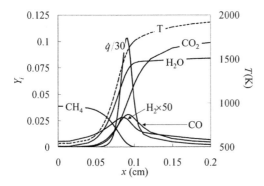

Figure 9.3 Computed major species mass fractions, heat release rate (kW/cm^3), and temperature (right axis) profiles through a CH$_4$/air flame ($\phi = 0.7, T^u = 533$ K, $p = 1$ atm, GRI 3.0 mechanism).

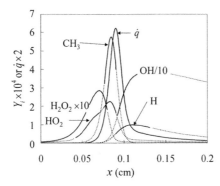

Figure 9.4 Computed minor species mass fractions and heat release rate (kW/cm^3) profiles through a CH$_4$/air flame ($\phi = 0.7, T^u = 533$ K, $p = 1$ atm, GRI 3.0 mechanism). Chemical production rate is positive for the solid line, negative for the dashed lines.

and "low-temperature" chemistry associated with HO$_2$ and H$_2$O$_2$ radicals. This point has been illustrated in Figure 9.4 by using the line type to indicate regions of production (solid) and destruction (dashed). The figure shows that the H radical is present for $x < 0.08$ cm, even though its production rate is negative in this region, illustrating the role of diffusion in bringing it from regions farther downstream where it is produced. The figure also shows chemical activity in the region that would be indicated as the "preheat zone" based on heat release rate, such as rising levels of HO$_2$ and H$_2$O$_2$ early in the flame.

This figure clearly shows that for this mixture and these conditions, the region of significant heat release is approximately one-half to one-third of the overall flame thickness and that there is a blurring of the boundaries between preheat and reaction zones. For example, the HO$_2$ radical is present throughout the majority of the preheat zone, due to the upstream diffusion of the H radical. In addition, other hydrocarbon fragments and oxidation products, such as CH$_3$, are present in this zone. Finally, the figure shows the presence of a "post-flame chemistry zone" associated with recombination of intermediate species, such as CO, H$_2$, OH, and H. Some of these recombination processes are slow, and consequently the recombination zone can be quite long.

9.1.2 Premixed Flame Dependencies

This section discusses calculations of flame burning rates, burning velocities, and flame thicknesses for H$_2$/CO, CH$_4$, and C$_3$H$_8$/oxidant mixtures. H$_2$/CO mixtures are the primary constituents of gasified fuels, as well as the products of rich combustion flames. They are also an important part of the chemistry of more complex fuels. Methane is considered because of its practical importance, and propane provides a useful introduction to the chemistry of larger fuel molecules. These calculations were performed using GRI 3.0 (CH$_4$) [5], San Diego (C$_3$H$_8$) [6], and Davis et al. (H$_2$/CO) [7] mechanisms.[1]

Due to the high overall activation energy of combustion reactions, $\dot{m}_F''$ is very sensitive to both the reactant temperature and adiabatic flame temperature. This latter sensitivity is indirectly illustrated in Figure 9.5 by the "fixed oxidizer" line, which shows the calculated dependence of methane burning velocity, $s^{u,0}$, on fuel/air ratio with standard air.

The flame speed and adiabatic flame temperature share a similar dependence on the equivalence ratio. The flame thickness, which is quantified using Eq. (9.6), has values on the order of 1 mm, with a fuel/air ratio dependence that is the inverse that of the flame speed,

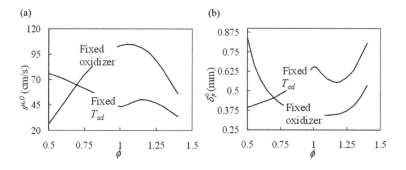

Figure 9.5 Dependence of the (a) unstretched flame speed, $s^{u,0}$, and (b) flame thickness, δ_{F^0}, for a methane-fueled flame on the equivalence ratio at $p = 1$ atm, $T^u = 533$ K. "Fixed oxidizer" denotes standard air with fixed oxygen content (21%/79% O$_2$/N$_2$) and varying T_{ad}. "Fixed T_{ad}" denotes 1900 K fixed flame temperature, with varying O$_2$/N$_2$ fraction.

---

[1] Some laminar flame speed computations presented here are performed at preheat temperatures and pressures extrapolated outside of the experimental conditions used to build these mechanisms. The Davis and GRI 3.0 mechanism parameters have been optimized using flame speed data targets obtained with room temperature reactants and pressures up to 15 and 20 atm, respectively. The San Diego mechanism (46 species, 325 reactions), is not built by optimization procedures, but by selection of relevant reactions. Reasonable comparisons of predicted propane/air flame speeds with measurements have been obtained from measurements at elevated temperatures and pressures [8, 9].

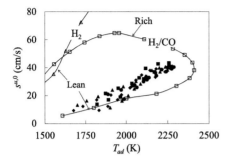

Figure 9.6 Measured dependence of the unstretched flame speed, $s^{u,0}$, on calculated adiabatic flame temperature for several n-alkane and H_2 fuel/air mixtures (♦ CH_4 [10], ■ C_2H_6 [11], ▲ C_3H_8 [10], ● C_7H_{16} [12], H_2/air [13], H_2/CO (5/95 by volume [14]). Data obtained at STP by varying fuel/air ratio.

$$\delta_F = \frac{(T^b - T^u)}{dT/dx|_{max}}. \tag{9.6}$$

In order to further illustrate this point about thermal effects, Figure 9.6 plots the dependence of measured $s^{u,0}$ values on calculated adiabatic flame temperatures, T_{ad}, for four different n-alkane fuels over a range of fuel/air ratios. The figure shows that the key n-alkane flame speed dependencies are captured with the flame temperature. However, other factors to be discussed further below also influence the flame speed, such as molecular fuel structure, as indicated in this plot by including two H_2 results.

In addition to adiabatic flame temperature, initial reactant temperatures are also important. For example, Figure 9.7 plots fixed adiabatic flame temperature results for CH_4, C_3H_8, and H_2/CO mixtures at several pressures, indicating that $s^{u,0}$ increases by a factor of roughly 10 as T^u increases from 300 K to 1000 K. This is accompanied by the flame becoming slightly thinner for the most part. Since these calculations were performed at a fixed $T_{ad} = 2000$ K value, this sensitivity is not due to a change in flame temperature. With the exception of the lower-pressure H_2/CO curves, these lines all indicate a very similar temperature dependence.

We next consider nonthermal effects [15]. Returning to Figure 9.6, the figure shows that significant nonthermal effects also influence the burning velocity. The H_2 and H_2/CO fuels exhibit very different characteristics than the n-alkane fuels, which can be attributed to the differences in their molecular weights and structure. In addition, the H_2 flame speed peaks at much richer stoichiometries than its flame temperature does, due to diffusive effects. Figure 9.5 provides another manifestation of nonthermal effects, by plotting the flame speed sensitivity to fuel/air ratio at constant flame temperature, obtained by adjusting N_2 diluent levels. The complete inversion of the flame speed dependence on fuel/air ratio relative to the "fixed oxidizer" result illustrates additional kinetic effects. This behavior is due to the corresponding variation of reactant O_2 mole fraction required to maintain constant T_{ad}. The O_2 content of the reactants is a minimum at the same location as the minimum in flame speed [16].

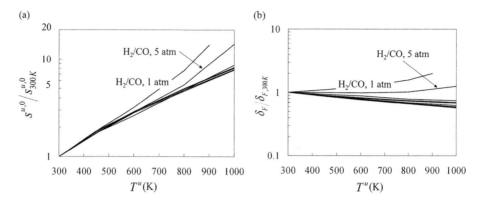

Figure 9.7 Dependence of the (a) normalized burning velocity and (b) flame thickness on preheat temperature at a fixed $T_{ad} = 2000$ K (obtained by varying stoichiometry at each T^u value). Cluster of lines obtained for CH_4/air and C_3H_8/air at $p = 1, 5,$ and 20 atm and a 50/50 H_2/CO blend at 20 atm.

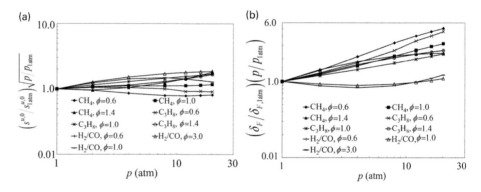

Figure 9.8 Calculated dependence of the (a) normalized burning velocity and (b) flame thickness on pressure for various fuel/air blends ($T^u = 300$ K).

Consider next pressure effects [17, 18]. Figure 9.8 plots calculated burning velocities and flame thicknesses for methane/air. These results show that the burning velocity, to first approximation, scales as $\sim p^{-1/2}$ and mass burning rates as $p^{1/2}$. Also, the flame thickness decreases roughly linearly with p. Both of these dependencies reflect the fact that the overall reaction order is less than 2, as can be seen from Eq. (9.4). Exercise 9.3 includes further results for pressure and preheat temperature effects.

Interested readers are referred to focused combustion texts [2, 4] for further discussion of how these sensitivities are controlled by diffusive and kinetic processes.

9.2 Premixed Combustion in Inhomogeneous, Autoigniting Mixtures

This section considers reaction wave propagation in situations where the mixture is inhomogeneous (e.g., temperature or composition are varying spatially) and residence times are long enough for autoignition to occur; i.e., $\tau_{ign}/\tau_{res} \sim O(1)$ [19–23]. It is important to differentiate autoignition as a process controlling flame stabilization (such as depicted in Figure 8.1) and controlling bulk reactant consumption – it is the latter case which is the focus of this section.

As noted in Section 8.2.2.4, even in the absence of a propagating deflagration wave, an autoignition wave propagates through a mixture with a velocity of $v_{ign} = 1/|\nabla \tau_{ign}|$. In the limiting case, the autoignition wave velocity is completely independent of diffusion. However, once an autoignition-induced reaction occurs, diffusive processes are inherently present and coupled reactive–diffusive waves (i.e., flames) will have a tendency to propagate away from the ignition site. In the laminar case, this flame propagation speed is given by the displacement speed, s_d, defined in Sections 7.1 and 9.3.4. In addition, at locations where the flame is stabilized by a pilot flame, or recirculating hot gases, the flame will propagate away from these stabilization locations.

Consider the local conditions around an autoignition site. If $s_d > v_{ign} = 1/|\nabla \tau_{ign}|$ then the flame propagates faster than the autoignition wave and reactant consumption will locally be controlled by flame propagation, even though the entire mixture may be under conditions where autoignition can occur. This ratio of the flame speed and autoignition wave speed is known as the Sankaran parameter, Sa [24, 25],

$$Sa = s_d |\nabla \tau_{ign}|. \tag{9.7}$$

For a mixture where $s_d |\nabla \tau_{ign}| \gg 1$, reactant consumption rates are flame propagation dominated; conversely, where $s_d |\nabla \tau_{ign}| \ll 1$, reactant consumption is autoignition dominated.

The ignition time gradient is, in turn, a function of gradients in temperature or mixture composition and can be written as

$$\nabla \tau_{ign}(T, \phi, \ldots) = \frac{\partial \tau_{ign}}{\partial T} \nabla T + \frac{\partial \tau_{ign}}{\partial \phi} \nabla \phi + \cdots. \tag{9.8}$$

This indicates that, more generally, $s_d |\nabla \tau_{ign}|$ is a function of a new family of kinetic parameters, $s_d \frac{\partial \tau_{ign}}{\partial T}$ or $s_d \frac{\partial \tau_{ign}}{\partial \phi}$, that are a property of the mixture, and configuration/turbulence specific parameters, like $|\nabla T|$ or $|\nabla \phi|$. These kinetic parameters can have sensitivities that differ from both s_d and τ_{ign} [26]. To illustrate, Figures 9.9 and 9.10 overlay isolines of $s_d \frac{\partial \tau_{ign}}{\partial T}$, s_d, and τ_{ign} as a function of pressure and temperature for methane and heptane. At a fixed pressure, the spacing between the isolines indicates the temperature sensitivity, showing that for methane, the temperature sensitivity of s_d is the smallest, while $s_d \frac{\partial \tau_{ign}}{\partial T}$ and τ_{ign} have comparable values. Both increasing pressure and temperature cause $\left| s_d \frac{\partial \tau_{ign}}{\partial T} \right|$ to decrease, showing that the role of autoignition relative to flame propagation increases. The heptane results are more complex, due to the NTC region (see Figure 8.6).

Outside of NTC regions, the role of autoignition relative to flame propagation increases with pressure and temperature. However, while increasing temperature generally increases the role of autoignition, the effect is not as strong as manifested in ignition times alone; i.e., $s_d \frac{\partial \tau_{ign}}{\partial T}$ generally has a weaker temperature sensitivity than

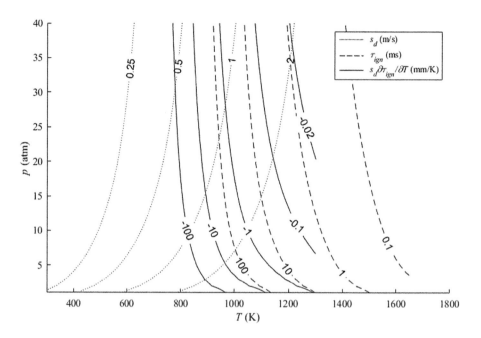

Figure 9.9 Isolines of flame speed, s_d, ignition time, τ_{ign}, and $s_d \frac{\partial \tau_{ign}}{\partial T}$ as a function of reactant pressure and temperature for methane at $\phi = 0.8$. Calculations courtesy of Minh Bau Louong and Hong Im.

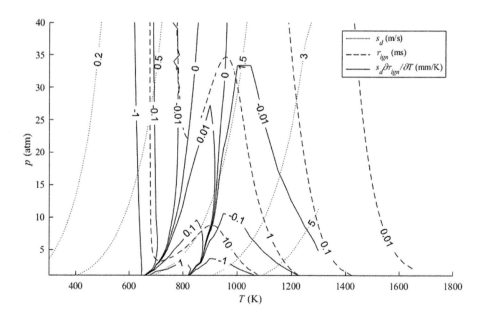

Figure 9.10 Isolines of flame speed, s_d, ignition time, τ_{ign}, and $s_d \frac{\partial \tau_{ign}}{\partial T}$ as a function of reactant pressure and temperature for n-heptane at $\phi = 0.8$. Calculations courtesy of Minh Bau Louong and Hong Im.

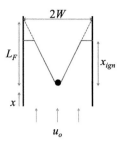

Figure 9.11 Example of a flow with a gradient in residence time before reaction, showing transition from flame propagation dominated to autoignition dominated.

τ_{ign} because of the competing influence of s_d. For example, calculations for lean syngas mixtures show that at 10 atm, increasing the temperature from 900 K to 1100 K decreases τ_{ign} by a factor of ~150, while $\left| s_d \frac{\partial \tau_{ign}}{\partial T} \right|$ decreases by about 20 [27]. These sensitivities are closer in the above methane results, but the same point is true. At the same conditions, τ_{ign} decreases by a factor of ~35, while, $s_d \frac{\partial \tau_{ign}}{\partial T}$ decreases by about 30. To repeat, while increases in preheat temperature increase the importance of autoignition relative to flame propagation, the effect is weaker than would be expected based on the temperature sensitivity of τ_{ign}.

In flows with gradients in residence time as well, the relative roles of the autoignition and deflagration wave in consuming reactants is controlled by gradients in their ratio, τ_{ign}/τ_{res}. To illustrate, Figure 9.11 shows a pedagogical laminar flow example, where the mixture is homogeneous with fixed ignition time delay; however, the flame spreading from a stabilization point implies that there is *stratification in residence time*. The flame position and burning rate/volume is controlled by the laminar burning velocity near the stabilization point. Farther downstream, however, the mixture reacts through autoignition.

The relative significance of flame propagation and autoignition can be worked out by simple scaling arguments. The flame length in the absence of autoignition is given by $L_F/W = u_o/s_d$. Similarly, the axial location of autoignition, x_{ign}, is given by $x_{ign}/u_o = \tau_{ign}$. Combining these two expressions leads to the following scaling for the location along the propagating flame front where autoignition processes become dominant:

$$\frac{x_{ign}}{L_F} = \frac{\tau_{ign} s_d}{W}. \tag{9.9}$$

Assuming $\tau_{ign} \sim p^{-1}$ and that $s_d \sim p^{-\frac{1}{2}}$, this expression shows that the importance of autoignition grows as $p^{-\frac{3}{2}}$. The temperature effect is more complex, as it increases flame speeds and decreases ignition times. Finally, this expression shows that autoignition effects grow with facility size, W.

9.3 Premixed Flame Stretch and Extinction

In this section, we generalize the premixed flame discussion from Section 9.1, which focused on flat flames in uniform flow fields, to more general configurations of curved flames in inhomogeneous flow fields. We show that these situations introduce a variety of new physics. Understanding these effects is critical to understanding flame propagation in shear layers or turbulent, premixed flames. The reader will find it useful to review the kinematic discussion of flame stretch in Section 7.4 before reading this section.

9.3.1 Overview

By definition, diffusive and convective fluxes are aligned in a one-dimensional premixed flame. An important additional degree of freedom we will explore in this section is the effects of their misalignment. To illustrate, consider the two geometries shown in Figure 9.12, which depict the preheat and reaction zones of "stretched" premixed flames. For the time being, we return to the simplified, high activation energy, one-step kinetic description of flames.

Start with the curved flame on the left. Note that with Fickian diffusion, diffusive heat and species fluxes occur in a direction normal to their respective isosurfaces. Thus, the curved flame shows heat flux normal to the isotherms, illustrating the "focusing" of thermal energy into the control volume. Similarly, reaction intermediates that have a high concentration in the reaction zone diffuse normal to their isoconcentration lines, also focusing into the control volume. In contrast, reactant species diffuse outward and are "defocused." Now, consider a control volume that coincides with the two indicated streamlines and the front and back of the flame. There are convective fluxes through the front and back of the control volume, but not through the sides. If the convective and diffusive fluxes are aligned, then this control volume is adiabatic; although there may be heat transfer internal to the control volume, no net heat is transferred out of it (assuming no external heat loss for the time being). In contrast, the control volume is not adiabatic if these convective and diffusive thermal energy transport processes are not aligned. Likewise, mass transport can lead to reactant fluxes in or out of the control volume, and an associated chemical energy flux. Similarly, important intermediate radicals can diffuse into or out of the control volume.

Misalignment of diffusive and convective fluxes also occurs if the flame is flat, but the flow streamlines curve ahead of the flame, as shown in the right figure. Again, drawing a control volume coinciding with the streamlines, note that these curved

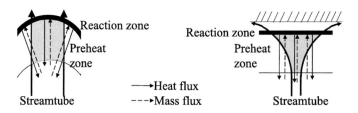

Figure 9.12 Illustration of the internal structure of two stretched premixed flames, showing misalignment of convective and diffusive fluxes. The gray region denotes the control volume.

streamlines cause the control volume to be nonadiabatic and for there to be net diffusion of species into and out of it.

To further analyze these effects, we will consider the diffusivities of heat and species; e.g., within the one-step kinetic description, these diffusivities are α, $\mathcal{D}_{Fuel}$, and $\mathcal{D}_{Ox}$. We will show next that stretch effects are critically dependent on the Lewis number, $Le = \alpha/\mathcal{D}$ (where $\mathcal{D}$ denotes some reference diffusivity, often taken as that of the deficient reactant), and the mass diffusivity ratio, $\mathcal{D}_{Fuel}/\mathcal{D}_{Ox}$. Lewis number effects can be understood by considering the energy equation and the effects of heat/mass transfer on the energy balance in the control volume. Following Eq. (1.29), neglecting viscous effects and body forces, the steady energy equation can be written as

$$\rho \vec{u} \cdot \nabla h_T = -\nabla \cdot \vec{q}. \tag{9.10}$$

This equation describes the balance between convective (left-hand side) and diffusive (right-hand side) fluxes of stagnation enthalpy and energy. Neglecting radiative heat transfer and the DuFour effect, and assuming Fickian diffusion, this heat flux vector is given by

$$\vec{q} = -k_T \nabla T - \rho \sum_{i=1}^{N} h_i \mathcal{D}_i \nabla Y_i. \tag{9.11}$$

This equation shows that energy flux occurs through both heat and mass diffusion. Thus, for the curved flame shown in Figure 9.12, energy flows in through the sides of the control volume through heat transfer, but also flows out through the sides through mass transfer of chemical energy associated with the reactants. If the heat and mass transfer diffusivities are the same, i.e., $Le = 1$, then these effects cancel for weakly curved flames and the energy balance in the control volume is not disturbed by these diffusive fluxes. These effects do not balance, however, if $Le \neq 1$, leading to an increase or decrease in thermal energy in the control volume. For this curved flame example, a $Le > 1$ value leads to net energy flux into the control volume, causing an increase in temperature and mass burning rate. Conversely, if the flame is curved in the opposite direction, a $Le > 1$ mixture leads to a decrease in flame temperature and burning rate. The opposite behavior occurs if $Le < 1$. Similarly for the stagnation flame shown in Figure 9.12, $Le > 1$ leads to a decrease in local temperature and burning rate. Thus, these results show that curvature/flow strain can lead to burning rate increases or decreases, depending on both the direction of curvature/strain and the Lewis number.

Next, consider differential diffusion effects. To illustrate, consider a lean CH_4/air mixture in the curved flame configuration in Figure 9.12. In this case, both oxygen and CH_4 diffuse through the sides of the control volume. However, the net loss of CH_4 is higher, since it is lighter and more diffusive than O_2. This shows that the mixture stoichiometry entering the reaction zone in the control volume is leaner than that of the approach flow reactive mixture. This leads to a decrease in flame temperature and burning rate. Conversely, if the mixture is rich, differential diffusion effects move the stoichiometry closer to unity, causing an increase in flame temperature and burning rate. The opposite behavior occurs for fuels that are heavier than air and/or if the direction of curvature is inverted. This behavior is illustrated in Figure 9.13, which plots the measured tip temperature of a Bunsen flame normalized by the calculated adiabatic flame temperature for CH_4/air and C_3H_8/air mixtures.

The figure clearly shows that the lighter than air fuel, CH_4, has a flame temperature that is less than and greater than T_{ad} for $\phi \lesssim 1.1$ and $\phi \gtrsim 1.1$, respectively. The C_3H_8 blend has the opposite trend. Both results are consistent with this discussion.

9.3.2 Expressions for Flame Stretch

In this section we relate the processes described above to the concepts of flame stretch which were considered from a flame sheet perspective in Section 7.4. Recall that flame stretch describes the increase in flame area by tangential velocity gradients along the flame sheet, κ_a, as well as motion of the curved flame surface, κ_b [28], as shown in Figure 9.14. The flame stretch rate, κ, is

$$\kappa = \frac{\partial u_{t,1}}{\partial t_1} + \frac{\partial u_{t,2}}{\partial t_2} + (\vec{v}_F \cdot \vec{n})(\nabla \cdot \vec{n}) = \underbrace{\nabla_t \cdot u_t}_{\kappa_a} + \underbrace{(\vec{v}_F \cdot \vec{n})(\nabla \cdot \vec{n})}_{\kappa_b}, \qquad (9.12)$$

where

$$\nabla_t \cdot u_t = -\vec{n} \cdot \nabla \times (\vec{u} \times \vec{n}) \qquad (9.13)$$

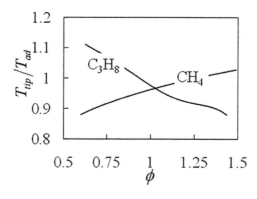

Figure 9.13 Measured dependence of the temperature at the curved tip of a Bunsen flame, normalized by calculated adiabatic flame temperature for CH_4/air and C_3H_8/air mixtures [2].

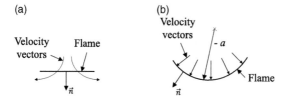

Figure 9.14 Illustration of two-dimensional flames showing positive stretching of a flame surface by (a) tangential velocity gradients along the flame and (b) motion of a curved flame. Flame normal is pointing into the products.

and $-\nabla \cdot \vec{n}$ equals $1/a$ and $2/a$ for cylindrical and spherical flame surfaces, respectively, where a is the radius of curvature.

The first term in this expression, κ_a, describes hydrodynamic flame stretch, which is the stretching of the flame sheet by the tangential velocity components. From the continuity equation, the tangential velocity variation leading to stretch can be alternatively thought of as a variation in normal velocity in the direction normal to the flame (since, just upstream of the flame, the relation $\nabla_t \cdot \vec{u}_t = \frac{\partial u_{t_1}}{\partial t_1} + \frac{\partial u_{t_2}}{\partial t_2} = -\frac{\partial u_n}{\partial n}$ holds). Thus, positive stretch associated with κ_a is also directly related to flow deceleration normal to the flame. The second term, κ_b, describes the effects of motion of a curved flame; such a term would be positive, for example, in the expanding spherical flame shown in Figure 9.14.

Both examples in Figure 9.12 illustrate the effects of hydrodynamic flame stretch. The stagnation flame is stretched because the tangential velocity of the flow varies along the flame. The curved flame is stretched because the angle of the flame changes with respect to the flow. Thus, a flame can experience hydrodynamic flame stretch even in a flow with no flow strain, $\underline{S} = \frac{1}{2}(\nabla \vec{u} + \nabla \vec{u}^T) = 0$, since a variation in the angle of the flame causes a change in tangential velocity with respect to the flame. Moreover, because both examples in Figure 9.12 were stationary flames, the flame motion term in Eq. (9.12), κ_b, is zero. In other words, curvature in a stationary flame induces flame stretch through the $\nabla_t \cdot \vec{u}_t$ term, not the $(\vec{v}_F \cdot \vec{n})(\nabla \cdot \vec{n})$ term. This immediately shows that stationary curved flames are only stretched if the flow is misaligned with the flame normal, i.e., $u_t \neq 0$. A stationary curved flame with $u_t = 0$ is unstretched, such as a stationary cylindrical flame fed by a central source.

While Eq. (9.12) is physically revealing in terms of fundamental mechanisms that cause stretch, we can rearrange it in other ways that more explicitly illustrate the role of flow nonuniformity and flame curvature (see Exercise 9.4):

$$\kappa = \underbrace{-\vec{n}\vec{n} : \nabla \vec{u} + \nabla \cdot \vec{u}}_{\kappa_S} \underbrace{-s^u(\nabla \cdot \vec{n})}_{\kappa_{curv}}. \tag{9.14}$$

Note that the $-\vec{n}\vec{n} : \nabla \vec{u} + \nabla \cdot \vec{u}$ term can be written as $-n_i n_j \partial u_i / \partial x_j + \partial u_i / \partial x_i$ in tensor form. We have again written the stretch as the sum of two terms. The first term, κ_s, is nonzero only if the flow has spatial gradients. This term incorporates the hydrodynamic flame stretch terms, κ_a, and part of the curved flame motion term, κ_b, from Eq. (9.12). The second term, κ_{curv}, quantifies how flame curvature in a uniform approach flow, as considered in Figure 9.12, leads to flame stretch.

9.3.3 Weak Stretch Effects

This section describes the sensitivity of flames to weak stretch. The discussion in Section 9.3.1 suggests that various flame characteristics, such as mass burning rate, burning velocity, and flame thickness, are functions of curvature and flow nonuniformity. Indeed, asymptotic analyses based on high activation energy, one-step kinetics show that these effects can be encapsulated into the sensitivity of the flame to stretch, κ [29]. Moreover, for steady, weakly stretched flames, the flame's sensitivity to the various stretch contributors, such as κ_a, κ_b, κ_s, or κ_{curv} are the same, so that we can write these stretch sensitivities simply as $s^u = s^u(\kappa)$. For small stretch values, it is useful to expand these expressions in a Taylor series, which to first order is given by

$$s^u(\kappa) = s^u|_{\kappa=0} + \frac{\partial s^u}{\partial \kappa}|_{\kappa=0} \kappa = s^{u,0} - \delta^u_M \kappa, \quad (9.15)$$

where, again, the superscript 0 denotes the value of the unstretched value of the quantity. This expression also defines the Markstein length, δ_M, which is the negative of the flame speed stretch sensitivity. The superscript u or b denotes the sensitivity of the burning velocity with respect to reactants or products, e.g.,

$$s^b(\kappa) = s^{b,0} - \delta^b_M \kappa, \quad (9.16)$$

where $\delta^b_M \neq \delta^u_M$; indeed, they can have opposite signs. We define the normalized flame stretch sensitivity, called the Markstein number, Ma, as

$$Ma^u = \frac{\delta^u_M}{\delta^0_F}. \quad (9.17)$$

In addition, we can define a normalized stretch rate, referred to as the Karlovitz number,[2] Ka, as

$$Ka = \frac{\delta^0_F \kappa}{s^{u,0}}. \quad (9.18)$$

Substituting these definitions into Eq. (9.15) leads to

$$\frac{s^u}{s^{u,0}} = 1 - Ma^u Ka, \quad (9.19)$$

The discussion in Section 9.3.1 suggests that Ma should be a function of the Lewis number, the relative diffusivities of fuel and oxidizer, and the fuel/air ratio. For example, the differential diffusion arguments suggest that lean mixtures of lighter-than-air fuels, such as methane or hydrogen, should have enhanced flame speeds for the positively stretched stagnation flame and decreased speeds for the negatively stretched curved flame in Figure 9.12; i.e., that $Ma < 0$. These arguments also suggest that $Ma > 0$ for rich fuel/air blends of these fuels, and that these trends should be inverted for fuels that are heavier than air, such as propane.

These very intuitive results have been confirmed by experimental results.[3] Figure 9.15 plots the measured dependence of propane/air flame speeds on normalized

---

[2] There are multiple ways to define the Karlovitz number, based on stretched and unstretched values, as well as burned or unburned values, so it is important to understand the definition being used in any particular study.
[3] Some caution should be used when analyzing stretch sensitivity results, as results for quantities such as Ma can vary significantly depending on how quantities are defined and what reference surface is being used to calculate the necessary quantities, because of the gradients in the flowfield both upstream and through the flame.

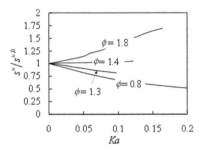

Figure 9.15 Measured dependence of C_3H_8/air flame speed on Karlovitz number, Ka, showing the sensitivity of the flame speed to stretch [30].

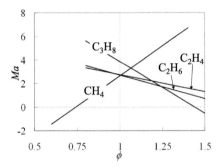

Figure 9.16 Measured dependence of the Markstein number on equivalence ratio for several n-alkane/air fuels with reactants at STP [30].

stretch rate, Ka. As suggested above, it shows that the flame speed of rich propane–air blends increases with stretch, with the $Ma = 0$ point occurring slightly rich, at $\phi \sim 1.4$.

Correlations for the Markstein number for several hydrocarbons are summarized in Figure 9.16 as a function of equivalence ratio. The general slopes of these curves with respect to fuel molecular weight can be anticipated from the discussion above. However, these results also show that the point of zero stretch sensitivity, $Ma = 0$, does not occur at $\phi = 1$, but at some higher or lower value depending on the fuel. This reflects coupled Lewis number, differential diffusion, and multicomponent chemistry effects.

In addition to the Markstein length's dependencies on stoichiometry and reactant specie diffusivity, Ma also exhibits sensitivities to other quantities. For example, quantities influencing flame thickness strongly influence δ_M, as it is ultimately the fact that the flame has finite thickness that makes it stretch sensitive. This can be seen by comparing the stretch sensitivity of the burning velocity at several pressures in Figure 9.17, noting that the unstretched flame thickness scales inversely with pressure. Note the monotonically decreasing dimensional stretch sensitivity with increasing pressure. Some, but not all, of this sensitivity is eliminated when the stretch rate is normalized and plotted as a Karlovitz number, as defined in Eq. (9.18).

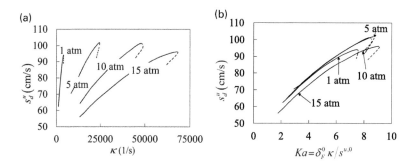

Figure 9.17 Calculated dependence of s_d^u (defined in Section 9.3.4) on (a) dimensional and (b) dimensionless stretch rate for a 30/70 H_2/CO blend at $T^u = 300$ K. Each mixture is at a different stoichiometry so that the unstretched flame speed stays constant at $s^{u,0} = 34$ cm/s; $\phi = 0.61$ (1 atm), $\phi = 0.75$ (5 atm), $\phi = 0.84$ (10 atm), $\phi = 0.89$ (15 atm).

9.3.4 Strong Stretch Effects, Consumption and Displacement Speeds, and Extinction

We have now seen what stretch is, why flames are sensitive to it, and provided some examples quantifying this sensitivity for weak stretch, $Ka \ll 1$ (recall that the previous section was based on a weak stretch expansion, see Eq. (9.15)). In this section, we expand this treatment to larger stretch values, $Ka \sim O(1)$ [31].

It is helpful to review a few points on the burning velocity, s, and burning rate, ρs, for flat, unstretched flames. The flame speed definition used up to this point in the text has been the speed of the flow with respect to the flame, which we will refer to more specifically as the *displacement speed*, s_d. Because of gas expansion through the flame, the displacement speed with respect to the unburned and burned flow, s_d^u and s_d^b, is different. These definitions can be generalized to refer to the speed of the flow with respect to any chosen temperature or concentration isosurface in the flame (see Section 7.4). This dependence of flame speed on chosen flame isosurface is not too problematic for unstretched or weakly stretched flames as, for example, the approach flow velocity varies weakly upstream of the flame. Moreover, the mass burning rate is constant through the flat, unstretched flame, since $\rho^u s^{u,0} = \rho^b s^{b,0}$.

In highly stretched flames, the mass flux itself varies significantly through the flame and the flow velocity gradients may occur over length scales that are on the order of the flame thickness [29]. Therefore, the displacement speed value becomes a strong function of the choice of isosurface. Moreover, the displacement speed definition for flame speed – i.e., the speed of the flow with respect to some flame isosurface – can lead to counterintuitive results. For example, in highly stretched flames, the displacement speed can become negative, i.e., $s_d < 0$. This seemingly nonsensical result is actually quite real; the bulk fluid convection occurs "backward" through the reference isosurface. However, the mass burning rate is positive, as it must be. This situation occurs in flames with very high concentration gradients where the flame is supplied with reactants by diffusive fluxes which are strong enough to counteract the bulk convection in the opposite direction. This situation can be realized in highly curved flames [32] or with a counterflow burner, by stagnating premixed reactants against hot products/inert gases. At very high strain rates, the flame moves across the stagnation surface into the inert side and is supplied fuel/oxidizer by diffusive flux [29]. Finally, displacement speeds can be negative in flames with oscillatory thicknesses (e.g., in an oscillatory stretch field), where the position of the reference isosurface moves.

An alternative flame speed definition can be defined based on spatially integrated chemical production rates, referred to as the consumption speed, s_c [33–35]. The consumption speed is basically the mass or sensible heat production rate integrated along a streamline through the flame, normalized by the unburned density or sensible enthalpy, respectively. To motivate the consumption speed definition, consider the steady-state, one-dimensional form of the energy equation, Eq. (1.35), neglecting body forces and viscosity. Integrating this equation over $-\infty < x < \infty$ leads to

$$\int_{-\infty}^{\infty} \frac{d(\rho u_x h_{sens})}{dx} dx = \int_{-\infty}^{\infty} \left(\dot{q} - \frac{d\dot{q}_x}{dx} + \frac{d}{dx}\left(\rho \sum_{i=1}^{N} h_{f,i}^0 Y_i u_{x,D,i} \right) \right) dx. \quad (9.20)$$

Assuming no spatial gradients at $\pm \infty$ leads to the following thermal consumption speed definition:

$$\rho^u s_c^u = \rho^b s_c^b = \frac{\int_{-\infty}^{\infty} \dot{q} dx}{(h_{sens,\infty} - h_{sens,-\infty})}. \quad (9.21)$$

The consumption speed based on a species consumption rate can be derived in a similar manner, starting with Eq. (1.25). Carrying out a similar analysis, the following expression is obtained for the species consumption speed:

$$s_c = \frac{\int_{-\infty}^{\infty} \dot{w}_i dx}{\rho(Y_{i,\infty} - Y_{i,-\infty})}. \quad (9.22)$$

The displacement and consumption speeds are identical for unstretched flames, i.e., $s_c^0 = s_d^0$, but differ for nonzero κ values. To illustrate, Figure 9.18 plots the calculated dependence of the displacement and consumption speed for a H_2/CO mixture on stretch rate.

Note that different results are obtained for the species (H_2) and enthalpy based definitions of s_c, both of which also differ from s_d. Moreover, the slopes of these curves in the low-stretch limits are different, implying that we can define both consumption- and displacement-based Markstein lengths.

Also, recall from the prior section that the flame's sensitivity to different stretch processes, e.g., κ_a, κ_b, κ_s, and κ_{curv}, is identical at low stretch rates, leading to a single Markstein length describing the flame's sensitivity to total stretch, κ. At high stretch rates, this is no longer the case, e.g., the burning rate sensitivity to hydrodynamic stretch, κ_H, and curvature stretch, κ_{curv}, are different [36]. To illustrate this point, Figure 9.19 presents the calculated dependence of the consumption speed on stretch rate for three different configurations – expanding cylindrical flames with three

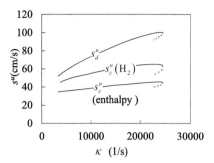

Figure 9.18 Calculated dependence of the displacement speed, s_d^u, and several consumption speeds, s_c^u, on stretch rate for a symmetric stagnation burner (s_d defined at the point of minimum velocity upstream of flame) for a 30/70 H$_2$/CO flame at $\phi = 0.75$, $T^u = 300$ K, $p = 5$ atm (2 cm nozzle separation distance, Davis kinetic mechanism [7]).

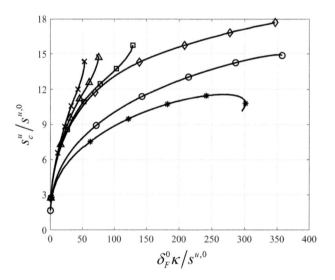

Figure 9.19 Calculated sensitivity of the consumption speed on normalized stretch rate for a planar counterflow flame (*), tubular counter flowflame (○), and expanding counter flow flame with different initial ignition radius ($a_i/\delta_F^0 = 0.25$ "◇", $a_i/\delta_F^0 = 0.5$ "□", $a_i/\delta_F^0 = 0.75$ "△", $a_i/\delta_F^0 = 1$ "×") [38].

different ignition kernel radii, a stagnation flame, and a tubular flame [37], which is essentially a curved stagnation flame that incorporates both tangential flame stretching and curvature.

In addition, there is a maximum stretch rate, κ_{ext}, that a flame can withstand before extinguishing. This extinction stretch rate can be inferred from the plots in Figures 9.17 and 9.18; note that this plot shows the top branch and part of the middle, unstable branch of the S-curve response of the flame to stretch. For example, Figure 9.17 shows that κ_{ext} monotonically increases with pressure, largely due to the flame thickness effect. In other words, what constitutes a "large stretch" rate as

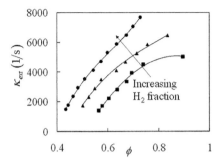

Figure 9.20 Measured dependence of the extinction stretch rate on equivalence ratio for 69/31, 83/17, and 100/0 CH_4/H_2 mixtures by volume ($T^u = 573$ K, $p = 1$ atm) [39].

measured by the flame depends on the flame's chemical time scale. The extinction stretch rate is sensitive to other parameters as well; for example, Figure 9.20 plots the significant sensitivity of κ_{ext} to fuel/air ratio and level of H_2 addition.

The extinction stretch rate is a useful quantity with which to scale flame characteristics, such as blowoff [40]. However, the dependence of κ_{ext} on the experimental configuration in which it is measured should be emphasized. For example, there is extensive data available for κ_{ext} for symmetric, opposed-flow flames. However, as noted above, the flame has different sensitivities to different stretch processes. Moreover, κ_{ext} depends on the details of the flow velocity profile through the flame, as the flow gradients occur over the same length scales as the flame thickness. Thus, for example, varying the distance between nozzles or the nozzle exit velocity profile for stagnation burners has an influence on κ_{ext} [41]. Section 9.10 discusses the quenching of flames by vortices, which introduces additional physics of flame interactions with the strain field.

9.4 Premixed Flames: Unsteady Effects

The preceding sections have treated several important dependencies of the burning rate on quantities such as pressure, temperature, and stretch rate. In this section, we consider unsteady effects associated with small fluctuations in these quantities. By expanding the burning rate in a Taylor series, $s = s_0 + s_1(t) + s_2(t) + \cdots$ (see Section 2.1), the first-order perturbation for the consumption speed can be written as

$$\frac{s_{c,1}}{s_{c,0}} = Ma_{c,curv}\frac{\kappa_{curv,1}}{(1/\delta_F)} + Ma_{c,s}\frac{\kappa_{s,1}}{(1/\delta_F)} + \frac{\partial(s_c/s_{c,0})}{\partial(p/p_0)}\bigg|_0 \frac{p_1}{p_0} + \frac{\partial(s_c/s_{c,0})}{\partial(\phi/\phi_0)}\bigg|_0 \frac{\phi_1}{\phi_0} + \cdots,$$

(9.23)

The partial derivatives on the right-hand side of Eq. (9.23) denote the frequency-dependent sensitivity of the consumption speed to flame curvature, hydrodynamic stretch, pressure, and equivalence ratio perturbations, respectively. The burning velocity sensitivities to curvature-induced stretch and hydrodynamic strain-induced stretch exhibit different frequency dependencies and, hence, are separated out in the expansion above.

Insight into the low-frequency characteristics of these sensitivity derivatives (such as $\partial(s_c/s_{c,0})/\partial(\phi/\phi_0)$), can be deduced directly from their steady-state tendencies. To illustrate this point, consider Figure 9.5(a), which plots the dependence of unstretched flame speed on reactant equivalence ratio. For hydrocarbon–air mixtures, the burning velocity attains its maximum value at a near-stoichiometric equivalence ratio, $\phi_{max\{s^{u,0}\}}$, implying that $\partial(s_c/s_{c,0})/\partial(\phi/\phi_0)$ is zero at this value of the fuel/air ratio. For $\phi < \phi_{max\{s^{u,0}\}}$, this sensitivity derivative is positive, while for $\phi > \phi_{max\{s^{u,0}\}}$, it is negative. The growing value of this sensitivity as ϕ_0 decreases has been argued to be a reason for the significance of equivalence ratio oscillations in exciting self-excited instabilities in lean, premixed systems [42].

These sensitivities become frequency dependent at higher frequencies where the internal flame structure cannot adjust fast enough relative to the external perturbations [43, 44]. Section 9.9 treats these non-quasi-steady effects in greater detail. In general, the flame response time scales as $\tau_{chem} \sim \delta_F^0/s^{u,0}$. This introduces a dimensionless time scale parameter, St_F, defined as

$$St_F = \frac{f\delta_F^0}{s^{u,0}}. \tag{9.24}$$

Typical values of this chemical time are plotted in Figure 9.21 as a function of pressure and temperature, calculated assuming that the pressure and temperature are isentropically related. For example, the lean methane/air flame results show that non-quasi-steady effects occur for $f \gtrsim 200$ Hz at 1 atm and $f > 1$ kHz at 20 atm. An important implication of this result is that non-quasi-steady effects may be highly significant in atmospheric, lab-scale devices, but negligible in the high-pressure system they may be attempting to emulate.

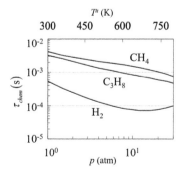

Figure 9.21 Calculated dependence of the characteristic chemical time, $\tau_{chem} = \delta_F^0/s_d^{u,0}$, on pressure/temperature for isentropically compressed C_3H_8/air, CH_4/air, and H_2/air mixtures at a fixed $T_{ad} = 1800$ K (ϕ value varies at each point to maintain constant T_{ad}; initial temperature before compression is 300 K). Isentropic temperature/pressure relation varies slightly between each blend; indicated temperature is for CH_4/air.

9.5 Nonpremixed Flame Overview

Nonpremixed flames occur at the interface between fuel and oxidizer. In contrast to premixed flames, they do not propagate; see also Section 7.1. Rather, fuel and oxidizer diffuse into the reaction zone in stoichiometric proportions. Furthermore, the burning rate of nonpremixed flames in the fast-chemistry limit is controlled completely by the rates at which fuel/oxidizer diffuse into the flame – it is independent of kinetic rates, and is only a function of the spatial gradients of fuel and oxidizer, their diffusivities, and the stoichiometric fuel/oxidizer ratio. In contrast, the burning rate of premixed flames is controlled by both kinetic and diffusive rates. In reality, finite-rate kinetic effects in nonpremixed flames do influence burning rates and some reactants do diffuse through the flame without burning. However, this effect acts as a perturbation to the infinite fast-chemistry limit, until the point where the spatial gradients become large enough that the flame extinguishes. This section overviews the nonpremixed flame problem in the fast-chemistry, single-step kinetics limit. Finite-rate effects are discussed in Section 9.6.

As in the premixed flame overview, the reader is referred to standard combustion texts for more detailed presentations on nonpremixed flames. The objective of this section is to summarize some basic ideas that will be utilized later. Consider the idealized, one-dimensional, nonpremixed flame in Figure 9.22 [2]. This one-dimensional geometry is realized by situating the flame between two boundaries where the fuel and oxidizer concentrations are kept constant at $Y_{Ox}(x=0) = Y_{Ox,a}$ and $Y_{Fuel}(x=L) = Y_{Fuel,b}$, at the same temperature, $T_a = T_b$. Combustion products are also absorbed at the boundaries so that $Y_{Prod}(x=0) = Y_{Prod}(x=L) = 0$.

In the single-step, fast-chemistry limit, there is negligible leakage of fuel and oxidizer across the flame. For $Le = 1$, and assuming constant values for the molecular transport coefficients, the spatial distributions of the fuel and oxidizer are given by [2]

$$x < x_F : \quad \frac{Y_{Ox}}{Y_{Ox,a}} = 1 - \frac{x}{x_F}, \tag{9.25}$$

$$\frac{Y_{Fuel}}{Y_{Fuel,b}} = 0; \tag{9.26}$$

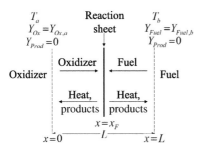

Figure 9.22 Illustration of the one-dimensional, nonpremixed flame configuration analyzed in the text.

$$x > x_F : \quad \frac{Y_{Fuel}}{Y_{Fuel,b}} = \frac{x - x_F}{L - x_F}, \qquad (9.27)$$

$$\frac{Y_{Ox}}{Y_{Ox,a}} = 0, \qquad (9.28)$$

where x_F denotes the flame location, given by

$$\frac{x_F}{L} = \frac{Y_{Ox,a}}{Y_{Ox,a} + \varphi_{Ox} Y_{Fuel,b}}, \qquad (9.29)$$

and φ_{Ox} denotes the stoichiometric mass ratio of oxidizer to fuel, respectively. These equations describe linear variations of the mass fractions between their values at the boundary and a value of zero at the flame location. Note that at the flame location, x_F, the diffusive fluxes of fuel and oxidizer are in stoichiometric proportion, as can be seen by the following calculation:

$$\frac{\left|\rho \mathscr{D} \frac{\partial Y_{Ox}}{\partial x}\right|_{x_F}}{\left|\rho \mathscr{D} \frac{\partial Y_{Fuel}}{\partial x}\right|_{x_F}} = \left| \frac{-(Y_{Ox,a} + \varphi_{Ox} Y_{Fuel,b})\left(1 - \frac{Y_{Ox,a}}{Y_{Ox,a} + \varphi_{Ox} Y_{Fuel,b}}\right)}{Y_{Fuel,b}} \right| = \varphi_{Ox}. \qquad (9.30)$$

Thus, the flame location, x_F, is biased towards the specie with the larger stoichiometric mass ratio, enabling a higher diffusive flux of this specie because of the higher concentration gradient. Assuming constant c_p, the temperature distribution is

$$x < x_F : \quad \frac{T}{T_a} = 1 + \frac{Y_{Fuel,b} Q}{c_p T_a} \left(\frac{x}{L}\right), \qquad (9.31)$$

$$x > x_F : \quad \frac{T}{T_a} = 1 + \frac{Y_{Ox,a} Q}{c_p \varphi_{Ox} T_a} \left(1 - \frac{x}{L}\right), \qquad (9.32)$$

where Q is defined in Eq. (8.16). Finally, the mass burning rate is given by

$$\dot{m}''_F = \dot{m}''_{Ox} + \dot{m}''_{Fuel} = \left|\rho \mathscr{D} \frac{\partial Y_{Ox}}{\partial x}\right|_{x_F} + \left|\rho \mathscr{D} \frac{\partial Y_{Fuel}}{\partial x}\right|_{x_F}, \qquad (9.33)$$

$$\dot{m}''_F = (1 + \varphi_{Ox})\left|\rho \mathscr{D} \frac{\partial Y_{Fuel}}{\partial x}\right|_{x_F} = \frac{(1 + \varphi_{Ox})}{\varphi_{Ox}} (Y_{Ox,a} + \varphi_{Ox} Y_{Fuel,b}) \left(\frac{\rho \mathscr{D}}{L}\right). \qquad (9.34)$$

This equation shows that burning rates are maximized when diffusivities and gradients (proportional to $1/L$) are high, and when the oxidizer/fuel ratios at the boundaries, $Y_{Ox,0}/Y_{Fuel,L}$, are stoichiometric.

Section 1.4 defined a mixture fraction, Z, as the mass fraction of material originating from the fuel side, defined by

$$Z \equiv Y_{Fuel} + \left(\frac{1}{\varphi_{Ox}+1}\right) Y_{Prod} = Y_{Fuel} + \left(\frac{1}{\varphi_{Ox}+1}\right)(1 - Y_{Ox} - Y_{Fuel}) = \frac{\varphi_{Ox} Y_{Fuel} - Y_{Ox} + Y_{Ox,a}}{\varphi_{Ox} Y_{Fuel,b} + Y_{Ox,a}}. \qquad (9.35)$$

For this problem, the mixture fraction and spatial coordinate are related by the simple relation

$$Z \equiv \frac{x}{L}. \tag{9.36}$$

The mixture fraction is a useful variable as it can be used in certain problems to reduce the number of independent variables, by replacing spatial coordinates by the mixture fraction [45]; i.e., the species mass fractions and temperature can be written as

$$T = T(Z, t), \tag{9.37}$$

$$Y_i = Y_i(Z, t). \tag{9.38}$$

For example, if x/L is replaced by Z in the above equations, it turns out that the above solutions for species mass fractions and temperature are generally true for more complex geometries, such as jet flames (given the same assumptions of fast chemistry, equal mass diffusivities, and a Lewis number of unity); i.e., $Y(Z)$ and $T(Z)$ are identical to Eqs. (9.25)–(9.28) and (9.31), respectively, while the spatial distribution of Z itself, $Z(x, y, z, t)$, is a function of the specific geometry.

9.6 Finite-Rate Effects in Nonpremixed Flames

Section 9.5 described the structure of nonpremixed flames in the fast-chemistry limit, emphasizing that in this limit, the burning rate is diffusion controlled, a function of geometry (parameterized by length scale, L) and independent of kinetic rates. This section considers finite-rate effects on nonpremixed flames and their internal structure in more detail. Finite-rate effects are directly linked to species concentrations gradients – high gradients cause fast diffusive time scales that can approach kinetic time scales. As such, in this section we emphasize the influence of species concentration gradients as controlling deviations of the nonpremixed flame from its fast-chemistry structure. Specifically, we will parameterize finite-rate chemistry effects through the scalar dissipation rate, χ, which naturally arises in these problems for reasons to be discussed shortly. It is defined as

$$\chi = 2\mathscr{D}|\nabla Z|^2. \tag{9.39}$$

Note that χ has units of 1/s, analogous to flame stretch rate. Recall that Section 9.3 emphasized the role of tangential velocity gradients, or more generally flame stretch, in controlling perturbations of the premixed flame from its one-dimensional structure. Flame stretch and species concentration gradients (one of fundamental significance to premixed flames, the other to nonpremixed flames) are sometimes, but not always, related, as shown in several examples in this section and in Section 11.3.

As discussed in the context of Eq. (9.37), it is often convenient to make a change of variables using the mixture fraction, Z, as a coordinate. In other words, rather than writing species mass fractions and temperature as $Y_i = Y_i(x, y, z, t)$ or

$T = T(x, y, z, t)$, we write them as $Y_i = Y_i(Z, \ell_1, \ell_2, t)$ or $T = T(Z, \ell_1, \ell_2, t)$, where ℓ_1 and ℓ_2 are local coordinates tangential to iso-Z surfaces, as shown in Figure 7.1 [29]. Assuming weak curvature of iso-Z surfaces, the species in Eq. (1.25) in this transformed coordinate system becomes [46]:

$$\rho\left(\frac{\partial Y_i}{\partial t} + \vec{u}_\ell \cdot \nabla_\ell Y_i\right) = \dot{w}_i + \nabla_\ell \cdot (\rho \mathscr{D} \nabla_\ell Z) + \rho \mathscr{D} |\nabla Z|^2 \frac{\partial^2 Y_i}{\partial Z^2}. \tag{9.40}$$

Assuming that tangential gradients are small relative to normal gradients in the vicinity of the reaction sheet, this expression simplifies to

$$\rho \frac{\partial Y_i}{\partial t} = \dot{w}_i + \frac{1}{2}\rho\chi \frac{\partial^2 Y_i}{\partial Z^2}. \tag{9.41}$$

This equation shows how the scalar dissipation rate, χ, naturally appears in the analysis. An analogous equation can be written for the temperature. Clearly, this equation is not valid in regions where tangential gradients are not negligible, such as for highly curved flames [46] or edge flames (Section 9.7).

In the fast-chemistry limit, the reaction rate is zero everywhere except in a very thin layer. Outside of this layer, the steady-state species profiles are described by the equation

$$\frac{\partial^2 Y_i}{\partial Z^2} = 0. \tag{9.42}$$

Note that the linearly varying species concentration profiles that are solutions of this equation and presented in Section 9.5 naturally fall out of this equation.

The scalar dissipation rate can be directly related to the flow strain rate in certain flow fields. For example, consider a potential, Hiemenz-type stagnation flow field [47] where fuel is stagnated against an oxidizer stream, such as illustrated in Figure 8.2(c). The centerline axial velocity profile in this configuration is given by

$$\frac{du_y}{dy} = -\kappa, \tag{9.43}$$

where $y = 0$ denotes the $u_y = 0$ stagnation point. In this configuration, the length scale over which species profiles vary (analogous to the length scale, L, in Figure 9.22) is given by $L \propto \sqrt{\mathscr{D}/\kappa}$. It can be shown that κ and χ are related through the expression [45]:

$$\chi = \frac{\kappa}{\pi} \exp\left(-2\left[\text{erf}^{-1}(1 - 2Z)\right]^2\right). \tag{9.44}$$

This equation shows the linear correspondence between strain rate and scalar dissipation rate at a given Z surface. Their ratio, χ/κ, is plotted in Figure 9.23. For example, at the $Z = 0.055$ surface (corresponding to Z_{st} for a methane–air flame), $\chi \approx 0.025\kappa$.

This stagnation flame problem serves as an important canonical problem for understanding finite-rate effects in practical devices by parameterizing flame structure as a function of χ. It can be used, for example, to understand high scalar dissipation

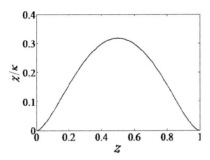

Figure 9.23 Ratio of the scalar dissipation rate, χ, and flow strain rate, κ, in a constant-density, opposed oxidizer–fuel flow, as a function of mixture fraction, Z.

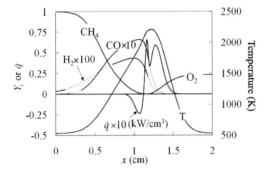

Figure 9.24 Calculated spatial profiles of the major species and heat release rate for a nonpremixed counterflow flame of methane stagnating against air ($T^u = 533$ K, $p = 1$ atm). The chemical production rate for H_2 and CO is positive for solid lines and negative for dashed lines.

rate effects in nonpremixed flames rolled up by vortices, or situated at the end of the splitter plate experiment shown in Figure 8.2(b). We next consider more realistic flame structures and finite-rate chemistry effects by integrating the conservation equations with a detailed chemical kinetic mechanism using this counterflow geometry. Figure 9.24 was obtained from a methane–air flame calculation, and plots the dependence of major species concentration as a function of spatial coordinate. As in premixed flames, the flame structure is more complex than that suggested by single-step, fast kinetic considerations. Note first the broadened heat release zone, and second, the region of negative heat release on the fuel side. This is due to the endothermic decomposition of CH_4 into H_2 and CO, both of which have significant mass fraction values over large parts of the fuel side.

Figure 9.25 plots intermediate radical profiles, also showing their presence over significant spatial regions. Regions of chemical production and destruction are demarcated by solid and dashed lines respectively, which can be used to infer whether a radical is present due to chemical production or diffusion. For example, the negative production rate of CH_3 on the far left side of the flame indicates that it exists at those locations through diffusion.

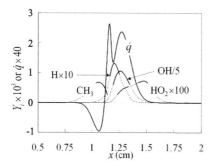

Figure 9.25 Calculated spatial profiles of the minor species and heat release rate for a nonpremixed counterflow flame of methane stagnating against air ($T^u = 533$ K, $p = 1$ atm). The chemical production rate is positive for solid lines and negative for dashed lines.

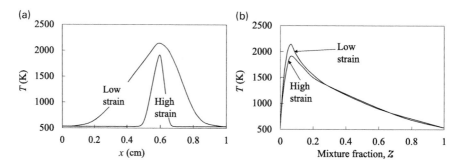

Figure 9.26 Temperature dependence on (a) spatial and (b) mixture fraction coordinate for a nonpremixed stagnation flame ("low strain" velocities are 20 cm/s at both fuel and air nozzle exits; "high strain" velocities are 480 cm/s and 275 cm/s on the fuel side and air side, respectively).

In order to demonstrate finite-rate kinetic effects, Figure 9.26 overlays temperature profiles plotted at two nozzle exit velocities. These results are plotted in both spatial and mixture fraction coordinates, to enable a comparison of how these results look in these two different representations.

In mixture fraction space, Z, the deviation of the higher strain case from the low strain rate line is a manifestation of finite-rate kinetic effects. In physical space, the thickness of the elevated temperature region is narrower with the higher strain case, as spatial gradients scale inversely with nozzle velocity/strain rate (as $1/\sqrt{\kappa}$ [2]).

Equation (9.33) shows that in the fast-chemistry limit, the mass burning rate increases as the gradients in the flow increase. For the one-dimensional flame in Figure 9.22, this is achieved by decreasing L, and for the stagnation flame this corresponds to increasing κ. In reality, this burning rate increase is moderated by finite-rate effects, as indicated by Figure 9.26. Some fuel and oxidizer inevitably diffuses across the flame without reaction, since the reaction rate is finite. This has the effect of decreasing the maximum flame temperature and increasing certain intermediate radicals [48]. For high activation energy chemistry, this leads to the familiar S-shaped dependence between temperature/reaction rate and strain rate (see Section 8.2.2.1). As leakage increases from zero, the reaction rate decreases, and then abruptly drops to near zero at some critical value of scalar dissipation rate [49, 50].

It is of interest to compare the extinction conditions for premixed and nonpremixed flames. Although the fundamental parameters leading to extinction are different – flame stretch and scalar dissipation rate – we can use Eq. (9.44) to relate the two for the specific stagnation flame configuration. These calculations show that the strain rate, κ, required to extinguish a nonpremixed flame is significantly lower than that required to extinguish a premixed flame [51, 52].

The reader is referred to more detailed references of finite-rate effects when coupled with differential diffusion of species, and strong gradients in and out of plane directions, such as due to strong flame curvature [46].

9.7 Edge Flames and Flame Spreading

9.7.1 Overview

The previous sections gave overviews of the structure of unstretched and stretched flames. An important feature of these flames is their one-dimensional or quasi-one-dimensional structure. In reality, most real flames have "edges." These edges are important for nonpremixed flame stabilization at fuel/oxidizer interfaces, as shown in Figure 9.27(a). They are also important for premixed flame stabilization in shear layers, see Figure 9.27(b), and for understanding the propagation of an ignition front, or in the vicinity of flame holes after extinction [53], as shown in Figure 9.27(c) and (d) respectively.

The edge of a flame can remain stationary, advance into fresh gases as an ignition wave, or retreat as a "failure wave." However, it is important to note the frame of

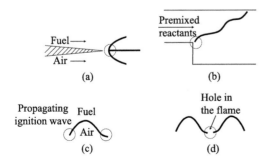

Figure 9.27 Illustration of several edge flame configurations showing (a) fuel/oxidizer interface with quasi-one-dimensional nonpremixed flame downstream, (b) premixed flame in shear layer with quasi-one-dimensional premixed flame downstream, (c) ignition front in a nonpremixed mixture, (d) flame holes induced by local extinction.

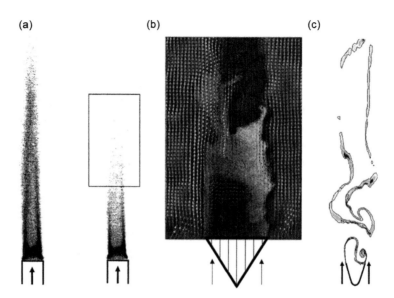

Figure 9.28 Images of several statistically stationary, retreating edge flames in premixed systems. (a) Piloted Bunsen flame without (left) and with (right) extinction near tip, (b) measured OH PLIF and PIV image, and (c) computed reaction rates of bluff body stabilized flames near blowoff. Images courtesy of (a) R. Rajaram [55], (b) Chaudhuri et al. [56], and (c) C. Smith et al. [57].

reference, because the reaction front is propagating with respect to the local tangential flow velocity. For example, an advancing edge flame may appear stationary in a lab frame, as the edge velocity (not to be confused with the flame speed) balances the local tangential flow field. Similarly, a retreating failure wave may also appear stationary in a lab frame because of flow convection in the opposite direction [54]. Finally, a stationary edge may move downstream as it is convected with the flow.

Several examples of flames with edges are depicted in Figure 9.28. Figure 9.28(a) shows a line-of-sight image of a piloted Bunsen flame at two fuel/air ratios [55]. The flow velocity is quite high, and the flames would blow off without the pilot. The flame on the left burns all of the fuel, but as the fuel/air ratio is reduced, the tip region of the flame on the right is partially extinguished. The region of visible reaction monotonically moves back toward the burner as the fuel/air ratio decreases, but the flame never blows off because of the pilot. This downstream point of vanishing luminosity is an example of a statistically stationary retreating edge flame, where the retreat velocity of the edge equals the tangential flow speed on average. Figure 9.28(b) shows instantaneous OH PLIF and PIV fields of a similar phenomenon for a bluff body stabilized flame very near blowoff [56]. Farther from blowoff, the flame front is clearly situated along the shear layer all the way downstream and no edges are observed. As blowoff is approached, local extinction occurs and the luminous flame region retracts back toward the bluff body with decreasing fuel/air ratio [40], as also shown in Figure 10.23. Figure 9.28(c) shows computed reaction rates of a bluff body flame near blowoff, similarly showing flame edges associated with local extinction [57].

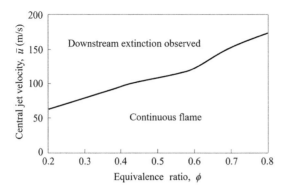

Figure 9.29 Relationship between core jet velocity and equivalence ratio at which downstream extinction of a premixed natural gas/air mixture occurs. Co-flow temperature and velocity are 1500 K and 0.8 m/s, respectively. Data obtained from Dunn et al. [61].

As discussed in the context of Figure 9.28(a), an important application of retreating edge flames occurs in piloted, high-velocity flows. In these cases, while vigorous reaction can be forced by an external pilot, reaction ceases at some point downstream if the flow strain is too high [58–60]. To illustrate, Figure 9.29 plots data from a Cabra burner, where a premixed core flow is piloted by a hot vitiated co-flow. The figure shows the dependence of the core flow velocity at which a given equivalence ratio mixture begins to extinguish downstream.

Important questions to be considered for edge flames are the conditions under which edge flames advance or retreat, and the velocity with which they do so. In addition, we will discuss the effects of flame stretch, operating conditions, and heat losses on these dependencies.

Edge flames have a unique structure compared to premixed or nonpremixed flames because of the high gradients in both the flame-normal and flame-tangential directions, which introduce fundamentally new physics that are not present for continuous fronts [62]. For nonpremixed flames, although the flame does not propagate, its edge does. For premixed flames, the advance/retreat velocity of the edge, v_F, while often proportional to the laminar burning velocity, can achieve both positive and negative values, with an absolute value that can substantially exceed the laminar burning velocity.

9.7.2 Buckmaster's Edge Flame Model Problem

Many important features of edge flames are illustrated by a model problem from Buckmaster, which is discussed in this section [63, 64]. Consider Figure 9.30, which generalizes the one-dimensional nonpremixed flame, discussed in Section 9.5, by including the edge of the flame. The flame structure asymptotes toward that of a continuous nonpremixed flame as $z \to +\infty$. Here, v_F denotes the flame edge velocity. The temperatures and mass fractions of the fuel and oxidizer at some transverse distance, $L/2$, from the front are given by $T_b = T_a$, $Y_{Fuel,b}$, and $Y_{Ox,a}$, as shown in

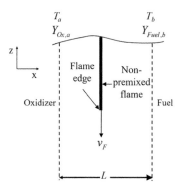

Figure 9.30 Model problem geometry used to illustrate edge flame concepts.

the figure. This transverse dimension, L, characterizes the length scale over which gradients occur normal to the flame, such as due to strain.

A simplified two-dimensional energy equation can be written in terms of the temperature (see Eq. (1.42)) as

$$\rho c_p \frac{\partial T}{\partial t} - k_T \frac{\partial^2 T}{\partial z^2} = \left[k_T \frac{\partial^2 T}{\partial x^2} - \rho c_p u_x \frac{\partial T}{\partial x} \right] - Q \dot{w}_{Fuel}, \tag{9.45}$$

where Q is defined in Eq. (8.16). We will assume equal mass diffusivities and $Le = 1$, which results in a linear relationship between temperature and the fuel and oxidizer mass fractions. The two bracketed terms on the right-hand side of Eq. (9.45) denote transverse fluxes. In order to develop a one-dimensional equation for the flame edge, these terms are modeled as conductive heat losses that are proportional to the difference between the local temperature and the ambient:

$$\left[k_T \frac{\partial^2 T}{\partial x^2} - \rho c_p u_x \frac{\partial T}{\partial x} \right] \equiv -\frac{k_T (T - T_b)}{L^2}. \tag{9.46}$$

Next, we use the following model for the reaction rate, which can be written in a similar form to Eq. (8.13):

$$\frac{\dot{w}_{Fuel}}{MW_{Fuel}} = -\mathcal{A}^* \frac{\rho Y_{Ox}}{MW_{Ox}} \frac{\rho Y_{Fuel}}{MW_{Fuel}} e^{-E_a/\mathcal{R}_u T} = -\mathcal{B} \frac{\rho Y_{Fuel}}{MW_{Fuel}} Y_{Ox} e^{-E_a/\mathcal{R}_u T}. \tag{9.47}$$

Analogous to the well-stirred reactor problem and Eq. (8.19), define $\tau_{chem} = 1/\mathcal{B}$ and $\tau_{flow} = \rho c_p L^2 / k_T$, leading to the following Damköhler number definition:

$$Da = \mathcal{B} \frac{\rho c_p L^2}{k_T}. \tag{9.48}$$

We will seek solutions where the flame edge propagates with a constant speed, v_F. Reverting to a coordinate system that is attached to the edge, the dimensionless temperature equation is given by (see Exercise 9.5):

9.7 Edge Flames and Flame Spreading

$$\tilde{v}_F \frac{d\tilde{T}}{d\tilde{z}} - \frac{d^2\tilde{T}}{d\tilde{z}^2} = F(\tilde{T}, Da), \tag{9.49}$$

where

$$F(\tilde{T}, Da) = 1 - \tilde{T} + Y_{Fuel,b} Da \frac{(1 - \tilde{T} + \tilde{Q})^2}{\tilde{Q}} \exp\left(-\frac{\tilde{E}}{\tilde{T}}\right), \tag{9.50}$$

$$\tilde{v}_F = \frac{v_F \rho c_p L}{k_T}, \tag{9.51}$$

and $\tilde{T} = T/T_b$, $\tilde{Q} = Y_{Fuel,b} Q / c_p T_b$, and $\tilde{E} = E_a / \mathscr{R}_u T_b$. Note that the function in Eq. (9.50) is identical to that derived in Section 8.2.2.1 for the well-stirred reactor, with the exception of the reaction here being second order in species concentrations.

A useful starting point for understanding the solution characteristics of Eq. (9.49) is to consider the steady-state solutions of this equation with no z-direction variation. In this case, the solution for the temperature is given by $F(\tilde{T}, Da) = 0$. A typical solution for this expression is plotted in Figure 9.31, showing that it traces out the familiar S curve for high activation energies, where the "ignition" and "extinction" Damköhler values are denoted by Da_{II} and Da_{I}, respectively. For this particular problem, $Da_{I} = 534$ and $Da_{II} = 12\,070$. As discussed earlier, the middle solution branch is unstable.

We are particularly interested in the $Da_{I} < Da < Da_{II}$ region, where it is possible to have either a self-sustaining flame, with temperature $\tilde{T}_{high}$, or no flame, with temperature, $\tilde{T}_{low}$. Next, consider the two-dimensional edge flame problem where the solution asymptotes to the vigorously reacting solution at $z = +\infty$ (with $\tilde{T} = \tilde{T}_{high}$), and the frozen, nonreacting solution at $z = -\infty$ (with $\tilde{T} = \tilde{T}_{low}$). Multiply Eq. (9.49) by $d\tilde{T}/d\tilde{z}$ and integrate over z to obtain

$$v_F = \frac{\int_{\tilde{T}_{low}}^{\tilde{T}_{high}} F(\tilde{T}, Da) d\tilde{T}}{\int_{-\infty}^{\infty} \left(\frac{d\tilde{T}}{d\tilde{z}}\right)^2 dz}. \tag{9.52}$$

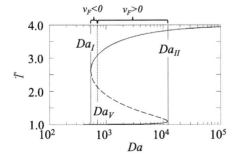

Figure 9.31 Dependence of the flame temperature on the Damköhler number, obtained from solving $F(\tilde{T}, Da) = 0$, for the case where there is no variation in z-direction properties ($\tilde{E} = 14$, $\tilde{Q} = 3$).

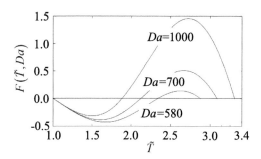

Figure 9.32 Temperature dependence of $F(\tilde{T}, Da)$ for three Da values (see also Figure 9.31).

Since the denominator is always positive, the sign of v_F is controlled by the sign of the numerator. To understand the factors influencing its sign, Figure 9.32 plots the temperature dependence of $F(\tilde{T}, Da)$ for three intermediate values between Da_{I} and Da_{II}, showing that it has negative and positive values for low and high $\tilde{T}$ values, respectively. Note that the zero crossings correspond to the steady solutions plotted in Figure 9.31.

These three Da values correspond to negative, zero, and positive values of the numerator in Eq. (9.52), $\int_{\tilde{T}_{low}}^{\tilde{T}_{high}} F(\tilde{T}, Da) d\tilde{T}$. We can define a critical Da value, Da_V, for which $\int_{\tilde{T}_{low}}^{\tilde{T}_{high}} F(\tilde{T}, Da) d\tilde{T} = 0$. These results imply that $v_F < 0$ for $Da < Da_V$ and $v_F > 0$ for $Da > Da_V$, also indicated on Figure 9.31. A stationary edge flame is only possible for one Damköhler number, $Da = Da_V$. Particularly interesting is the $Da_{\mathrm{I}} < Da < Da_V$ region, where a steady, one-dimensional solution is possible (corresponding to $534 < Da < 700$ for this model problem). However, this solution shows that if the flame forms any edges, they will retreat since $v_F < 0$ and lead to eventual extinction of the whole flame!

To summarize, the flame edge acts as an ignition source for the unreacted gases at large Da values, specifically where $Da > Da_V$. At low Da values, the unreacted gases quench the flame. Note also that, in contrast to the propagation speed of premixed flames, the edge velocity can attain a continuum of values ranging from negative to positive.

9.7.3 Edge Flame Velocities

In this section, we discuss the factors influencing the edge velocity, v_F. An advancing edge flame in a nonpremixed flow (an ignition front, $v_F > 0$) often has a triple flame structure consisting of a nonpremixed flame connected at a triple point to a rich and lean premixed flame [65]. Retreating nonpremixed flames (an extinction front, $v_F < 0$) generally consist of a single edge, as do both advancing and retreating premixed flames. Premixed flame edges generally have a significant bend, or 'hook-like structure' [66]. To illustrate, Figure 9.33 plots computed reaction rate contours for advancing, slightly advancing, and retreating nonpremixed flames (note that the definition of Da is different than used in Section 9.7.2), as well as two experimental images for edge flames in a low- and high-stretch environment. Figure 9.34 plots experimental and computed images of a premixed edge flame.

Figure 9.33 (a) Reaction rate contours of adiabatic nonpremixed edge flames, reproduced from Daou et al. [67], where Da is equated to the inverse of their dimensionless flame thickness squared. (b) Images of an edge flame in a low-stretch (left) and high-stretch (right) environment, reproduced from Cha and Ronney [68].

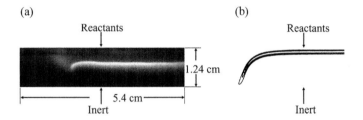

Figure 9.34 Experimental image (a) and computed reaction rate contours (b) of a premixed edge flame in a counterflow burner of reactants stagnating against inert. Reproduced from Liu and Ronney [66] and Vedarajan and Buckmaster [69], respectively.

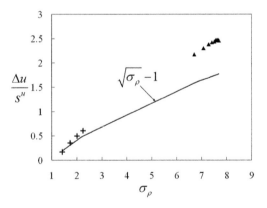

Figure 9.35 Illustration of the approach flow deceleration, Δu, induced by the triple flame on the flame density ratio. Obtained from calculations by Ruetsch et al. [70] (+) and Fernandez-Tarrazo et al. [71] (▲).

In this section we discuss the factors influencing the edge velocity, v_F, with a focus on density ratio, Damköhler number, and heat loss effects. The edge speed is a function of the density jump across the flame, σ_ρ, which leads to alterations of the flow field near and across the flame because of its convex orientation relative to the reactants, as shown in Figure 9.27(a). Figure 9.35 plots the approach flow deceleration in front of this triple flame, showing that the approach flow velocity in front of the flame monotonically decreases with σ_ρ. Consequently, the flame leading edge can situate itself in a higher velocity region of the flow than would be expected based on

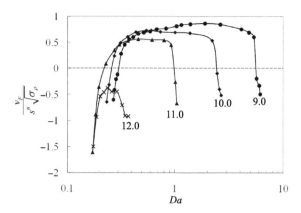

Figure 9.36 Measured nonpremixed flame edge speed dependence on Damköhler number, obtained in a counter flow burner. Fuel is $CH_4/O_2/N_2$ with ratios of $1/2/N$, where N is indicated in the figure. Data obtained from Cha and Ronney [68], where Da is equated to the inverse of their dimensionless flame thickness squared.

considerations of the nominal flow field in the absence of the flame. The physical processes responsible for this behavior are discussed in Section 7.5 (see Figures 7.17 and 7.18). For this reason, a typical edge velocity scaling used for advancing, adiabatic, nonpremixed flames at high Damköhler numbers is [70]:

$$v_F|_{Da\to\infty} \propto s^u \sqrt{\sigma_p}. \tag{9.53}$$

The edge speed is also a strong function of Damköhler number [72], as illustrated by the model problem in Section 9.7.2. Experimental data illustrating the dependence of the normalized edge velocity on Da is presented in Figure 9.36 [68].

Note the nonmonotonic dependence of v_F on Damköhler number. Moreover, negative edge speeds are observed not only at low Da values, as expected from the above discussion, but also at high values. These high Da, negative v_F trends are due to heat losses [67, 68, 73, 74], which introduce new physics not considered in the model problem in Section 9.7.2. Heat losses are well known to significantly alter unstretched flame properties [2], but are generally of secondary importance in highly stretched flames near extinction. In this latter case, diffusive losses associated with strong, stretch-induced gradients are the dominant loss process so the flame can usually be treated as effectively adiabatic. Edge flames near boundaries, such as a flame stabilized in a shear region near a boundary, see Figure 9.27(b), are an important application where heat loss effects are significant, even in highly stretched flames. Calculations suggest that heat loss can cause flame extinction, even with edge flames whose edge speed, v_F, is positive [74]. To further illustrate these points, Figure 9.37 plots edge speeds at several heat loss rates, using results from a one-step kinetic model. In this result, heat loss is modeled as a volumetric loss term, given by $\kappa_{loss}(T - T_b)$, where κ_{loss} is a heat loss coefficient and T_b is the ambient temperature. This plot shows that high heat losses or high stretch can lead to negative edge flame propagation.

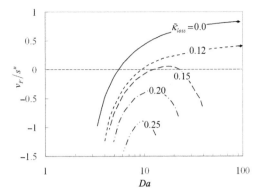

Figure 9.37 Calculated edge speed dependence on Damköhler number (based on the ratio of stretch rate to chemical time; chemical term estimated from stoichiometric laminar flame speed) where Da is equated to the inverse of the dimensionless flame thickness squared. $\tilde{\kappa}_{loss}$ is a dimensionless heat loss coefficient, given by: $\tilde{\kappa}_{loss} = \frac{E_a(T_{ad}-T_b)}{\mathcal{R}_u T_{ad}^2} \frac{\alpha}{(s^{u,0})^2} \kappa_{loss}$. The $s^{u,0}$ value is computed at $\phi = 1$. Calculation results obtained from Daou et al. [74].

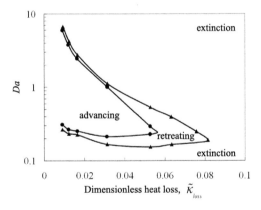

Figure 9.38 Effects of the Damköhler number and heat loss on regimes where edge flames advance, retreat, or where the entire flame extinguishes. Data reproduced from Cha and Ronney [68], where Da is equated to the inverse of their dimensionless flame thickness squared.

Experimental results summarizing the conditions under which the flame edge advances, retreats, or where no flame is possible at all (extinction) are plotted in Figure 9.38.

In addition to the flame density ratio and Damköhler number, edge velocities are also a function of Lewis number, ratios of diffusivities, and stoichiometric mixture fraction, Z_{st}. The reader is referred to the references for further details on these dependencies [68, 71].

9.7.4 Conditions at the Flame Edge

In this section we consider in more detail the conditions at the flame edge in a flow with a spatial gradient in stretch or scalar dissipation rate. Consider a continuous,

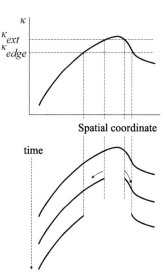

Figure 9.39 Illustration of a flame hole opening in a flow with spatially varying stretch rate.

premixed flame front with no holes subjected to a spatial stretch rate gradient at $t = 0$, such as shown in Figure 9.39.

This continuous flame will extinguish at points where the local stretch rate, κ, exceeds κ_{ext} (where κ_{ext} here denotes the extinction stretch rate of a continuous flame sheet). However, the resulting flame edge does not correspond to the point in the flow where $\kappa = \kappa_{ext}$, i.e., the stretch rate at the edge of the hole, $\kappa_{edge} \neq \kappa_{ext}$. Thus, Figure 9.39 shows that once a hole is initiated in a region of high stretch, it leads to flamelet extinction at adjoining points that would otherwise not have extinguished at the local conditions (neglecting for now the mixing of hot products with reactants that would inevitably occur at the hole). For example, simulations of the quenching of a nonpremixed flame by a vortex show the initial flame hole formation and then the recession of the flame edge due to negative edge speeds once the edge exists [75]. The flame edge then advances to close the hole once the vortex has passed, due to positive edge speeds.

Using the model problem considered in Section 9.7.2, the following relationship can be developed between κ_{edge} and κ_{ext} [64, 76]:

$$\frac{\kappa_{edge}}{\kappa_{ext}} = \frac{e^2}{(1 - T_0/T_{ad})} \frac{\mathcal{R}_u T_{ad}}{E_a}, \qquad (9.54)$$

where $e \approx 2.718$. Thus, assuming dimensionless activation energies of $E_a/(\mathcal{R}_u T_{ad}) = 10 - 20$ and $T_0/T_{ad} = 0.2$ leads to $\kappa_{edge}/\kappa_{ext} \sim 0.45 - 0.9$. This value is consistent with measurements, which indicate ranges of $0.5 < \kappa_{edge}/\kappa_{ext} < 1$ [66, 76–78]. While most measurements have been obtained in cases where $\kappa_{edge} < \kappa_{ext}$, there are also indications that κ_{edge} is greater than κ_{ext} in cases with tangential flow to the flame edge; this may be seen in a retreating edge where hot gases from the flame flow over the edge [79].

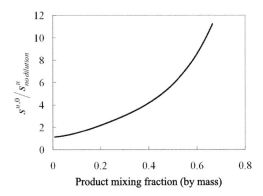

Figure 9.40 Computed laminar flame speeds of a $\phi = 0.58$ methane/air flame diluted with varying levels of equilibrium products, assuming adiabatic product mixing and an unburned reactant temperature of $T^u = 450$ K. Image courtesy of M.K. Bobba.

As flame holes develop in premixed flames, the opportunity exists for transport of mass and energy between reactants and products. This leads to dilution, but also preheating and radical introduction into the reactants. This has favorable influences on the flame. To indirectly illustrate this point, Figure 9.40 plots calculations of the unstretched laminar flame speed of methane/air flames diluted with varying levels of equilibrium products, assuming adiabatic mixing [80]. These results show that the flame speed increases with reactant dilution. This increase is primarily due to faster diffusivities associated with the higher-temperature reactants, as the peak heat release rate actually changes very little across these different dilution levels. This result suggests that the initial development of flame holes is partially self-correcting, as the burning rate adjacent to the hole is augmented [81]. Moreover, at very high dilution/preheating levels, the flame loses its characteristic ignition/extinction character [82] that plays such a key role in the edge flame dynamics discussed in the previous sections, as discussed further in Section 10.2.3. In other words, $\tilde{E}$ becomes small such that the S-curve behavior is lost; see Figure 8.8.

9.7.5 Implications on Flame Spread after Ignition

Returning to the introductory comments on ignition in Section 8.1, it is clear that the presence of edges on a reaction kernel after a successful ignition does not imply a self-sustaining, propagating flame. This is best illustrated from the model problem in Section 9.7.2 and Figure 9.31. In particular, an ignition front on a nonpremixed flame must have an edge. Note that the Damköhler number range over which a propagating flame with no edges is possible, $Da > Da_I$, is broader than the range over which a flame with edges will have a positive edge velocity, $Da > Da_V$. Thus, this argument suggests that although a successful ignition event can be achieved in the $Da_I < Da < Da_V$ range, the kernel will not successfully expand if it has an edge; this can only be achieved for $Da > Da_V$. However, if a flame can be successfully

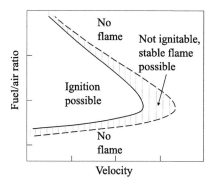

Figure 9.41 Data illustrating range over which ignition and stable flames are possible for a given combustor and ignitor. Adapted from figure 5.29 in Lefebvre and Ballal [83].

initiated and stabilized in this $Da > Da_V$ range, then Da can then be decreased into the $Da_I < Da < Da_V$ range (such as by decreasing the fuel/air ratio when lean or increasing velocity) and the flame will stay stable. The main point here is that the parameter space over which ignition leads to flame stabilization is narrower than that over which a stable flame can occur. Equation (9.54) predicts that this contraction in Damköhler space is given by $Da_I/Da_V \sim 0.5 - 0.9$. This contraction is well known experimentally [83]. For example, Figure 9.41 illustrates typical measured results on the ranges in velocity–fuel/air ratio space where a flame can be stabilized, and where it can be successfully ignited. In order to generate data in the "not ignitable, stable flame possible" region, the flame is initiated in the "ignition possible" region and then the operating conditions are moved toward the "not ignitable" region. Also notice that the "not ignitable" region is strongly dependent on the given combustor and ignitor; e.g., it is a function of ignitor location and power, as well as combustor flow field.

9.8 Intrinsic Flame Instabilities

Consider a nominally flat flame in a uniform flow field, such as analyzed in Section 7.3. Perturbations in either flow velocity or burning rate cause movement of the flame front of the form $\xi_1 = \mathcal{A}_\xi e^{iky} e^{-i\omega t}$ (see Eq. (7.71)). Even in the absence of external forcing, flames are often unsteady because of intrinsic instabilities [29, 84–87]. Premixed flame instabilities with one-step chemistry can be grouped into several basic categories [29, 88]: body force (Rayleigh–Taylor), hydrodynamic, diffusive–thermal, and Saffman–Taylor. In addition, multistep chemistry introduces additional instabilities of purely kinetic origin [89, 90].

The hydrodynamic, or Darrieus–Landau, instability stems from alteration of the approach flow by flame wrinkling. The underlying instability mechanism is due to the density drop across the flame and can be understood from the discussion in Section 7.5. As shown in Figure 7.17, any bulge of the flame front into the reactants causes the flow in front of the flame to decelerate, causing the bulge to advance deeper into the

9.8 Intrinsic Flame Instabilities

reactants. Similarly, any trough on the flame leads to approach flow acceleration, causing the trough to grow larger. The dispersion relation relating the growth rate of the instability to the size of the disturbance in an infinite domain can be obtained from Section 7.8, by setting $\mathscr{A}_2^u = 0$. The resulting homogeneous equations lead to the following dispersion relation between disturbance growth rate and wavenumber:

$$\frac{\omega}{ks^{u,0}} = \frac{i\sigma_\rho(\sqrt{1+\sigma_\rho - 1/\sigma_\rho} - 1)}{1+\sigma_\rho}. \tag{9.55}$$

This equation shows that ω is purely imaginary, illustrating that all disturbance wavelengths, k, grow exponentially in time through this mechanism, within the constant burning velocity assumption. The disturbance growth rate is linearly proportional to the wavenumber, k, and laminar burning velocity, $s^{u,0}$. The dependence of the growth rate on density ratio across the flame scales as $\omega/(ks^{u,0}) \propto i(\sqrt{\sigma_\rho} - 1)$ and $i(\sigma_\rho - 1)/2$ for large and small gas expansion across the flame, respectively. Of course, this dispersion relation only accounts for hydrodynamic effects. The overall flame stability dispersion relation is also influenced by gravity, viscosity, stretch, and other pertinent effects.

We will now move to another instability mechanism, the body force instability. This mechanism is equivalent to the classical Rayleigh–Taylor buoyant mechanism where a heavy fluid resting above a lighter one is destabilized by the action of gravity (see also Section 3.8). In the same way, flames propagating upward divide a higher- and lower-density region and, thus, are unstable. As discussed later, similar instabilities can be induced by acceleration of the flame sheet [91], either through a variation of the burning velocity or an externally imposed flow perturbation, such as the acoustic velocity field.

The above two instability mechanisms are due to the alteration of the flowfield by perturbations in flame position. In contrast, the thermal–diffusive instability mechanism arises from alteration of the burning velocity by flame wrinkling. This sensitivity to flame curvature can be either stabilizing (for $Ma > 0$) or destabilizing (for $Ma < 0$). Thus, spontaneous cellular structures arise on lean H_2–air or rich, large hydrocarbon flames and can make it difficult to obtain smooth, laminar flames in the laboratory. Figure 9.42 illustrates several images of thermodiffusively unstable flames.

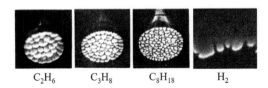

Figure 9.42 Images of cellular structures on thermal-diffusively unstable flames. C_2H_6, C_3H_8, and C_8H_{18} images are photographs of fuel-rich, laminar flames and reproduced from Markstein [92]. H_2 image obtained from OH PLIF image from fuel-lean, turbulent, low-swirl burner; reproduced from Cheng [93].

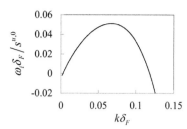

Figure 9.43 Predicted relationship between the instability growth rate and the disturbance wavenumber for a lean propane–air flame. Adapted from Clanet and Searby [87].

The overall stability of the flame is controlled by the interaction of all of these instabilities interacting simultaneously. To illustrate, Figure 9.43 plots the predicted dispersion relationship between the instability growth rate, ω_i, and wavenumber of disturbance, k, for a downward-propagating, thermodiffusively stable flame. The long wavelengths (low k values) are stabilized by gravity, the midrange wavenumbers are destabilized by gas expansion effects, and the short wavelengths are stabilized by thermal diffusive processes.

Note that the peak instability growth rate has values on the order of $\omega_i \sim O(0.1\, s^{u,0}/\delta_F)$. Thus, assuming a value of $\delta_F/s^{u,0} \sim 0.005\,\text{s}$ (typical values for methane or propane at 1 atm, as shown in Figure 9.21), the instability grows by a factor of e^1 in ~50 ms. For reference, the wrinkle development time on a 10 cm-long flame with a 10 m/s flow velocity is 10 ms. For such a flame, the instability would have little time to develop. In contrast, in a lower-velocity flow, such as a laminar burner, a 50 ms residence time is very typical and one would correspondingly expect to see significant spontaneous wrinkling of the flame through this instability.

These baseline instability characteristics are influenced by a variety of factors, such as stretch, heat loss, flow acceleration, flame edges [63], and confinement. For example, time-averaged positive flame stretch generally has a stabilizing influence on cellular instabilities, which can be conceptually understood through the smoothing effect that positive stretch has on flame disturbance features [2]. Moreover, confinement restricts the allowable wavenumbers possible on the flame and adds a wall boundary condition. The Saffman–Taylor instability arises because of confinement and the jump in viscosity across the flame, which alters the pressure balance across the flame, but is important only for flames propagating in narrow gaps [94]. Oscillatory acceleration of the flow in the direction normal to the flow also has important influences on these instabilities, as detailed in Section 11.6.

Nonpremixed flames also exhibit intrinsic instabilities [95], manifested as stationary or time-varying cellular flame structures, or even bulk flame oscillations. As discussed above, fluid mechanics and the coupling of the flame and flow by gas expansion play important roles in premixed flame instability. In contrast, nonpremixed flame oscillations appear to be driven only by thermal–diffusive effects, with gas expansion processes causing quantitative influences on stability boundaries, but not introducing new instability mechanisms. These instabilities are associated with differential diffusion of mass and heat, leading to regions of excess or shortage of enthalpy. Instability mechanisms are closely tied to the fact that nonunity Lewis number or mass diffusivity ratios lead to nonsimilar temperature and fuel/oxidizer concentration fields. In addition, heat loss is generally destabilizing for similar reasons, namely causing nonsimilarity in thermal and concentration profiles [95].

9.9 Aside: Unsteady Flame Response Effects

In this section we consider the effects of fluctuations in equivalence ratio, stretch rate, and pressure on the flame response at frequencies where the flame response is not quasi-steady [96]. Equation (9.23) is reproduced below, illustrating the effects of small fluctuations in these quantities on the burning velocity:

$$\frac{s_{c,1}}{s_{c,0}} = Ma_{c,curv} \frac{\kappa_{curv,1}}{(1/\delta_F^0)} + Ma_{c,s} \frac{\kappa_{s,1}}{(1/\delta_F^0)} + \left.\frac{\partial(s_c/s_{c,0})}{\partial(p/p_0)}\right|_0 \frac{p_1}{p_0} + \left.\frac{\partial(s_c/s_{c,0})}{\partial(\phi/\phi_0)}\right|_0 \frac{\phi_1}{\phi_0} + \cdots. \tag{9.56}$$

In order to simplify the notation, we will replace the pressure and equivalence ratio sensitivity coefficients by $s^u_{c,(\ )} = \frac{\partial(s_c/s^u_{c,0})}{\partial(\ )/(\ )_0}$. In addition, it is convenient to write the frequency domain equivalent of this expression, as we will specifically consider these sensitivities for harmonic oscillations:

$$\frac{\hat{s}_{c,1}}{s_{c,0}} = \hat{M}a_{c,curv} \frac{\hat{\kappa}_{curv,1}}{(1/\delta_F^0)} + \hat{M}a_{c,s} \frac{\hat{\kappa}_{s,1}}{(1/\delta_F^0)} + \hat{s}_{c,p} \frac{\hat{p}_1}{p_0} + \hat{s}_{c,\phi} \frac{\hat{\phi}_1}{\phi_0} + \cdots. \tag{9.57}$$

Each of these sensitivities has its own peculiarities, which are discussed further in this section. First, the unsteady flame response is not the same for hydrodynamic stretch as it is for curvature-induced stretch, i.e., $\hat{M}a_{c,curv} \neq \hat{M}a_{c,s}$. Second, the instantaneous displacement and consumption speeds are not the same [97–101], similar to steady-state, highly stretched flames (see Section 9.3.4). Finally, the flame responds as a low-pass filter and phase shifter to temporal fuel/air ratio and hydrodynamic strain fluctuations [102]. The response to curvature-induced stretch or pressure fluctuations is quite different and discussed later in this section. Because of this low-pass filter character, unsteady flames can be operated at fuel/air ratios that are instantaneously below their flammability limit or hydrodynamic stretch rates that are instantaneously above κ_{ext} [99]. As such, quantities such as κ_{ext} are functions of excitation frequency and amplitude, in addition to the time-averaged quantities already discussed.

To illustrate, Figure 9.44 plots computations of the dependence of the instantaneous burning velocity on instantaneous equivalence ratio and hydrodynamic stretch rate, κ_s. The quasi-steady equivalence ratio and stretch sensitivities are indicated as dashed lines on the plots for reference. Note how the magnitude of the burning velocity fluctuations decreases with increase in frequency. The low-frequency forcing cases follow the quasi-steady response line, but the flame response lags in phase and drops in gain as the frequency increases.

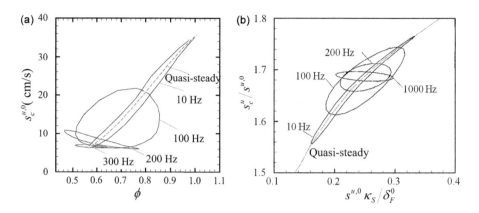

Figure 9.44 Dependence of the instantaneous flame consumption speed, s_c, on the instantaneous (a) equivalence ratio and (b) stretch rate, κ_s, at several frequencies of oscillation. Images reproduced from Sankaran and Im [102] and Im and Chen [98] (a CH$_4$/air flame at $\kappa = 300$ 1/s, b H$_2$/air flame, $\phi = 0.4$).

Note that these plots display the time variation of the consumption speed. The corresponding unburned displacement speed can exhibit very different characteristics, can become negative, and can even exhibit increasing gain with frequency [98]. This behavior reflects the fact that fluctuations in flow velocity at the leading edge of the preheat zone are not the same as those at the reaction zone in cases where the flame response is non-quasi-steady.

This unsteady response to fuel/air ratio oscillations can be approximately modeled by replacing the instantaneous fuel/air ratio value with its average over the thermal thickness of the flame, δ_F^0, with a phase shift equal to the convective propagation time over this same distance [103]. To illustrate, consider a harmonic perturbation in equivalence ratio expressed as

$$\phi(t) = \phi_0[1 + \varepsilon \cos(2\pi f t)]. \tag{9.58}$$

The "effective" equivalence ratio, $\tilde{\phi}(t)$, may then be evaluated as

$$\tilde{\phi}(t) = \phi_0\left[1 + \varepsilon \frac{\sin(\pi St_F)}{\pi St_F} \cos\left(2\pi\left(\frac{St_F}{2} + ft\right)\right)\right], \tag{9.59}$$

where $St_F = \frac{f \delta_F^0}{s^{u,0}}$. Thus, the instantaneous flame response is approximately equal to the quasi-steady response of a flame that is excited by the equivalent fuel/air ratio fluctuation $\tilde{\phi}$. On accounting for non-quasi-steady effects in such a manner, the flame speed sensitivity is diminished and phase shifted as [104]:

$$\hat{s}_{c,\phi} = \frac{\sin(\pi St_F)\exp(-i\pi St_F)}{\pi St_F} \hat{s}_{c,\phi}\big|_{quasi-steady}. \tag{9.60}$$

Note the low-pass filter character, as this sensitivity rolls off with frequency as $1/St_F$. This behavior, as well as computed results for hydrodynamic stretch oscillations, are plotted in Figure 9.45.

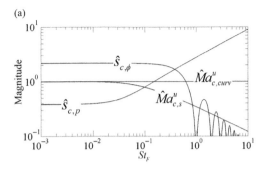
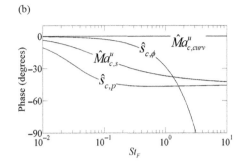

Figure 9.45 Predicted (a) magnitude and (b) phase sensitivities of the flame speed to curvature, hydrodynamic stretch, equivalence ratio, and pressure disturbances for a flame with $\sigma_\rho = 1$, $Le = 1$, $\gamma = 1.3$, $\tilde{E} = 10$.

The sensitivity of the flame to unsteadiness in flame curvature, κ_{curv}, is quite different. Models based on high activation energy asymptotics and single-step chemistry suggest that the flame's frequency response remains almost invariant over a wide range of frequencies [88, 105]; an illustrative calculation result is also plotted in Figure 9.45. From Joulin [88], the expression for the Markstein number for curvature has the form $Ma^u_{c,curv} = (1 + f^+(Le, St_F))/(1 + f^-(Le, St_F))$, where the complex functions $f^\pm(Le, St_F)$ have a maximum magnitude of 1 for $St_F = 0$ and decrease approximately at the same rate with increasing St_F, thereby keeping the magnitude of $Ma^u_{c,curv}$ of order unity.

Finally, consider the flame response to isentropic pressure fluctuations, such as due to an acoustic wave. Assuming that the pressure fluctuation is isentropic implies that the pressure, density, and temperature are actually all oscillating (i.e., not just the pressure) and so the "pressure sensitivity" discussed here rolls all these effects into a single number. This pressure response is unique from the other sensitivities in that it has been predicted to *increase* with frequency until the reaction zone itself becomes non-quasi-steady [106–111]. For example, for a flat flame responding to harmonic pressure disturbances, the following expression for the mass burning rate response was derived using single-step, high-activation chemistry [106]:

$$\hat{S}_{c,p} = \frac{-2\pi i St_F(1 - \gamma^{-1})\tilde{E}\left(2\sigma_\rho^{-1}F_a - (\sigma_\rho^{-1} - 1)\right)}{\left(Le\left(F_a - \frac{1}{2}\right) + \left(\frac{1}{2} - F_b\right)\right)\tilde{E}(\sigma_\rho^{-1} - 1) - 4F_a\left(\frac{1}{2} - F_b\right)} - \gamma^{-1}, \quad (9.61)$$

where

$$F_a = \frac{1}{2}(1 - 8\pi i Le\, St_F)^{1/2}, \quad F_b = \frac{1}{2}(1 - 8\pi i\, St_F)^{1/2}, \quad (9.62)$$

and $\tilde{E} = E_a/\mathscr{R}_u T^u$.

These sensitivities are summarized in Figure 9.45. Note that all of the curves are flat for small St_F values, with values asymptoting toward their respective, quasi-steady sensitivities. The high-frequency sensitivities in the $Le = 1$, $\sigma_\rho = 1$ limits are

$$\hat{M}a^u_{c,curv} = 1, \tag{9.63}$$

$$\left|\hat{M}a^u_{c,s}\right| \sim St_F^{-1/2}, \tag{9.64}$$

$$\left|\hat{s}_{c,p}\right| \sim St_F^{1/2}, \tag{9.65}$$

$$\left|\hat{s}_{c,\phi}\right| \sim St_F^{-1}, \tag{9.66}$$

showing a range of different frequency exponents. We close this section by noting that more work is needed on this topic for realistic chemistry, particularly to assess the curvature and pressure sensitivities which were calculated from single-step, high-activation descriptions of the flame.

9.10 Aside: Flame Extinction by Vortices

Section 9.3.4 showed how high levels of flame stretch lead to flame extinction using calculation obtained from a stagnation flow field. In practice, these high flame stretch levels are often associated with flame stabilization in separating shear layers and during vortex–flame interactions, as also shown in Exercise 7.4. While the steady, stagnation flow model problem is useful for understanding basic extinction concepts, vortex–flame interactions involve additional effects due to unsteadiness and consideration of the vortex length scale with respect to that of the flame [112, 113]. In this aside, we discuss the extinction of flames by vortices.

Quasi-steady, infinitely thin flame concepts suggest that for a given vortex rotational velocity, u_θ, smaller vortices (parameterized here by their length scale, L_v) are more effective at quenching flames than larger vortices, since the hydrodynamic stretch rate scales as u_θ/L_v. In reality, this effect is moderated by at least two effects since small vortices on the order of the flame thickness (1) are less effective at wrinkling the flame, and thus introducing curvature-induced stretch effects, and (2) diminish significantly in strength in the flame preheat zone by dilatation and diffusion effects (recall that the kinematic viscosity can increase by an order of magnitude across the flame). To illustrate, Figure 9.46 plots a flame-quenching correlation obtained from measurements of a toroidal, counter-rotating vortex incident on a

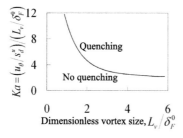

Figure 9.46 Flame quenching dependence on Ka and vortex size, obtained from measurements of a toroidal vortex incident on a $\phi = 0.55$ methane/air flame [114].

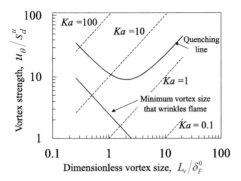

Figure 9.47 Dependence of flame quenching on vortex strength and vortex size. The plot also indicates the minimum size and strength of a vortex required to wrinkle the flame. Obtained from measurements of a toroidal vortex incident on a $\phi = 0.55$ methane/air flame [114, 115].

premixed flame [114]. Specifically, this figure shows that the Karlovitz number at quenching has values on the order of 2–3 for cases where $L_v \gg \delta_F^0$, but sharply increases to values that exceed 10 when $L_v/\delta_F^0 \lesssim 2$. In other words, the effectiveness of a vortex of a given rotational velocity, u_θ, at quenching the flame drops sharply as its size approaches the flame thickness.

An alternative way to plot these results is shown in Figure 9.47, which plots the quenching line, as well as a line indicating the minimum vortex strength/size required to significantly wrinkle the flame. Also plotted are lines of constant Karlovitz number, defined as $Ka = (u_\theta/s_d^u)/(L_v/\delta_F^0)$.

Note the nonmonotonic dependence of the vortex strength, u_θ/s_d^u, required to quench the flame on vortex size. This is due to the fact that for large vortices, where $L_v/\delta_F^0 \gg 1$, decreasing L_v/δ_F^0 leads to increases in flame stretch, implying a reduction in the vortex strength required to achieve the same Karlovitz number value. In contrast, the ineffectiveness of very small vortices in quenching the flame cause the required vortex strength to rise for $L_v/\delta_F^0 < 1$.

9.11 Aside: Wave Speeds of Reaction–Diffusion Equations

This section presents several results for one-dimensional reaction–diffusion equations of the form

$$\frac{\partial \tilde{c}}{\partial t} = \mathscr{D} \frac{\partial^2 \tilde{c}}{\partial x^2} + \omega(\tilde{c}), \tag{9.67}$$

where $0 < \tilde{c}(x) < 1$ denotes a normalized progress variable, such as $\tilde{c}(x) = (T(x) - T^u)/(T^b - T^u)$. While this expression is simplified, its solution provides insights into the processes controlling combustion wave speeds that can be used as a conceptual framework for analyzing more complex situations. We seek steady-state, traveling-wave solutions for the reactive–diffusive wave of the form

$$\tilde{c}(x, t) = \tilde{c}(x - s_d t), \tag{9.68}$$

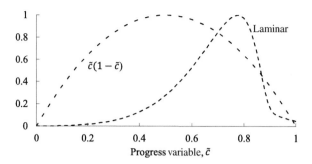

Figure 9.48 Dependence of reaction rate on temperature for a laminar flame (replotting the methane/air calculation from Figure 9.3), and for a common turbulent reaction rate closure, $\tilde{c}(1-\tilde{c})$, that satisfies the criterion in Eq. (9.70).

where s_d is the wave speed. Reaction–diffusion equations of this form, similar to the energy and species equations for combustion applications, arise in a host of other applications, and a large body of work exists on their general properties. We will not consider questions of stability of the solution or transients, but refer the reader to focused treatments for more detailed discussions [116–120].

We will assume that the diffusivity $\mathscr{D} > 0$ and that $\omega(\tilde{c} = 0) = \omega(\tilde{c} = 1) = 0$, and consider two functional dependencies for $\omega(\tilde{c})$ behaviors that are of particular interest for combustion applications, both plotted in Figure 9.48. For laminar, premixed flames with high activation energy chemistry, reaction rates are negligible for most $\tilde{c}$ values, except near $\tilde{c} \sim 1$, and $\omega(\tilde{c})$ has an inflection point in the profile. This can be seen from the figure, which replots the laminar profile from the methane/air calculation shown in Figure 9.3, plotting the heat release rate as a function of normalized temperature.

A second functional dependency of interest applies to the averaged behavior of turbulent, premixed combustion, where heat release occurs more uniformly over the entire turbulent flame brush. A common closure for average reaction rate through the turbulent flame brush is of the form $\tilde{c}(1-\tilde{c})$ [33], which is also plotted in the figure. This dependency arises in the flamelet case, such as shown in Figure 12.19(a), as the flame can be instantaneously located at different locations in the flame brush at different instants, with maximum probability of occurrence near the middle ($\tilde{c} \sim 0.5$), and lowest probabilities at the very front ($\tilde{c} \sim 0$) and back of the brush ($\tilde{c} \sim 1$).

If the reaction rate profile has an inflection point and is concentrated in $\tilde{c}$ space, the steady-state wave speed asymptotes to

$$s_{d,ZFK} = \sqrt{2\mathscr{D}\int_0^1 \tilde{\omega}(\tilde{c})d\tilde{c}}, \qquad (9.69)$$

where ZFK denotes Zeldovich–Frank-Kamenetskii wave speed [121, 122]. Note that s_d is controlled by the integral of the reaction rate profile and is independent of its local values, or the location of the maximum.

9.11 Wave Speeds of Reaction–Diffusion Equations

Another important limiting case occurs when the profile does not have an inflection point. Then, the maximum value of $d\tilde{\omega}/d\tilde{c}$ occurs at the leading edge of the wave, $\tilde{c} \to 0$; i.e.,

$$\left.\frac{d\tilde{\omega}}{d\tilde{c}}\right|_{\tilde{c}\to 0} = \max_{0\le\tilde{c}\le 1} \left.\frac{d\tilde{\omega}}{d\tilde{c}}\right|_{\tilde{c}}. \tag{9.70}$$

Then, the wave speed is given by

$$s_{d,KPP} = 2\sqrt{\mathscr{D}\left.\frac{d\tilde{\omega}}{d\tilde{c}}\right|_{\tilde{c}\to 0}}, \tag{9.71}$$

where KPP denotes Kolmogorov–Petrovsky–Piskunov [116]. This result shows that the wave speed is only a function of quantities at the leading edge of the wave and independent of the functional profile of $\omega(\tilde{c})$. For this reason, such waves are called "pulled fronts." While laminar flames with high activation energy chemistry will not satisfy the KPP criterion, the time- or Favre-averaged version of a turbulent flame could, as the average reaction rate consists of an ensemble of thin flamelets distributed spatially, such as in Figure 12.19(a).

To summarize, the reaction–diffusion equation admits at least two fundamentally different families of solutions, one controlled by integrated reaction rate values and the other by reaction rates at the leading edge. The reader is encouraged to review Section 11.2.3.3, which examines the propagation speed of a propagating front in a flow with space–time-varying velocity. While this is a completely different problem, there are analogies between the local/integrated attributes of the KPP/ZFK limits and the local/global dependencies for that solution, such as encapsulated in Figure 11.29.

A natural question to ask is the s_d behaviors between these two limiting functional dependencies of $\tilde{c}$. This transition can be explored numerically or with simplified source functions [118, 123]. For example, consider the reaction rate profile

$$\tilde{\omega}(\tilde{c}) = A(1-\tilde{c})\left(e^{-\beta(1-\tilde{c})} - e^{-\beta}\right), \tag{9.72}$$

where $A = \beta^2/\left(1 - e^{-\beta}\left(1 + \beta + \frac{\beta^2}{2}\right)\right)$, so that $\int_0^1 \tilde{\omega}(\tilde{c})d\tilde{c} = 1$; i.e., the β parameter redistributes the heat release but not its overall amount. As the parameter β increases, the source function changes from being uniformly distributed over the domain to being narrow and concentrated close to $\tilde{c} = 1$, as shown in Figure 9.49.

Figure 9.50 plots the computed dependence of s_d on β. Also shown are the corresponding $s_{d,ZFK}$ and $s_{d,KPP}$ values, computed using Eqs. (9.69) and (9.71). Note how the exact solution asymptotes to these two limits for small and large β, respectively, and smoothly transitions between them for intermediate values.

We close by briefly considering the constraint on $\mathscr{D} > 0$. This constraint is satisfied for Fickian diffusion in laminar flames. However, the turbulent diffusivity may be negative, as combustion can lead to counter-gradient diffusion [124, 125].

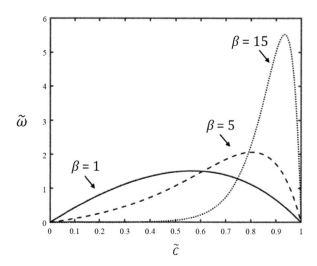

Figure 9.49 Source function given by Eq. (9.72) for increasing values of β.

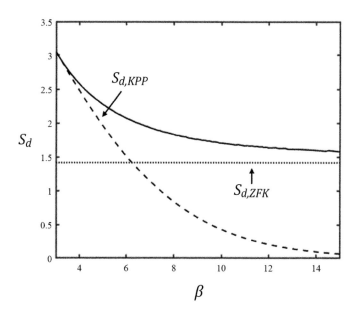

Figure 9.50 Wave speed computed numerically (solid line) showing a transition from KPP value (dashed line) to ZFK value (dotted line) with increasing β [118, 123].

It has been suggested that turbulent diffusivity is of the gradient type if $u'/s_d \gtrsim (\sigma_p/2)$ [33], showing that the $\mathscr{D} > 0$ criterion may not be met if considering averaged reaction–diffusion conditions for low turbulence intensity flames with significant levels of gas expansion.

Exercises

9.1 Asymptotic premixed flame structure. As discussed in the text, the reaction zone is much thinner than the preheat zone in the single-step chemistry, high activation energy limit. Work out the scaling of Eq. (9.1) relating the flame reaction zone and preheat zone thickness.

9.2 Mass burning rate scaling for a premixed flame. Work out the scaling of Eq. (9.3) for the mass burning rate of a premixed flame. For this purpose it is sufficient to utilize the simplified energy equation,

$$\dot{m}_F'' c_p \frac{dT}{dx} = \lambda \frac{d^2 T}{dx^2} + \dot{w} Q, \tag{9.82}$$

subject to the boundary conditions

$$T(-\infty) = T^u, \quad T(+\infty) = T^{b,0},$$
$$\left(\frac{dT}{dx}\right)_{-\infty} = \left(\frac{dT}{dx}\right)_{+\infty} = 0, \tag{9.83}$$

where the thermal conductivity, λ, and the specific heat, c_p, are constant and the heat of reaction per unit mass, Q, is given by

$$Q \sim c_p \left(T^{b,0} - T^u\right). \tag{9.84}$$

9.3 Burning velocity sensitivities. Following the calculations of the pressure and preheat temperature sensitivities in Section 9.1.1, perform a series of calculations and plot the pressure and temperature sensitivity of the burning velocity of CH$_4$/air over 300 K $< T^u <$ 1000 K and 1 atm $< p <$ 20 atm. At each point, adjust the mixture stoichiometry to maintain a fixed adiabatic flame temperature of 2000 K. Comment on your results.

9.4 As shown in the text, flame stretch can be formulated in different ways. Equation (9.12) illustrates the role of tangential gradients in tangential velocity and flame front motion, while Eq. (9.14) explicitly brings out the role of hydrodynamic stretching and flame curvature. Demonstrate the equivalency of these two forms.

9.5 Edge flames. Derive the dimensionless temperature equation given in Eqs. (9.49) and (9.50).

References

1. Neophytou A., Mastorakos E., and Cant R., DNS of spark ignition and edge flame propagation in turbulent droplet-laden mixing layers. *Combustion and Flame*, 2010, **157**(6): pp. 1071–1086.
2. Law C.K., *Combustion Physics*. 2006, Cambridge University Press.
3. Turns S.R., *An Introduction to Combustion: Concepts and Applications*. 2nd ed. 2001, MacGraw-Hill.

4. Glassman I. and Yetter R.A., *Combustion*. 4th ed. 2008, Academic Press.
5. Smith G.P., Golden D.M., Frenklach M., Moriarty N.W., Eiteneer B., Goldenberg M., Bowman T., Hanson R.K., Song S., Gardiner W.C.J., Lissianski V.V., and Qin Z., GRI-Mech. Available from: http://combustion.berkeley.edu/gri-mech/.
6. Mechanical and Aerospace Engineering (Combustion Research), *Chemical-Kinetic Mechanisms for Combustion Applications*, San Diego Mechanism web page 2012. Available from: https://web.eng.ucsd.edu/mae/groups/combustion/mechanism.html/.
7. Davis S.G., Joshi A.V., Wang H., and Egolfopoulos F., An optimized kinetic model of H_2/CO combustion. *Proceedings of the Combustion Institute*, 2005, **30**(1): pp. 1283–1292.
8. Lowry W., de Vries J., Krejci M., Petersen E., Serinyel Z., Metcalfe W., Curran H., and Bourque G., Laminar flame speed measurements and modeling of pure alkanes and alkane blends at elevated pressures. *Journal of Engineering for Gas Turbines and Power*, 2011, **133**: p. 091501.
9. Kochar Y.N., Vaden S.N., Lieuwen T.C., and Seitzman J.M., Laminar flame speed of hydrocarbon fuels with preheat and low oxygen content, in *48th AIAA Aerospace Sciences Meeting Including the New Horizons Forum and Aerospace Exposition*. 2010.
10. Vagelopoulos C., Egolfopoulos F., and Law C., Further considerations on the determination of laminar flame speeds with the counterflow twin-flame technique. *Symposium (International) on Combustion*, 1994, **25**(1): pp. 1341–1347.
11. Vagelopoulos C.M. and Egolfopoulos F.N., Direct experimental determination of laminar flame speeds, *Symposium (International) on Combustion*, 1998, **27**(1): pp. 513–519.
12. Smallbone A., Liu W., Law C., You X., and Wang H., Experimental and modeling study of laminar flame speed and non-premixed counterflow ignition of *n*-heptane. *Proceedings of the Combustion Institute*, 2009, **32**(1): pp. 1245–1252.
13. Dowdy D.R., Smith D.B., Taylor S.C., and Williams A., The use of expanding spherical flames to determine burning velocities and stretch effects in hydrogen/air mixtures. *Symposium (International) on Combustion*, 1991, **23**(1): pp. 325–332.
14. McLean I.C., Smith D.B., and Taylor S.C., The use of carbon monoxide/hydrogen burning velocities to examine the rate of the CO + OH reaction. *Symposium (International) on Combustion*, 1994, **25**(1): pp. 749–757.
15. Qiao L., Kim C., and Faeth G., Suppression effects of diluents on laminar premixed hydrogen/oxygen/nitrogen flames. *Combustion and Flame*, 2005, **143**(1–2): pp. 79–96.
16. Amato A., Hudak B., D'Carlo P., Noble D., Scarborough D., Seitzman J., and Lieuwen T., Methane oxycombustion for low CO_2 cycles: Blowoff measurements and analysis. *Journal of Engineering for Gas Turbines and Power*, 2011, **133**: p. 061503-1.
17. Burke M.P., Chaos M., Dryer F.L., and Ju Y., Negative pressure dependence of mass burning rates of $H_2/CO/O_2$/diluent flames at low flame temperatures. *Combustion and Flame*, 2010, **157**(4): pp. 618–631.
18. Burke M.P., Dryer F.L., and Ju Y., Assessment of kinetic modeling for lean $H_2/CH_4/O_2$/diluent flames at high pressures. *Proceedings of the Combustion Institute*, 2011, **33**(1): pp. 905–912.
19. Chen J.H., Hawkes E.R., Sankaran R., Mason S.D., and Im H.G., Direct numerical simulation of ignition front propagation in a constant volume with temperature inhomogeneities. I. Fundamental analysis and diagnostics. *Combustion and Flame*, 2006, **145**(1): pp. 128–144.

20. Hawkes E.R., Sankaran R., Pébay P.P., and Chen J.H., Direct numerical simulation of ignition front propagation in a constant volume with temperature inhomogeneities. II. Parametric study. *Combustion and Flame*, 2006, **145**(1): pp. 145–159.
21. Zhang H., Hawkes E.R., Chen J.H., and Kook S., A numerical study of the autoignition of dimethyl ether with temperature inhomogeneities. *Proceedings of the Combustion Institute*, 2013, **34**(1): pp. 803–812.
22. Yoo C.S., Lu T., Chen J.H., and Law C.K., Direct numerical simulations of ignition of a lean *n*-heptane/air mixture with temperature inhomogeneities at constant volume: Parametric study. *Combustion and Flame*, 2011, **158**(9): pp. 1727–1741.
23. Schulz O. and Noiray N., Combustion regimes in sequential combustors: Flame propagation and autoignition at elevated temperature and pressure. *Combustion and Flame*, 2019, **205**: pp. 253–268.
24. Sankaran R., Im H.G., Hawkes E.R., and Chen J.H., The effects of non-uniform temperature distribution on the ignition of a lean homogeneous hydrogen–air mixture. *Proceedings of the Combustion Institute*, 2005, **30**(1): pp. 875–882.
25. Mansfield A.B. and Wooldridge M.S., High-pressure low-temperature ignition behavior of syngas mixtures. *Combustion and Flame*, 2014, **161**(9): pp. 2242–2251.
26. Luong M.B., Hernández Pérez F.E., and Im H.G., Prediction of ignition modes of NTC-fuel/air mixtures with temperature and concentration fluctuations. *Combustion and Flame*, 2020, **213**: pp. 382–393.
27. Pal P., Mansfield A.B., Arias P.G., Wooldridge M.S., and Im H.G., A computational study of syngas auto-ignition characteristics at high-pressure and low-temperature conditions with thermal inhomogeneities. *Combustion Theory and Modelling*, 2015, **19**(5): pp. 587–601.
28. Groot G., van Oijen J., de Goey L., Seshadri K., and Peters N., The effects of strain and curvature on the mass burning rate of premixed laminar flames. *Combustion Theory and Modelling*, 2002, **6**(4): pp. 675–695.
29. Williams F.A., *Combustion Theory: The Fundamental Theory of Chemically Reacting Flow Systems*. 1994, Perseus Books.
30. Tseng L.K., Ismail M., and Faeth G.M., Laminar burning velocities and Markstein numbers of hydrocarbon/air flames. *Combustion and Flame*, 1993, **95**(4): pp. 410–426.
31. Aggarwal S.K., Extinction of laminar partially premixed flames. *Progress in Energy and Combustion Science*, 2009, **35**(6): pp. 528–570.
32. Gran I.R., Echekki T. and Chen J.H., Negative flame speed in an unsteady 2-D premixed flame: a computational study, *Symposium (International) on Combustion Science*, 1996, **26**(1): pp. 323–329.
33. Poinsot T. and Veynante D., *Theoretical and Numerical Combustion*. 2005, RT Edwards, Inc.
34. Sankaran R. and Im H.G., Effects of hydrogen addition on the Markstein length and flammability limit of stretched methane/air premixed flames. *Combustion Science and Technology*, 2006, **178**(9): pp. 1585–1611.
35. Poinsot T., Echekki T., and Mungal M., A study of the laminar flame tip and implications for premixed turbulent combustion. *Combustion Science and Technology*, 1992, **81**(1): pp. 45–73.
36. Wang P., Hu S., Wehrmeyer J.A., and Pitz R.W., Stretch and curvature effects on flames, in *42nd AIAA Aerospace Sciences Meeting and Exhibit*. 2004, Reno, NV: AIAA.
37. Wang P., Wehrmeyer J.A., and Pitz R.W., Stretch rate of tubular premixed flames. *Combustion and Flame*, 2006, **145**(1–2): pp. 401–414.

38. Amato A., Day M., Cheng R.K., Bell J., Dasgupta D., and Lieuwen T., Topology and burning rates of turbulent, lean, H_2/air flames. *Combustion and Flame*, 2015, **162**(12): pp. 4553–4565.
39. Jackson G.S., Sai R., Plaia J.M., Boggs C.M., and Kiger K.T., Influence of H_2 on the response of lean premixed CH_4 flames to high strained flows. *Combustion and Flame*, 2003, **132**(3): pp. 503–511.
40. Shanbhogue S.J., Husain S., and Lieuwen T., Lean blowoff of bluff body stabilized flames: Scaling and dynamics. *Progress in Energy and Combustion Science*, 2009, **35**(1): pp. 98–120.
41. Egolfopoulos F. Geometric and radiation effects on steady and unsteady strained laminar flames. *Symposium (International) on Combustion*, 1994, **25**(1): pp. 1375–1381.
42. Lieuwen T., Neumeier Y., and Zinn B., The role of unmixedness and chemical kinetics in driving combustion instabilities in lean premixed combustors. *Combustion Science and Technology*, 1998, **135**(1): pp. 193–211.
43. Welle E.J., Roberts W.L., Carter C.D., and Donbar J.M., The response of a propane-air counter-flow diffusion flame subjected to a transient flow field. *Combustion and Flame*, 2003, **135**(3): pp. 285–297.
44. Sung C., Liu J., and Law C., Structural response of counterflow diffusion flames to strain rate variations. *Combustion and Flame*, 1995, **102**(4): pp. 481–492.
45. Peters N., *Turbulent Combustion*. 1st ed. 2000, Cambridge University Press.
46. Xuan Y., Blanquart G., and Mueller M.E., Modeling curvature effects in diffusion flames using a laminar flamelet model. *Combustion and Flame*, 2014, **161**(5): pp. 1294–1309.
47. Kee R.J., Coltrin M.E., and Glarborg P., *Chemically Reacting Flow: Theory and Practice*. 1st ed. 2003: John Wiley & Sons.
48. Barlow R., Dibble R., Chen J.Y., and Lucht R., Effect of Damköhler number on super-equilibrium OH concentration in turbulent nonpremixed jet flames. *Combustion and Flame*, 1990, **82**(3–4): pp. 235–251.
49. Sutton J.A. and Driscoll J.F., Imaging of local flame extinction due to the interaction of scalar dissipation layers and the stoichiometric contour in turbulent non-premixed flames. *Proceedings of the Combustion Institute*, 2007, **31**(1): pp. 1487–1495.
50. Won S.H., Dooley S., Dryer F.L., and Ju Y., Kinetic effects of aromatic molecular structures on diffusion flame extinction. *Proceedings of the Combustion Institute*, 2011, **33**: pp. 1163–1170.
51. Mastorakos E., Taylor A., and Whitelaw J., Scalar dissipation rate at the extinction of turbulent counterflow nonpremixed flames. *Combustion and Flame*, 1992, **91**(1): pp. 55–64.
52. Peters N., *Length Scales in Laminar and Turbulent Flames. Numerical Approaches to Combustion Modeling* (A 92-16977 04-25). Washington, DC, American Institute of Aeronautics and Astronautics, Inc., 1991, pp. 155–182.
53. Ihme M., Cha C.M., and Pitsch H., Prediction of local extinction and re-ignition effects in non-premixed turbulent combustion using a flamelet/progress variable approach. *Proceedings of the Combustion Institute*, 2005, **30**(1): pp. 793–800.
54. Carnell W.F. and Renfro M.W., Stable negative edge flame formation in a counterflow burner. *Combustion and Flame*, 2005, **141**(4): pp. 350–359.
55. Rajaram R. and Lieuwen T., Parametric studies of acoustic radiation from premixed flames. *Combustion Science and Technology*, 2003, **175**(12): pp. 2269–2298.

56. Chaudhuri S., Kostka S., Tuttle S.G., Renfro M.W., and Cetegen B.M., Blowoff mechanism of two dimensional bluff-body stabilized turbulent premixed flames in a prototypical combustor. *Combustion and Flame*, 2011, **158**: pp. 1358–1371.
57. Smith C., Nickolaus D., Leach T., Kiel B., and Garwick K., LES blowout analysis of premixed flow past v-gutter flameholder, in *45th AIAA Aerospace Sciences Meeting and Exhibit*. 2007, Reno NV: AIAA.
58. Barlow R., Ozarovsky H., Karpetis A., and Lindstedt R., Piloted jet flames of CH_4/H_2/air: experiments on localized extinction in the near field at high Reynolds numbers. *Combustion and Flame*, 2009, **156**(11): pp. 2117–2128.
59. Dunn M., Masri A., Bilger R., Barlow R., and Wang G.H., The compositional structure of highly turbulent piloted premixed flames issuing into a hot coflow. *Proceedings of the Combustion Institute*, 2009, **32**(2): pp. 1779–1786.
60. Lindstedt R., Ozarovsky H., Barlow R., and Karpetis A., Progression of localized extinction in high Reynolds number turbulent jet flames. *Proceedings of the Combustion Institute*, 2007, **31**(1): pp. 1551–1558.
61. Dunn M.J., Masri A.R., and Bilger R.W., A new piloted premixed jet burner to study strong finite-rate chemistry effects. *Combustion and Flame*, 2007, **151**(1–2): pp. 46–60.
62. Owston R. and Abraham J., Structure of hydrogen triple flames and premixed flames compared. *Combustion and Flame*, 2010, **157**(8): pp. 1552–1565.
63. Buckmaster J., Edge-flames. *Progress in Energy and Combustion Science*, 2002, **28**(5): pp. 435–475.
64. Buckmaster J., Edge-flames and their stability. *Combustion Science and Technology*, 1996, **115**(1–3): pp. 41–68.
65. Im H. and Chen J., Structure and propagation of triple flames in partially premixed hydrogen-air mixtures. *Combustion and Flame*, 1999, **119**(4): pp. 436–454.
66. Liu J.B. and Ronney P.D., Premixed edge-flames in spatially-varying straining flows. *Combustion Science and Technology*, 1999, **144**(1–6): pp. 21–46.
67. Daou R., Daou J. and Dold J., Effect of volumetric heat loss on triple-flame propagation. *Proceedings of the Combustion Institute*, 2002, **29**(2): pp. 1559–1564.
68. Cha M.S. and Ronney P.D., Propagation rates of nonpremixed edge flames. *Combustion and Flame*, 2006, **146**(1–2): pp. 312–328.
69. Vedarajan T. and Buckmaster J., Edge-flames in homogeneous mixtures. *Combustion and Flame*, 1998, **114**(1–2): pp. 267–273.
70. Ruetsch G., Vervisch L., and Liñán A., Effects of heat release on triple flames. *Physics of Fluids*, 1995, **7**(6): pp. 1447–1454.
71. Fernandez-Tarrazo E., Vera M., and Linan A., Liftoff and blowoff of a diffusion flame between parallel streams of fuel and air. *Combustion and Flame*, 2006, **144**(1–2): pp. 261–276.
72. Santoro V.S., Liñán A., and Gomez A., Propagation of edge flames in counterflow mixing layers: experiments and theory. *Proceedings of the Combustion Institute*, 2000, **28**(2): p. 2039–2046.
73. Daou R., Daou J., and Dold J., The effect of heat loss on flame edges in a non-premixed counterflow within a thermo-diffusive model. *Combustion Theory and Modelling*, 2004, **8**(4): pp. 683–699.
74. Daou R., Daou J., and Dold J., Effect of heat-loss on flame-edges in a premixed counterflow. *Combustion Theory and Modelling*, 2003, **7**(2): pp. 221–242.

75. Favier V. and Vervisch L., Edge flames and partially premixed combustion in diffusion flame quenching. *Combustion and Flame*, 2001, **125**(1–2): pp. 788–803.
76. Shay M.L. and Ronney P.D., Nonpremixed edge flames in spatially varying straining flows. *Combustion and Flame*, 1998, **112**(1–2): pp. 171–180.
77. Cho S.J. and Takita K., Numerical study of premixed twin edge flames in a counterflow field. *Combustion and Flame*, 2006, **144**(1–2): pp. 370–385.
78. Takita K., Sado M., Masuya G., and Sakaguchi S., Experimental study of premixed single edge-flame in a counterflow field. *Combustion and Flame*, 2004, **136**(3): pp. 364–370.
79. Carnell Jr W. and Renfro M., Influence of advective heat flux on extinction scalar dissipation rate and velocity in negative edge flames. *Combustion Theory Modelling*, 2006, **10**: pp. 815–830.
80. Bobba M.K., *Flame Stabilization and Mixing Characteristics in a Stagnation Point Reverse Flow Combustor*. 2007, Georgia Institute of Technology.
81. Driscoll J., Turbulent premixed combustion: flamelet structure and its effect on turbulent burning velocities. *Progress in Energy and Combustion Science*, 2008, **34**: pp. 91–134.
82. Mastorakos E., Taylor A.M.K.P. and Whitelaw J.H., Extinction of turbulent counterflow flames with reactants diluted by hot products. *Combustion and Flame*, 1995, **102**: pp. 101–114.
83. Lefebvre A.H. and Ballal D.R., *Gas Turbine Combustion: Alternative Fuels and Emissions*. 2010, Taylor and Francis.
84. Clavin P., Dynamic behavior of premixed flame fronts in laminar and turbulent flows. *Progress in Energy and Combustion Science*, 1985, **11**(1): pp. 1–59.
85. Aldredge R. and Killingsworth N., Experimental evaluation of Markstein-number influence on thermoacoustic instability. *Combustion and Flame*, 2004, **137**(1–2): pp. 178–197.
86. Bychkov V. and Liberman M.A., Dynamics and stability of premixed flames. *Physics Reports*, 2000, **325**(4–5): pp. 115–237.
87. Clanet C. and Searby G., First experimental study of the Darrieus–Landau instability. *Physical Review Letters*, 1998, **80**(17): pp. 3867–3870.
88. Joulin G., On the response of premixed flames to time-dependent stretch and curvature. *Combustion Science and Technology*, 1994, **97**(1): pp. 219–229.
89. Park Y. and Vlachos D., Isothermal chain-branching, reaction exothermicity, and transport interactions in the stability of methane/air mixtures. *Combustion and Flame*, 1998, **114**(1–2): pp. 214–230.
90. Kalamatianos S., Park Y., and Vlachos D., Two-parameter continuation algorithms for sensitivity analysis, parametric dependence, reduced mechanisms, and stability criteria of ignition and extinction. *Combustion and Flame*, 1998, **112**(1–2): pp. 45–61.
91. Bauwens C.R.L., Bauwens L., and Wierzba I., Accelerating flames in tubes, an analysis. *Proceedings of the Combustion Institute*, 2007, **31**(2): pp. 2381–2388.
92. Markstein G.H., *Non-Steady Flame Propagation*. AGARD-AR-75. 1964. Originally published by the Advisory Group for Aeronautical Research and Development, North Atlantic Treaty Organization (AGARD/NATO).
93. Cheng R.K., Turbulent combustion properties of premixed syngas, in *Synthesis Gas Combustion: Fundamentals and Applications*, T. Lieuwen, V. Yang, and R. Yetter, eds. 2010, CRC Press, pp. 129–168.
94. Aldredge R., Saffman–Taylor influence on flame propagation in thermoacoustically excited flow. *Combustion Science and Technology*, 2004, **177**(1): pp. 53–73.
95. Matalon M., Intrinsic flame instabilities in premixed and nonpremixed combustion. *Annual Review of Fluid Mechanics*, 2007, **39**: pp. 163–191.

96. Zambon A. and Chelliah H., Acoustic-wave interactions with counterflow single-and twin-premixed flames: Finite-rate kinetics, heat release and phase effects. *Proceedings of the Combustion Institute*, 2007, **31**(1): pp. 1247–1255.
97. Saitoh T. and Otsuka Y., Unsteady behavior of diffusion flames and premixed flames for counter flow geometry. *Combustion Science and Technology*, 1976, **12**(4): pp. 135–146.
98. Im H.G. and Chen J.H., Effects of flow transients on the burning velocity of laminar hydrogen/air premixed flames. *Proceedings of the Combustion Institute*, 2000, **28**(2): pp. 1833–1840.
99. Im H., Bechtold J., and Law C., Response of counterflow premixed flames to oscillating strain rates. *Combustion and Flame*, 1996, **105**(3): pp. 358–372.
100. Huang Z., Bechtold J., and Matalon M., Weakly stretched premixed flames in oscillating flows. *Combustion Theory and Modelling*, 1999, **2**(2): pp. 115–133.
101. Egolfopoulos F.N. *Dynamics and structure of unsteady, strained, laminar premixed flames. Symposium (International) on Combustion*, 1994, **25**(1): pp. 1365–1373.
102. Sankaran R. and Im H.G., Dynamic flammability limits of methane/air premixed flames with mixture composition fluctuations. *Proceedings of the Combustion Institute*, 2002, **29**(1): pp. 77–84.
103. Lauvergne R. and Egolfopoulos F.N., Unsteady response of C_3H_8/air laminar premixed flames submitted to mixture composition oscillations. *Proceedings of the Combustion Institute*, 2000, **28**: pp. 1841–1850.
104. Shreekrishna and Lieuwen T. High frequency response of premixed flames to acoustic disturbances, in *15th AIAA/CEAS Aeroacoustics Conference*. 2009, Miami, FL.
105. Clavin P. and Joulin G., High-frequency response of premixed flames to weak stretch and curvature: a variable-density analysis. *Combustion Theory and Modelling*, 1997, **1**(4): pp. 429–446.
106. McIntosh A.C., Pressure disturbances of different length scales interacting with conventional flames. *Combustion Science and Technology*, 1991, **75**: pp. 287–309.
107. Ledder G. and Kapila A., The response of premixed flame to pressure perturbations. *Combustion Science and Technology*, 1991, **76**: pp. 21–44.
108. van Harten A., Kapila A., and Matkowsky B.J., Acoustic coupling of flames. *SIAM Journal of Applied Mathematics*, 1984, **44**(5): pp. 982–995.
109. Peters N. and Ludford G.S.S., The effect of pressure variation on premixed flames. *Combustion Science and Technology*, 1983, **34**: pp. 331–344.
110. Keller D. and Peters N., Transient pressure effects in the evolution equation for premixed flame fronts. *Theoretical and Computational Fluid Dynamics*, 1994, **6**: pp. 141–159.
111. McIntosh A.C., Deflagration fronts and compressibility. *Philosophical Transactions of the Royal Society of London:A*, 1999, **357**: pp. 3523–3538.
112. Hancock R.D., Schauer F.R., Lucht R.P., Katta V.R., and Hsu K.Y., Thermal diffusion effects and vortex–flame interactions in hydrogen jet diffusion flames. *Symposium (International) on Combustion*, 1996, **26**(1): pp. 1087–1093.
113. Patnaik G. and Kailasanath K., A computational study of local quenching in flame-vortex interactions with radiative losses. *Symposium (International) on Combustion*, 1998, **27**(1): pp. 711–717.
114. Roberts W.L., Driscoll J.F., Drake M.C., and Goss L.P., Images of the quenching of a flame by a vortex to quantify regimes of turbulent combustion. *Combustion and Flame*, 1993, **94**(1–2): pp. 58–69.

115. Roberts W.L. and Driscoll J.F., A laminar vortex interacting with a premixed flame: Measured formation of pockets of reactants. *Combustion and Flame*, 1991, **87**(3–4): pp. 245–256.
116. Kolmogorov A.N., A study of the equation of diffusion with increase in the quantity of matter, and its application to a biological problem. *Moscow University Bulletin of Mathematics*, 1937, **1**: pp. 1–25.
117. Aldushin A., Zel'dovich Y.B., and Khudyaev S., Numerical investigation of flame propagation in a mixture reacting at the initial temperature. *Combustion, Explosion and Shock Waves*, 1979, **15**(6): pp. 705–710.
118. Clavin P. and Liñán A., Theory of gaseous combustion, in *Nonequilibrium Cooperative Phenomena in Physics and Related Fields*, M.G. Velarde, ed. 1984, Springer, pp. 291–338.
119. Aronson D.G. and Weinberger H.F., Nonlinear diffusion in population genetics, combustion, and nerve pulse propagation, in *Partial Differential Equations and Related Topics*, J.A. Goldstein, ed. 1975, Springer, pp. 5–49.
120. Ebert U. and van Saarloos W., Front propagation into unstable states: universal algebraic convergence towards uniformly translating pulled fronts. *Physica D: Nonlinear Phenomena*, 2000, **146**(1–4): pp. 1–99.
121. Zeldovich Y.B. and Barenblatt G., Theory of flame propagation. *Combustion and Flame*, 1959, **3**: pp. 61–74.
122. Zeldovich Y.B. and Frank-Kamenetskii D., A theory of thermal propagation of flame. *Acta physiochimica URSS*, 1938, **9**: pp. 341–350.
123. Benguria R., Cisternas J., and Depassier M., Variational calculations for thermal combustion waves. *Physical Review E*, 1995, **52**(4): p. 4410.
124. Sabelnikov V.A. and Lipatnikov A.N., Transition from pulled to pushed premixed turbulent flames due to countergradient transport. *Combustion Theory and Modelling*, 2013, **17**(6): pp. 1154–1175.
125. Corvellec C., Bruel P., and Sabelnikov V., Turbulent premixed flames in the flamelet regime: burning velocity spectral properties in the presence of countergradient diffusion. *Combustion and Flame*, 2000, **120**: pp. 585–588.

10 Flame Stabilization, Flashback, Flameholding, and Blowoff

This chapter initiates the third section of the text, discussing transient and time-harmonic combustor phenomena. This particular chapter focuses on the transient phenomena of flashback, flame stabilization, and blowoff. Chapters 11 and 12 then focus on time-harmonic and broadband flame forcing.

This chapter is divided into two main sections. Section 10.1 treats flame flashback. It shows that there are multiple mechanisms through which a flame can propagate upstream into premixed reactants, each of which has different sensitivities to the flowfield and operating conditions. We also show that the processes controlling the initiation of flashback, and those controlling its behavior once it has begun to propagate upstream, are quite different. Section 10.2 then treats flame stabilization and blowoff. This chapter starts with the classical treatment of the problem by considering the relative balance between flame speed and flow velocity in the shear layer. However, flames are strongly affected by stretch effects near the stabilization point, as they lie in regions of high shear. As such, we then work out the scalings for flame stretch rate in a shear layer and show that quite different results are possible, depending on the configuration. We also discuss the effects of flow recirculation on flame stabilization and the processes leading to blowoff.

10.1 Flashback and Flameholding

Flashback describes the upstream propagation of a premixed flame into a region not designed for the flame to exist. It is undesirable for the same reasons that autoignition in premixing passages is, although the fundamental processes controlling flashback and autoignition are quite different. Of most interest are situations where the average axial flow velocity exceeds the laminar and/or turbulent flame speed (if this is not the case, the flashback mechanism is trivial). Even if this criterion is met, however, several mechanisms can lead to flashback: (1) flame propagation in the high-velocity core flow (see Figure 10.1(a)), (2) flashback due to combustion instabilities, and (3) flashback in the boundary layer (see Figure 10.1(b)) [1–3]. The relative significance of these mechanisms is a strong function of fuel composition, operating conditions, and fluid mechanics.

Combustion instability induced flashback is associated with the fact that the instantaneous axial flow velocity can drop to quite low values, or even negative ones, during large-amplitude oscillations, as shown in Figure 6.15 [5]. In fact, flashback is

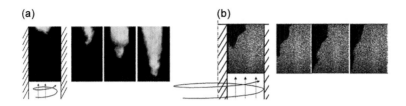

Figure 10.1 Experimental images of flame flashback in (a) the core flow and (b) the boundary layer. Images adapted from (a) Kröner et al. [2] and (b) Heeger et al. [4].

often the mechanism through which combustion instabilities damage combustion systems. Combustion instabilities are discussed in Chapter 6 and not treated further in this chapter. The subsequent discussion focuses on the controlling processes for flashback in the core flow or boundary layer.

In closing this section, it is important to differentiate the phenomena of flashback and flameholding. Flashback, defined earlier as upstream propagation of a premixed flame into a region not designed for it to exist, is usually of secondary interest to designers of hardware that must be robust to disturbances, such as an air flow transient. Rather, a designer is interested in the more severe criterion of whether a flame, introduced into either the premixing section core flow or boundary layer, will be pushed out by the flow. In other words, will a system under "normal operation" push a flame back out if flashback had occurred due to some unexpected disturbance? We will use the term *flameholding* to describe situations that do not meet this criterion.[1] As we will discuss next, there are many hysteretic elements to the problem, rendering the need to distinguish between flashback and flameholding. Note that flameholding physics involves elements of both flashback and blowoff; i.e., once a flame has flashed back and is in a premixing passage, will it blow off if the flow conditions are returned to the nominal operating point?

10.1.1 Flame Propagation in the Core Flow

In this section we consider the upstream propagation of a flame through the high-velocity core flow. In a nonswirling flow, analysis of this problem essentially comes to determining whether any locations in the premixer have local axial flow velocities that are lower than the turbulent displacement speed. From a design point of view, this requires a flowfield without any significant velocity deficits, such as wakes from premixing hardware. Indeed, this is one instance where the distinction between flashback and flameholding is important, as even if a low-velocity wake feature has smoothed out at the combustor entrance so as to prevent flashback, a flame may still stabilize in it if it is somehow introduced there.

---

[1] Note that "flameholding" is sometimes used to more generally describe flame stabilization in the literature. "Flameholding" in this text refers to flame stabilization only in the premixing sections, not in the combustor itself.

10.1 Flashback and Flameholding

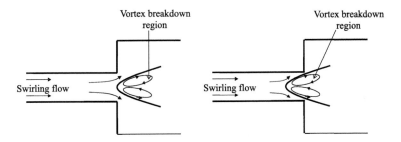

Figure 10.2 Flashback due to combustion-induced vortex breakdown, showing the upstream propagation of the flame and vortex breakdown bubble [7].

Additional physics are present in swirling flows. In this case, the interactions of the flame with the approach flow field may lead to flashback, even if the axial flow velocity *in the absence of the flame* everywhere exceeds the laminar or turbulent displacement speed. For example, Section 7.3.4.2 showed that azimuthal flow rotation leads to baroclinic vorticity generation and an induced flow velocity in the direction of flame propagation. In other words, the local flow velocity in front of the flame is different than it would be if the flame were not there. This can be seen from Figure 7.13, which shows the monotonic increase of v_F with azimuthal velocity, u_θ.

At higher swirl numbers, heat release effects alter the vortex breakdown characteristics. Following Sattelmayer, this phenononmenon is referred to as *combustion-induced vortex breakdown* [2, 6–9], shown in Figure 10.2.

As discussed in Section 4.4.2, vortex breakdown refers to the formation of a stagnation point and axially recirculating flow in a system with sufficient swirl. The subsequent low or negative velocity allows for upstream flame propagation into a flow with an axial velocity that is nominally too high for upstream propagation. The basic phenomenon is that the flame contributes to vortex breakdown (because of the adverse pressure gradient and radial divergence that the concave flame imposes on the reactants, see Section 7.5), and therefore generates a region of low or negative flow velocity ahead of it. The flame advances forward, causing the location of the vortex breakdown region to advance farther upstream into the mixing zone. This process continues as the flame proceeds farther and farther upstream. In this case, flashback occurs in the core of swirling flows even if the flame displacement speed is everywhere less than the flow velocity in the isothermal case. Moreover, increasing core flow velocity may not reduce flashback proclivity [1].

The phenomenon of combustion-induced vortex breakdown is intimately linked to the bistable nature of vortex breakdown. Consider the stability of the premixer flow to vortex breakdown. As illustrated in Figure 4.39, no breakdown occurs in flows for low swirl numbers, $S < S_A$, and only vortex breakdown states are present for high swirl numbers, $S > S_B$. As such, for low enough swirl numbers breakdown is never possible in the premixer, while for high swirl numbers it will always occur. In the latter case, the flame would always stabilize in the premixer. Thus, it seems that the combustion-induced vortex breakdown phenomenon occurs because of the existence of an intermediate hysteresis regime ($S_A < S < S_B$) in the nonreacting flow, where either flow state is possible. Nominally, no breakdown occurs in the nozzle; rather, it would occur at the combustor inlet where a rapid area expansion causes radial flow divergence, forcing vortex breakdown. However, heat release can also provide the finite-amplitude perturbation required to move the system from a nonbreakdown state to a breakdown state.

10.1.2 Boundary Layer Flashback

Clearly, the axial flow velocity in premixing sections must drop to zero at the combustor walls, which introduces the possibility of flame flashback in the boundary layer. Section 10.1.2.1 summarizes the basic processes associated with boundary layer flashback for zero heat release, stretch-insensitive flames. We then discuss heat release effects, which alter the approach flow in the boundary layer, and stretch effects in Section 10.1.2.2. This section also treats the flameholding problem, discussing the role of additional processes, such as heat release induced boundary layer separation, that occur once the flame has flashed back.

10.1.2.1 Basic Considerations Influencing Flashback Limits

Flashback in laminar boundary layers is a classical topic [10–12]. Figure 10.3 presents the physical picture put forward by these studies, which basically treat the propagation of a constant-density, stretch-insensitive front.

As shown in the figure, the flow velocity increases from zero at the wall. Similarly, the laminar flame speed varies spatially and reactions completely quench at the location δ_q from the wall because of heat losses. A leading-order description of the phenomenon, then, is that flashback occurs if the displacement speed exceeds the flow velocity at this point near δ_q,

$$u_x(y = \delta_q) = s_d^u(y = \delta_q). \tag{10.1}$$

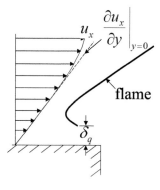

Figure 10.3 Illustration of a premixed flame in a laminar boundary layer.

Expanding u_x in a Taylor series about $y = 0$ and retaining the leading-order term leads to

$$u_x(y = \delta_q) = u_x(y = 0) + \frac{\partial u_x}{\partial y}\bigg|_{y=0} \delta_q = g_u \delta_q, \qquad (10.2)$$

where we have defined the wall velocity gradient as

$$g_u = \partial u_x/\partial y|_{y=0}. \qquad (10.3)$$

Combining Eqs. (10.1) and (10.2) leads to the flashback condition

$$\frac{g_u \delta_q}{s_d^u} = 1. \qquad (10.4)$$

Furthermore, if we assume that the quenching distance, δ_q, and flame thickness, δ_F, are proportional, we can define the flashback Karlovitz number for correlating data with:

$$Ka = \frac{g_u \delta_F}{s_d^u}. \qquad (10.5)$$

This basic idea is well validated as a suitable approach for correlating flashback limits, and a number of references have measured critical velocity gradients for different fuel types and operating conditions. To illustrate, Figure 10.4 plots measured critical velocity gradient values at flashback for methane/air flames at several preheat temperatures. It shows that the critical gradient is highest for near-stoichiometric mixtures where the flame speed peaks. It also increases with preheat temperature for the same reason.

For the same reasons, flames flash back more easily as pressure increases since the flame thickness, and therefore the quenching distance, are inversely proportional to pressure (see the scaling in Section 9.1.2) [14]. This is because a thinner flame can exist closer to the wall where flow velocities are lower. Hydrogen and hydrogen

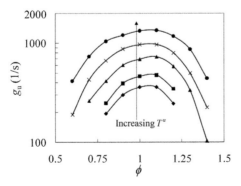

Figure 10.4 Measured dependence of the critical velocity gradient, g_u, on equivalence ratio for methane/air flames at reactant temperatures of $T^u = 300$ K (♦), 348 K (■), 423 K (▲), 473 K (×), and 523 K (•) [13].

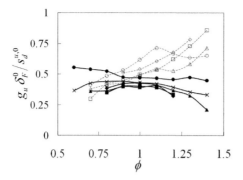

Figure 10.5 Normalized critical gradient for methane/air and propane/air blends. Flashback data obtained from Refs. [13] and [15]. Values of δ_F^0 and $s_d^{u,0}$ calculated with GRI-MECH 3.0 [16] for CH_4 and USC-MECH [17] for C_3H_8. Solid and dashed lines denote CH_4 and C_3H_8, respectively (300 K (◆), 348 K (■), 423 K (▲), 473 K (×), and 523 K (•) for CH_4; 306 K (◊), 422 K (□), and 506 K (Δ) for C_3H_8).

blends have a particularly high susceptibility to flashback, due to their high flame speeds and small quenching distances.

Figure 10.5 plots the same data shown in Figure 10.4, along with several propane/air blends, but with the critical gradient normalized to form a Karlovitz number, following Eq. (10.5). Note the good collapse of each fuel across the range of equivalence ratios and T^u values. The systematic differences between the propane and methane mixtures is likely a manifestation of the stretch sensitivity of these blends, discussed further in Section 10.1.2.2.

Flashback boundaries are also sensitive to wall temperatures and, therefore, the heat transfer characteristics of the boundaries. This has been illustrated by experimental measurements of the dependence of flashback boundaries on burners with different wall materials; one measurement showed that g_u values in a stainless steel burner were more than double those of a copper burner [18]. In this case, this variation can be directly correlated with the wall material heat transfer coefficients. The copper burner has a much lower wall temperature due to its high thermal conductivity relative to stainless steel. This point also illustrates that burners that do not flash back under steady-state operation could flash back during a transient. For example, consider the scenario where wall temperatures are proportional to thermal heat loading, i.e., reactant flow rates. A burner may not flash back at the high flow velocity, high wall temperature or low flow velocity, low wall temperature steady states. However, a sudden drop in flow rate could lead to a situation with low reactant velocities and high wall temperatures, leading to flashback.

Flashback in turbulent boundaries has the added complexity of the multizone turbulent boundary layer structure, as discussed in Section 4.6. Neglecting the influence of heat release on the boundary layer instabilities for the moment, the basic ideas embodied in Eq. (10.1) should hold. However, whether the Taylor series expansion used in Eq. (10.2) is a good approximation for $u_x(y = \delta_q)$ depends on the relative

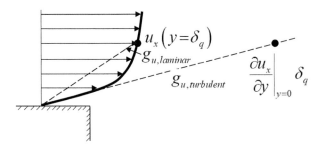

Figure 10.6 Time-averaged velocity profile in a turbulent boundary layer, illustrating the increase in the critical velocity gradient from turbulent to laminar for the same $u_x(y = \delta_q)$ value.

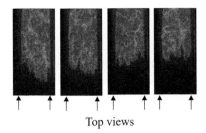

Top views

Figure 10.7 Several instantaneous images of a flame in a two-dimensional diffuser channel, viewed from the top. Image courtesy of C. Eichler [21].

thicknesses of the quenching distance and the laminar sublayer of the turbulent boundary layer, δ_v. If $\delta_q < \delta_v$, then the basic scaling described in Eq. (10.5) should hold. In contrast, if $\delta_q > \delta_v$, then $u_x(y = \delta_q) < \frac{\partial u_x}{\partial y}\big|_{y=0}\delta_q$, as shown in Figure 10.6.

While more detailed characterization of this issue is an active area of research [19], it appears that the latter scenario holds for the majority of the data in the literature as most data on critical flashback gradients in turbulent boundary layers show similar trends to laminar ones, but with $g_{u,turbulent} \sim 3 g_{u,laminar}$ [11, 14, 20]. However, this relationship should not be treated as generally applicable since, as mentioned above, flashback limits are a function of the structure of the turbulent boundary layer.

The flashback process in turbulent flows also exhibits significant space–time variation. To illustrate, Figure 10.7 shows several top views of a flame in a nominally two-dimensional diffuser channel, showing the flame "fingers" that have propagated forward the farthest. These are likely manifestations of coherent structures in the turbulent boundary layer.

10.1.2.2 Heat Release and Stretch Effects

The preceding discussion treated the burning velocity and approach flow velocity as known quantities which could be specified. In reality, both of these critical velocities are influenced by heat release and stretch effects, so that their values cannot be specified, and must be calculated as part of the problem (see also Sections 7.5 and

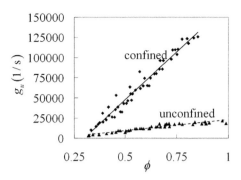

Figure 10.8 Measured dependence of the critical flashback velocity gradient on equivalence ratio for a H_2/air flame [22].

9.3.3). These effects have particular significance for the flameholding problem. This is the more practical flashback problem because, as noted in an earlier section, a designer is interested in the more severe criterion of whether a flame that is introduced into the premixing section boundary layer will be pushed out by the flow if the system is returned to normal operation. As we discuss in this section, this is a much more severe test, since it is more difficult to push a flame back out of a premixing section once it has entered it. This point can be seen from the data in Figure 10.8, which shows the measured dependence of the flashback velocity gradient for an open ("unconfined") flame, similar to the data shown in the previous section. Data are also shown for the flashback boundaries of a confined flame that is nominally stabilized by a small step inside a tube, simulating the conditions of a flame that is flashing back. Note the factor of ~5 increase in critical velocity gradient in the confined case. This illustrates that once a flame has flashed back, it is much more difficult to expel it from inside the burner.

Understanding this phenomenon requires consideration of two additional processes – flame stretch and gas expansion [23]. We discuss flame stretch first. Note that a flame stabilized in the shear layer at a burner exit will inevitably be stretched due to tangential flow nonuniformities along the sheet (the κ_s term in Eq. (9.14)). However, once it begins to flash back, it will attain very strong positive curvature, as illustrated in Figure 10.3, that scales as $(\partial u_x/\partial y)/s_d^u$ [24]. This can cause $s_d^u(y = \delta_q) \gg s_d^{u,0}$ for reactants where $Ma < 0$, such as lean H_2/air flames as shown in Figure 9.17. Therefore, once a $Ma < 0$ flame enters the channel and achieves this strong positively curved orientation, scaling its displacement speed with $s_d^{u,0}$ is an underestimate of its true displacement speed.

Second, gas expansion across the flame leads to $u_x(y = \delta_q)$ values that can differ significantly from their values when the flame is absent. As shown in Figures 7.17 and 7.18, positive flame curvature leads to flow deceleration and an adverse pressure gradient ahead of the flame. Obviously, this makes flashback easier since the flow velocity is lower. To illustrate, Figure 10.9 replots the ratio of confined and unconfined velocity gradient data shown in Figure 10.8 as a function of density ratio across the flame (bottom x-axis). Note the monotonic increase in this ratio with temperature ratio. For reference, lines proportional to $\sqrt{\sigma_\rho}$ and σ_ρ, as suggested by Figure 9.35, are also included. Note that this increase in $g_{u,confined}/g_{u,unconfined}$ is likely not only a function of σ_ρ but also Markstein length, whose calculated values are indicated along the secondary x-axis on the top of the graph.

This influence of the flame on the approach flow can be dramatically amplified in cases where the adverse pressure gradient is strong enough to cause the boundary layer to separate. In this case, a phenomenon analogous to combustion-induced vortex breakdown occurs, where the reverse flow in the separated boundary layer ahead of the flame literally "sucks" the flame back into the nozzle. The separated flow region ahead of a flame that is flashing back is clearly visible in the images shown in Figure 10.10. These images show simultaneous flame positions and flow fields for a flame propagating down the boundary layer of a centerbody in an annular, swirling flow. There are elements of both combustion-induced vortex breakdown and boundary

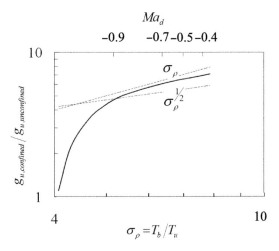

Figure 10.9 Dependence of the critical velocity gradient ratio between confined and unconfined flames on calculated flame temperature ratio for a H_2/air mixture [22]. Markstein numbers calculated with the Davis mechanism [25] using a premixed stagnation flame.

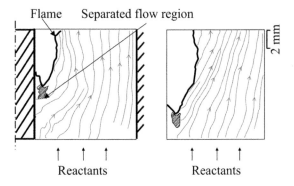

Figure 10.10 Two measured instantaneous flame positions and flowfields in the vicinity of a flame flashing back showing boundary layer separation ahead of the flame (denoted by hatched region). Adapted from Heeger et al. [4].

layer flashback in this particular event. These data illustrate the flow streamlines ahead of the flame, clearly indicating the influence of the flame on the approach flow. Also indicated by the hatched region is an area of reverse flow ahead of the flame where the boundary layer has separated.

Finally, once the flame is in contact with hot metal, back conduction of heat through the nozzle structure causes heating of the approach flow boundary layer. As shown in the context of Figure 4.57 and Exercise 4.2, boundary layer heating introduces an inflection point in the velocity profile, causing the boundary layer to thicken and possibly separate through an entirely different mechanism than discussed in the previous paragraph.

10.2 Flame Stabilization and Blowoff

Because the flow velocity in any real combustor exceeds the flame speed, special consideration is needed to determine the conditions required for flame stability. This section analyzes flame stabilization processes and the limit conditions under which no stabilization points exist. Determining the conditions where the flame cannot be stabilized and is convected downstream by the flow – referred to here as blowoff – is an important issue in any practical device [26–30] as it sets fundamental operational boundaries on the combustor. A photograph of the immediate aftermath of flame blowoff in both of an SR-71's engines is shown in Figure 10.11, where the flames can be seen some distance behind the aircraft!

In addition to blowoff, determining flame stabilization locations is also an important problem, as flames can exhibit fundamentally different shapes and lengths in

Figure 10.11 Flame blowoff in the SR-71 during a high-acceleration turn, reproduced from Campbell and Chambers [31].

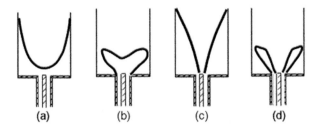

Figure 10.12 Illustration of basic flame configurations possible for an annular, swirling flow geometry.

situations where multiple stabilization points exist. For example, Figure 10.12 illustrates four possible premixed flame configurations in an annular, swirling combustor arrangement, previously discussed in Figure 4.41. In a low-velocity flow, configuration (d) is typically observed. However, the flame can exhibit one of the other three shapes if it cannot stabilize in both the inner and outer shear layers. For example, if the flame cannot anchor in the outer shear layer, configuration (d) will bifurcate to configuration (c). If it then cannot anchor in the inner shear layer, configuration (c) may bifurcate to configuration (a). These shifts in flame location can be thought of as a sequence of local blowoff events. Blowoff occurs when no stabilization locations are possible.

The location/spatial distribution of the flame in a combustion chamber is a fundamental problem that has important ramifications for combustor operability, durability, and emissions. For example, flame location has an important influence on combustion instability boundaries, as detailed by the flame response analysis in Chapter 12. Combustor stability limits are controlled by the time delay between when a fuel/air ratio disturbance or vortex is created and when it reaches the flame. This time delay is much longer for configurations (a) and (c) than (b) and (d). This also illustrates that discontinuous changes in combustor stability behavior may occur when the flame abruptly bifurcates from one stabilization location to another.

Special methods are usually required to stabilize the flame in high-velocity flows. If the flame finds a stabilization point, it can then spread downstream with a location controlled by the kinematic balance between flame and flow speed for premixed flames, or where fuel and oxidizer meet in stoichiometric proportions for nonpremixed flames. Pilot flames are often used for flame stabilization. In addition, flames are often stabilized in low-velocity, high-shear regions, such as locations of flow separation, as shown in images (a) and (b) of Figure 10.13. Alternatively, flames can stabilize in regions where the flow velocity is aerodynamically decelerated without flow separation, e.g., near the forward stagnation point of a vortex breakdown bubble, or in a rapidly diverging jet (such as a low-swirl burner) [32], such as shown in image (c) of Figure 10.13. Figure 10.13(d) shows a photograph taken through a combustor side window, where multiple annular nozzles are situated around the combustion chamber (similar to the arrangement shown in Figure 10.12). The prominent flame in the center of the window is stabilized in the shear layer of an axisymmetric bluff body.

10.2.1 Basic Effects in Premixed Flames: Kinematic Balance Between Flow and Burning Velocities [10]

Fundamentally, blowoff occurs when the upstream propagation velocity of the combustion wave is everywhere less than the flow velocity. In simple flows, this involves a balance between laminar burning velocity and the flow, or in more complex flows it involves upstream propagation speeds of edge flames or autoignition waves. To start the discussion, it is useful to summarize the classical treatment of flame stabilization that follows from consideration of relative values of the laminar displacement speed, s_d^u, and the flow velocity. To illustrate, consider Figure 10.14, which is similar to the sketch used in Figure 10.3 to illustrate boundary layer flashback. This figure sketches the radial dependence of the laminar burning velocity near the burner lip, showing that it has values of s_d^u in the core of the flow, but drops to zero within the quenching distance, δ_q, from the lip. Similarly, the flow velocity is assumed to increase linearly from zero at the wall with a velocity gradient of g_u. As also noted in Section 10.1.2.2, the presence of the flame actually alters the character of the approach flow velocity, so that this classical description is essentially an isothermal treatment of a propagating front.

The figure shows three different velocity gradients, $g_{u,a}$, $g_{u,b}$, and $g_{u,c}$. Note that the flow velocity for line a everywhere exceeds s_d^u. This flame will blow off – not due to flame extinction, but simply because the burning velocity is everywhere less than the flow velocity. For line c, there is a region where $s_d^u > u_x$ so the flame will propagate upstream. For line b, there is a single point where $s_d^u = u_x$, which acts as the stabilization point from which the flame propagates into the core. Analogous to the

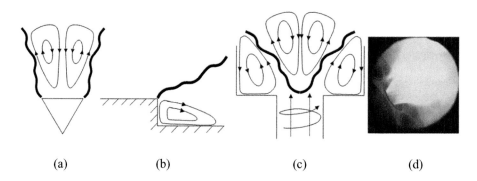

(a) (b) (c) (d)

Figure 10.13 Illustration of flames stabilized by (a) a bluff body, (b) a backward-facing step, and (c) a swirling flow. (d) The view through the side window of a multinozzle combustor showing a centerbody-stabilized flame (image courtesy of GE Energy).

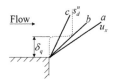

Figure 10.14 Illustration used to demonstrate blowoff concepts, showing the transverse dependence of the laminar burning velocity and flow velocity.

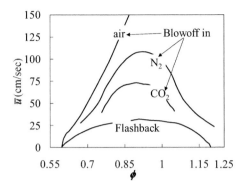

Figure 10.15 Dependence of the critical blowoff and flashback velocity of natural gas/air mixtures (exiting into the environment specified in the figure) on equivalence ratio, based on figure 92 in Lewis and von Elbe [10].

development in Section 10.1.2.1, these considerations lead to the scaling of blowoff limits with the velocity gradient or with a blowoff Karlovitz number,

$$Ka = \frac{g_u \delta_F^0}{s_d^u}. \tag{10.6}$$

Figure 10.15 illustrates typical dependencies of measured blowoff velocities, which are related to g_u, on equivalence ratio [10, 13, 15]. For reference, it also shows flashback values, indicating that g_u at flashback is less than at blowoff, as expected. The region between the flashback and blowoff curves defines the operability window of the burner where a stationary, stable flame is possible. Note that the blowoff boundaries depend on the gas composition of the surrounding environment into which the reactants discharge, for reasons to be discussed later.

The criterion that u_x must exactly equal the burning velocity and never less seems like a very special condition that will very rarely be met. Indeed, it suggests that stable flames must be a rare occurrence and that most premixed flames will either blow off or flash back! However, intrinsic stabilizing mechanisms exist so that flames find an anchoring point that adjusts itself over a range of velocities and fuel/air ratios. Consider curve $g_{u,a}$ in Figure 10.14. As the flame moves downstream, wall quenching decreases, causing the flame speed to increase in the low-velocity boundary layer. Thus, it may find a stable location further downstream. Similarly, if the flame is propagating upstream, quenching effects increase, causing the flame speed to drop near the wall. The end result is that the flame can often find a stable point over a range of conditions where the flow velocity and flame speed match.

The fact that the flame stabilizes downstream of the burner lip [11] implies that the reactants are diluted with the ambient gas – typically air for open flames and recirculated products for confined flames. This is the reason that the blowoff velocity for air increases monotonically with equivalence ratio in Figure 10.15. In other words, as a fuel-rich flame lifts off, the fuel/air ratio is decreased by mixing with the ambient air toward stoichiometric values where the flame speed is higher. In contrast, the rich side

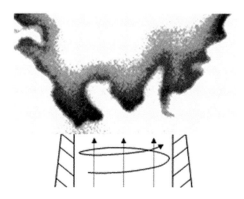

Figure 10.16 OH PLIF image of a flame stabilized in the low-velocity core region in a low-swirl burner. Image reproduced from Petersson et al. [34].

mirrors the flashback curve if the ambient does not contain O_2, such as when the ambient environment is N_2 or CO_2. In this case, entrainment of ambient gases has an equally adverse impact on burning velocities, whether the flame is lean or rich. Moreover, CO_2 has a higher specific heat than N_2 and an inhibiting impact on chemistry, which is why Figure 10.15 shows that the flame in a CO_2 ambient is the easiest to blow off of the different gases shown.

A related mechanism also controls stabilization of nonpremixed flames. As discussed in Section 9.7.3, high Damköhler number nonpremixed flames are preceded by lean and rich premixed wings. These premixed flames propagate into the unburned mixture at the relevant burning velocity. Moreover, as shown in Figure 9.35, the curved triple flame causes the approach flow to decelerate through gas expansion effects, which allows the flame to stabilize at a value of the upstream flow velocity that is higher than the burning velocity.

While the preceding introductory comments focused on the laminar burning velocity, the question of whether the laminar or turbulent displacement velocity, or edge flame velocity, is more appropriate for scaling blowoff is an important one. In shear layer stabilized flames, the flame attachment point is generally fixed in space and so flame wrinkling is minimal. Thus, although the flame may become significantly wrinkled farther downstream, a stretched laminar burning velocity or edge velocity scaling is most appropriate near the attachment point. In contrast, flames stabilized away from a fixed spatial region and in a low-velocity region in the jet core, such as shown in Figure 10.16, are highly wrinkled and move around substantially. In these cases, the turbulent displacement speed is the appropriate velocity scale. In fact, this configuration is a recommended approach for measurements of turbulent displacement speeds [33].

10.2.2 Stretch Rates for Shear Layer Stabilized Flames

Flames stabilized in shear layers reside in regions of very strong flow gradients and are highly stretched. In this section we analyze these stretch rate characteristics. Section

10.2 Flame Stabilization and Blowoff

10.2.1 illustrated the basic principles of flame stabilization, which fundamentally come down to a kinematic balance between a flame velocity (be it a laminar or turbulent burning velocity, or edge flame velocity) and the flow velocity. It also showed that quenching of the flame was key to understanding flame stabilization in shear layers, as there always exist regions of low flow velocities, due to the flow velocity becoming zero at the wall. However, in high-velocity flows, the critical quenching phenomenon may not be due to wall quenching, or at least not exclusively. Rather, the high rates of flow shear can lead to significant stretching of the flame [35]. For example, the time-averaged shearing rate component of the flow strain tensor, $\underline{\underline{S}}$, in a 50 m/s flow with a 1 mm boundary layer is of the order of $du_z/dx \approx 50\,000$ s^{-1}. This is a very large number relative to typical premixed flame extinction stretch rates, such as shown in Figure 9.20.

However, flow strain, $\underline{\underline{S}}$, is not the same as flame stretch, κ. Moreover, the rollup of the shear layer through the Kelvin–Helmholtz instability can lead to curvature-induced flame stretch, as shown in Figure 10.17. This figure illustrates three realizations of the instantaneous vorticity and position of a flame stabilized in the shear layers of an annular gap.

It is useful to revisit Eqs. (9.12) and (9.14) in order to understand how flame stretch, κ, is related to flow strain, $\underline{\underline{S}}$, in a shear layer. The coordinate system used for this discussion is shown in Figure 10.18, which illustrates a flame anchored on a centerbody.

Assuming a two-dimensional, steady flame and an incompressible flow upstream of the flame sheet, the hydrodynamic flame stretch term may be written as

$$\kappa_s = -n_x^2 \frac{\partial u_x^u}{\partial x} - n_z^2 \frac{\partial u_z^u}{\partial z} - n_x n_z \frac{\partial u_x^u}{\partial z} - n_z n_x \frac{\partial u_z^u}{\partial x}, \qquad (10.7)$$

or, using the continuity equation and grouping terms,

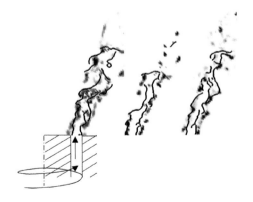

Figure 10.17 Three instantaneous flame fronts and isovorticity contours in a flame stabilized in an annular swirl flow, showing flame stabilization in high-shear regions and rollup of the shear layers into concentrated regions of high vorticity [36]. Image courtesy of Q. Zhang.

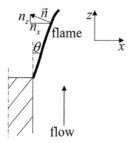

Figure 10.18 Flow and flame coordinate system for a centerbody-stabilized, two-dimensional flame.

(a) Shearing flow velocity profile

(b) Decelerating flow velocity profile

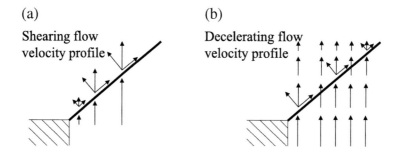

Figure 10.19 Illustration of the manner in which (a) flow shearing and (b) flow deceleration cause flame stretch through variation in the tangential flow velocity along the flame sheet.

$$\kappa_s = \left(n_x^2 - n_z^2\right)\frac{\partial u_z^u}{\partial z} - n_x n_z \left(\frac{\partial u_x^u}{\partial z} + \frac{\partial u_z^u}{\partial x}\right). \tag{10.8}$$

From Eq. (10.8), we can explicitly identify the flame stretch terms arising from normal (symmetric) and shear (antisymmetric) flow strain as

$$\kappa_{s,normal} = \left(n_x^2 - n_z^2\right)\frac{\partial u_z^u}{\partial z}, \tag{10.9}$$

$$\kappa_{s,shear} = -n_x n_z \left(\frac{\partial u_x^u}{\partial z} + \frac{\partial u_z^u}{\partial x}\right). \tag{10.10}$$

The $\kappa_{s,shear}$ term quantifies the manner in which shearing flow strain translates into flame stretch. Similarly, the $\kappa_{s,normal}$ term describes the impact of flow acceleration and deceleration on flame stretch. Flow deceleration typically occurs simultaneously with shear in flame stabilization regions, as the approach flow boundary layer separates into an increase in cross-sectional flow area. The manner in which these normal and shear flow strain terms translate into flame stretch is illustrated in Figure 10.19. In both cases, flame stretch occurs due to variations in tangential flow velocity along the flame sheet.

Figure 10.19(a) depicts the effect of shear strain (assumed to be dominated by shearing in the separating boundary layer). This image shows that shearing flow strain leads to positive stretching of the flame. This occurs because the tangential flow velocity increases along the flame. Figure 10.19(b) shows the effect of normal strain, assuming that the axial flow is decelerating. This image shows that compressive normal strain leads to negative stretching, or compression, of the flame. This occurs because the tangential flow component along the flame decreases along the flame. The opposite would happen if the bulk flow were accelerating, such as due to gas expansion in a confined channel.

We next analyze the two flame stretch terms in more detail, starting with the shear contribution, given by Eq. (10.10). To analyze this term, assume that $\partial u_x^u/\partial z \ll \partial u_z^u/\partial x$ and that $\theta \ll 1$, then rewrite Eq. (10.8), neglecting terms of $O(\theta^2)$ or higher:

$$\kappa_{s,shear} \approx \theta \frac{\partial u_z^u}{\partial x}. \tag{10.11}$$

Thus, this equation shows that shearing flow strain and flame stretch are related by the flame angle, θ. The angle in which the flame spreads from the stabilization/ignition point is given by

$$\theta \approx \frac{s_d^u}{u_z^u}. \tag{10.12}$$

Combining these expressions, we obtain

$$\kappa_{s,shear} \approx \frac{s_d^u}{u_z^u} \frac{\partial u_z^u}{\partial x}. \tag{10.13}$$

This expression suggests that flame stretch introduced by flow shear scales as

$$\kappa_{s,shear} \sim \frac{s_d^u}{u_z^u} \frac{u_z^u}{\delta} = \frac{s_d^u}{\delta}, \tag{10.14}$$

where δ is the shear layer thickness. This equation is quite significant and also not intuitive as it suggests that the flame stretch rate due to shear is independent of flow velocity! The reason for this flow velocity independence follows from two cancelling effects in Eq. (10.11). Note, first, that if $\theta = 0$ then the flame is not stretched by shear. In practical situations, θ is small but nonzero and a function of the flow velocity. Increases in flow velocity increase the flow shearing term, $\partial u_z^u/\partial x$, but decrease the flame angle, θ. These two effects cancel each other, leading to Eq. (10.14). In reality, flow velocity does exert an influence through the shear layer thickness, δ. For example, for a laminar boundary layer, $\delta \propto L/Re_L^{1/2} = (\nu_\nu L/u_z^u)^{1/2}$, where L is a characteristic geometric dimension, implying that

$$\kappa_{s,shear} \sim s_d^u (u_z^u/\nu_\nu L)^{1/2}. \tag{10.15}$$

We can also write a scaling law for the corresponding Karlovitz number associated with shear, showing that it is proportional to the relative thicknesses of the flame, δ_F, and shear layer, δ:

$$Ka_{s,shear} \sim \frac{s_d^u}{\delta} \frac{\delta_F}{s_d^u} = \frac{\delta_F}{\delta}. \qquad (10.16)$$

We next turn to scaling flame stretch from the normal strain term, shown in Eq. (10.9). Again, assuming that θ is small leads to the conclusion that $(n_x^2 - n_z^2) = 1 - O(\theta^2)$, so that

$$\kappa_{s,deceleration} = \frac{\partial u_z^u}{\partial z} \sim \frac{u_z^u}{L}, \qquad (10.17)$$

where L is a characteristic combustor geometric dimension. Comparing Eqs. (10.17) and (10.15) suggests that two fundamentally different flame stretch rate scalings exist. We next turn to the question of which contribution to flame stretch is dominant in shear layers. Using the small-θ approximation, the full expression for κ_s is

$$\kappa_s \approx \underbrace{\frac{\partial u_z^u}{\partial z}}_{\text{Axial Flow Deceleration}} + \underbrace{\theta \frac{\partial u_z^u}{\partial x}}_{\text{Transverse Flow Shearing}}. \qquad (10.18)$$

In the shear layer, we can expect both velocity gradient terms in this equation to be nonzero, and for the shear gradient to be much larger than the axial flow deceleration, i.e., $\partial u_z^u/\partial x \gg \partial u_z^u/\partial z$. However, the much larger shearing term is multiplied by the small flame angle, θ. Thus, it can be seen that neither term obviously dominates, indicating that the near-field flame stretch could be either positive or negative. Moreover, κ_s is a sum of two numbers of opposite sign, but potentially similar magnitudes. This implies that relatively small changes in flame angle, axial flow deceleration, or shear layer characteristics could fundamentally alter the value of the local stretch from positive to negative values, or vice versa. For example, measurements of these two terms at the centerbody of an annular swirling flow found that the flow deceleration term was dominant for that flow [37]. However, even though the axial flow deceleration is particularly strong in swirling flows, this conclusion may not hold for other separating flows.

The flame will locally extinguish in the low-velocity stabilization region if the flame stretch rate exceeds κ_{ext}. As such, these results have important implications on blowoff scaling of shear layer stabilized flames,[2] such as the sensitivity of blowoff limits to velocity. In order to illustrate measured velocity sensitivities of the calculated extinction stretch rate at blowoff, Figure 10.20 reproduces several data sets from Foley et al.'s [38] compilation. As expected, the figures show that increasing flow velocities shift blowoff conditions to points associated with higher extinction stretch rates. Blowoff scaling is discussed further in the next section.

10.2.3 Product Recirculation Effects on Flame Stabilization and Blowoff

The preceding sections discussed basic blowoff concepts by considering first a purely kinematic balance between flame propagation and flow velocity, and then adding

---

[2] However, as emphasized in the next section, blowoff and local extinction are distinct processes.

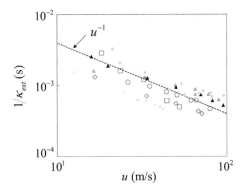

Figure 10.20 Dependence of computed extinction stretch rate at blowoff on flow velocity for several C_3H_8-fueled, bluff body data sets [39–43]. Chemical times computed using the San Diego mechanism [44].

concepts of stretch-induced flame extinction. Many flames in high-velocity flows are supported by recirculating hot gases, which adds additional dynamics to the problem and makes blowoff a "process" and not a discontinuous event [45–47].

The precise mechanisms through which product recirculation promotes flame stabilization is a topic of active research. First, product recirculation transfers heat back to metal conductive surfaces, which reduces heat losses from flame edges and causes boundary layer heating (see also the discussion in the context of Figure 4.57). For example, after a cold combustor is lit, it is known that blowoff boundaries gradually retract as the combustor bulkhead heats up [48]. The larger effect, however, likely comes from reactant/product mixing, which occurs where holes/edges in the flame exist – this can occur at the base of the flame where reactions are quenched near the wall, or farther downstream across holes in the flame sheet. Both the flame speed and extinction stretch rate increase substantially with hot product dilution, as discussed in the context of Figure 9.40.

Furthermore, at high dilution/preheating levels, the flame does not "extinguish." Note that increases in reactant temperature are equivalent to a reduction in dimensionless activation energy, $\tilde{E} = E_a/\mathscr{R}_u T^u$. For example, the "S curve" is stretched out and does not have an ignition/extinction character in Figure 8.8 for $\tilde{E} < 8$ values (this $\tilde{E}$ value is derived in Exercise 8.2). Similarly, calculations and measurements in stretched flames show that flames do not extinguish above a certain reactant temperature [49]. To illustrate, Figure 10.21 plots the dependence of the calculated burning velocities of a reactant stream stagnated into hot product gases, the configuration shown in Figure 8.2(c). This is an approximate model for the flame processes in the high-shear region of a recirculation zone supported flame.

For this example, the consumption and displacement speeds both abruptly drop to zero when the stretch rate exceeds $\kappa_{ext} = 176$ s^{-1} when the stagnation gas temperature is 1350 K. However, for larger counterflow gas temperatures, the flame does not extinguish. Rather, the consumption speed monotonically drops with increasing stretch

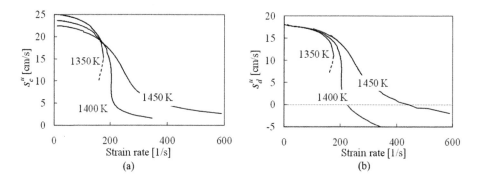

Figure 10.21 Calculated dependence of the (a) consumption and (b) displacement speeds on the stretch rate of a CH_4/air flame stagnating against hot products, whose temperature is indicated on the plot ($\phi = 0.6$, $T^u = 366$ K). Consumption speed determined from Eq. (9.21). Displacement speed equals the flow velocity at the location of 5% of maximum CH_4 consumption, multiplied by the ratio of T^u and the temperature at that point.

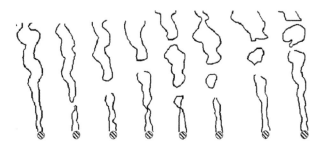

Figure 10.22 Sequence of experimental flame images, 10 ms apart, taken at $\phi/\phi_{LBO} = 1.10$, showing the presence of flame holes. Image courtesy of S. Nair [50].

rate. Moreover, at high stretch rates the reaction zone moves into the hot product side and exhibits a negative displacement speed, as illustrated in Figure 10.21(b). For these reasons, blowoff of recirculation zone stabilized flames cannot be understood from consideration of flame processes based on reactant properties only. Furthermore, blowoff probably cannot be understood from a simple consideration of flame propagation and stretch-induced extinction at the stabilization point.

We will focus our discussion on bluff body stabilized flames in the rest of this section, as this configuration has been analyzed most extensively [45–47]. Empirically, it is observed that blowoff is preceded by the development of local extinction on the flame sheet downstream of the stabilization zone, manifested as holes in the flame sheet, as shown in Figures 10.22 and 9.28(c). Figure 10.22 illustrates a flame near blowoff, with a sequence of images showing the initiation of a hole in the flame and its convection downstream. In this experiment, such holes were observed once the equivalence ratio approached about 10% of the blowoff value.

These holes are apparently initiated in high-stretch regions. However, holes are not observed near the bluff body, where the stretch rate is highest! This is likely

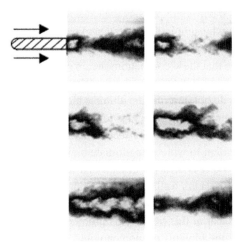

Figure 10.23 Instantaneous line-of-sight images of near-blowoff flames (not sequential), showing the presence of reaction near a bluff body and extinction farther downstream. Images courtesy of D. Noble.

a reflection of the fact that at the stabilization point, consisting of reactants diluted with hot products, the flame can withstand these high stretch levels but farther downstream it cannot. For these reasons, near-blowoff flames often do not extinguish near the stabilization point, but downstream of it, as shown in Figures 10.23 and 9.28(b). These images both show reaction occurring immediately downstream of the bluff body, but quenching of these reactions farther downstream. The downstream location where extinction occurs moves monotonically upstream toward the bluff body as blowoff is approached, such as by decreasing fuel/air ratio. This is the opposite behavior observed for nonpremixed jet flames, whose liftoff height monotonically increases as the flame approaches blowoff (see Section 10.2.4).

Extinction events occur near blowoff, and grow in frequency of occurrence as blowoff is approached, but do not immediately lead to blowoff [47]. The flame can apparently persist indefinitely and is simply unsteady under these conditions. This has been referred to as "stage 1" of the near-blowoff flame dynamics, where the overall position and qualitative dynamics of the flame, in the instances where it is continuous, appear essentially the same as those of the flame under stable conditions [47]. This first stage is distinct from blowoff; blowoff and local extinction must not be confused.

The actual blowoff event is a complex phenomenon that not only involves local stretch-induced extinction, but also wake/recirculation zone feedback and probably multiple chemical kinetic time scales involving stretch-induced extinction, ignition, and flame propagation. The blowoff event is preceded by changes of the wake zone dynamics, referred to as "stage 2" of the blowoff process [47]. These large-scale changes in the recirculating flow dynamics reflect significant alteration of the wake structure because of entrainment of reactants into the wake followed by ignition.

Given that blowoff involves much more physics than simply local stretch-induced extinction, a natural question is why simple Karlovitz/Damköhler number scalings do a reasonable job in correlating blowoff boundaries, such as shown in Figure 10.20. The reason for this appears to be that these correlations are, in essence, correlations for the first pre-blowoff stage where holes in the flame occur. That is, from a fundamental point of view, such correlations do not describe the ultimate blowoff physics itself, but rather the condition at which flame extinction begins to occur. As such, the ability of Karlovitz/Damköhler number correlations to describe/predict the actual blowoff condition is directly linked to the extent to which the ultimate blowoff event is correlated with these extinction events [47]. It is clear from data that the flame can withstand some extinction, but that blowoff occurs well before the majority of the flame sheet is extinguished. However, determination of what constitutes this critical level of flamelet disruption and extinction is unclear at this point. This problem, which remains to be solved, is key to understanding the blowoff phenomenology.

10.2.4 Nonpremixed Flame Liftoff and Blowoff

This section considers the blowoff problem for nonpremixed flames. The distinction between extinction and blowoff, emphasized in earlier sections, is equally significant in nonpremixed jet flames. Consider an experiment where a fuel jet discharges into ambient air at some velocity, $\bar{u}_x$. We can define two limiting velocities, $\bar{u}_{x,L}$ and $\bar{u}_{x,B}$. For $\bar{u}_x < \bar{u}_{x,L}$, the flame anchors on the burner lip; for $\bar{u}_{x,L} < \bar{u}_x < \bar{u}_{x,B}$, the flame lifts off from the burner with an average height H_{lift}; and for $\bar{u}_x > \bar{u}_{x,B}$, the flame blows off. There is hysteresis in these flame liftoff velocities [51]. The liftoff height, H_{lift}, is empirically known to scale as [52–54]:

$$H_{lift} \propto \frac{\bar{u}_x}{(\max(s^{u,0}))^2}, \tag{10.19}$$

where $\max(s^{u,0})$ denotes the maximum laminar flame speed for a given reactive mixture. The liftoff height is empirically observed to be independent of burner diameter. To illustrate, Figure 10.24 reproduces methane jet liftoff heights in still air.

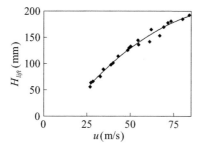

Figure 10.24 Measured dependence of the flame liftoff height, H_{lift}, on flow velocity for a methane jet flame. Burner diameters vary from 4 to 10 mm. Reproduced from Kalghatgi [52].

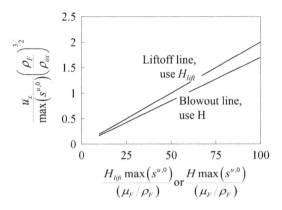

Figure 10.25 Correlation of blowoff velocity and liftoff height for a range of hydrocarbon fuels, following Kalghatgi [52, 55].

The blowoff velocity has a similar relationship, given by

$$\bar{u}_{x,B} \propto H\left(\max\left(s^{u,0}\right)\right)^2, \tag{10.20}$$

where H is the empirically determined distance from the burner exit to the location at which the average fuel/air ratio is stoichiometric along the burner axis [55]. These correlations have additional dependencies on the molecular transport coefficients and ratio of oxidizer to fuel density, as shown in Figure 10.25, which summarizes Kalghatgi's correlations for liftoff and blowoff, obtained for a range of hydrocarbon fuels [52, 55]. Putting the liftoff height and blowoff correlations together shows that blowoff occurs when the liftoff height $H_{lift} \sim 0.7H$ [52].

Significant questions still remain on exactly what processes control the flame liftoff height and blowoff conditions. The initial liftoff of the flame clearly involves non-premixed flame extinction processes, and appears to be associated with scalar dissipation rates that exceed the extinction value (see Section 9.6). However, once the flame lifts off, some premixing occurs upstream of the flame, so that the flame leading edge can propagate into the premixed reactants at the local flame speed. For example, measurements at the instantaneous flame base indicate that the fuel/air ratio lies within the flammability limit and that the local scalar dissipation rate falls well below the extinction value, indicating that the flame leading edge is not controlled by nonpremixed flame extinction phenomena [56]. An approximate criterion, then, is that the flame liftoff height corresponds to the point of correspondence between local axial flow velocity and displacement flame speed. Similarly, blowoff occurs when the maximum laminar or turbulent flame speed at a given axial location falls below the corresponding axial flow velocity.

A more complete picture of the intrinsically dynamic character of the leading edge of lifted nonpremixed flames involves both premixed and nonpremixed flame concepts, large-scale structures that arise from the underlying instability of the jet (see Section 4.3) [57], and entrainment of hot combustion products [51, 58]. For example, as the leading edge of the flame bounces axially back and forth, it evolves from a propagating triple flame farther downstream to a nonpremixed flame closer to the jet exit [51, 59]. Large-scale structures (as well as strain rate/scalar dissipation rate) cause the flow field that this leading edge experiences to be spatiotemporally unsteady. In addition, these structures cause the premixed reactant composition/temperature to be altered by back-mixing of hot combustion products. As such, these dynamics involve elements of premixed flame propagation, extinction of the nonpremixed flames as the leading edge moves into high scalar dissipation rate regions (e.g., vortex braids), and possibly ignition of reactants by entrainment of hot products.

Finally, we note that an important generalization of the above results is the effect of co-flow both along and perpendicular to (i.e., a jet in cross flow) the jet axis. Even low levels of co-flow can exert nonnegligible influences on blowoff velocities [51, 60]. Liftoff of reacting jets in cross flow was also discussed in Section 4.3.4.2.

References

1. Kröner M., Fritz J., and Sattelmayer T., Flashback limits for combustion induced vortex breakdown in a swirl burner. *Journal of Engineering for Gas Turbines and Power*, 2003, **125**: p. 693.
2. Kröner M., Sattelmayer T., Fritz J., Kiesewetter F., and Hirsch C., Flame propagation in swirling flows: Effect of local extinction on the combustion induced vortex breakdown. *Combustion Science and Technology*, 2007, **179**(7): pp. 1385–1416.
3. Lieuwen T., McDonell V., Santavicca D., and Sattelmayer T., Burner development and operability issues associated with steady flowing syngas fired combustors. *Combustion Science and Technology*, 2008, **180**(6): pp. 1169–1192.
4. Heeger C., Gordon R., Tummers M., Sattelmayer T., and Dreizler A., Experimental analysis of flashback in lean premixed swirling flames: Upstream flame propagation. *Experiments in Fluids*, 2010, **49**: pp. 1–11.
5. Thibaut D. and Candel S., Numerical study of unsteady turbulent premixed combustion: Application to flashback simulation. *Combustion and Flame*, 1998, **113**(1–2): pp. 53–65.
6. Tangermann E., Pfitzner M., Konle M., and Sattelmayer T., Large-eddy simulation and experimental observation of combustion-induced vortex breakdown. *Combustion Science and Technology*, 2010, **182**(4): pp. 505–516.
7. Kiesewetter F., Konle M., and Sattelmayer T., Analysis of combustion induced vortex breakdown driven flame flashback in a premix burner with cylindrical mixing zone. *Journal of Engineering for Gas Turbines and Power*, 2007, **129**: p. 929.
8. Nauert A., Petersson P., Linne M., and Dreizler A., Experimental analysis of flashback in lean premixed swirling flames: Conditions close to flashback. *Experiments in Fluids*, 2007, **43**(1): pp. 89–100.
9. Blesinger G., Koch R., and Bauer H., Influence of flow field scaling on flashback of swirl flames. *Experimental Thermal and Fluid Science*, 2010, **34**(3): pp. 290–298.
10. Lewis B. and von Elbe G., *Combustion, Flames and Explosions of Gases*. 3rd ed. 1987, Academic Press.
11. Wohl K. Quenching, flash-back, blow-off-theory and experiment. *Symposium (International) on Combustion*, 1952, **4**(1): pp. 68–89.

12. Putnam A.A. and Jensen R.A. Application of dimensionless numbers to flash-back and other combustion phenomena. *Symposium on Combustion, Flame, and Explosion Phenomena*. 1948, **3**(1): pp. 89–98.
13. Grumer J. and Harris M.E., Temperature dependence of stability limits of burner flames. *Industrial and Engineering Chemistry*, 1954, **46**(11): pp. 2424–2430.
14. Fine B., The flashback of laminar and turbulent burner flames at reduced pressure. *Combustion and Flame*, 1958, **2**(3): pp. 253–266.
15. Dugger G.L., Flame stability of preheated propane–air mixtures. *Industrial and Engineering Chemistry*, 1955, **47**(1): pp. 109–114.
16. Smith G.P., Golden D.M., Frenklach M., Moriarty N.W., Eiteneer B., Goldenberg M., Bowman T., Hanson R.K., Song S., Gardiner W.C.J., Lissianski V.V., and Qin Z., GRI-Mech. Available from: http://combustion.berkeley.edu/gri-mech/.
17. Wang H., You X., Joshi A.V., Davis S.G., Laskin A., Egolfopoulos F., and Law C., *High-Temperature Combustion Reaction Model of $H_2/CO/C_1$–C_4 Compounds*. USC Mech Version II. 2007; Available from: http://ignis.usc.edu/Mechanisms/USC-Mech%20II/USC_Mech%20II.htm.
18. Bollinger L.E. and Edse R., Effect of burner-tip temperature on flash back of turbulent hydrogen–oxygen flames. *Industrial and Engineering Chemistry*, 1956, **48**(4): pp. 802–807.
19. Eichler C. and Sattelmayer T., Experiments on flame flashback in a quasi-2D turbulent wall boundary layer for premixed methane–hydrogen–air mixtures. *Journal of Engineering for Gas Turbines and Power*, 2011, **133**: p. 011503.
20. Khitrin L., Moin P., Smirnov D., and Shevchuk V., Peculiarities of laminar-and turbulent-flame flashbacks. *Symposium (International) on Combustion*, 1965, **10**(1): 1285–1291.
21. Eichler C. and Sattelmayer T., Premixed flame flashback in wall boundary layers studied by long-distance micro-PIV. *Experiments in Fluids*, 2012, **52**: pp. 347–360.
22. Eichler C., Baumgartner G., and Sattelmayer T., Experimental investigation of turbulent boundary layer flashback limits for premixed hydrogen–air flames confined in ducts, in *ASME Turbo Expo*, 2011. Vancouver, Canada.
23. Lee S.T. and T'ien J., A numerical analysis of flame flashback in a premixed laminar system. *Combustion and Flame*, 1982, **48**: pp. 273–285.
24. Kurdyumov V., Fernandez E., and Lián A., Flame flashback and propagation of premixed flames near a wall. *Proceedings of the Combustion Institute*, 2000, **28**(2): pp. 1883–1889.
25. Davis S.G., Joshi A.V., Wang H., and Egolfopoulos F., An optimized kinetic model of H_2/CO combustion. *Proceedings of the Combustion Institute*, 2005, **30**(1): pp. 1283–1292.
26. Blust J., Ballal D., and Sturgess G., Fuel effects on lean blowout and emissions from a well-stirred reactor. *Journal of Propulsion and Power*, 1999, **15**(2): pp. 216–223.
27. Stoehr M., Boxx I., Carter C., and Meier W., Dynamics of lean blowout of a swirl-stabilized flame in a gas turbine model combustor. *Proceedings of the Combustion Institute*, 2011, **33**(2): pp. 2953–2960.
28. Kim W.W., Lienau J.J., Van Slooten P.R., Colket III M.B., Malecki R.E., and Syed S., Towards modeling lean blow out in gas turbine flameholder applications. *Journal of Engineering for Gas Turbines and Power*, 2006, **128**(1): pp. 40–48.
29. Eggenspieler G. and Menon S., Combustion and emission modelling near lean blow-out in a gas turbine engine. *Progress in Computational Fluid Dynamics, An International Journal*, 2005, **5**(6): pp. 281–297.

30. Katta V., Forlines R., Roquemore W., Anderson W., Zelina J., Gord J., Stouffer S., and Roy S., Experimental and computational study on partially premixed flames in a centerbody burner. *Combustion and Flame*, 2011, **158**(3): pp. 511–524.
31. Campbell J. and Chambers J., *Patterns in the Sky: Natural Visualization of Aircraft Flow Fields*. 1994, National Aeronautics and Space Administration.
32. Chan C., Lau K., Chin W., and Cheng R., Freely propagating open premixed turbulent flames stabilized by swirl. *Symposium (International) on Combustion*. 1992, **24**(1): 511–518.
33. Shepherd I. and Cheng R., The burning rate of premixed flames in moderate and intense turbulence. *Combustion and Flame*, 2001, **127**(3): pp. 2066–2075.
34. Petersson P., Olofsson J., Brackman C., Seyfried H., Zetterberg J., Richter M., Aldén M., Linne M.A., Cheng R.K., and Nauert A., Simultaneous PIV/OH-PLIF, Rayleigh thermometry/OH-PLIF and stereo PIV measurements in a low-swirl flame. *Applied Optics*, 2007, **46**(19): pp. 3928–3936.
35. Dunn M.J., Masri A.R., Bilger R.W., and Barlow R.S., Finite rate chemistry effects in highly sheared turbulent premixed flames. *Flow, Turbulence and Combustion*, 2010, **85**(3–4): pp. 621–648.
36. Zhang Q., Sanbhogue S.J., and Lieuwen T., Dynamics of premixed H_2/CH_4 flames under near blowoff conditions. *Journal of Engineering for Gas Turbines and Power*, 2010, **132** (11): pp. 111502(1)–111502(8).
37. Zhang Q., Shanbhogue S.J., Lieuwen T., and O Connor J., Strain characteristics near the flame attachment point in a swirling flow. *Combustion Science and Technology*, 2011, **183** (7): pp. 665–685.
38. Foley C.W., Seitzman J., and Lieuwen T. Analysis and scalings of blowoff limits of 2D and axisymmetric bluff body stabilized flames, in *ASME Turbo Expo*, 2012, Copenhagen, Denmark. Paper # GT2012-70048, 2012.
39. Fetting F., Choudhury A., and Wilhelm R., Turbulent flame blow-off stability, effect of auxiliary gas addition into separation zone. *Symposium (International) on Combustion*, 1958, **7**(1): pp. 621–632.
40. Filippi F. and Fabbrovich-Mazza L., Control of bluff-body flameholder stability limits. *Symposium (International) on Combustion*, 1961, **8**(1): pp. 956–963.
41. Löblich K.R., Semitheoretical consideration on scaling laws in flame stabilization. *Symposium (International) on Combustion*, 1963, **9**(1): pp. 949–957.
42. Williams G.C., Basic studies on flame stabilization. *Journal of the Aeronautical Sciences*, 1949, **16**(12): pp. 714–722.
43. DeZubay E., Characteristics of disk-controlled flame. *Aero Digest*, 1950, **61**(1): pp. 54–56.
44. Mechanical and Aerospace Engineering (Combustion Research), *Chemical-Kinetic Mechanisms for Combustion Applications*, San Diego Mechanism web page 2012. Available from: https://web.eng.ucsd.edu/mae/groups/combustion/mechanism.html/.
45. Chaudhuri S., Kostka S., Renfro M.W., and Cetegen B.M., Blowoff dynamics of bluff body stabilized turbulent premixed flames. *Combustion and Flame*, 2010, **157**(4): pp. 790–802.
46. Chaudhuri S., Kostka S., Tuttle S.G., Renfro M.W., and Cetegen B.M., Blowoff mechanism of two dimensional bluff-body stabilized turbulent premixed flames in a prototypical combustor. *Combustion and Flame*, 2011, **158**: pp. 1358–1371.
47. Shanbhogue S.J., Husain S., and Lieuwen T., Lean blowoff of bluff body stabilized flames: scaling and dynamics. *Progress in Energy and Combustion Science*, 2009, **35**(1): pp. 98–120.

48. Chterev I., Foley C.W., Noble D.R., Ochs B.A., Seitzman J.M., and Lieuwen T.C., Shear layer flame stabilization sensitivities in a swirling flow, in *Proceedings of ASME Turbo Expo*, 2012. Paper # GT2012–68513.
49. Mastorakos E., Taylor A.M.K.P., and Whitelaw J.H., Extinction of turbulent counterflow flames with reactants diluted by hot products. *Combustion and Flame*, 1995, **102**: pp. 101–114.
50. Nair S., and Lieuwen, T., Near-blowoff dynamics of a bluff-body stabilized flame. *Journal of Propulsion and Power*, 2007, **23**(2): pp. 421–427.
51. Lyons K.M., Toward an understanding of the stabilization mechanisms of lifted turbulent jet flames: experiments. *Progress in Energy and Combustion Science*, 2007, **33**(2): pp. 211–231.
52. Kalghatgi G.T., Lift-off heights and visible lengths of vertical turbulent jet diffusion flames in still air. *Combustion Science and Technology*, 1984, **41**(1): pp. 17–29.
53. Vanquickenborne L. and Van Tiggelen A., The stabilization mechanism of lifted diffusion flames. *Combustion and Flame*, 1966, **10**(1): pp. 59–69.
54. Pitts W.M. Assessment of theories for the behavior and blowout of lifted turbulent jet diffusion flames. *Symposium (International) on Combustion*, 1989, **22**(1): 809–816.
55. Kalghatgi G.T., Blow-out stability of gaseous jet diffusion flames. Part I: In still air. *Combustion Science and Technology*, 1981, **26**(5): pp. 233–239.
56. Schefer R., Namazian M., and Kelly J., Stabilization of lifted turbulent-jet flames. *Combustion and Flame*, 1994, **99**(1): pp. 75–86.
57. Dahm W.J.A. and Mayman A.G., Blowout limits of turbulent jet diffusion flames for arbitrary source conditions. *AIAA Journal*, 1990, **28**, pp. 1157–1162.
58. Chao Y.C., Chang Y.L., Wu C.Y., and Cheng T.S., An experimental investigation of the blowout process of a jet flame. *Proceedings of the Combustion Institute*, 2000, **28**(1): pp. 335–342.
59. Kelman J., Eltobaji A., and Masri A., Laser imaging in the stabilisation region of turbulent lifted flames. *Combustion Science and Technology*, 1998, **135**(1): pp. 117–134.
60. Kalghatgi G.T., Blow-out stability of gaseous jet diffusion flames. Part II: Effect of cross wind. *Combustion Science and Technology*, 1981, **26**(5): pp. 241–244.

11 Forced Response I: Flamelet Dynamics

The final two chapters treat the response of flames to forced disturbances, both time-harmonic and random. This chapter focuses on local flame dynamics; i.e., on characterizing the local space–time fluctuations in flame position. Chapter 12 treats the resulting heat release induced by disturbances, as well as sound generation by heat release fluctuations. These two chapters particularly stress the time-harmonic problem, with more limited coverage of flames excited by stochastic disturbances. This latter problem is essentially the focus of turbulent combustion studies, a topic which is the focus of dedicated treatments [1–3].

These unsteady flame–flow interactions involve kinetic, fluid mechanic, and acoustic processes over a large range of scales. Fundamentally different physical processes may dominate in different regions of the relevant parameter space, depending on the relative magnitudes of various temporal/spatial scales. Section 11.1 starts the chapter by reviewing the key length and time scales involved with flame–flow interactions. Then, Sections 11.2 and 11.3 analyze premixed and nonpremixed flame dynamics, respectively.

11.1 Overview of Length/Time Scales

A variety of length and time scales exist in unsteady combustor flow fields. First, the broadband turbulent fluctuations consist of a continuum of length scales, ranging from the integral length scale, L_{11}, down to the Kolmogorov length scale, L_η, where the two are related by the scaling $L_{11}/L_\eta \sim Re_L^{3/4}$ in fully developed turbulence, where $Re_L = (u_0 L_{11})/\nu_\nu$. Similarly, separating shear layers or boundary layers are characterized by high velocity gradient regions over length scales of δ. The corresponding time and velocity scales associated with these processes also span many orders of magnitude. In contrast, narrowband fluctuations are characterized by a discrete time scale(s), $\mathcal{T}$, and by convective and acoustic length scales, $u_0 \mathcal{T}$ and $c_0 \mathcal{T}$, respectively.

Reactive processes also have a variety of different scales, associated with coupled diffusive, convective, and kinetic processes. For example, the overall flame thickness, δ_F, and the reaction zone thickness, δ_{rxn}, are natural length scales for premixed flames, arising from convective–diffusive and diffusive–reactive balances, respectively. Similarly, the laminar burning velocity, s^u, is a natural velocity scale. Additional scales can be defined for various internal or post-flame processes, such as H consumption layers, heat release regions, or the post-flame OH recombination region, to name just a few [4]. The rest of this section is divided into two subsections. The first will treat flame interactions with broadband turbulent disturbances, and the second with narrowband disturbances.

11.1.1 Premixed Flame Interactions with Broadband Disturbance Fields

To start the discussion on broadband disturbances it is useful to consider premixed flames and review the Borghi combustion regime diagram shown in Figure 11.1 [1, 2]. This plot illustrates the different regimes of interaction between a premixed combustion process and broadband turbulent fluctuations. Similar diagrams could also be generated for other cases, e.g., turbulent nonpremixed flames [1] as discussed briefly at the end of this section, flame propagation in laminar shear layers, or post-flame NO_x production. The axes indicate the ratio of the integral length scale to flame thickness, L_{11}/δ_F, and turbulence intensity to laminar burning velocity, u_{rms}/s^u. For $u_{rms}/s^u < 1$ the flame is weakly wrinkled, but becomes highly wrinkled and multiconnected for $u_{rms}/s^u > 1$. In addition, the range of flame front wrinkling length scales increases monotonically with the ratio L_{11}/δ_F.

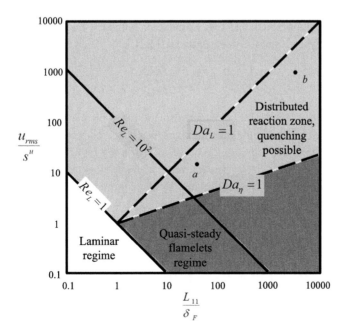

Figure 11.1 Turbulent combustion diagram illustrating different combustion regimes as a function of L_{11}/δ_F (integral length scale/flame thickness) and u_{rms}/s^u (turbulent velocity fluctuations/flame speed). Points a and b are used in Figure 11.3.

Lines of constant turbulent Reynolds number, $Re_L = u_{rms} L_{11}/\nu_\nu$, are also indicated in the figure. Assuming equal mass and momentum diffusivities, the Reynolds number can be related to u_{rms}/s^u and L_{11}/δ_F through the relation (see Exercise 11.1):

$$Re_L = \left(\frac{u_{rms}}{s^u}\right)\left(\frac{L_{11}}{\delta_F}\right). \tag{11.1}$$

Thus, the lower left region describes laminar flames. Increasing the Reynolds number corresponds to moving upwards and to the right in the figure, and is associated with a broader range of turbulent length scales interacting with the flame, since $L_{11}/L_\eta \sim Re_L^{3/4}$.

Next, lines of constant Damköhler number, i.e., the ratio of flow and flame time scales $Da = \tau_{flow}/\tau_{chem}$, are discussed. We define two Damköhler numbers based on flow time scales associated with integral and Kolmogorov length scales:

$$Da_L = \frac{\tau_{flow}|_{Large\ scales}}{\tau_{chem}} = \left(\frac{L_{11}}{u_{rms}}\right) \bigg/ \left(\frac{\delta_F}{s^u}\right) = \left(\frac{s^u}{u_{rms}}\right)\left(\frac{L_{11}}{\delta_F}\right), \tag{11.2}$$

$$Da_\eta = \frac{\tau_{flow}|_{Kolmogorov\ scales}}{\tau_{chem}} = \left(\frac{L_\eta}{u_\eta}\right) \bigg/ \left(\frac{\delta_F}{s^u}\right) = \left(\frac{s^u}{u_{rms}}\right)^{3/2}\left(\frac{L_{11}}{\delta_F}\right)^{1/2}, \tag{11.3}$$

where u_η denotes the velocity fluctuations associated with the Kolmogorov scale eddies. Separate Damköhler numbers can also be defined using the preheat thickness and the reaction zone thicknesses, to differentiate between disturbance effects on the preheat and reaction zone. The two Damköhler numbers, Da_L and Da_η, are related through the Reynolds number (see Exercise 11.2):

$$Re_L = \frac{Da_L^2}{Da_\eta^2}. \tag{11.4}$$

Lines of $Da_L = 1$ and $Da_\eta = 1$ are also indicated on the figure. An alternative interpretation of these Damköhler numbers is as the reciprocal of their respective Karlovitz numbers, defined as the ratio of a flow strain rate to an extinction stretch rate, e.g., $Da_\eta = 1/Ka_\eta$. Note that the ratio of flame thickness to the Kolmogorov length scale, δ_F/L_η, and Da_η are related by the following expression, as shown in Exercise 11.3:

$$Da_\eta \left(\frac{\delta_F}{L_\eta}\right)^2 = 1, \tag{11.5}$$

showing that $\delta_F/L_\eta = 1$ when $Da_\eta = 1$. This is an important expression as it shows that the flame and Kolmogorov time scales are the same, i.e., $Da_\eta = 1$, when the flame thickness equals the Kolmogorov length scale, $\delta_F/L_\eta = 1$. It should be emphasized, however, that this relationship between time and length scales for disturbance interactions is not true in general, as will be noted in the context of acoustic–flame interactions later. Consider the limit of $\delta_F/L_\eta \ll 1$, where the flame essentially appears as a discontinuity to the flow. Thus, the flame structure is unaltered by

11.1 Overview of Length/Time Scales

turbulent fluctuations. This also means that $Da_\eta \gg 1$, so the flame responds in a quasi-steady manner to the flow disturbances. For this reason, this region is labeled "quasi-steady flamelets" in the turbulent combustion diagram in Figure 11.1. In other words, the flame's internal structure can be understood at each instant in time by an analysis of the corresponding steady-state laminar flame with the same local conditions, topics covered in Chapter 9. In essence, the flame has a locally laminar structure that becomes increasingly wrinkled in amplitude with increases in u_{rms}/s^u, over an increasing range of length scales with L_{11}/δ_F. The local propagation velocity of the front may vary from location to location due to flow straining and flame curvature in a time-varying, but quasi-steady, manner.

Consider next the case where $Da_\eta \sim O(1)$ and, therefore, δ_F/L_η and Ka_η are $\sim O(1)$ as well. In this situation, the flame is highly stretched, its response is not quasi-steady, and the flow length scales are the same as the flame thickness. Hence, the descriptor "distributed reaction zone" is used in the combustion diagram. Laminar flames under quasi-steady conditions extinguish when $Ka \sim O(1)$. However, the turbulent flame does not necessarily extinguish because the flame is not quasi-steady, nor does it have the same structure as a laminar flame. In this regime, laminar flamelet concepts described in Chapter 9 cannot be directly applied. For example, laminar flames under non-quasi-steady conditions can withstand instantaneous Karlovitz numbers that exceed the steady-state extinction value (see Section 9.9). Moreover, the internal flame structure itself is altered from its laminar value, such as the correlations between temperature and species concentrations. Some indication of this alteration can be seen by plotting the averaged relationship between local fuel consumption rate and the temperature through a turbulent hydrogen flame, shown in Figure 11.2, whose adiabatic flame temperature, $T_{ad} = 1426$ K. For reference, this relationship is also plotted for a laminar unstretched flame and for a stretched flame, very near the extinction stretch rate, κ_{ext}, whose peak temperature falls below T_{ad}. Results are shown for two different Da_η values, one corresponding to weak and the other very intense turbulence.

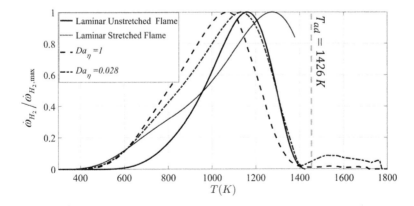

Figure 11.2 Averaged dependence of local fuel consumption rate on temperature in a hydrogen flame under weak and strongly turbulent conditions. For reference, laminar unstretched and near-critically stretched flame results are also shown. Image courtesy of D. Dasgupta [5].

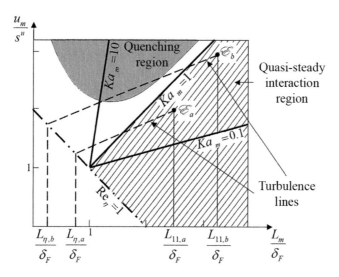

Figure 11.3 Spectral diagram showing the range of length and velocity scales interacting with the flame, following Poinsot and Veynante [2]. The "turbulence lines" in the graph correspond to the single points a and b in Figure 11.1.

For the turbulent flame results, two items are evident. First, there is enhanced reactivity at lower flame temperatures relative to the laminar calculation, presumably due to enhanced turbulent diffusivity. Second, there is the "hump" in reaction rates above T_{ad}, between ~1400 K and 1800 K. These correspond to stretched regions of the flame, where local flame temperatures exceed the adiabatic value.

While helpful, the diagram in Figure 11.1 characterizes turbulence by a single velocity and length scale. In order to better visualize flame–turbulence interactions over the full range of length scales, it is helpful to use a spectral diagram [2], as shown in Figure 11.3. The x-axis of this figure plots the entire range of length scales, L_m, between the integral and Kolmogorov length scales, $L_\eta < L_m < L_{11}$. The y-axis plots the normalized velocity fluctuations, u_m, associated with the corresponding eddies of length scale L_m.

For a given operating point on Figure 11.3, L_m and u_m are not independent. For example, in the inertial subrange, the turbulent dissipation rate, $\mathscr{E}$, is a constant that is independent of scale, given by

$$\mathscr{E} = u_m^3 / L_m. \tag{11.6}$$

This point is indicated by the "turbulence line" in Figure 11.3. Thus, each *point* in Figure 11.1 corresponds to a *line* in Figure 11.3 that describes the spectrum of length and velocity scales associated with a given Reynolds number and dissipation rate. The $\mathscr{E}_a$ line corresponding to the point a in Figure 11.1 indicates the range of flow length scales existing in this flow between the integral, $L_{11,a}$, and Kolmogorov, $L_{\eta,a}$, length scales. Also shown in this figure is a "Kolmogorov line" corresponding to $Re_\eta = 1$. The intersection of this Kolmogorov line with the spectral line denotes the corresponding Kolmogorov length scale associated with that given $\mathscr{E}$ value.

11.1 Overview of Length/Time Scales

A Karlovitz number can be defined for each length scale, L_m, as

$$Ka_m = \frac{u_m/L_m}{s^u/\delta_F}. \tag{11.7}$$

The $Ka_m = 1/Da_m = 1$ line is indicated in Figure 11.3. This line divides the plot into two regions where the flame response is and is not quasi-steady with respect to the given turbulence length scale. Quasi-steady, laminar flamelet considerations suggest that flame quenching can occur in the region of the spectral diagram corresponding to $Ka_m > 1$. In reality, this "quenching line" is displaced to higher Ka_m values, a topic also discussed in Section 9.10 for a single vortex, and curves upward at small length scales, as shown in Figure 11.3 [2, 6].

As shown in Figure 9.46, the fact that the single vortex quenching line curves upward at small scales suggests that the smallest vortices, which have the highest corresponding cold flow hydrodynamic strain rates, $\underline{S}$, are actually not effective in quenching the flame. Non-quasi-steadiness, gas expansion, and viscous dissipation are important factors for this behavior [7]. Thus, u_m is not necessarily the appropriate velocity to parameterize the fluctuations associated with eddies of scale size L_m in the flame. For example, a vortex that is thinner than the flame cannot cause flame wrinkling, and its magnitude is reduced by gas expansion on crossing the preheat zone (see Section 7.3.4). In addition, increased viscous diffusion encountered at the higher gas temperatures in the preheat zone causes the vorticity magnitude to decrease. This latter effect can be appreciated by considering the ratio of the viscous diffusion time of a Kolmogorov-scale eddy, L_η^2/ν_ν, to the preheat zone residence time, δ_F/s^u:

$$\frac{L_\eta^2/\nu_\nu}{\delta_F/s^u} = \left(\frac{L_\eta}{\delta_F}\right)^2. \tag{11.8}$$

This equation shows that the time scale required to "diffuse out" an eddy of length scale comparable to the flame thickness, $L_\eta/\delta_F \sim O(1)$, is equal to its preheat zone residence time. Thus, vortices smaller than the flame thickness are, in fact, acted on substantially by viscous diffusion in the preheat zone. Combining these results shows that the quasi-steady extinction condition $Ka_\eta = 1/Da_\eta = 1$ is necessary, but not sufficient, for flame extinction by turbulent eddies. Flame quenching potentially occurs for turbulent fields with sufficiently high dissipation rates over an intermediate range of turbulent length scales that cross the quenching line. As such, although points a and b both lie in the $Da_\eta < 1$ zone in Figure 11.1, only point b crosses the quenching region in Figure 11.3.

Consider next nonpremixed flames. A key challenge is that the nonpremixed flame does not have a clearly defined characteristic velocity or length scale. We can define a diffusion layer thickness, which is not a function of some internal flame process but is imposed on the flame by the outer flow, by [2]:

$$\delta = 1/|\nabla Z|_{Z=Z_{st}}. \tag{11.9}$$

Unlike the premixed flame, the diffusion layer thickness for a nonpremixed flame is not controlled by kinetic processes and is geometry dependent. Similarly, a characteristic time scale can be defined by $1/\chi$, where χ is the scalar dissipation rate (see Section 9.6) which, again, is imposed on the flame by the outer flow. This time scale, in conjuction with a characteristic chemical time, τ_{chem}, can be used to define a Damköhler number. The reader is referred to focused texts for further elaboration on the different nonpremixed combustion regimes and their dependencies on key flame/flow length and time scales [1, 2].

11.1.2 Flame Interactions with Narrowband Velocity Disturbance Fields

In this section, we consider narrowband disturbances, characterized by the time scale, $\mathcal{T} = 1/f$, and by both convective and acoustic length scales, $\lambda_c = u_c \mathcal{T}$ and $\lambda_\Lambda = c_o \mathcal{T}$, where u_c denotes the disturbance phase speed, such as the convection speed of a vortex. These length scales are referred to as convective and acoustic wavelengths, respectively.

Disturbances influence both the local internal flame structure, such as the burning rate, and its global geometry, such as the flame area [8]. As shown in Figure 11.4, we will consider time and length scales associated with both the overall flame size, L_F, and its thickness, δ_F, to characterize the noncompactness[1] and non-quasi-steadiness characteristics that are required to understand these physics.

In order to understand the dynamics associated with the overall flame position, it is necessary to first consider the velocity with which disturbances on the flame sheet propagate downstream – this is not the same as the flow velocity. This is given by the phase speed of $\xi(y,t)$, the instantaneous location of the flame with respect to the flame base, as shown in Figure 11.4. We will use the variables $u_{\xi,0}$ and $u_{\xi x,0}$ to denote this velocity along the flame sheet and its component along the axial coordinate,

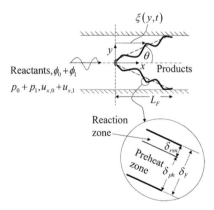

Figure 11.4 Illustration of the key length scales associated with narrowband flame–flow interactions.

---

[1] Compactness and noncompactness imply that the ratio of flame to disturbance length scales is small (i.e., the disturbance has a much longer length scale than the flame) or large, respectively.

respectively. Since nonpremixed flames have no flame propagation mechanism, the wrinkles propagate axially at the flow speed to leading order [9]. Because premixed flames propagate normal to themselves, $u_{\xi,0}$ equals the magnitude of the vector sum of the flow velocity and burning velocity normal to the flame. These processes are discussed more generally in Section 11.2.2.2, but for this section we assume a spatially uniform axial velocity field, $u_{x,0}$, a constant burning velocity, and no transverse flow field, so that the speed of a wrinkle on the flame and the flow speed are related by $u_{\xi,0} = u_{x,0} \cos \theta$ and $u_{\xi x,0} = u_{x,0} \cos^2 \theta$. Thus, note the distinction between the bulk flow velocity, $u_{x,0}$, the flow disturbance phase speed, u_c, and the flame wrinkle phase speed, $u_{\xi x,0} = u_{x,0} \cos^2 \theta$. We will show later that the ratio of flow disturbance to flame wrinkle velocity, given by the parameter k_c, has a significant influence on the space–time dynamics of the flame:

$$k_c = \frac{u_c}{u_{\xi x,0}} = \frac{u_c}{u_{x,0} \cos^2 \theta}. \tag{11.10}$$

We define three dimensionless time ratios to characterize global flame geometry and internal flame structure processes: (1) St_2, the dimensionless ratio of the flame wrinkle convection time from flame base to tip, to the acoustic period; (2) St_{L_F}, the dimensionless ratio of the flow convection time from flame base to tip, to the acoustic period; and (3) St_F, the ratio of the chemical time to the acoustic period:

$$St_2 = \frac{L_F/u_{x,0} \cos^2 \theta}{\mathcal{T}} = f L_F / (u_{x,0} \cos^2 \theta), \tag{11.11}$$

$$St_{L_F} = \frac{f L_F}{u_{x,0}}, \tag{11.12}$$

$$St_F = \frac{\delta_F/s^u}{\mathcal{T}} = f \delta_F / s^u. \tag{11.13}$$

St_2 and St_{L_F}, which naturally arise for premixed or nonpremixed flames (see Section 11.4), respectively, are *global* non-quasi-steadiness parameters – the overall flame shape adjusts in a quasi-steady manner to disturbances for $St_2 \ll 1$, as shown in Figure 11.5(a). Global quasi-steadiness implies that the flame position at each instant

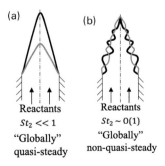

Figure 11.5 Illustration of instantaneous flame shapes at two time instances for flow disturbances when (a) $St_2 \ll 1$ and (b) $St_2 \sim O(1)$.

of time is the same as its steady-state position for the same instantaneous conditions. In contrast, the flame shape is not quasi-steady for St_{L_F} or $St_2 \sim O(1)$. Flames excited by disturbances at these higher Strouhal numbers generally have significant wrinkling, as shown in Figure 11.5(b). St_2 can also be defined as the ratio of the actual length of the flame, $L_F/\cos\theta$, to the convective wavelength of wrinkles on the flame front, λ_{ℓ}, where $\lambda_{\ell} = u_{\ell,0}/f = u_{x,0}\cos\theta/f$.

Consider next the time scale ratio, St_F, defined in Eq. (11.13) for a premixed flame, which is the ratio of the flame response time scale to forcing time scale and is an "internal" non-quasi-steadiness parameter. It is the harmonic forcing analogue to Da_η and Ka discussed in Section 11.1.1. Sections 9.4 and 9.9 show that St_F naturally appears in analyzing the internal flame response to fluctuations in fuel/air ratio, flame strain rate, and pressure. This time scale is related to the relaxation of the preheat zone of the flame to imposed disturbances. Very high frequency interactions also influence the reaction zone, δ_{rxn} [10–12].

Since the flame is much longer than its thickness, $L_F \gg \delta_F$, a flame becomes globally non-quasi-steady at much lower frequencies than it becomes internally non-quasi-steady. Operating conditions also influence these two non-quasi-steadiness parameters differently. For example, the flame length typically exhibits less sensitivity to pressure than the flame thickness. Then, two flames at low and high pressure would exhibit global non-quasi-steadiness at similar frequencies, but internal non-quasi-steadiness at very different ones. Moreover, in this example, the low- and high-pressure flame would exhibit similar low-frequency forced response characteristics, but very different response characteristics at higher frequencies.

Having considered time scale ratios, we next consider length scale ratios, which include the following:

$$He = L_F/\lambda_\Lambda, \tag{11.14}$$

$$L_F/\lambda_C = fL_F/u_C = St_C, \tag{11.15}$$

$$He_\delta = (\delta_F/\lambda_\Lambda \sin\theta), \tag{11.16}$$

$$(\delta_F/\lambda_C \sin\theta) = (u_{x,0}/u_C)(f\delta_F/s^u) = St_F(u_{x,0}/u_C). \tag{11.17}$$

The Helmholtz number, $He = L_F/\lambda_\Lambda$, is an acoustic compactness parameter – i.e., it equals the ratio of the flame length and an acoustic wavelength. When $He \ll 1$, the spatially integrated heat release, $\dot{Q}(t)$, controls the system's thermoacoustic stability and combustion noise characteristics (see Sections 6.5 and 12.4.3). In this case, the distribution of unsteady heat release, $\dot{q}(x,t)$, is unimportant.

The ratios L_F/λ_C and $\delta_F/\lambda_C \sin\theta$ are convective compactness parameters, i.e., the ratio of the flame length or thickness, respectively, to the length scale of an imposed convecting disturbance. Note that the disturbance length scale is multiplied by $\sin\theta$ for the internal flame parameter to obtain the convective length scale in the direction normal to the flame, see Figure 11.4. Additionally, note the correspondence between length and time scales, e.g., $L_F/\lambda_C = St_C$. Furthermore, notice the close relationship, but not exact equality, between the convective compactness parameter, L_F/λ_C, and global quasi-steadiness parameter, St_2, since $k_c(L_F/\lambda_C) = St_2$. Similar considerations apply for the ratios of the flame thickness and convective wavelength, since $(\delta_F/\lambda_C \sin\theta) = St_F(u_{x,0}/u_C)$.

Last, consider the ratio of flame thickness and acoustic wavelength, He_δ. For acoustic disturbances, the criteria for non-quasi-steadiness and noncompactness are not the same. Rather, for low Mach number flows, the flame response becomes non-quasi-steady at much lower frequencies than when the flame becomes acoustically noncompact (in contrast, non-quasi-steadiness and noncompactness occur at similar conditions for convecting disturbances, as noted previously). The ratio of acoustic and convective length scales is given by

$$\frac{\lambda_C}{\lambda_\Lambda} \sim \frac{u_C/f}{c_0/f} = M_0, \qquad (11.18)$$

indicating that $\lambda_C \ll \lambda_\Lambda$ at low flow Mach numbers.

11.2 Dynamics of Premixed Flame Sheets

Section 7.2.1 derived an expression for the dynamics of premixed flame sheets, denoted as the G-equation. This section further analyzes the space–time dynamics of premixed flames, using this formulation.

11.2.1 Model Problems for Two-Dimensional Configurations [13]

It is helpful to consider a simplified situation in order to study the solution properties of propagating fronts. Assume a two-dimensional flame front whose location is a single-valued function, ξ, of the coordinate y, see Figure 11.6. The reader is referred to Section 11.2.2.4 for treatment of the flame response to three-dimensional disturbances, such as convecting helical disturbances.

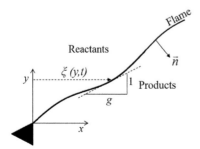

Figure 11.6 Schematic of model problem of the top half of a two-dimensional flame whose position is a single-valued function of the coordinate, y.

We define $G(x, y, t) \equiv x - \xi(y, t)$. Inserting this expression into Eq. (7.9) leads to

$$\frac{\partial \xi}{\partial t} - u_x^u + u_y^u \frac{\partial \xi}{\partial y} = -s_d^u \sqrt{1 + \left(\frac{\partial \xi}{\partial y}\right)^2}. \tag{11.19}$$

We will refer to this equation as the "ξ-equation," and note that ξ can be defined as a function of either transverse or axial coordinates [14] – the coordinate system is generally picked based on the configuration such that the flame is a single-valued function.

Note that the flow velocities are evaluated on the upstream side of the flame sheet, i.e., $u_x^u(y, t) = u_x(\xi(y, t), y, t)$. We will also consider the evolution equation for the flame slope, $g = \partial \xi / \partial y$, obtained by differentiating Eq. (11.19) with respect to y:

$$\frac{\partial g}{\partial t} - \frac{\partial u_x^u}{\partial y} + \frac{\partial u_y^u}{\partial y} g + u_y^u \frac{\partial g}{\partial y} = -\frac{\partial}{\partial y}\left(s_d^u \sqrt{1 + g^2}\right). \tag{11.20}$$

The next subsections work through a series of model problems to understand the dynamics described by the ξ-equation. In general, the flow velocity and flame position are coupled, as shown in Section 7.8, because of gas expansion across the flame. However, in order to obtain analytical solutions, the velocity will be prescribed for these problems.

11.2.1.1 Flat Flames, Flashback, and Blowoff

Consider a flat flame in a uniform flow field. The solutions are straightforward and can be derived by inspection, but are useful for building up familiarity with use of the level set equation for analysis of flame dynamics. Consider a flow with uniform axial velocity, u_x, no transverse velocity, $u_y = 0$, and a flame with constant flame speed, s_d^u. In an infinite domain where g is a continuous function of y (flames actually tend to form discontinuities in the slope, g, a generalization to be discussed later in this section), a family of solutions of Eq. (11.19) is

$$\xi(y, t) = \xi_{init} + \left(u_x - s_d^u \sqrt{1 + g^2}\right) \cdot t, \tag{11.21}$$

where ξ_{init} is the initial flame front location at $t = 0$. In the case of a flat flame (i.e., $g = 0$), the solution describes a flame either propagating upstream (flashback) or moving downstream (blowoff). The velocity with which the front is moving upstream or downstream is given by

$$\frac{\partial \xi(t)}{\partial t} = v_F = u_x^u - s_d^u. \tag{11.22}$$

As such, we can see that the following conditions describe flashback and blowoff in this model problem, as also shown in Figure 11.7:

$$\begin{aligned} u_x^u < s_d^u &: \text{flashback}, \\ u_x^u > s_d^u &: \text{blowoff}. \end{aligned} \tag{11.23}$$

Note also that a steady-state solution is only possible when $u_x^u = s_d^u$.

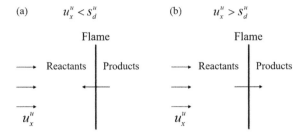

Figure 11.7 Illustration of (a) flashback and (b) blowoff of a flat flame.

11.2.1.2 Attached Steady-State Flames

We next consider a semi-infinite domain $(x, y \geq 0)$ with a steady uniform velocity field where $u_x^u > s_d^u$, and specify the following boundary condition at $y = 0$:

$$\xi(y = 0, t) = 0. \tag{11.24}$$

This boundary condition simulates flame attachment. Understanding this attachment condition requires consideration of the internal flame structure and edge flames, topics considered in Chapter 9. Note that a steady-state solution is possible for all $u_x^u \geq s_d^u$ values:

$$\xi(y) = \left(\left(\frac{u_x^u}{s_d^u}\right)^2 - 1\right)^{\frac{1}{2}} \cdot y. \tag{11.25}$$

This solution describes the familiar result that the flame equilibrates at an angle where the component of the velocity normal to the front $\vec{u}^u \cdot \vec{n}$ is equal to the displacement speed, s_d^u. As such, the flame spreading angle, θ, is given by

$$\sin \theta = \frac{s_d^u}{u_x^u}. \tag{11.26}$$

11.2.1.3 Attached Transient Flames

Consider an unsteady problem where there is a step change in axial velocity over the entire domain from $u_{x,a}$ to $u_{x,b}$, both of which exceed s_d^u:

$$u_x^u(t) = \begin{cases} u_{x,a} & t < 0, \\ u_{x,b} & t \geq 0. \end{cases} \tag{11.27}$$

We will show that the flame relaxation process consists of a "wave" that propagates along the flame in the flow direction. The governing equation for the flame slope, obtained from Eq. (11.20), is

$$\frac{\partial g}{\partial t} + \frac{\partial}{\partial y}\left(s_d^u \sqrt{1 + g^2}\right) = 0. \tag{11.28}$$

Note that this equation is a special case of the more general equation [15]:

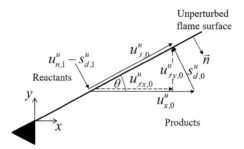

Figure 11.8 Illustration of the top half of an attached flame, indicating the velocity components tangential and normal to the flame.

$$\frac{\partial g}{\partial t} + c(g)\frac{\partial g}{\partial y} = 0, \qquad (11.29)$$

where $c(g)$ is the component of the disturbance propagation velocity in the y-direction, given for this problem by $s_d^u g/\sqrt{1+g^2}$ (equivalent to $u_{ty,0}^u$ in Figure 11.8 for the unforced problem), or in general

$$c(g) \equiv \left(\vec{u}^u - s_d^u \,\vec{n}\right) \cdot \vec{e}_y. \qquad (11.30)$$

This equation can develop discontinuities in the solution, analogous to shock waves, when $u_{x,b} > u_{x,a}$, as shown by the solution developed in Exercise 11.4:

$$g(y,t) = \begin{cases} \sqrt{\left(\dfrac{u_{x,b}}{s_d^u}\right)^2 - 1} & y \leq c_{shock} \cdot t, \\[2mm] \sqrt{\left(\dfrac{u_{x,a}}{s_d^u}\right)^2 - 1} & y > c_{shock} \cdot t, \end{cases} \qquad (11.31)$$

where

$$c_{shock} = \frac{\left(s_d^u\right)^2}{u_{x,a} + u_{x,b}} \cdot \left[\sqrt{\left(\frac{u_{x,a}}{s_d^u}\right)^2 - 1} + \sqrt{\left(\frac{u_{x,b}}{s_d^u}\right)^2 - 1}\right]. \qquad (11.32)$$

Several instantaneous pictures of the solution showing the flame relaxation transient are shown in Figure 11.9.

Note the "wavelike" nature of the flame response. The change in slope from the initial steady-state value of $\left(u_{x,a}^2/(s_d^u)^2 - 1\right)^{1/2}$ to the final steady-state value of $\left(u_{x,b}^2/(s_d^u)^2 - 1\right)^{1/2}$ is initiated at the flame attachment point. This slope discontinuity propagates along the flame front at a velocity projected along the y-coordinate at c_{shock}. This "shock" propagation velocity lies between the wave propagation velocities of the initial condition and final steady-state solutions, i.e.,

$$c(g^+) = s_d^u\left(1 - (s_d^u)^2/u_{x,a}^2\right)^{1/2} < c_{shock} < c(g^-) = s_d^u\left(1 - (s_d^u)^2/u_{x,b}^2\right)^{1/2}. \qquad (11.33)$$

11.2 Dynamics of Premixed Flame Sheets

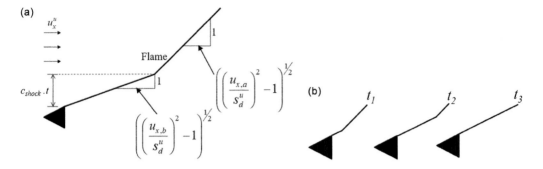

Figure 11.9 (a) Illustration of the top half of the instantaneous flame position for a problem where the axial flow velocity is changed from $u_{x,a}$ to $u_{x,b}$ at $t = 0$. (b) Evolution of the flame position at different times, $t_1 < t_2 < t_3$.

Note the analogies to the shock propagation velocity in gas dynamics. The flame equilibrates at its new steady-state position after a time τ, which physically corresponds to the time for the discontinuity to reach the end of the flame. Readers are referred to Exercise 11.5 for the case of $u_{x,b} < u_{x,a}$, which leads to an "expansion wave" solution for the flame dynamics.

11.2.2 Linearized Dynamics of Constant Burning Velocity Flames

Because of the nonlinearity of the ξ-equation, exact solutions are possible only for a limited set of problems. General solutions can be derived for the linearized ξ-equation – i.e., the spatiotemporal dynamics of small-amplitude disturbances on the flame. This section works through these dynamics in detail for laminar flames without stretch effects. Stretch effects and harmonic forcing of turbulent flames are discussed in Sections 11.4 and 11.5.

11.2.2.1 Linearized Formulations

Some insight into the transient flame dynamics can be obtained from the problem considered in the previous section. If the velocity change from $u_{x,a}$ to $u_{x,b}$ is infinitesimal, the wave propagation speed along the flame is given by

$$c(g_0) = \frac{s_d^u g_0}{\sqrt{1 + g_0^2}}. \tag{11.34}$$

This propagation speed is equal to the y-component of the mean flow velocity in the tangential direction along the flame, $u_{ty,0}$, as illustrated in Figure 11.8, and is discussed further below. We expand the flame position and flow variables at the flame as (see Section 2.1):

$$\begin{aligned}
\xi &= \xi_0(y) + \xi_1(y, t), \\
\vec{u}^u &= \vec{u}_0^u(y) + \vec{u}_1^u(y, t), \\
s_d^u &= s_{d,0}^u(y) + s_{d,1}^u(y, t).
\end{aligned} \tag{11.35}$$

A note on notation: $s_{d,0}^u$ denotes the unperturbed displacement flame speed, as defined by the expansions in Eq. (2.2), while $s_d^{u,0}$ denotes the unstretched displacement speed, as discussed in Chapter 9. Inserting this expansion into Eq. (11.19) and retaining only the first-order terms leads to

$$\frac{\partial \xi_1}{\partial t} + \underbrace{\left(u_{y,0}^u + s_{d,0}^u \cdot \frac{\partial \xi_0/\partial y}{\sqrt{1+(\partial \xi_0/\partial y)^2}} \right)}_{u_{\chi,0}^u \cdot \sin \theta} \cdot \frac{\partial \xi_1}{\partial y} = \underbrace{u_{x,1}^u - u_{y,1}^u \frac{\partial \xi_0}{\partial y} - s_{d,1}^u \sqrt{1+(\partial \xi_0/\partial y)^2}}_{(u_{n,1}^u - s_{d,1}^u)/\sin \theta},$$

(11.36)

where $u_{\chi,0}^u$ represents the mean tangential velocity along the flame, $u_{n,1}^u$ denotes the fluctuating velocity component in the direction normal to the unperturbed flame front, and $s_{d,1}^u$ denotes the fluctuating displacement speed. The $\sin \theta$ term on both sides comes from the choice of the flame position coordinate; i.e., if ξ is written instead as a function of x, then a $\cos \theta$ arises; alternatively, if the coordinate system is aligned with the unperturbed flame position, this term does not appear at all. This equation can be written in the physically more revealing form

$$\frac{\partial \xi_1}{\partial t} + u_{\chi,0}^u \sin \theta \frac{\partial \xi_1}{\partial y} = \frac{v_{F,n1}}{\sin \theta}, \qquad (11.37)$$

where

$$v_{F,n1} = u_{n,1}^u - s_{d,1}^u. \qquad (11.38)$$

The left-hand side of Eq. (11.37) is an advection operator and describes the propagation of wrinkles along the flame with a velocity projected in the y-direction given by $u_{\chi,0}^u \sin \theta = u_{y,0}^u$ (see Figure 11.8). The right-hand side describes excitation of flame disturbances, showing that they are due to velocity (in the normal direction to the flame) and flame speed disturbances. Both of these terms are illustrated in Figure 11.8.

11.2.2.2 Flame Dynamics with Tangential Flow

This section describes the linearized dynamics of flames with tangential flow, i.e., flows where $u_{\chi,0}$ is not equal to 0, and illustrates the effects of flame attachment, flow forcing, and advection processes on the flame response. Notice that in the absence of tangential flow the linearized flame dynamics are described by Langevin's equation, which is an ordinary differential equation (see Section 11.2.2.5). In this case, the unsteady flame motion at each point is a function of the local disturbance field, $v_{F,n1}$. In contrast, flame dynamics with tangential flow are described by a partial differential equation, usually introducing an additional spatial boundary condition to the problem and making the local flame position a nonlocal function of the disturbance field.

We will use the following conventions for solutions in this section. First, y_s and t_s will be used to denote the position and time at which the flame position/slope is being evaluated, in order to distinguish them from differentiation with respect to t or y.

Second, the notation $\left[\frac{\partial F}{\partial t}(y,t)\right]_{y=\sigma, t=\tau}$ will be used to denote differentiation of a function F with respect to t and then substituting the arguments σ and τ in for y and t.

The flame position at a given (y_s, t_s) is a superposition of wrinkles from all upstream points that were generated at earlier times and have propagated along the flame front at the velocity $u^u_{\chi,0}$. This can be seen from the solution of Eq. (11.36). To simplify the presentation, we assume that the mean velocity field is spatially uniform. Assuming the flame boundary condition at $y = 0$ of $\xi(y = 0, t) = \xi_b(t)$, the general solution of the ξ-equation is

$$\xi_1(y_s, t_s) = \xi_b \left(t_s - \frac{y_s}{u^u_{\chi,0} \sin\theta} \right) + \frac{1}{u^u_{\chi,0} \sin\theta} \int_0^{y_s} \frac{1}{\sin\theta} v_{F,n1}\left(s, t_s - \frac{y_s - s}{u^u_{\chi,0,y}} \right) ds, \tag{11.39}$$

$$g_1(y_s, t_s) \equiv \frac{\partial \xi_1(y_s, t_s)}{\partial y} = -\frac{1}{u^u_{\chi,0} \sin\theta} \left[\frac{d\xi_b}{dt}(t)\right]_{t=t_s - \frac{y_s}{u^u_{\chi,0} \sin\theta}}$$

$$+ \frac{1}{u^u_{\chi,0} \sin\theta} \left[\int_0^{y_s} \frac{1}{\sin\theta} \left[\frac{\partial v_{F,n1}}{\partial y}(y, t)\right]_{y=y_s, t=t_s - \frac{y_s - s}{u^u_{\chi,0} \sin\theta}} ds + \frac{1}{\sin\theta} v_{F,n1}\left(0, t_s - \frac{y_s}{u^u_{\chi,0} \sin\theta} \right) \right]. \tag{11.40}$$

The first terms in the expressions for ξ_1 and g_1 on the right-hand sides of Eqs. (11.39) and (11.40), respectively, describe the homogeneous solution of the ξ-equation – i.e., the solution in the absence of external forcing. The remaining terms describe the additional effects of forcing from the $v_{F,n1}$ term. The physical significance of these homogeneous and particular solutions can be understood by considering a series of model problems, which are analyzed next.

The simplest example is the case without flow forcing, i.e., where $v_{F,n1} = 0$. In that case, only the first terms on the right-hand sides of Eqs. (11.39) and (11.40), respectively, remain. These physically represent fluctuations at the boundary, such as due to an oscillating flameholder location, which leads to wrinkle propagation along the flame at the speed of $u^u_{\chi,0}$. The resulting solutions are

Moving flameholder, no flow forcing:

$$\xi_1(y_s, t_s) = \xi_b \left(t_s - \frac{y_s}{u^u_{\chi,0} \sin\theta} \right), \tag{11.41}$$

$$g_1(y_s, t_s) = -\frac{1}{u^u_{\chi,0} \sin\theta} \left[\frac{d\xi_b}{dt}(t)\right]_{t=t_s - \frac{y_s}{u^u_{\chi,0} \sin\theta}}. \tag{11.42}$$

In other words, even though the flame is only perturbed at the base, tangential convection causes the entire downstream portion of the flame to exhibit wrinkles. The downstream propagation of waves excited by flameholder motion can be seen in the experimental data shown in Figure 11.10.

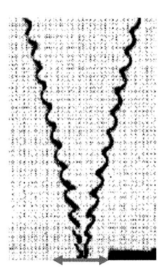

Figure 11.10 Photograph of flame excited by a transversely oscillating flameholder, showing downstream propagation of flame wrinkles. Reproduced from Petersen and Emmons [16].

The second example is a spatially uniform disturbance, i.e., one where $v_{F,n1}$ is not a function of spatial coordinates, and therefore $\partial v_{F,n1}/\partial y = 0$. This velocity disturbance does not excite the second term in Eq. (11.40). If the flame base moves at the same velocity as the velocity fluctuations, i.e., $\frac{d\xi_b(t)}{dt} = \frac{v_{F,n1}(y=0,t)}{\sin\theta}$, then the last two terms cancel in the equation for g_1. In this case, g_1 is identically zero for all y and t, and the entire flame simply moves up and down in a bulk motion without any change in its shape or area [17]:

Spatially uniform disturbance field, flame base moves with flow:

$$\xi_1(y_s, t_s) = \xi_b(t_s), \tag{11.43}$$

$$g_1(y_s, t_s) = 0. \tag{11.44}$$

If a flame-anchoring boundary condition is imposed, $\frac{d\xi_b(t)}{dt} = 0$, the flow disturbance excites a flame front disturbance that originates at the boundary [18], leading to:

Spatially uniform disturbance field, flame base fixed:

$$\xi_1(y_s, t_s) = \frac{1}{u^u_{x,0}\sin^2\theta}\int_0^{y_s} v_{F,n1}\left(0, t_s - \frac{y_s - s}{u^u_{x,0}\sin\theta}\right)ds, \tag{11.45}$$

$$g_1(y_s, t_s) = \frac{1}{u^u_{x,0}\sin^2\theta} \cdot v_{F,n1}\left(0, t_s - \frac{y_s}{u^u_{x,0}\sin\theta}\right). \tag{11.46}$$

An experimental image of a harmonically excited flame that was forced in this manner is shown in Figure 11.11.

11.2 Dynamics of Premixed Flame Sheets

Figure 11.11 Experimental image showing a Bunsen flame excited by a longitudinal acoustic wave. Image courtesy of S.K. Thumuluru.

If the forcing term is spatially nonuniform, i.e., $\partial v_{F,n1}/\partial y \neq 0$, then the second term on the right-hand side of Eq. (11.40) is excited. This results in waves originating at the spatial location(s) of flow nonuniformity that also propagate along the flame at $u^u_{x,0} \sin \theta$. Since the propagation speed of the flame wrinkle and the flow disturbance exciting the wrinkle are not necessarily the same, the resulting flame fluctuation at a given (t_s, y_s) is a convolution of flame wrinkles excited at all upstream locations, $0 < y \leq y_s$, and times $t_s - y_s/(u^u_{x,0} \sin \theta) < t \leq t_s$. As such, the flame exhibits "memory," and the fluctuations at a given point are nonlocal; i.e., they are not only a function of the disturbance at that space–time point, but also all prior locations on the flame front. The next two subsections work through this problem in detail, first considering the harmonically forced problem and then the randomly forced one.

11.2.2.3 Example: Attached Flame Excited by a Harmonically Oscillating, Convecting Disturbance [19]

In this section we work through the response characteristics of the attached flame illustrated in Figure 11.8 to a convecting, decaying disturbance field. The fluctuating velocity component normal to the flame is assumed to convect axially at a phase speed of u_c and decay at a rate of ζ. The disturbance field at the flame surface is given by

$$\frac{v_{F,n1}}{u_{x,0}}\bigg|_{x=\xi(y,t)} = \varepsilon_n \cdot \left[e^{-\zeta x} \cos\left(2\pi f(t - x/u_c)\right)\right]_{x=\xi(y,t)}. \tag{11.47}$$

This disturbance can be due to axial or transverse velocity fluctuations, fuel/air ratio fluctuations, or any other disturbance leading to flame speed disturbances. Since the level set equation in Eq. (11.37) has been posed with y as the independent variable, we replace ξ with $(y/\tan\theta + \xi_1)$. Then, Eq. (11.47) is written as

$$\frac{v_{F,n1}(y,t)}{u_{\chi,0}^u} = \varepsilon_n \cdot e^{-\zeta y/\tan\theta} \cos(2\pi f(t - y/(u_c \cdot \tan\theta))) + O(\varepsilon_n \cdot \xi_1). \quad (11.48)$$

Although $v_{F,n1}$ is shown as a function of y, it does not imply that the flow is convecting and decaying in the y-direction. This is simply a reflection of the choice of coordinate system. We will further consider this same problem in Section 11.2.3.2, showing how nonlinear effects modify these results. The solution of Eq. (11.37) for the flame position and slope for this disturbance field is (see Exercise 11.7):

$$\frac{\xi_1}{u_{\chi,0}^u/f} = \text{Real}\left\{\frac{-i\cdot\varepsilon_n/\sin\theta}{2\pi\left(u_{\chi,0}^u \cos\theta/u_c - 1\right) + i\zeta\, u_{\chi,0}^u \cos\theta/f} \times \left[e^{-\zeta y/\tan\theta}e^{i2\pi f(y/(u_c \tan\theta)-t)} - e^{i2\pi f(y/(u_{\chi,0}^u \sin\theta)-t)}\right]\right\}, \quad (11.49)$$

$$\frac{\partial(\xi_1/(u_{\chi,0}^u/f))}{\partial y} = \text{Real}\left\{\frac{-i\cdot\varepsilon_n/\sin\theta}{2\pi\left(u_{\chi,0}^u \cos\theta/u_c - 1\right) + i\zeta\, u_{\chi,0}^u \cos\theta/f} \times \begin{bmatrix}\left(-\dfrac{\zeta}{\tan\theta} + \dfrac{i2\pi f}{u_c \tan\theta}\right)\cdot e^{-\zeta y/\tan\theta}e^{i2\pi f(y/(u_c \tan\theta)-t)} \\ -\dfrac{i2\pi f}{u_{\chi,0}^u \sin\theta}\cdot e^{i2\pi f(y/(u_{\chi,0}^u \sin\theta)-t)}\end{bmatrix}\right\}.$$

$$(11.50)$$

Note that the solution consists of two parts, as discussed in Section 11.2.2.2. The first term is the particular solution that is excited by the unsteady velocity field at all positions upstream of the position, y. The second term is the homogeneous solution, which is determined by the zero response of the flame at the anchoring point to the nonzero velocity disturbance at that point. Mirroring the underlying perturbation velocity field, the particular solution decays downstream and propagates with the phase velocity of the excitation, u_c. In contrast, the homogeneous solution has a constant axial magnitude and propagates at the flame wrinkle advection velocity, $u_{\chi y,0}$. Note that these two disturbances propagate at the same phase velocity when $k_c = 1$, or equivalently, $u_{\chi,0}^u \cos\theta = u_c$. This equation shows that the flame response is controlled by two parameters, $u_{\chi,0}^u \cos\theta/u_c$ and $\zeta u_{\chi,0}^u \cos\theta/f$, whose influence is discussed next. These parameters can be alternatively cast in terms of the length scales $\lambda_c = u_c/f$, $\lambda_{\chi} = u_{\chi,0}^u \cos\theta/f$, and $1/\zeta$, respectively.

Figure 11.12 plots gain and phase results for the flame position and slope at several values of the decay parameter, $\zeta u_{\chi,0}^u \cos\theta/f$. Note that the flame position, $|\xi_1|$, and slope, $|\partial\xi_1/\partial y|$, are directly related to the amplitude of flame flapping and area fluctuations at each axial position. First, notice that all three solutions for $|\xi_1|$ converge to a common solution near $y = 0$. In fact, the following general solution for the flame

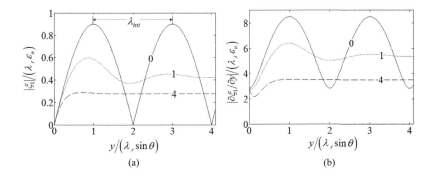

Figure 11.12 Dependence of (a) flame position, $|\xi_1|$, and (b) gradient, $|\partial \xi_1/\partial y|$, on the transverse coordinate for different values of the velocity decay rate parameter, $\zeta u_{\chi,0} \cos\theta/f$ ($k_c = 2.0$, $\theta = 45°$).

response near $y = 0$ can be worked out for an arbitrary velocity field where $|\xi_1(y=0,t)| = 0$, as shown in Exercise 11.6:

$$\frac{\partial |\xi_1|}{\partial y} = \left|\frac{\partial \xi_1}{\partial y}\right| = \frac{1}{\sin^2\theta} \frac{|v_{F,n1}|}{u_{\chi,0}^u}. \tag{11.51}$$

This equation shows that the fluctuation of the flame sheet in the direction normal to the mean flame front is equal to the ratio of the fluctuating perturbation field normal to the flame front and mean velocity tangential to the flame front. Returning to Figure 11.12, note next that for the no-decay case, $\zeta u_{\chi,0} \cos\theta/f = 0$, the gain in flame response grows, reaches a local maximum, and then oscillates with constant local maximum value. This is due to interference, which occurs when $k_c \neq 1$, due to differing propagation speeds of the two waves along the flame sheet. For decaying disturbance fields, $\zeta u_{\chi,0} \cos\theta/f > 0$, this local maximum decays axially until settling at a constant magnitude given by the homogeneous solution.

Figure 11.12 shows that interference processes create a new length scale, λ_{int}, over which $|\xi_1|$ and $|\partial \xi_1/\partial y|$ oscillate spatially, given by

$$\lambda_{int}/(\lambda_A \sin\theta) = \frac{1}{|1/k_c - 1|}. \tag{11.52}$$

This shows that faster-propagating disturbances (such as acoustic waves) with longer wavelengths lead to shorter-wavelength undulations in flame response [20], a somewhat counterintuitive result. The limiting case is an axially uniform disturbance, corresponding to $\lambda_c = \infty$, where $\lambda_{int}/(\lambda_A \sin\theta) = 1$. These interference patterns are often observed experimentally. To illustrate, Figure 11.13 plots two measured flame response curves.[2] In Figure 11.13(a), three interference maxima are evident, with an

---

[2] Note that the measurements quantify the flame position along the axial direction, while the analysis uses the transverse direction. These can be linearly scaled to relate one to the other in the linear amplitude regime.

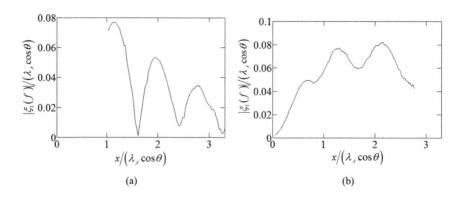

Figure 11.13 Two different measured dependencies of $|\xi_1|$ on axial coordinate showing flame front interference patterns. Images courtesy of (a) S. Shanbhogue and D.H. Shin [21] and (b) V. Acharya et al. [22].

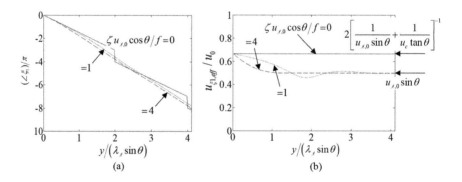

Figure 11.14 Calculated dependence of (a) phase and (b) phase slope (where $u_{\xi1,eff} = 2\pi f/(\partial \angle \xi_1/\partial y)$) of flame position fluctuation, ξ_1, on axial coordinate for the model problem, using different values of velocity decay rate parameter, $\zeta u_{x,0} \cos\theta/f$ ($k_c = 2.0$, $\theta = 45°$).

overall wavelength of $\lambda_{int}/(\lambda_\chi \sin\theta) \sim 1$, suggesting that the phase speed of the disturbance is fast relative to the flow field, probably an acoustic disturbance. In Figure 11.13(b), two length scales of flame wrinkling are evident, a longer-wavelength pattern on which a shorter-wavelength undulation is present. This flame data corresponds to the velocity field data shown in Figure 2.25, which showed the simultaneous presence of a convecting vortical disturbance and an acoustic disturbance. As the vortical disturbance amplitude is larger than the acoustic one, the longer-wavelength flame wrinkle is the dominant feature, with the lower-amplitude, shorter-wavelength wrinkle causing undulations in this overall pattern. Other data is also shown in Figure 11.24, showing a single large interference peak.

Returning to the calculation results for the model velocity profile, consider next the corresponding calculated phase of the flame wrinkle in Figure 11.14(a), which rolls off monotonically with transverse coordinate. Only the flame position, ξ_1, is shown as its phase characteristics are quite similar to $\partial \xi_1/\partial y$. An alternative way to illustrate these phase dependencies is through the local slope of the phase curve in (b), which can be interpreted as a local phase velocity through the relationship $-2\pi f/(\partial\angle\xi_1/\partial y)$. In cases where $\zeta > 0$, the velocity excitation source is zero far downstream, implying that the phase speed in the y-direction should asymptote to $u_{z,0}\sin\theta$. In the near-field, the phase characteristics are influenced by the superposition of convecting velocity disturbances and flame wrinkle propagation. As shown in Exercise 11.8, the normalized phase slope at $y=0$ is given by $2[1/(u_{z,0}\sin\theta) + 1/(u_c\tan\theta)]^{-1}$ and $[1/(u_{z,0}\sin\theta) + 1/(u_c\tan\theta)]^{-1}$ for ξ_1 and $\partial\xi_1/\partial y$, respectively [23].

Finally, consider the $k_c = 1$ case. Note that special care is required in evaluating Eq. (11.49) as this is a "resonant" case, where the disturbance exciting wrinkles on the flame and the wrinkle convection speed exactly match. This limit of Eq. (11.49) for $\zeta = 0$ is given by the following solution:

$$\xi_1(y,t) = \frac{y}{\sin^2\theta} \text{Real}\left[\varepsilon_n e^{i2\pi f\left(\frac{y}{u_{z,0}^{u}\sin\theta} - t\right)}\right] \tag{11.53}$$

In this case, the flame wrinkle disturbance grows monotonically with downstream distance, as there is no interference; indeed, the interference length scale, λ_{int}, becomes infinite.

11.2.2.4 Example: Flame Response to Harmonic Oscillating, Three-Dimensional Disturbances [24]

The previous examples have considered two-dimensional configurations. An important generalization of this is round configurations, where the flame is subjected to three-dimensional disturbances, such as helical modes. Consider an axisymmetric geometry as shown in Figure 11.15, where a swirling premixed flame is stabilized on a centerbody. We will use θ to denote the polar angle in this section, and ψ for the flame angle (note that θ denotes flame angle for the earlier two-dimensional cases in this chapter, such as Figure 11.8).

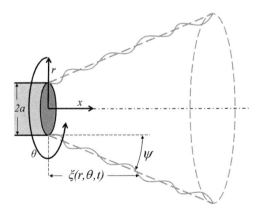

Figure 11.15 Schematic of the axisymmetric flame configuration.

If an axisymmetric flame is subjected to axisymmetric disturbances, then swirl has no influences on flame dynamics and its local response can be understood from discussions in earlier sections. However, when the flame is subjected to nonaxisymmetric disturbances, swirl convects the flame wrinkles azimuthally, leading to a complex axial and azimuthal dependence of flame patterns. To illustrate, consider the mean flow

$$u_{x,0} = u_0,$$
$$u_{r,0} = 0, \quad (11.54)$$
$$u_{\theta,0} = \Omega r,$$

where the base flow swirl rate is parameterized by Ω. We will consider a velocity disturbance along the mean flame surface that mirrors the profile used in the previous section, with the addition of an azimuthal component:

$$\frac{v_{F,n1}(r,\theta,t)}{u_{\chi,0}} = \varepsilon_n \cdot e^{-\zeta r/\tan\psi} \cos\left(2\pi f(t - r/(u_c \cdot \tan\psi)) - m\theta\right). \quad (11.55)$$

The flame response is given by

$$\frac{\xi_1(r,\theta,t)}{u^u_{\chi,0}/f} = \frac{\varepsilon_n}{\sin\psi} \text{Real} \left\{ \left(\frac{e^{\left(\frac{i2\pi f k_{c,m}}{u^u_{\chi,0}\cos\psi}\right)\frac{r}{\tan\psi}} - e^{-\zeta\frac{r}{\tan\psi}}}{i2\pi k_{c,m} + \frac{\zeta u^u_{\chi,0}\cos\psi}{f}} \right) e^{i2\pi f(r/(u_c\tan\psi)-t)+im\theta} \right\}. \quad (11.56)$$

Here,

$$k_{c,m} = \left(1 - \frac{u^u_{\chi,0}\cos\psi}{u_c} + \frac{m\Omega}{2\pi f}\right) = \left(1 - \frac{1}{k_c} + \frac{m\Omega}{2\pi f}\right). \quad (11.57)$$

This expression describes flame wrinkling in the axial and azimuthal directions. In this case, interference effects are possible in both the axial direction, such as the discussion in Section 11.2.2.3, and in the azimuthal direction.

In addition, there is a given helical mode which causes maximum flame wrinkling for a given velocity disturbance, denoted $m_{\xi max}$. This helical mode is the one with minimal interference, analogous to the $k_c = 1$ case discussed in the context of Eq. (11.53). This helical mode is given by the integer closest to

$$m_{\xi max} = \frac{2\pi f}{\Omega}\left(\frac{1}{k_c} - 1\right). \quad (11.58)$$

An analogous result can be derived for nonpremixed flames [9]. Finally, there are also disturbance modes, m, which lead to maximum spatially integrated heat release fluctuations, $m_{\dot{Q}max}$ (Section 12.3.1.3), or sound emissions, m_{pmax} (Section 12.4.3.2), which are generally not the same as $m_{\xi max}$.

11.2.2.5 Turbulent Flow Disturbances Exciting Flame with No Tangential Flow

In this section we work through a different class of example problems, focusing on the response of a flat flame to convecting broadband disturbances [25–28], illustrated in

11.2 Dynamics of Premixed Flame Sheets

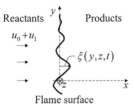

Figure 11.16 Illustration of model problem showing a flat flame perturbed by random flow disturbances.

Figure 11.16. The combined effect of harmonic and broadband forcing are treated in Section 11.5. Before proceeding, the reader may wish to review the definition of the ensemble average in Section 2.8, which is used extensively in the next two examples.

In the ensuing analysis, we will assume that the disturbances are spatially homogeneous and isotropic, characterized by their root mean square value, $u_{rms}^2 = u_{x,1}^2 + u_{y,1}^2 + u_{z,1}^2$, and longitudinal integral length scale, L_{11}. We will solve for two quantities in this section, the turbulent burning rate and the turbulent flame brush thickness. The turbulent burning rate for a constant laminar burning velocity flame is directly proportional to the ensemble-averaged increase in local burning area due to flame surface wrinkling. This increase in burning rate is quantified using the "turbulent consumption speed", $s_{T,c}$ (discussed further in Section 12.4.1), which is related to flame areas for a constant laminar burning velocity flame as

$$\frac{s_{T,c}(y,z)}{s^{u,0}} = \lim_{\Delta y \Delta z \to 0} \frac{\langle A(y,z,t; \Delta y, \Delta z) \rangle}{\Delta y \Delta z}, \qquad (11.59)$$

where A is the instantaneous area of the flame surface within a tube of dimensions Δy and Δz normal to the x-axis and centered around the point (y, z). While this expression is formally correct, a deeper question is whether the turbulent flame speed increases because the area increases, or whether the area increases because the turbulent flame speed increases. This question of causality is touched on further in Section 11.2.3.3.

Given a flame whose location can be written as a single-valued function of y and z, this implies that $\frac{s_{T,c}}{s^{u,0}} = \left\langle \sqrt{1 + \left(\frac{\partial \xi}{\partial y}\right)^2 + \left(\frac{\partial \xi}{\partial z}\right)^2} \right\rangle$. Expanding this expression to leading order yields

$$\frac{s_{T,c}}{s^{u,0}} = 1 + \frac{1}{2}\left\{\left\langle \left(\frac{\partial \xi_1}{\partial y}\right)^2 \right\rangle + \left\langle \left(\frac{\partial \xi_1}{\partial z}\right)^2 \right\rangle\right\} + O\left(\left(\frac{\partial \xi_1}{\partial y}, \frac{\partial \xi_1}{\partial z}\right)^4\right). \qquad (11.60)$$

Thus, the leading-order local consumption speed depends on the mean-squared fluctuations in local flame surface slope. Note, however, that this asymptotic tendency is crucially dependent on whether $\partial \xi_0 / \partial y$ is or is not zero; see the related discussion in Section 11.2.3.3. The flame brush thickness is defined as $\langle \xi_1^2 \rangle^{1/2}$ and is the random analogue of the envelope, $|\xi_1|$, that was characterized in Section 11.2.2.3. Thus, expressions are needed for the mean squared flame position fluctuation, ξ_1, and slopes, $\partial \xi_1 / \partial y$ and $\partial \xi_1 / \partial z$, in order to solve for the flame brush thickness and turbulent flame speed.

The linearized ξ-equation, Eq. (11.19), takes the form

$$\frac{\partial \xi_1}{\partial t} = u_{x,1}. \qquad (11.61)$$

Similarly, the flame slope is described by the expression

$$\frac{\partial}{\partial t}\left(\frac{\partial \xi_1}{\partial y}\right) = \frac{\partial u_{x,1}}{\partial y}. \qquad (11.62)$$

Integrating both of these expressions with respect to time at a point (y,z) and using the initial condition that $\xi_1(x, t=0) = \partial u_{x,1}/\partial y(x, t=0) = 0$ yields

$$\xi_1(t) = \int_0^t u_{x,1}(t')dt', \qquad (11.63)$$

$$\frac{\partial \xi_1(t)}{\partial y} = \int_0^t \frac{\partial u_{x,1}(t')}{\partial y} dt'. \qquad (11.64)$$

Using these expressions, the mean squared values of $\langle \xi_1(t)^2 \rangle^{1/2}$ and $\langle (\partial \xi_1(t)/\partial y)^2 \rangle$ are

$$\langle \xi_1^{\,2}(t) \rangle = \left\langle \int_0^t \int_0^t u_{x,1}(t_1) u_{x,1}(t_2) dt_1 dt_2 \right\rangle, \qquad (11.65)$$

$$\left\langle \left(\frac{\partial \xi_1(t)}{\partial y}\right)^2 \right\rangle = \left\langle \int_0^t \int_0^t \frac{\partial u_{x,1}(t_1)}{\partial y} \frac{\partial u_{x,1}(t_2)}{\partial y} dt_1 dt_2 \right\rangle. \qquad (11.66)$$

These integrals depend on the correlation function of the velocity, defined for a homogeneous and stationary velocity field as

$$\wp_{corr, u_x u_x}(\Delta_x, \Delta_y, \Delta_z, \Delta_t) = \frac{\langle u_{x,1}(x,y,z,t) \cdot u_{x,1}(x+\Delta_x, y+\Delta_y, z+\Delta_z, t+\Delta_t)\rangle}{\langle (u_{x,1}(x,y,z,t))^2 \rangle}, \qquad (11.67)$$

where Δ_x, Δ_y, Δ_z, and Δ_t represent the spatial separation distances and separation time, respectively. The correlation function for $\partial u_{x,1}/\partial y$ can be written as [29]:

$$\wp_{corr, \frac{\partial u_x}{\partial y}\frac{\partial u_x}{\partial y}}(\Delta_x, \Delta_y, \Delta_z, \Delta_t) = -\frac{\langle u_{x,1}^{\,2}\rangle}{\langle (\partial u_{x,1}/\partial y)^2\rangle}\frac{\partial^2}{\partial \Delta_y^{\,2}} \wp_{corr, u_x u_x}(\Delta_x, \Delta_y, \Delta_z, \Delta_t). \qquad (11.68)$$

The value of $\langle (\partial u_{x,1}/\partial y)^2 \rangle$ in this expression is used to define the transverse Taylor microscale, L_{taylor}, as [30]:

$$\langle (\partial u_{x,1}/\partial y)^2 \rangle = \frac{2\langle u_{x,1}^{\,2} \rangle}{(L_{taylor})^2}. \qquad (11.69)$$

Then, as detailed in Exercise 11.9, the flame brush thickness and turbulent flame speed are

$$\langle \xi_1^2(t) \rangle = \langle u_{x,1}^2 \rangle \int_0^t 2(t-\tau) \ell_{corr, u_x u_x}(\tau) d\tau, \qquad (11.70)$$

$$\frac{s_{T,c}(t)}{s^{u,0}} = 1 + \int_0^t (t-\tau) \left[\left\langle \left(\frac{\partial u_{x,1}}{\partial y}\right)^2 \right\rangle \ell_{corr, \frac{\partial u_x}{\partial y} \frac{\partial u_x}{\partial y}}(\tau) + \left\langle \left(\frac{\partial u_{x,1}}{\partial z}\right)^2 \right\rangle \ell_{corr, \frac{\partial u_x}{\partial z} \frac{\partial u_x}{\partial z}}(\tau) \right] d\tau. \qquad (11.71)$$

While the general forms of these solutions depend on the particular correlation, results can be derived for small or large time (for the latter case, recalling the caveat of the linear approximation in Eq. (11.61)):

$$t \ll \tau_{int}: \quad \langle \xi_1(t)^2 \rangle^{1/2} \approx \frac{u_{rms}}{\sqrt{3}} \cdot t, \qquad (11.72)$$

$$t \ll \tau_{int}: \quad \frac{s_{T,c}(t)}{s^{u,0}} \approx 1 + \frac{2}{3}\left(\frac{u_{rms}}{L_{taylor}}\right)^2 \cdot t^2, \qquad (11.73)$$

$$t \gg \tau_{int}: \quad \langle \xi_1(t)^2 \rangle^{1/2} \approx \sqrt{\frac{2}{3}(u_{rms})^2 \tau_{int} t}, \qquad (11.74)$$

$$t \gg \tau_{int}: \quad \frac{s_{T,c}(t)}{s^{u,0}} \approx 1 + \frac{4}{3}\left(\frac{u_{rms}}{L_{taylor}/\tau_{int}}\right)^2 \cdot \frac{t}{\tau_{int}} \qquad (11.75)$$

where τ_{int} is the integral time scale, defined as:[3]

$$\tau_{int} = \int_0^\infty \ell_{corr, u_x u_x}(\tau) d\tau = \int_0^\infty \ell_{corr, \frac{\partial u_x}{\partial y}\frac{\partial u_x}{\partial y}}(\tau) d\tau = \int_0^\infty \ell_{corr, \frac{\partial u_x}{\partial z}\frac{\partial u_x}{\partial z}}(\tau) d\tau \qquad (11.76)$$

These show that the flame brush thickness grows linearly with time initially, as the turbulent motions are correlated for $t \ll \tau_{int}$. For large time, the brush grows as $t^{1/2}$, analogous to a diffusion process or random walk, with a growth rate that is proportional to the integral time scale, $\tau_{int}^{1/2}$. In addition, the turbulent flame speed exhibits a quadratic dependence on turbulence intensity for weak turbulence, a result we will return to in Section 12.4.1 where we consider higher turbulence intensities.

Using Taylor's hypothesis [30], the integral time and length scales can be related as $\tau_{int} = L_{11}/u_{x,0}$. The above expressions can then be alternatively written as:

$$t \gg \tau_{int}: \quad \langle \xi_1^2(t) \rangle^{1/2} \approx \varepsilon_T L_{11} \sqrt{\frac{2}{3}\frac{t}{\tau_{int}}}, \qquad (11.77)$$

---

[3] This expression assumes that the space–time dependence of the correlation functions are separable. If they are not, then different integral time scales need to be defined for the velocity and its gradient.

432 Forced Response I: Flamelet Dynamics

$$t \gg \tau_{int}: \quad \frac{s_{T,c}(t)}{s^{u,0}} \approx 1 + \frac{4}{3}\varepsilon_T^2 \left(\frac{L_{11}}{L_{taylor}}\right)^2 \cdot \frac{t}{\tau_{int}}, \quad (11.78)$$

where

$$\varepsilon_T = \frac{u_{rms}}{u_{x,0}}. \quad (11.79)$$

11.2.2.6 Turbulent Flow Disturbances Exciting Flame with Tangential Flow [26–28]

This section analyzes an anchored flame with tangential flow as shown in Figure 11.8, subjected to random turbulent fluctuations. This problem is analogous to the harmonically forced problem considered in Section 11.2.2.3, with the velocity model replaced by a random function. As shown in Figure 11.17, we will use a coordinate system fixed to the unforced flame position, (t, n, z), as opposed to the previous sections where (x, y, z) was used. The derivations are quite lengthy, so we only present the results and refer the reader to Hemchandra [26] for details. We will assume Taylor's hypothesis in order to relate space and time correlations and explicitly specify the following longitudinal velocity correlation function:

$$\rho_{corr,LL}(r) = \exp\left(-\frac{\pi}{4}\left(\frac{r}{L_{11}}\right)^2\right). \quad (11.80)$$

The solutions for $\langle \xi_1^2 \rangle^{1/2}$ and $s_T/s^{u,0}$ are:[4]

$$\langle \xi_1(t)^2 \rangle^{1/2} = \left\{ \frac{4L_{11}^2}{\pi} \left[\exp\left(-\frac{\pi}{4}\left(\frac{t\tan\theta}{L_{11}}\right)^2\right) - 1 \right] + 2tL_{11}\tan\theta \cdot \text{erf}\left(\frac{\sqrt{\pi}t\tan\theta}{2L_{11}}\right) \right\}^{1/2} \frac{u_{rms}}{\sqrt{3}s^{u,0}}, \quad (11.81)$$

$$\frac{s_{T,c}(t)}{s^{u,0}} = 1 + \left\{ \frac{8\left[\exp\left(-\frac{\pi}{4}\left(\frac{t\tan\theta}{L_{11}}\right)^2\right) - 1\right]}{L_{11}} + \frac{4t\pi\tan\theta}{L_{11}} \cdot \text{erf}\left(\frac{\sqrt{\pi}t\tan\theta}{2L_{11}}\right) + \tan^2\theta \right\} \cdot \frac{u_{rms}^2}{6(s^{u,0})^2}, \quad (11.82)$$

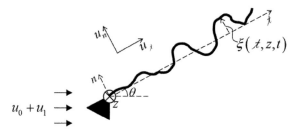

Figure 11.17 Schematic of an anchored turbulent flame with tangential flow showing the coordinate system used for the model problem.

---

[4] The author is grateful to D. H. Shin for deriving the flame brush result and finding an error in the flame speed expressions in Ref. [27], which is corrected here.

11.2 Dynamics of Premixed Flame Sheets

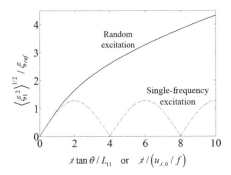

Figure 11.18 Comparison of the axial dependence of the flame brush thickness of a flame subjected to harmonic $(\xi_{ref} = (v_{F,n1})_{rms}/f, k_c = 0.8)$ and random $(\xi_{ref} = (u_{rms}/(\sqrt{3}s^{u,0}))L_{11})$ disturbances.

whose leading-order contribution for small $\not{t}/L_{11}$ values is given by

$$\not{t}/L_{11} \ll 1: \quad \langle \xi_1^1 \rangle^{1/2} = \frac{u_{rms} \tan \theta}{\sqrt{3}s^{u,0}} \not{t} + O(\not{t}^3), \tag{11.83}$$

$$\not{t}/L_{11} \ll 1: \quad \frac{s_{T,c}(\not{t})}{s^{u,0}} = 1 + \left(\frac{u_{rms} \tan \theta}{s^{u,0}}\right)^2 \left\{\frac{\pi}{3}\left(\frac{\not{t}}{L_{11}}\right)^2 + \frac{1}{6}\right\}. \tag{11.84}$$

This shows that the flame brush thickness and turbulent flame speed grow linearly and quadratically with downstream distance, $\not{t}$, respectively. The downstream growth in flame brush thickness is a well-known experimental result for turbulent flames [31].

Figure 11.18 compares these flame brush calculations with the solutions for $|\xi_1|$ for the harmonically forced case where the decay parameter, ζ, is set to zero. Note the identical behavior of both curves near the attachment point (given by Eq. (11.51)), but then the sharp difference farther downstream, with interference effects dominating the behavior of the harmonic case and turbulent diffusion in the random case.

We will return to this problem in Section 11.2.3.2, discussing the growing role that nonlinear effects have in modifying this behavior with increasing downstream distance.

11.2.3 Nonlinear Flame Front Dynamics

Linearization of the ξ-equation provides important insights into the propagation of disturbances, but excludes physics associated with convective nonlinearities and kinematic restoration [1]. "Convective nonlinearities" refers to the propagation of flame wrinkles by flow perturbations; an example of the way in which they influence results is modifying the time-averaged convection speed of disturbances along the flame. "Kinematic restoration" refers to the smoothing of wrinkles by flame propagation normal to itself. This latter effect significantly alters (generally destroys) flame wrinkles, as illustrated in Figure 11.19. The top figure shows a wrinkled flame front at

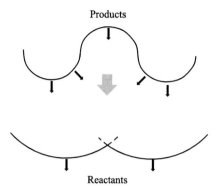

Figure 11.19 Illustration showing branches of a wrinkled flame front propagating toward each other, leading to destruction of the flame area [32].

time $t = 0$ in a quiescent flow field. The flame moves downward as it propagates into the reactants. In addition, the flame wrinkles are smoothed out because of flame propagation normal to itself, and there is annihilation of the flame area (given by the dashed lines) as the wrinkles propagate into each other. For context, recall from the solutions in, for example, Eq. (11.41) that flame wrinkles propagate with constant amplitude along the flame front in the linear limit for constant burning velocity flames.

While this figure provides a qualitative description of this phenomenon, describing it quantitatively is difficult because of its inherently nonlinear nature. The next section considers model problems that enable analytical insight. Then, Section 11.2.3.2 considers nonlinear effects on anchored flames. Section 11.3.3 also considers kinematic restoration effects on premixed flame rollup by a vortex.

11.2.3.1 Kinematic Restoration: Overview

We will consider a class of problems where the flame propagates into a quiescent flow, $u_x = u_y = 0$. In reality, the presence of a wrinkled, moving flame induces a flow field ahead of it (see Section 7.5), so this assumption implicitly assumes a low-density jump across the flame, $\sigma_p \to 1$. The governing equation for the flame position is

$$\frac{\partial \xi}{\partial t} = -s_d^u \sqrt{1 + \left(\frac{\partial \xi}{\partial y}\right)^2}. \tag{11.85}$$

The general solution of Eq. (11.85) is described by the Hopf–Lax formula [33]:

$$\xi(y, t) = \min_{y - s_d^u t \ < \ y^* \ < \ y + s_d^u t} \left(\xi_{init}(y) - \sqrt{(s_d^u t)^2 - (y - y^*)^2} \right), \tag{11.86}$$

where $\xi(y, t = 0) = \xi_{init}(y)$ denotes the initial condition. Though analytically involved, the basic physical processes described by this equation are quite straightforward. The equation simply states that the flame propagates normal to itself at the local displacement speed, s_d^u, also referred to as "Huygens propagation," as shown in

Figure 11.20 Illustration of Huygens propagation and flame surface annihilation described by the Hopf–Lax solution for flame propagation.

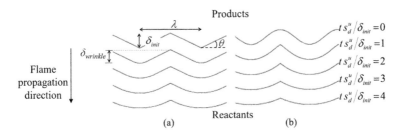

Figure 11.21 Calculations showing the time evolution of (a) a triangular and (b) a sinusoidal flame propagating at a constant flame speed ($\delta_{init}/\lambda = 0.25$). Image courtesy of D. H. Shin.

Figure 11.20. This figure shows the instantaneous flame location at two instances of time. The dashed lines indicate points of constant radial distance from the flame position. The flame position at later times can be constructed geometrically by drawing such circles about every point along the flame at each time, as shown in the figure. If the flame is curved, the direction of propagation varies with position. Moreover, in regions where multiple points on the flame propagate toward each other, annihilation of flame area occurs, such as the concave feature in the center. This annihilation is reflected in the solution, Eq. (11.86), being the "minimum" of

$$\xi_{init}(y) - \sqrt{(s_d^u t)^2 - (y - y^*)^2}.$$

To illustrate this point, consider the time evolution of a flame with a periodic triangular shape, shown in Figure 11.21(a), with wrinkle "wavelength" λ and height δ_{init}. Although this example is contrived because the regions of discontinuity in flame slope would be smoothed out or extinguished by flame stretch, the constant value of flame slope allows for a simple solution that can be put together by geometric construction. Note that the slope discontinuity at the trailing edge of the flame propagates downward at a velocity of $s_d^u / \cos\theta$. For steep wrinkles, this trace velocity can be arbitrarily fast. The leading-edge arcs propagate forward at a speed of s_d^u. Thus, the height of the wrinkle progressively decreases with time. Moreover, for times $t > \lambda/(2s_d^u \sin\theta)$ the flame position is completely governed by the leading edge (note the analogy with pulled fronts, discussed in Section 9.11, and leading points in Section 11.2.3.3) and forms a collection of arcs whose centers originate at the leading edges. After this time, the flame shape is completely independent of the initial wrinkling amplitude, δ_{init}, and is only a function of the wrinkling wavelength, λ.

Consider a second example, shown in Figure 11.21(b), this time where the sheet has an initially sinusoidal shape, similarly parameterized by wavelength λ and amplitude δ_{init} [32]:

$$\xi_{init}(y) = \frac{\delta_{init}}{2} \cos(2\pi y/\lambda). \tag{11.87}$$

The solution character here is more complex because of the variation in flame slope along the flame. Several computed instantaneous snapshots of the flame position are shown in Figure 11.21(b). Note, first, the formation of a cusp at the trailing edge of the flame – i.e., a discontinuity in flame slope, even though the initial slope distribution was continuous – this propensity towards cusp formation is a generic property of premixed flame fronts. In reality, this cusp region will be either smoothed or extinguished by flame stretch, and analysis of the flame shape in this regime requires consideration of curvature sensitivity of the flame speed, discussed in Section 9.3 [32]. At large enough times, the solution for the flame position becomes identical to the triangular shape shown in (a), as solutions become independent of initial conditions, are controlled by arcs propagating from the leading edge, and are only a function of λ (see Exercise 11.10).

The destruction of flame wrinkles by kinematic restoration can be quantified by the distance between the leading and trailing edges of the flame, denoted $\delta_{wrinkle}$. Figure 11.22 plots the time dependence of $\delta_{wrinkle}$ for both the triangular and sinusoidally wrinkled flames. For the latter case, note that all of the curves remain constant at $\delta_{wrinkle} = \delta_{init}$ for a duration corresponding to the time required for formation of a cusp at the trailing edge. As time increases, the two initially different flame shapes asymptote to the same solution, as discussed above. Also, these results show that the wrinkle size decreases with time at a progressively faster rate with increasing

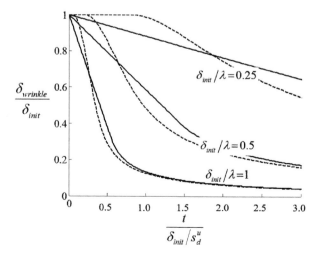

Figure 11.22 Dependence of the flame wrinkle height on time for an initially triangular (solid) and sinusoidal (dashed) flame. Calculations courtesy of D. H. Shin.

δ_{init}/λ. This shows the coupled effect of wrinkle height, δ_{init}, and length scale, λ, in controlling wrinkle destruction time scales. For larger times, the wrinkle destruction rate becomes independent of δ_{init} for a fixed λ.

11.2.3.2 Nonlinear Effects on Harmonically Forced, Anchored Flames with Tangential Flow

Many of the above concepts can be directly applied to anchored flames with tangential flow. To illustrate, Figure 11.23 plots a series of instantaneous images of a harmonically forced Bunsen flame. Note the sinusoidal wrinkling of the flame base near the attachment point and the cusped wrinkles farther downstream. This behavior is the spatial analogue of the temporal results shown in Figure 11.21.

Nonlinear effects accumulate with distance downstream (or time in Figure 11.21) and so become increasingly evident with increasing axial position. The roughly linear (in amplitude) behavior of the flame position near the flame attachment point is clearly evident in data. Figure 11.24 reproduces data from an experiment where a bluff body stabilized flame was excited by harmonically oscillating acoustic disturbances at three different amplitudes. The left image plots the axial dependence of the flame position fluctuation magnitude, $|\hat{\xi}'(x,f_0)|$, showing that it grows with amplitude of forcing, as expected. The right image shows this same result but with $|\hat{\xi}'(f_0)|$ normalized by the forcing amplitude. Note the collapse of these data to a common curve near the attachment point, demonstrating the linearity of the near-field flame dynamics. The curves diverge with downstream distance, demonstrating the growing significance of nonlinear effects.

Similar results can be seen from a sample calculation, by returning to the fully nonlinear level set equation and forcing it with the velocity field $u_{x,1} = v_{F,n1}/\sin\theta$ using the $v_{F,n1}$ forcing in Eq. (11.47). Recall that Section 11.2.2.3 analyzed the flame response to this disturbance field using the linearized flame response equation. We will now retain the nonlinear terms in order to illustrate the effect of disturbance

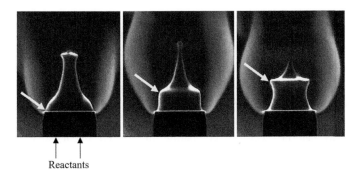

Reactants

Figure 11.23 Image of a harmonically forced Bunsen flame, showing the concave portion of a flame wrinkle becoming progressively more cusped as it moves downstream. Note how this action mirrors the results shown in Figure 11.19, with the added dynamic of tangential wrinkle convection. Reproduced from Ducruix et al. [34].

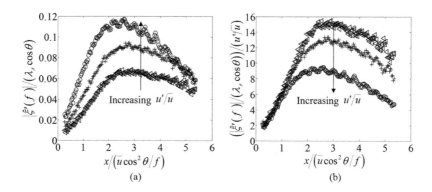

Figure 11.24 (a) Measured axial amplitude dependence of flame wrinkle amplitude of an acoustically excited, bluff body stabilized flame at three forcing amplitudes at the forcing frequency. (b) Same data, but normalized by perturbation velocity magnitude. Reproduced from Shanbhogue et al. [35].

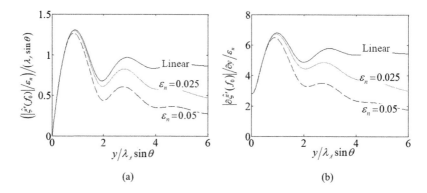

Figure 11.25 Calculated dependence of (a) $|\hat{\xi}'|$ and (b) $|\partial \hat{\xi}'/\partial y|$ on transverse location at several excitation amplitudes; $\theta = 45°$, $\zeta(\bar{u}_{x,0} \cos \theta/f) = 3$, $k_c = 2$. Calculations courtesy of D.H. Shin.

amplitude on the flame response characteristics. Figure 11.25 plots $|\hat{\xi}'(f_0)|$ and $|\partial \hat{\xi}'(f_0)/\partial y|$, normalized by the amplitude of excitation, in order to explicitly illustrate nonlinear effects. Note how all three curves collapse to a single line near $y = 0$. Moving downstream, the reduction in interference-generated oscillations in flame response, as well as the decreasing magnitude of flame wrinkling, is clearly evident in the higher-amplitude cases. This dissipation in flame wrinkling is amplitude dependent, as can be seen by comparing the scaled linear solution with the $\varepsilon_n = 0.025$ and 0.05 cases. Similar points are evident from the scaled data in Figure 11.24(b), showing faster decay of the flame wrinkle with increasing amplitude. The corresponding phase is not shown, as all three amplitude results are virtually identical for this particular calculation.

This amplitude-dependent smoothing of flame wrinkles is particularly evident in cases where flames are perturbed by strong vortices, such as the images in

Figure 12.12. These images clearly show the rollup of the flame by the vortex near the flameholder and the progressive smoothing of the flame front with downstream distance.

11.2.3.3 Example: Flame Propagation through a Periodic Shear Flow

In this section we consider a flame with constant laminar displacement speed, s_d^u, propagating through periodic shear flow, and consider the factors influencing its propagation speed. Section 11.2.3.1 analyzed a periodically perturbed flame in a homogeneous flow where its propagation speed ultimately relaxes back to the laminar speed. In this problem, the average flame propagation speed, denoted here by s_T, reaches a steady-state value $s_T > s_d^u$, because of the persistence of the flow oscillations. This example raises very interesting questions about the factors that are ultimately responsible for this augmentation of flame speed/burning rate, showing that two different interpretations of the results arise.

As shown in Figure 11.26, we consider a velocity profile, $u_x(y) = f(y)$, that is periodic over length scale λ with a constant tangential flow $u_y = u_t$, closely following Refs. [36, 37]. The mean tangential flow can be interpreted as a parameter of the "unsteadiness," since in a reference frame attached to the tangential flow, the shear flow is time dependent. In this new reference frame, the shear is therefore acting on the flame with a time scale given by λ/u_t. Without tangential flow, the flame settles into a steady position with the leading edge of the flame located at the point of maximum velocity. Tangential flow continuously translates the flame transversely, so that, in the moving reference frame, it is continuously evolving in a changing velocity field. In the stationary reference frame, this also causes the leading edge of the flame to not align with the point of maximum velocity.

First, consider the case where $u_t = 0$. Then, it can be shown that

$$s_T = s_d^u + \max(f(y)). \tag{11.88}$$

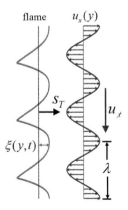

Figure 11.26 Depiction of flame propagating into a periodic shear flow.

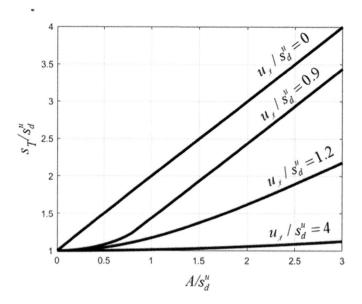

Figure 11.27 Dependence of s_T on the velocity disturbance amplitude, A, at different values of the unsteadiness parameter, u_t [36].

In other words, the turbulent flame speed is completely independent of the length scale, λ, or spatial profile, $f(y)$, of the velocity field. Rather, it is completely controlled by its maximum value, which is the point where the leading edge of the flame propagates into the fresh mixture.

Next, consider the $u_t \neq 0$ case. In order to demonstrate results more quantitatively, we consider the periodic flowfield

$$u_x(y) = A \cos(2\pi y/\lambda). \tag{11.89}$$

Figure 11.27 shows the dependence of s_T on the velocity disturbance amplitude, A. When $u_t = 0$, $s_T = s_d^u + A$, i.e., it increases linearly with A. The plot also shows this dependence for different values of the "unsteadiness parameter," u_t, showing that s_T decreases monotonically for increasing u_t. This effect is very large – note how the $u_t/s_d^u = 4$ case shows very weak sensitivity to A. In addition, the dependence of s_T on A is completely different for $u_t \neq 0$ ($s_T \sim A^2$) and $u_t = 0$ ($s_T \sim A$) for $A \sim 0$. In all cases, $s_T \sim A$ for large A [36, 37].

The classical interpretation of these effects is in terms of flame area [38], i.e., increases in flame area are responsible for the increases in s_T. Indeed, it can readily be verified that the steady-state flame area is directly proportional to s_T [36], as it must be. However, the result in Eq. (11.88) clearly shows that the increase in s_T is due to the leading edge of the flame propagating into the point of maximum velocity, which then leads to the necessary creation of flame area to consume all the reactants. In other words, increases in flame area are the effect, not the cause, of the increase in s_T.

Further insight into the factors driving the flame dynamics comes from solving Eq. (11.19) backward in time from the steady-state flame position. Defining this

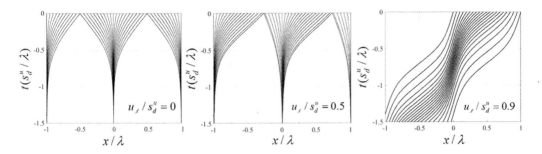

Figure 11.28 $x-t$ plots showing constant characteristic lines of solution for flame position, calculated backward in time from the steady-state solution at $t = 0$. Three plots are shown for three different values of $u_{\cancel{z}}/s_d^u$ for $A/s_d^u = 0.5$ [36].

steady-state flame position at $t = 0$, Figure 11.28 is an $x-t$ plot showing the location of the solution characteristic curves for earlier times.

Figure 11.28(a) shows that when $u_{\cancel{z}}/s_d^u = 0$, all the characteristic curves converge backward in time to a single point, which happens to coincide with the point on the flame aligned with the maximum in the velocity field. In other words, after some transient, this single point of the velocity field completely controls the subsequent position of the entire flame (note a similar discussion in the context of Figure 11.21). Figure 11.28(b), corresponding to $u_{\cancel{z}}/s_d^u = 0.5$, has a similar convergence back to a single point. Figure 11.28(c) shows completely different behavior – note that no curves converge. This means that the steady-state flame position is controlled by the full spatial profile of the shear flow; i.e., each part of the velocity profile influences a part of the flame position.

Thus, the relationship between disturbance velocity field and flame position exhibits two fundamentally different characters. The transition between these two behaviors is not smooth, but is demarcated by a bifurcation in solution space at a critical value of $u_{\cancel{z}}/s_d^u$. Figure 11.29 plots the $A/s_d^u - u_{\cancel{z}}/s_d^u$ combination where these two behaviors occur.

Figure 11.29 demonstrates that in a certain range of $A-u_{\cancel{z}}$ conditions, the front displacement speed is controlled by velocity field characteristics at discrete points on the flame. Under these conditions, any spatially periodic velocity fields with the same maximum value lead to identical front displacement speeds. The front speed is completely independent of the velocity field distribution or length scale. In contrast, there exists another set of behaviors in another range of $A-u_{\cancel{z}}$ conditions, where the front displacement speed is not controlled by discrete points, but rather by the entire spatial distribution of the velocity field. For these conditions, there is no single point in the velocity field that has special dynamical significance in controlling the front displacement speed. It is interesting to note that that there is not intermediate behavior in this problem. For example, there is not a monotonic growth in the region of influence from the discrete point to gradually encompass the whole velocity field. Rather, either a single point is controlling, or the entire velocity field, and there is a sharp bifurcation between these two behaviors.

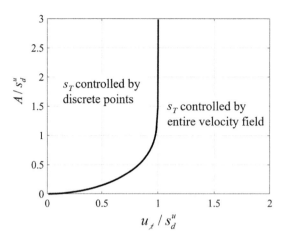

Figure 11.29 Regime map showing regions where characteristic curves converge or do not converge back to a single point [36].

Note the analogies of this problem with the the KPP solution for the speed of a reaction–diffusion wave in Section 9.11 for the average propagation speed of reaction–diffusion equations. It was similarly noted that there the leading edge $\tilde{c} \to 0$ of the turbulent flame brush controlled the turbulent flame speed under certain conditions, just as the leading edge of the flame controls this problem when $u_{\it{f}} = 0$.

A key question in combustion is "Why do velocity disturbances lead to increased averaged burning rates?" Taken together, these two problems show that the answer to this question has considerable nuance. A common explanation in the literature is that increasing turbulence intensity increases turbulent flame speeds because the velocity disturbances lead to flame wrinkling and increases in flame area. While certainly the flame area and flame speed must ultimately be related if all the reactants are consumed, these two examples show that the ultimate causality is more complex and that the common explanation cited above is, in fact, wrong in certain cases.

11.3 Dynamics of Nonpremixed Flame Sheets

This section treats the dynamics of nonpremixed flames in the thin reaction sheet limit, using the Z equation formulation from Section 7.2.2. As noted in Section 7.2, the nonpremixed problem is, by nature of the underlying equations, a geometry-specific one. As such, our treatment consists of analysis of several specific configurations: reacting mixing layer, stagnation flame, flame–vortex interaction, and harmonically forced jet flame.

11.3.1 Example Problem: Mixing Layer

In this section we consider two problems with the same solution characteristics that are illustrated in Figure 11.30. Figure 11.30(a) illustrates the temporal problem, which

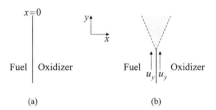

Figure 11.30 Illustration of (a) temporal and (b) spatial problems considered in this section.

consists of a one-dimensional interface that at time $t = 0$ separates two semi-infinite, quiescent $(\vec{u} = 0)$ regions of pure fuel $(x < 0)$ and pure oxidizer $(x > 0)$ [1, 2]. The premixed flame analogue of this problem, where the flame divided regions of reactants and products, was considered in Section 11.2.1.1. Neglecting density changes and assuming constant diffusivities, the mixture fraction is given by

$$\frac{\partial Z}{\partial t} = \mathscr{D} \frac{\partial^2 Z}{\partial x^2}. \tag{11.90}$$

Figure 11.30(b) illustrates the related spatial problem, consisting of a steady mixing layer with two streams of equal and uniform velocity, u_y, that mix as they convect in the y-direction downstream. Neglecting density changes and diffusion in the streamwise direction, and assuming constant diffusivities, the mixture fraction is given by

$$u_y \frac{\partial Z}{\partial y} = \mathscr{D} \frac{\partial^2 Z}{\partial x^2}. \tag{11.91}$$

For the spatial problem, neglect of streamwise diffusion is a good approximation for high-velocity streams except near $y = 0$, where there are significant axial and transverse gradients. This assumption is discussed further in Section 11.3.4. The initial and/or boundary conditions are identical for the two problems:

$$\begin{aligned} Z(x < 0, t \text{ or } y = 0) &= 1, \\ Z(x > 0, t \text{ or } y = 0) &= 0, \end{aligned} \tag{11.92}$$

$$\begin{aligned} Z(x \to -\infty, t \text{ or } y > 0) &\to 1, \\ Z(x \to \infty, t \text{ or } y > 0) &\to 0. \end{aligned} \tag{11.93}$$

This section presents the solution in terms of the temporal problem, but the corresponding spatial problem can be obtained by replacing t with y/u_y. This is a standard unsteady diffusion problem, whose solution is

$$Z = \frac{1}{2} \text{erfc}\left(x/\delta_{diff}(t)\right), \tag{11.94}$$

where erfc denotes the complimentary error function, and the time-varying thickness of the diffusive layer, $\delta_{diff}(t)$, is given by

$$\delta_{diff}(t) = 2\sqrt{\mathscr{D} t}, \tag{11.95}$$

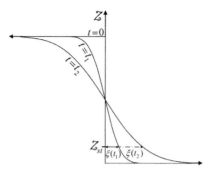

Figure 11.31 Spatial variation of mixture fraction at several instants of time.

showing that it grows as the square root of time. Note that there is no natural time/length scale, so the solution depends on the similarity variable $x/\sqrt{\mathscr{D} t}$, and not x or t independently. The solution for Z at several instants of time is illustrated in Figure 11.31, showing the progressive broadening of the mixing layer.

In the fast-chemistry, limit, the flame position, $\xi(t)$, and velocity, $v_F = d\xi/dt$, are given by the location where $Z = Z_{st}$:

$$\xi(t) = 2(\sqrt{\mathscr{D} t})\mathrm{erf}^{-1}(1 - 2Z_{st}), \tag{11.96}$$

$$v_F(t) = \frac{d\xi(t)}{dt} = (\sqrt{\mathscr{D}/t})\mathrm{erf}^{-1}(1 - 2Z_{st}) \tag{11.97}$$

Note that the velocity of the reaction sheet, v_F, is equivalent to the displacement speed, s_d (i.e., the speed of the surface with respect to the flow normal to itself, see Section 7.1) for this problem because there is no axial flow. This flame motion is simply a requirement of the fact that the reaction sheet must adjust its position so that the fuel and oxidizer diffusive flux rates remain in stoichiometric proportions. In order to interpret this solution, it is helpful to note that the inverse of the error function, $\mathrm{erf}^{-1}(x)$, has the same sign as its argument, x. Thus, the flame sheet remains stationary, i.e., $v_F = 0$ for $Z_{st} = 1/2$. This occurs because the problem is symmetric about $x = 0$ and the flame requires equal diffusive fluxes of fuel and oxidizer. For $Z_{st} < 1/2$, as is the case for hydrocarbon/air flames, more oxidizer than fuel is required and the reaction sheet drifts in the positive x-direction toward the oxidizer reservoir in order to maintain the higher oxidizer gradient needed to maintain stoichiometric mass fluxes. The velocity, v_F, increases without bound as $Z_{st} \to 0$ or 1.

The mass burning rate is given by

$$\dot{m}_F'' = \dot{m}_{Ox}'' + \dot{m}_{Fuel}'' = (1 + \varphi_{Ox})\rho \mathscr{D} \left. \frac{\partial Y_{Fuel}}{\partial x} \right|_{x=\xi(t)}, \tag{11.98}$$

which is given by [2]:

$$\dot{m}''_F = \frac{(1+\varphi_{Ox})\rho\mathscr{D}}{2\sqrt{\pi\mathscr{D}t}} \frac{\exp\left[-\left(\mathrm{erf}^{-1}\{1-2Z_{st}\}\right)^2\right]}{(1-Z_{st})}. \qquad (11.99)$$

This expression shows that the burning rate decreases as the square root of time, in contrast to the constant burning rate for stretch-insensitive premixed flames. Evaluation of the Z_{st} dependency also shows that the burning rate is minimized when $Z_{st} = 1/2$, where $\dot{m}''_F = (1+\varphi_{Ox})\rho\mathscr{D}/\sqrt{\pi\mathscr{D}t}$. It increases symmetrically for higher and lower values of Z_{st}, growing without bound as Z_{st} approaches 0 or 1. Finally, the scalar dissipation rate is given by

$$\chi_{st} = \frac{\exp\left[-2\left(\mathrm{erf}^{-1}\{1-2Z_{st}\}\right)^2\right]}{2\pi t}. \qquad (11.100)$$

This problem is, thus, an example of a flow with zero strain and nonzero scalar dissipation rate. As such, while strain and scalar dissipation are related in certain problems, such as the stagnation flame in Eq. (9.44), their distinction should be emphasized.

11.3.2 Example Problem: Transient Stagnation Flame [2]

The nonpremixed stagnation flame, where opposing fuel and oxidizer jets flowing in the positive and negative x-directions are stagnated against each other, was previously considered in Section 9.6 to illustrate finite chemistry effects in nonpremixed flames. In that section, we also showed that the flow strain rate and scalar dissipation rate at the flame were linearly related through Eq. (9.44). In this section, we return to this problem in the fast-chemistry limit, and consider the response of the flame to a step change in flow strain rate from κ_a to κ_b. It will be shown that the flame exhibits a lag in response associated with the limiting diffusive processes, even for infinitely fast chemistry. This problem is detailed in Poinsot and Veynante [2], and the solution is simply stated without proof. The steady-state solution for the mixture fraction of the stagnation flame is given by

$$Z = \frac{1}{2}\mathrm{erfc}\left[\sqrt{\frac{\kappa_a}{2\mathscr{D}}}x\right]. \qquad (11.101)$$

The time variation of the mixture fraction for the step change in strain rate is given by

$$Z(x,t) = \frac{1}{2}\mathrm{erfc}\left[\frac{\exp(\kappa_b t)\sqrt{\kappa_b/2\mathscr{D}}\,x}{\sqrt{\kappa_b/\kappa_a - 1 + \exp(2\kappa_b t)}}\right], \qquad (11.102)$$

showing the delayed response in Z. The time constant associated with the system response is a nonlinear function of κ_b/κ_a, but for small changes in strain rate is given by $1/\kappa_a$. Note the distinction in solution characteristics of this problem from the self-similar solution derived in the previous section, where the thickness of the diffusive layer grew in a self-similar manner as $\delta_{diff} \sim \sqrt{\mathscr{D}t}$. In this problem, the thickness of the diffusion layer is fixed in time and given by $\delta_{diff} \sim \sqrt{\mathscr{D}/\kappa}$. Finally, note the time-varying relationship between flow strain rate and scalar dissipation rate.

11.3.3 Example Problem: Isothermal Nonpremixed and Premixed Flame Rollup by a Vortex

This section considers the rollup of the flame by a vortex. While the rollup of a passive scalar interface was already analyzed in Section 7.4.3, the problem is modified in nonpremixed and premixed flames by diffusive and/or flame propagation terms, $\nabla \cdot (\mathscr{D} \nabla Z)$ and $s_d |\nabla G|$, respectively. In the nonpremixed flame, successive regions winding around each other cause flame layers to successively grow closer with time, driving up concentration gradients and, therefore, burning rates [39], and causing the vortex core to burn out. This rollup process is clearly illustrated in the time sequence of experimental images shown in Figure 11.32.

Premixed flame propagation normal to itself also has a smoothing action, as discussed in Section 11.2.3. These points are illustrated in Figure 11.33, which shows the rollup of a material line (using the solution from Section 7.4.3) and a flame by a vortex.

In this section, we will analyze the scaling for this burnout in the vortex core. The analysis is quite involved even for the isothermal problem, and so we will simply derive the scalings; the reader is referred to the references for more details [41–44].

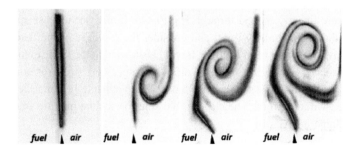

Figure 11.32 Image of nonpremixed flame–vortex interaction problem. Adapted from Cetegen and Basu [40].

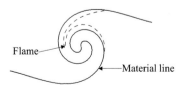

Figure 11.33 Illustration of the effects of diffusion or flame propagation on a flame wrapped up in vortex.

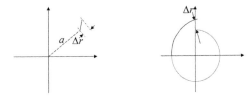

Figure 11.34 Illustration of the stretching of a small segment of the flame surface at two instants of time.

Consider two closely spaced points on the flame that are a distance a from the center of the vortex at time $t = 0$, and a distance Δr from each other.

As shown in Figure 11.34, if this were a nondiffusive material line, the two points separate from each other because of the straining action of the vortex, and the outer point spins an additional 360 degrees around the origin relative to the inner point after the time interval

$$[u_\theta(a + \Delta r) - u_\theta(a)]\Delta t = 2\pi a. \tag{11.103}$$

Expanding the first term in a Taylor series and solving leads to

$$\frac{du_\theta(a)}{dr}\Delta t \Delta r = 2\pi a. \tag{11.104}$$

In reality, premixed flame propagation or nonpremixed flame diffusion occurs simultaneously, which causes closely spaced windings to burn out. The time scale required for diffusive/flame propagation across the winding distance, Δr, is:

$$\text{premixed flame:} \quad \tau_{propagation} = \frac{\Delta r/2}{s_d^u}, \tag{11.105}$$

$$\text{nonpremixed flame:} \quad \tau_{diffusion} \propto \frac{(\Delta r/2)^2}{\mathscr{D}}. \tag{11.106}$$

Equating these time scales to Δt in Eq. (11.104) sets the condition over which the flame will consume the reactants between the windings in the required turnover time, allowing us to solve for the burned core radius:

$$\text{premixed flame:} \quad a \propto \frac{du_\theta(a)}{dr}\frac{(\Delta t)^2 s_d^u}{\pi}, \tag{11.107}$$

$$\text{nonpremixed flame:} \quad a \propto \frac{du_\theta(a)}{dr}\sqrt{\mathscr{D}\Delta t}. \tag{11.108}$$

These are implicit expressions for the product core size, a. Explicit results can be worked out for a potential vortex, whose azimuthal velocity is given by $u_\theta(r) = \Gamma/2\pi r$. Substituting this into the above expressions leads to:

$$\text{premixed flame:} \quad a \propto \left((s_d^u)^2 t^4 \Gamma^2\right)^{1/6}, \tag{11.109}$$

nonpremixed flame: $a \propto (\mathscr{D} t^3 \Gamma^2)^{1/6}$. (11.110)

Taking the ratio of these expressions,

$$\frac{a_{premixed}}{a_{nonpremixed}} \propto \left[\frac{(s_d^u)^2 t}{\mathscr{D}}\right]^{1/6} \propto \left[\frac{t}{\tau_{PF}}\right]^{1/6}, \quad (11.111)$$

where $\tau_{PF} = \delta_F^0 / s^{u,0}$. This scaling shows that the isothermal burnout rate of the premixed flame slightly exceeds that of the nonpremixed flame.

11.3.4 Example Problem: Harmonic Forcing of a Confined, Overventilated Flame [9]

We next consider the response of jet flames to harmonic bulk flow oscillations. This problem incorporates significant flow advection tangential to the flame sheet and some analogous phenomena to the convection of wrinkles that was emphasized in the premixed flame discussion. The geometry is shown in Figure 11.35.

Consider the case of a two-dimensional diffusion flame in a uniform axial flow field, $u_{x,0}$. At the inlet ($x = 0$), fuel and oxidizer flow into the domain. We will use the following inflow conditions:

$$Z(x = 0, y) = \begin{cases} 1 & 0 \le |y| < W_{II}, \\ 0 & W_{II} \le |y| < W_{I}. \end{cases} \quad (11.112)$$

Enforcing this boundary condition enables an analytical solution of the problem. However, in reality there is axial diffusion of fuel into the oxidizer supply and vice versa, so that the solution must actually be solved over a larger domain that includes the fuel/oxidizer supply systems. As such, the boundary condition in Eq. (11.112) implicitly neglects axial diffusion at $x = 0$. Assuming symmetry at $y = 0$ and no diffusion through the walls at $y = W_I$ leads to the following two additional boundary conditions:

$$\begin{aligned} \frac{\partial Z}{\partial y}(x, y = 0) &= 0, \\ \frac{\partial Z}{\partial y}(x, y = W_I) &= 0. \end{aligned} \quad (11.113)$$

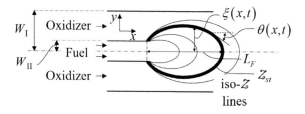

Figure 11.35 Illustration of the forced nonpremixed flame model problem.

11.3 Dynamics of Nonpremixed Flame Sheets

We will derive the solution in the limit of small perturbations, and so expand each variable as $()(x,t) = ()_0(x,y) + ()_1(x,y,t)$, as discussed in Section 2.1. Since the governing equation, Eq. (7.11), is linear, this procedure is not necessary in order to obtain an analytic solution. However, this assumption simplifies the temporal dynamics, as the first-order solution consists only of oscillations at the forcing frequency and not additional harmonics. Moreover, it enables an explicit analytic expression for the space–time dynamics of the flame position, $\xi_1(x,t)$. This is useful in analyzing various controlling features of the flame dynamics.

The mixture fraction field in the absence of forcing is obtained from Eq. (7.11):

$$u_{x,0}\frac{\partial Z_0}{\partial x} = \mathscr{D}\frac{\partial^2 Z_0}{\partial y^2} + \mathscr{D}\frac{\partial^2 Z_0}{\partial x^2}. \tag{11.114}$$

The solution of this equation is derived generally using separation of variables in Exercise 11.11. For the subsequent analysis we will present a simplified version of the solution that neglects axial diffusion, since we have already done so implicitly in formulating the boundary condition in Eq. (11.112). This solution is

$$Z_0(x,y) = \frac{W_{II}}{W_I} + \sum_{n=1}^{\infty}\frac{2}{n\pi}\sin(\mathcal{A}_n)\cos\left(\mathcal{A}_n\frac{y}{W_{II}}\right)\exp\left(-\mathcal{A}_n^2\frac{x}{PeW_{II}}\right), \tag{11.115}$$

where

$$\mathcal{A}_n = n\pi\frac{W_{II}}{W_I} \tag{11.116}$$

and the Peclet number, Pe, is given by

$$Pe = \frac{u_{x,0}W_{II}}{\mathscr{D}}. \tag{11.117}$$

The Peclet number physically corresponds to the relative time scales for convective and diffusive processes to transport mass over a distance W_{II}, $\frac{\tau_{diffusion}}{\tau_{convection}} = \frac{W_{II}^2/\mathscr{D}}{W_{II}/u_{x,0}} = \frac{u_{x,0}W_{II}}{\mathscr{D}}$. As such, the $Pe \gg 1$ limit physically corresponds to the limit where convective processes are much faster than diffusive ones in the axial direction. The solution in Eq. (11.115) is equivalent to the $Pe \gg 1$ limit derived in Exercise 11.11. The implicit solution for the flame sheet position, $\xi_0(x)$, can then be determined from the coordinates where $Z_0(x,\xi_0(x)) = Z_{st}$, yielding

$$Z_{st} = \frac{W_{II}}{W_I} + \sum_{n=1}^{\infty}\frac{2}{n\pi}\sin(\mathcal{A}_n)\cos\left(\mathcal{A}_n\frac{\xi_0(x)}{W_{II}}\right)\exp\left(-\mathcal{A}_n^2\frac{x}{PeW_{II}}\right). \tag{11.118}$$

The flame length is then given implicitly by

$$Z_{st} = \frac{W_{II}}{W_I} + \sum_{n=1}^{\infty}\frac{2}{n\pi}\sin(\mathcal{A}_n)\exp\left(-\mathcal{A}_n^2\frac{L_F}{PeW_{II}}\right). \tag{11.119}$$

Now, consider the effect of adding a uniform bulk fluctuation in flow velocity on this base flow,

$$u_{x,1} = \varepsilon u_{x,0} \exp[-i\omega t]. \tag{11.120}$$

From Eq. (7.11), the dynamical equation for Z_1 is given by

$$\frac{\partial Z_1}{\partial t} + u_{x,0}\frac{\partial Z_1}{\partial x} - \mathscr{D}\frac{\partial^2 Z_1}{\partial y^2} - \mathscr{D}\frac{\partial^2 Z_1}{\partial x^2} = -u_{x,1}\frac{\partial Z_0}{\partial x}. \tag{11.121}$$

Again, the general solution is derived in Exercise 11.12, but we will focus the subsequent analysis on the $Pe \gg 1$ limit:

$$Z_1 = \sum_{n=1}^{\infty}\left[\frac{i\varepsilon(\mathscr{A}_n)^2 \frac{2}{n\pi}\sin(\mathscr{A}_n)}{2\pi St_W Pe}\right]\cos\left(\mathscr{A}_n\frac{y}{W_{II}}\right)\exp\left(-\mathscr{A}_n^2\frac{x}{Pe\,W_{II}}\right)$$
$$\times\left\{1 - \exp\left(2\pi i St_W \frac{x}{W_{II}}\right)\right\}\exp[-i\omega t], \tag{11.122}$$

where the Strouhal number based on the half-width of the fuel lip is defined as

$$St_W = \frac{f W_{II}}{u_{x,0}}. \tag{11.123}$$

This expression can also be written without the summation by writing it in terms of Z_0 as

$$Z_1 = \left[\frac{-i\varepsilon L_F}{2\pi St_{L_F}}\right]\frac{dZ_0}{dx}\left\{1 - \exp\left(2\pi i St_{L_F}\frac{x}{L_F}\right)\right\}\exp[-i\omega t]. \tag{11.124}$$

These expressions for Z_0 and Z_1 provide solutions for the mixture fraction field over the entire domain. The reaction sheet location can be determined from the expression

$$Z_0(x, \zeta_0(x)) + Z_1(x, \zeta_0(x) + \zeta_1(x,t), t) = Z_{st}. \tag{11.125}$$

The temporal evolution of the flame position calculated from this expression is plotted in Figure 11.36. Note the bulk axial pulsing of the flame at lower Strouhal numbers, and the substantial wrinkling on the sheet at higher values.

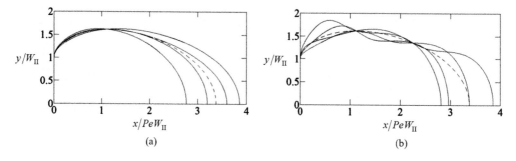

Figure 11.36 Snapshots showing four instantaneous positions of a forced nonpremixed flame at two different forcing frequencies using nominal values of $Z_{st} = 0.3$ and $Pe = 50$: (a) $St_{L_F} = 0.2$, (b) $St_{L_F} = 1.5$. Unforced flame indicated by dashed lines.

11.3 Dynamics of Nonpremixed Flame Sheets

We will next consider the dynamics of the flame position itself. Expanding Eq. (11.125) to first order leads to the following explicit expression for fluctuating flame position:

$$\xi_{1,n}(x,t) = -\frac{Z_1(x, y = \xi_0(x), t)}{\frac{\partial Z_0}{\partial n}(x, y = \xi_0(x))} = -\frac{Z_1(x, y = \xi_0(x), t)}{|\nabla Z_0|_{(x,y=\xi_0(x))}}, \qquad (11.126)$$

where $\xi_{1,n}$ is measured normal to the mean flame surface. We use this normal displacement measurement for quantifying flame motion in this section due to the substantial change in angle of the reaction sheet with axial location (in contrast, if flame motion were measured as radial displacement, its value would be infinite at the flame tip). Pulling these results together leads to

$$\xi_{1,n}(x,t) = -\left(\frac{dZ_0/dx}{|\nabla Z_0|}\right)_{(x,y=\xi_0(x))} \left[\frac{-i\varepsilon L_F}{2\pi St_{L_F}}\right]\left\{1 - \exp\left(2\pi i St_{L_F}\frac{x}{L_F}\right)\right\}\exp[-i\omega t]. \qquad (11.127)$$

The $\left(\frac{dZ_0/dx}{|\nabla Z_0|}\right)$ term can be written in terms of the local angle of the flame using the relations

$$|\nabla Z_0| = \sqrt{\left(\frac{dZ_0}{dx}\right)^2 + \left(\frac{dZ_0}{dy}\right)^2} = \frac{\frac{dZ_0}{dx}}{\sin\theta(x)}, \qquad (11.128)$$

where θ denotes the angle of the flame with respect to the axial coordinate (see Figure 11.35). Using these results, the solution for $\xi_{1,n}(x,t)$ can be written in the following very simple form:

$$\xi_{1,n}(x,t) = \frac{i\varepsilon u_{x,0}}{2\pi f} \sin\theta(x)\left\{1 - \exp\left(i2\pi f\frac{x}{u_{x,0}}\right)\right\}\exp[-i2\pi ft]. \qquad (11.129)$$

For reference, the corresponding fluctuations of an attached premixed flame with constant flame speed subjected to bulk flow oscillations are given by

$$\xi_{1,n}(x,t) = \frac{i\varepsilon u_{x,0}}{2\pi f}\sin\theta\left\{1 - \exp\left(i2\pi f\frac{x}{u_{x,0}\cos^2\theta}\right)\right\}\exp[-i2\pi ft], \qquad (11.130)$$

where the angle θ is a constant. Notice the similarities in the premixed and nonpremixed solutions, with the key difference lying in the spatial phase dependence, the $1 - e^{i2\pi f x/u_{x,0}}$ term. This difference reflects the influence of premixed flame propagation on wrinkle convection speeds. In contrast, the nonpremixed flame does not propagate and wrinkles convect downstream at a speed of $u_{x,0}$. In both cases, local maxima and minima arise through this term, $1 - e^{i2\pi f x/u_{x,0}} = 2\sin(\pi f x/u_{x,0})e^{i(\pi f x/u_{x,0} - \pi/2)}$, due to interference between wrinkles generated at the $x = 0$ boundary and disturbances excited locally. To summarize, Eq. (11.129) explicitly illustrates the effects of velocity fluctuations normal to the flame sheet in inducing flame wrinkles (through the $\sin\theta(x)$ term) and axial flow in convecting wrinkles downstream. Section 11.4 further discusses and compares these

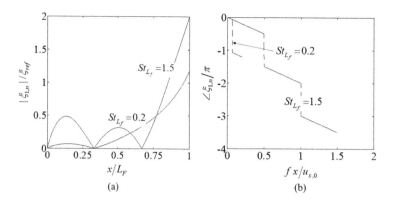

Figure 11.37 Axial dependence of the magnitude and phase of flame response, where $\zeta_{ref} = \frac{\varepsilon \cdot u_{x,0}}{Pe \cdot 2\pi f}$, and using nominal values of $Z_{st} = 0.3$, $Pe = 50$, and (a) $St_{L_F} = 0.2$, (b) $St_{L_F} = 1.5$.

wave propagation features for premixed and nonpremixed flames, incorporating finite Pe effects, showing that they lead to dissipation and dispersion of wrinkles.

These features are illustrated in Figure 11.37, which plots the magnitude and phase of the nonpremixed results. The nodes and local maxima and minima referred to in the previous paragraph are clearly evident in the figure. The phase rolls off linearly with axial distance, again reflecting the convection process described by the $1 - e^{i2\pi f x/u_{x,0}}$ term, and jumps 180 degrees across the nodes.

11.4 Aside: Dissipation and Dispersion of Disturbances on Premixed and Nonpremixed Flames

Sections 11.2.2.3 and 11.3.4 analyzed linear dynamics of anchored premixed and nonpremixed flames, showing that flame wrinkles were excited at the attachment point and convected downstream with constant amplitude and at constant phase speed. These effects are altered in nonpremixed flames through finite Pe effects and in premixed flames by stretch effects.

Consider the premixed case. In order to compare the result directly to the nonpremixed case, we will write the flame position as a function of x for this section, $\zeta(x,t)$. Returning to Eq. (11.36), perturbations in $v_{Fn,1}$ lead to unsteady stretch on the flame by causing fluctuations in the hydrodynamic, κ_s, and curvature, κ_{curv}, stretch terms; see Section 9.3.2. For thermodiffusively stable/unstable flames, curvature effects tend to smooth out/amplify these wrinkles through purely linear mechanisms[5] [45, 46]. To illustrate this point, consider the linearized modification of the flame displacement speed caused by flame curvature. This may be expressed in terms of derivatives of ζ_1 as

---

[5] In contrast, for stretch-insensitive flames with constant burning velocity, nonlinear processes are required to smooth out flame wrinkles, as discussed in Section 11.2.3.1.

11.4 Dissipation and Dispersion of Disturbances

$$\frac{s_{d,1}^u}{s_0^{u,0}} = -\frac{\delta_{M,d}^u \kappa_1}{s^{u,0}} = -\delta_{M,d}^u \cos^3\theta \cdot \frac{\partial^2 \xi_1}{\partial x^2}. \quad (11.131)$$

Substituting this expression into Eq. (11.37) leads to the following expression for the linearized flame dynamics:

$$\frac{\partial \xi_1}{\partial t} + u_{x,0} \cos\theta \frac{\partial \xi_1}{\partial x} = -\frac{u_{n,1}^u}{\cos\theta} + (s^{u,0} \cos^2\theta)\delta_{M,d}^u \frac{\partial^2 \xi_1}{\partial x^2}. \quad (11.132)$$

Equation (11.132) shows that the flame dynamics equation has a second-order spatial derivative term, which resembles a diffusive term. Define the dimensionless stretch sensitivity through the Markstein number [45],

$$Ma_a = \frac{\delta_{M,d}^u \sin\theta}{a}. \quad (11.133)$$

The solution is most easily interpretable for large Pe and small Ma_a. By following this procedure to $O(1/Pe^2)$ and $O(Ma_{a^2})$, we can develop the following result for premixed or nonpremixed flames subjected to axial bulk forcing, where $St_a = \omega a/u_{c,F}$, and $u_{c,F}$ is the flame wrinkle convection speed defined in Table 11.1:

$$\frac{\xi_{1,n}(x,t)}{a} = -\left[\frac{\varepsilon \sin\theta(x)e^{-i\omega t}}{iSt_a}\right]\left[1 - \exp\left(i\frac{\omega x}{u_{c,F}}\right)\exp\left(\frac{-\zeta x}{a}\right)\exp\left(i\beta\frac{\omega x}{u_{c,F}}\right)\right]$$

$$+ O\left(\frac{1}{Pe^3}, Ma_a^3\right). \quad (11.134)$$

The waveform term is parameterized by a flame wrinkle convection speed, $u_{c,F}$, axial dissipation rate, ζ, and dispersion term, β, as defined in Table 11.1.

Note the symmetry in the solutions between Ma_a and $1/Pe$. Consider the wave propagation term in Eq. (11.134). Consistent with the discussion in Section 3.2, the leading unity term derives from the particular solution of the equation, and lacks spatial dependence because of the nature of the assumed bulk forcing. The second term describes a decaying, dispersive traveling wave generated at the boundary, $x = 0$, resulting from the assumption of flame attachment, i.e., $\xi_{1,n}(x = 0, t) = 0$, or fixed mixture fraction at the burner outlet, $Z_1(x = 0, y, t) = 0$, for the premixed and nonpremixed cases, respectively.

The leading-order expansion of this expression, $1 - e^{i\omega x/u_{c,F}}$, was previously presented and discussed in Section 11.3.4. Consider next $O(1/Pe)$ or $O(Ma_a)$ terms

Table 11.1 Propagation, dissipation, and dispersion terms in Eq. (11.134).

	Premixed	Nonpremixed
Propagation speed, $u_{c,F}$	$u_{x,0} \cos^2\theta$	$u_{x,0} + O(1/Pe^2)$
Dissipation, ζ	$(St_a)^2 Ma_a$	$(St_a)^2\left(\frac{1}{Pe}\right)$
Dispersion, β	$-2(St_a)^2 Ma_a^2$	$-2(St_a)^2\left(\frac{1}{Pe}\right)^2$

which, as shown in Table 11.1, cause wrinkles to decay exponentially with downstream distance. This causes the interference effect discussed above to become imperfect, an effect that increases as St_a^2. The mechanisms for wave dissipation for the two flame types are different: for premixed flames, it is due to the dependence of the flame speed on flame surface curvature, which causes positive Markstein length flames to be thermodiffusively stable. For nonpremixed flames, it is due to the progressive smoothing by diffusion of the spatial variations in the Z field with downstream distance. Lastly, $O(1/Pe^2)$ and $O(Ma_a^2)$ effects introduce dispersion, as shown explicitly by the β term in Table 11.1, i.e., a frequency-dependent wave propagation speed.

11.5 Aside: Harmonic Forcing of Turbulent, Premixed Flames

Real flames inevitably exist in a turbulent flow environment, and so the flame is simultaneously disturbed by both spatiotemporally narrowband and broadband turbulence fluctuations. While the G or Z equation approaches, coupled with the triple decomposition in Section 2.8, can certainly be used to analyse the response of flames to these disturbances, such problems quickly become analytically intractable. Referring to the images in Figure 2.23, which show illustrative instantaneous and phase-averaged images, it is desirable to develop expressions for the phase-averaged flame position, $\langle \xi \rangle$, or transfer function, $\frac{\langle \hat{Q}_1 \rangle / Q_0}{\langle \hat{u}_1 \rangle / u_0}$, without actually having to solve the full turbulent problem and then phase averaging. Several approaches have been used in the literature. The "quasi-laminar" modeling approach essentially mirrors the treatment described earlier, replacing the nominal flame position, ξ_0, by its average, $\bar{\xi}$; the laminar flame speed, s_d^u, by an average turbulent flame speed, $s_{T,d}$; and the instantaneous velocity by its ensemble-averaged value (note the analogies of this approach to the mean flow stability analysis discussed in Section 3.5):

$$\frac{\partial \langle \xi \rangle}{\partial t} - \langle u_x \rangle + \langle u_y \rangle \frac{\partial \langle \xi \rangle}{\partial y} + \langle u_z \rangle \frac{\partial \langle \xi \rangle}{\partial z} = s_{T,d}(y,z,t) \left[1 + \left(\frac{\partial \langle \xi \rangle}{\partial z} \right)^2 + \left(\frac{\partial \langle \xi \rangle}{\partial y} \right)^2 \right]^{1/2}.$$

(11.135)

Empirically, this quasi-laminar approach has been found to yield reasonable results, using measured mean velocity fields and flame positions for $s_{T,d}$ [47].

Measurements and asymptotic theory have shown that nonlinear coupling effects introduce additional phenomena. First, not included in the above expression is a shift in the average convection speed of wrinkles along the flame, $\bar{u}_t$, due to correlations between flame wrinkles and turbulent velocity fluctuations [48]. Second, the phase-averaged turbulent burning rate exhibits coherent modulation at the harmonic forcing frequency. In other words, the turbulent flame speed, $s_{T,d}$, which is generally taken as an average of the flame speed/burning rate, exhibits coherent oscillations. An approximate fit to the data, such as shown in Figure 11.38, is given by

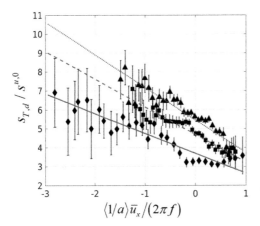

Figure 11.38 Dependence of the ensemble-averaged turbulent displacement speed on ensemble-averaged curvature for three turbulence intensities, $u'/\bar{u}_x = 15.7\%$ (solid line, diamonds), 29.5% (dashed line, squares), and 33.1% (dotted line, triangles) [49]. © 2017 Cambridge University Press, reprinted with permission.

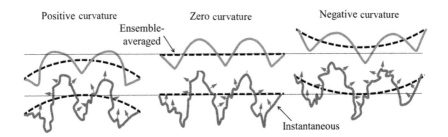

Figure 11.39 Schematic of the interaction of narrowband flame curvature with broadband turbulent wrinkling, showing how ensemble-averaged curvature influences the turbulent displacement speed [48]. © 2013 Cambridge University Press, reprinted with permission.

$$s_{T,D}(y,t) = s_{T,0}(y)\left(1 - \delta_{MT}(y)\left\langle\frac{1}{a(y,t)}\right\rangle\right), \tag{11.136}$$

where $a(y,t)$ is the radius of flame curvature, $\langle 1/a \rangle$ is the phase-averaged flame curvature, and δ_{MT} is the sensitivity of the turbulent flame speed to ensemble-averaged curvature.

In other words, the ensemble-averaged displacement speed increases in regions of negative curvature, and decreases in regions of positive curvature, as shown in Figure 11.39.

Note the clear analogy to the stretch sensitivity of laminar flames. For this reason, δ_{MT} is termed the "turbulent displacement Markstein length." However, this analogy should not be pressed too far, because this curvature sensitivity exists here for completely different reasons. For example, δ_{MT} is nonzero for stretch-insensitive laminar flames, and also appears to only take positive values, although only limited measurements of its characteristics exist.

11.6 Aside: Forced Response Effects on Natural Flame Instabilities

Section 9.8 showed that flames have natural instabilities that cause them to spontaneously wrinkle or pulsate. Oscillatory acceleration of the flow, and therefore the flame sheet, modifies the baseline flame stability characteristics by adding a periodic acceleration of the two different density regions across the flame [50–58]. To illustrate, Figure 11.40 shows calculated stability regions in disturbance wavenumber–amplitude space for a lean methane–air flame. The results at $u_1/s^{u,0} = 0$ correspond to the unforced case, whose stability characteristics are discussed in Section 9.8. Also, these calculations show $s^{u,0}$ as the amplitude normalization factor for a flame that is normal to the flow at low flow velocities. For flames with tangential flow, the values of the mean and fluctuating quantities normal to the flame should be used for normalization.

Note that the flame remains unstable through the Darrieus–Landau instability for low disturbance amplitudes, $u_1/s^{u,0} < 3.5$ for this particular calculation. Forcing amplitudes where $u_1/s^{u,0} > 3.5$ stabilizes the Darrieus–Landau instability [54, 57] – indeed, experiments have exploited this fact to create unwrinkled flames that would otherwise be unstable [60]. An approximate expression for this threshold amplitude required to suppress the Darrieus–Landau instability is

$$u_1/s^{u,0} = \left(2\sigma_p \frac{\sigma_p+1}{\sigma_p-1}\right)^{1/2} \text{ [57]}.$$

The figure also shows that higher forcing amplitudes, $u_1/s^{u,0} > 5$ for this calculation, lead to a new parametric instability. This instability occurs due to the oscillatory acceleration of the density interface. It is same type of instability that occurs when a variable-density region is subjected to acceleration, whether steady, impulsive, or time harmonic, and is discussed in Section 3.8. This parametric instability leads to cellular structures on the flame front that oscillate at half the frequency of the imposed excitation, $f_0/2$. While the analysis used to generate plots like Figure 11.40 was done

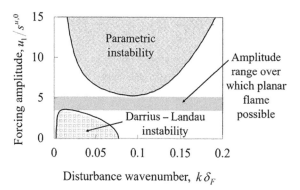

Figure 11.40 Calculated influence of the periodic flow oscillation amplitude, $u_1/s^{u,0}$, on methane–air flame stability ($\phi = 0.8$). Adapted from Aldredge and Killingsworth [59].

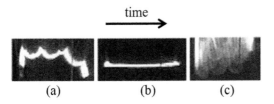

Figure 11.41 Sequence of images showing the flame as it propagates downward in a tube [56]. (a) Flame wrinkled by Darrieus–Landau instability mechanism. (b) "Planarization" of flame by low-amplitude velocity oscillations. (c) Parametric instability induced by large-amplitude acoustic oscillations. Images courtesy of R.C. Aldredge.

for laminar flames, the presence of this parametric instability has been noted in turbulent flow fields as well [61–63]. These different behaviors are illustrated in Figure 11.41, showing a sequence of images of a flame propagating downward in a tube with a thermoacoustic instability present, so that the amplitude of oscillations grows from left to right [56]. These images illustrate (a) the flame wrinkled by the Darrieus–Landau instability, (b) "planarization" of the flame by low-amplitude velocity oscillations, and (c) parametric instability induced by large-amplitude acoustic oscillations.

Exercises

11.1 The turbulent combustion diagram in Figure 11.1 shows lines of constant turbulent Reynolds number $Re_L = u_{rms}L_{11}/\nu_\nu$. Derive the relationship $Re_L = \left(\frac{u_{rms}}{s^u}\right)\left(\frac{L_{11}}{\delta_F}\right)$ between u_{rms}/s^u and L_{11}/δ_F assuming a unity Prandtl number.

11.2 The turbulent combustion diagram in Figure 11.1 shows lines of constant Damköhler number. Derive the relationship between the Reynolds number and Damköhler numbers, $Re_L = \frac{Da_L^2}{Da_\eta^2}$.

11.3 The Damköhler number based on the Kolmogorov scales, Da_η, and the ratio of flame thickness to the Kolmogorov length scale, δ_F/L_η, are related by the expression $Da_\eta \left(\frac{\delta_F}{L_\eta}\right)^2 = 1$. Derive this expression.

11.4 Derive Eq. (11.31) for the dynamics of the flame position in response to a step change in velocity from $u_{x,a}$ to $u_{x,b}$ for the case $u_{x,b} > u_{x,a}$. Use the Rankine–Hugoniot jump condition to relate the value of g before and after the discontinuity (see Section 11.2.1.3) [33].

11.5 Exercise 11.4 analyzed the flame response to a step increase in flow velocity and showed that this leads to a cusp, or "shock," on the flame. In this exercise we consider a step decrease in flow velocity, which leads to an "expansion wave" type solution. Derive the following solution for the dynamics of the flame position in response to a step change in velocity from $u_{x,a}$ to $u_{x,b}$ for the case $u_{x,b} < u_{x,a}$ (see Section 11.2.1.3):

$$g(y,t) = \begin{cases} \sqrt{\left(\dfrac{u_{x,b}}{s_d^u}\right)^2 - 1} & y < s_d^u\sqrt{1 - \left(s_d^u/u_{x,b}\right)^2}\cdot t, \\ \sqrt{\dfrac{y^2}{\left(s_d^u\right)^2 t^2 - y^2}} & s_d^u\sqrt{1 - \left(s_d^u/u_{x,b}\right)^2}\cdot t < y < s_d^u\sqrt{1 - \left(s_d^u/u_{x,a}\right)^2}\cdot t, \\ \sqrt{\left(\dfrac{u_{x,a}}{s_d^u}\right)^2 - 1} & s_d^u\sqrt{1 - \left(s_d^u/u_{x,a}\right)^2}\cdot t < y. \end{cases}$$

(11.153)

11.6 Derive Eq. (11.51), which is an expression describing the characteristics of an anchored flame response near $y = 0$ to disturbances for an arbitrary disturbance field.

11.7 The extended example in Section 11.2.2.3 analyzed the response characteristics of a flame subjected to a decaying, convecting disturbance. Derive the solution for the flame position, ξ_1, given in Eq. (11.49) from the linearized flame position equation, Eq. (11.39).

11.8 Using Eq. (11.37), derive asymptotic expressions for the flame position magnitude and phase near $y = 0$, subject to the boundary condition $\xi_1(y = 0, t) = 0$.

11.9 Derive the expressions for the flame brush thickness and turbulent consumption speed in Eq. (11.70) from Eq. (11.65).

11.10 Analyze the far-field behavior of flame wrinkles due to kinematic restoration effects. Consider a flame with periodic wrinkles of wavelength λ propagating into a quiescent medium, as discussed in Section 11.2.3.1. After a certain transient, the flame asymptotes to a far-field solution that is controlled by the leading edges of the flame, and has a shape which equals the locations of arcs radiating out from these leading edges, as shown in Figure 11.45. Write expressions for the position of these leading edge arcs, as well as the magnitude

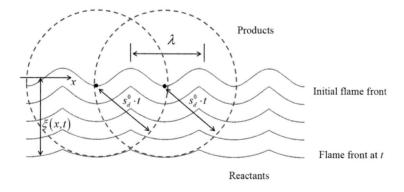

Figure 11.45 Sketch showing far-field behavior of flame with periodic wrinkle propagating into quiescent reactants. Image courtesy of D.H. Shin.

of the flame wrinkling at the wavenumber corresponding to the initial wavelength of wrinkling, $k = 2\pi/\lambda$.

11.11 Nonpremixed jet flame solution. Derive the solution for the unforced mixture fraction field in Eq. (11.115). This can be done using separation of variables.

11.12 Forced nonpremixed jet flame solution. Derive the solution for the forced mixture fraction field in Eq. (11.122). This can be done assuming separation of variables.

References

1. Peters N., *Turbulent Combustion*. 1st ed. 2000, Cambridge University Press.
2. Poinsot T. and Veynante D., *Theoretical and Numerical Combustion*. 2nd ed. 2005, R.T. Edwards, Inc.
3. Echekki T. and Mastorakos E., *Turbulent Combustion Modeling: Advances, New Trends and Perspectives*. Vol. 95. 2010, Springer.
4. Wang G.H., Clemens N.T., Varghese P.L., and Barlow R.S., Turbulent time scales in a nonpremixed turbulent jet flame by using high-repetition rate thermometry. *Combustion and Flame*, 2008, **152**(3): pp. 317–335.
5. Dasgupta D., *Turbulence–Chemistry Interactions for Lean Premixed Flames*. 2018, Georgia Institute of Technology.
6. Roberts W.L., Driscoll J.F., Drake M.C., and Goss L.P., Images of the quenching of a flame by a vortex – to quantify regimes of turbulent combustion. *Combustion and Flame*, 1993, **94**(1–2): pp. 58–69.
7. Driscoll J.F., Turbulent premixed combustion: flamelet structure and its effect on turbulent burning velocities. *Progress in Energy and Combustion Science*, 2008, **34**(1): pp. 91–134.
8. Clanet C., Searby G., and Clavin P., Primary acoustic instability of flames propagating in tubes: cases of spray and premixed combustion. *Journal of Fluid Mechanics*, 1999, **385**: pp. 157–197.
9. Magina N., Acharya V., and Lieuwen T., Forced response of laminar non-premixed jet flames. *Progress in Energy and Combustion Science*, 2019, **70**: pp. 89–118.
10. McIntosh A.C., The linearized response of the mass burning rate of a premixed flame to rapid pressure changes. *Combustion Science and Technology*, 1993, **91**: pp. 329–346.
11. McIntosh A.C., Batley G., and Brindley J., Short length scale pressure pulse interactions with premixed flames. *Combustion Science and Technology*, 1993, **91**: pp. 1–13.
12. Beardsell G. and Blanquart G., Impact of pressure fluctuations on the dynamics of laminar premixed flames. *Proceedings of the Combustion Institute*, 2019, **37**(2): pp. 1895–1902.
13. Markstein G., *Nonsteady Flame Propagation*. 1964. Published for and on behalf of Advisory Group for Aeronautical Research and Development, North Atlantic Treaty Organization by Pergamon Press.
14. Humphrey L., Acharya V., Shin D.-H., and Lieuwen T., Coordinate systems and integration limits for global flame transfer function calculations. *International Journal of Spray and Combustion Dynamics*, 2014, **6**(4): pp. 411–416.
15. Whitham G., *Linear and Nonlinear Waves*. Vol. 226. 1974, New York: Wiley.
16. Petersen R.E. and Emmons H.W., Stability of laminar flames. *Physics of Fluids*, 1961, **4**(4): pp. 456–464.

17. Lieuwen T., Modeling premixed combustion-acoustic wave interactions: A review. *Journal of Propulsion and Power*, 2003, **19**(5): pp. 765–781.
18. Fleifil M., Annaswamy A.M., Ghoneim Z.A., and Ghoniem A.F., Response of a laminar premixed flame to flow oscillations: A kinematic model and thermoacoustic instability results. *Combustion and Flame*, 1996, **106**(4): pp. 487–510.
19. Shin D., Shanbhogue S., and Lieuwen T., Premixed flame kinematics in an axially decaying, harmonically oscillating vorticity field, in *44th AIAA/ASME/ SAE/ASEE Joint Propulsion Conference*. 2008, pp. 1–16.
20. Boyer L. and Quinard J., On the dynamics of anchored flames. *Combustion and Flame*, 1990, **82**(1): p. 51–65.
21. Shin D.H., Romani M., and Lieuwen T.C., Interference effects in controlling flame response to harmonic flow perturbations, in *49th AIAA Aerospace Sciences Meetings including the New Horizons Forum and Aerospace Exposition*. 2011, Orlando, FL: AIAA.
22. Acharya V., Emerson B., Mondragon U., Brown C., Shin D.H., McDonell V., and Lieuwen T., Measurements and analysis of bluff-body flame response to transverse excitation, in *2011 ASME-IGTI Turbo Expo*. 2011.
23. Shin D., Plaks D.V., and Lieuwen T., Dynamics of a longitudinally forced, bluff body stabilized flame. *Journal of Propulsion and Power*, 2011, **27**(1): pp. 105–116.
24. Acharya V.S., Shin D.-H., and Lieuwen T., Premixed flames excited by helical disturbances: flame wrinkling and heat release oscillations. *Journal of Propulsion and Power*, 2013, **29**(6): pp. 1282–1291.
25. Aldredge R.C., The speed of isothermal-front propagation in isotropic, weakly turbulent flows. *Combustion Science and Technology*, 2006, **178**(7–9): pp. 1201–1216.
26. Hemchandra S., *Dynamics of Premixed Turbulent Flames in Acoustic Fields*. 2009, Atlanta, GA: Georgia Institute of Technology.
27. Hemchandra S. and Lieuwen T., Local consumption speed of turbulent premixed flames: an analysis of "memory effects." *Combustion and Flame*, 2010, **157**(5): pp. 955–965.
28. Hemachandra S., Shreekrishna, and Lieuwen T., Premixed flames response to equivalence ratio perturbations, in *Joint Propulsion Conference*. 2007, Cincinnatti, OH.
29. Bendat J.S. and Piersol A.G., *Random Data: Analysis and Measurement Procedures*. 2000, John Wiley & Sons, Inc.
30. Pope S.B., *Turbulent Flows*. 2000, Cambridge University Press.
31. Lipatnikov A. and Chomiak J., Turbulent flame speed and thickness: Phenomenology, evaluation, and application in multi-dimensional simulations. *Progress in Energy and Combustion Science*, 2002, **28**(1): pp. 1–74.
32. Law C.K. and Sung C.J., Structure, aerodynamics, and geometry of premixed flamelets. *Progress in Energy and Combustion Science*, 2000, **26**(4–6): pp. 459–505.
33. Evans L., *Partial Differential Equations*, Graduate Studies in Mathematics, Vol. 19. 1998, American Mathematics Society.
34. Ducruix S., Durox D., and Candel S., Theoretical and experimental determinations of the transfer function of a laminar premixed flame. *Proceedings of the Combustion Institute*, 2000, **28**(1): pp. 765–773.
35. Shanbhogue S.J., Seelhorst M., and Lieuwen T., Vortex phase-jitter in acoustically excited bluff body flames. *International Journal of Spray and Combustion Dynamics*, 2009, **1**(3): pp. 365–387.
36. Amato A. and Lieuwen T.C., Analysis of flamelet leading point dynamics in an inhomogeneous flow. *Combustion and Flame*, 2013, **161**: pp. 1337–1347.

37. Embid P.F., Majda A.J., and Souganidis P.E., Effective geometric front dynamics for premixed turbulent combustion with separated velocity scales. *Combustion Science and Technology*, 1994, **103**: pp. 85–115.
38. Damköhler G., *The Effect of Turbulence on the Flame Velocity in Gas Mixtures*. 1947, Washington, DC: National Advisory Committee for Aeronautics.
39. Lakshminarasimhan K., Ryan M.D., Clemens N.T., and Ezekoye O.A., Mixing characteristics in strongly forced non-premixed methane jet flames. *Proceedings of the Combustion Institute*, 2007, **31**(1): pp. 1617–1624.
40. Cetegen B.M. and Basu S., Soot topography in a planar diffusion flame wrapped by a line vortex. *Combustion and Flame*, 2006, **146**(4): pp. 687–697.
41. Marble F., Growth of a diffusion flame in the field of a vortex, in *Recent Advances in the Aerospace Sciences*, C. Casci and C. Bruno eds. 1985, New York: Plenum Press, pp. 395–413.
42. Peters N. and Williams F., Premixed combustion in a vortex. *Proceedings of the Combustion Institute*, 1988, **22**: p. 495.
43. Cetegen B.M. and Sirignano W.A., Study of mixing and reaction in the field of a vortex. *Combustion Science and Technology*, 1990, **72**: pp. 157–181.
44. Cetegen B.M. and Mohamad N., Experiments on liquid mixing and reaction in a vortex. *Journal of Fluid Mechanics*, 1993, **249**: pp. 391–414.
45. Wang H.Y., Law C.K., and Lieuwen T., Linear response of stretch-affected premixed flames to flow oscillations. *Combustion and Flame*, 2009, **156**(4): pp. 889–895.
46. Preetham T., Sai K., Santosh H., and Lieuwen T., Linear response of laminar premixed flames to flow oscillations: Unsteady stretch effects. *Journal of Propulsion and Power*, 2010, **26**(3): pp. 524–532.
47. Acharya V., Malanoski M., Aguilar M., and Lieuwen T., Dynamics of a transversely excited swirling, lifted flame: Flame response modeling and comparison with experiments. *Journal of Engineering for Gas Turbines and Power*, 2014, **136**(5): p. 051503.
48. Shin D.H. and Lieuwen T., Flame wrinkle destruction processes in harmonically forced, turbulent premixed flames. *Journal of Fluid Mechanics*, 2013, **721**: pp. 484–513.
49. Humphrey L., Emerson B., and Lieuwen T., Premixed turbulent flame speed in an oscillating disturbance field. *Journal of Fluid Mechanics*, 2018, **835**: pp. 102–130.
50. Kaskan W., *An Investigation of Vibrating Flames*. 1953, Elsevier.
51. Searby G., Acoustic instability in premixed flames. *Combustion Science and Technology*, 1992, **81**(4): pp. 221–231.
52. Clavin P., Pelcé P., and He L., One-dimensional vibratory instability of planar flames propagating in tubes. *Journal of Fluid Mechanics*, 1990, **216**(1): pp. 299–322.
53. Markstein G., Flames as amplifiers of fluid mechanical disturbances, in *Proceedings of the Sixth National Congress on Applied Mechanics*, 1970, pp. 11–33.
54. Searby G. and Rochwerger D., A parametric acoustic instability in premixed flames. *Journal of Fluid Mechanics*, 1991, **231**(1): pp. 529–543.
55. Pelcé P. and Rochwerger D., Vibratory instability of cellular flames propagating in tubes. *Journal of Fluid Mechanics*, 1992, **239**(1): pp. 293–307.
56. Vaezi V. and Aldredge R., Laminar-flame instabilities in a Taylor–Couette combustor. *Combustion and Flame*, 2000, **121**(1–2): pp. 356–366.
57. Bychkov V., Analytical scalings for flame interaction with sound waves. *Physics of Fluids*, 1999, **11**: pp. 3168–3173.
58. Bychkov V. and Liberman M., Dynamics and stability of premixed flames. *Physics Reports*, 2000, **325**(4–5): pp. 115–237.

59. Aldredge R. and Killingsworth N., Experimental evaluation of Markstein-number influence on thermoacoustic instability. *Combustion and Flame*, 2004, **137**(1–2): pp. 178–197.
60. Clanet C. and Searby G., First experimental study of the Darrieus–Landau instability. *Physical Review Letters*, 1998, **80**(17): pp. 3867–3870.
61. Vaezi V. and Aldredge R., Influences of acoustic instabilities on turbulent-flame propagation. *Experimental Thermal and Fluid Science*, 2000, **20**(3–4): pp. 162–169.
62. Savarianandam V.R. and Lawn C.J., Thermoacoustic response of turbulent premixed flames in the weakly wrinkled regime. *Proceedings of the Combustion Institute*, 2007, **31**(1): pp. 1419–1426.
63. Lawn C.J., Williams T.C., and Schefer R.W., The response of turbulent premixed flames to normal acoustic excitation. *Proceedings of the Combustion Institute*, 2005, **30**(2): pp. 1749–1756.
64. Shin D. and Lieuwen T., Flame wrinkle destruction processes in harmonically forced, laminar premixed flames, in *50th AIAA Aerospace Sciences Meeting*. 2012, Nashville, TN.

12 Forced Response II: Heat Release Dynamics

Chapter 11 described the dynamics of flamelets forced by velocity or burning rate oscillations and illustrated the key physics controlling the spatiotemporal dynamics of the flame position. This chapter focuses on the impacts of these disturbances on the mass burning rate and/or heat release rate itself. For example, a key quantity of interest for the thermoacoustic instability problem is the heat release fluctuations that are induced by the flame disturbances. Section 12.1 gives an overview of the basic mechanisms through which flow disturbances lead to heat release oscillations, and differentiates between velocity coupling, fuel/air ratio coupling, pressure coupling, and acceleration coupling. Section 12.2 treats the effects of the flame configuration on its sensitivity to these disturbances, such as geometry or reactant premixing.

Given these basic considerations, Section 12.3 specifically works through a number of problems for harmonically excited flames. The key topic addressed in this section is the gain and phase response of the unsteady heat release in response to disturbances. For example, given a disturbance velocity fluctuation of magnitude ε, what is the magnitude and phase shift of the resulting unsteady heat release, $\dot{Q}(t)$? This phase shift has profound implications on thermoacoustic instability limits in particular.

Section 12.4 then treats broadband flame excitation, turbulent flame speeds, and the generation of sound by turbulent flames. Section 12.4.1 discusses the influence of broadband fluctuations on the time-averaged burning rate (i.e., the turbulent flame speed), a key problem in turbulent combustion. Section 12.4.2 treats the spectrum of heat release fluctuations induced by broadband flow disturbances, an important problem for combustion noise applications. Finally, Section 12.4.3 treats the sound generated by unsteady heat release fluctuations.

12.1 Overview of Forced Flame Response Mechanisms

This section considers the factors influencing the heat release rate and how its fluctuations and time average are influenced by flow and thermodynamic disturbances. The heat release per unit surface area of the flamelet is given by

$$q(\vec{x}, t) = \dot{m}_F''(\vec{x}, t) \, h_R(\vec{x}, t), \tag{12.1}$$

where $\dot{m}''_F$ is the reactant mass consumption rate per unit area, and h_R is the heat release per unit mass of reactant consumed. This mass burning rate term is controlled by quite different processes for premixed and nonpremixed flames [1]. For premixed flames, the mass burning rate is given by

$$\text{Premixed flame:} \quad \dot{m}''_F = \rho^u s^u_c, \tag{12.2}$$

where ρ^u is the density and s^u_c the laminar consumption speed of the unburned reactant. For nonpremixed flames, it is given by [1]

$$\text{Nonpremixed flame:} \quad \dot{m}''_F = -\frac{(1+\varphi_{Ox})^2}{\varphi_{Ox}} \rho \mathscr{D} \left.\frac{\partial Z}{\partial n}\right|_{Z=Z_{st}}, \tag{12.3}$$

where n represents the direction normal to the flame surface into the oxidizer. For example, considering the two-dimensional flames shown in Figure 11.35, $\partial Z/\partial n$ is given by

$$\frac{\partial Z}{\partial n} = \frac{\partial Z}{\partial x}\sin\theta - \frac{\partial Z}{\partial y}\cos\theta. \tag{12.4}$$

The spatially integrated heat release, $\dot{Q}(t)$, for a flamelet is given by

$$\dot{Q}(t) = \int_A \dot{m}''_F h_R \, dA, \tag{12.5}$$

where the integral is performed over the flamelet surface. Equation (12.5) shows three fundamentally different mechanisms generating heat release disturbances: fluctuations in burning rate, heat of reaction, and flame surface area.

Start with fluctuations in mass burning rate. This quantity is sensitive to perturbations in pressure, temperature, stretch/scalar dissipation rate, and mixture composition. Pressure and temperature fluctuations are generated by acoustic and entropy perturbations, while stretch/scalar dissipation rate fluctuations are associated with acoustic or vortical velocity fluctuations. Section 12.3.1.2 will show that the mass burning rate sensitivities of premixed and nonpremixed flames are fundamentally different.

Flame area fluctuations are associated with disturbances in the position and orientation of the flame that, in turn, are generated by fluctuations in flow velocity. For premixed flames, they can also be generated by disturbances in local flame speed, e.g., in the linear limit, by the term $v_{F,n1} = u^u_{n,1} - s^u_{d,1}$ in Eq. (11.37). Finally, fluctuations in heat of reaction, h_R, are driven by variations in reactive mixture composition and, to a lesser extent, pressure and temperature [2]. Note also that the burning area fluctuations are nonlocal functions of $v_{F,n1}$ – i.e., perturbations in flame position and area are a function of the entire space–time history of the upstream disturbance field, see Section 11.2.2.2, while perturbations in ρ^u, s^u_c, and h_R are functions of the local spatial conditions.

As can be seen from the above discussion, there are different potential ways to classify flame coupling mechanisms. For example, they can be classified based on

12.1 Overview of Forced Flame Response Mechanisms

either fundamental flame processes or type of disturbance mode. In terms of the type of disturbance mode, heat release perturbations arise from acoustic, vortical, or entropy disturbances. In addition, they can be classified based on whether they modify the overall area of the flame or its internal structure [3]. Finally, they can be classified by the flow and thermodynamic variables leading to the oscillation, e.g., due to pressure [4–6], temperature [7, 8], velocity [9–14], fuel/air ratio [15–27], acceleration (see Section 11.6), or density fluctuations. In general, these flow/thermodynamic perturbations are related and coexist. For example, an acoustic wave is accompanied by pressure, temperature, density, acceleration, and flow velocity oscillations. The classification approach used here is to quantify the flame response through its sensitivity to acoustic pressure, flow velocity, fuel/air ratio, flow acceleration, and entropy. It should be emphasized that this classification approach is not unique, but is motivated by differentiating fundamentally different physical processes exciting the flame, as well as common classifications used in the past.

We start by considering the velocity-coupled mechanism, where the flame is perturbed by acoustic and/or vortical perturbations [28, 29] (the pressure and temperature fluctuations accompanying the acoustic velocity are treated in the acoustic pressure coupling section). Figure 12.1 illustrates the mechanism by which flow perturbations lead to heat release oscillations. As discussed later, burning rate perturbations, route 2, are the dominant factor leading to heat release oscillations in non-premixed flames. Conversely, the flame speed remains essentially constant at low forcing frequencies in premixed flames, rendering the heat release oscillations directly proportional to the fluctuating flame area, route 1.

However, the oscillating stretch along the flame, due to both hydrodynamic straining and curvature, grows in importance with frequency, causing oscillations in flame speed [30]. These flame speed oscillations perturb the heat release both directly (route 2a) and indirectly (route 2b) by affecting the burning area, and become an important contribution at higher frequencies [31, 32].

Finally, note that this velocity coupling mechanism could also be called "normal velocity coupling," as it is only the velocity fluctuations normal to the flame that wrinkle it, such as shown in Eq. (11.37). However, these normal velocity perturbations may themselves be a result of flow disturbances tangential to the flame. An important example of this tangential-to-normal velocity conversion occurs in swirl

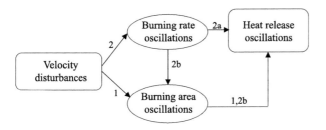

Figure 12.1 Physical mechanisms through which velocity oscillations lead to heat release oscillations.

flows, where acoustic oscillations through the swirler induce an oscillatory azimuthal velocity in the flow [13, 14, 33]. These azimuthal flow disturbances may be tangential to the flame and, consequently, do not disturb it. However, they also lead to axial flow divergence angle oscillations, leading to oscillations in the flow direction normal to the flame.

We next consider the fuel/air ratio oscillation mechanism [19–27]. In premixed systems, this would consist of oscillations in the local stoichiometry of the fuel/air mixture. For nonpremixed flames, it would consist of disturbances in composition of the fuel and/or oxidizer stream. The physical processes that lead to heat release oscillations are illustrated in Figure 12.2. This mechanism has been studied primarily in the context of premixed flames, so this discussion primarily focuses on them. In premixed flames, equivalence ratio perturbations cause fluctuations in local flame speed (route 2a) and heat of reaction (route 1) along the flame surface. These fluctuations in flame speed and mixture heat of reaction then cause the local heat release rate to oscillate. This is a direct route of influence. In addition, flame speed variations also excite flame wrinkles that propagate along the flame through the $v_{F,n1}$ term in Eq. (11.37). This leads to an oscillation in the burning area of the flame (route 2b), also causing the net heat release rate to oscillate. This indirect route of influence is also nonlocal; i.e., the flame area fluctuations at a given time and position are a convolution of the fuel/air ratio oscillations at all upstream locations at earlier times.

Finally, the wrinkling of the flame induced by these burning rate perturbations causes stretch rate oscillations that also perturb the flame speed, thereby establishing another route by which the flame speed fluctuates (route 2S). These fluctuations in flame speed can then disturb the heat release directly (route 2Sa) or indirectly by altering the burning area fluctuations (route 2Sb). Finally, the unsteady wrinkling on

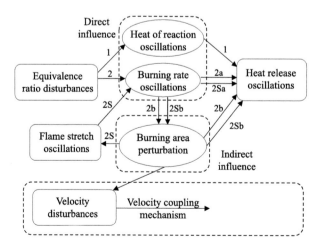

Figure 12.2 Physical mechanisms through which equivalence ratio oscillations lead to heat release oscillations [34]. Velocity coupling mechanism referred to at the bottom of the figure is detailed in Figure 12.1.

the flame introduces perturbations in gas velocity upstream of the flame, introducing the entire velocity coupling mechanism illustrated in Figure 12.1. This effect is due to gas expansion and, hence, scales with σ_ρ [34, 35].

Heat of reaction oscillations, route 1, are dominant for lean, premixed flames when $St_2 \ll 1$ (see Eq. (11.11)), as detailed later [25, 34]. Routes 1, 2a, and 2b are all of comparable importance when $St_2 \sim O(1)$ [25, 34]. Finally, flame stretch becomes important through unsteady curvature at higher frequencies, introducing routes 2Sa and 2Sb [30].

We next consider the acoustic pressure coupled mechanism or, more specifically, *pressure–temperature–density* coupling, as we assume that the three are isentropically related in the acoustic perturbation field. The routes through which these disturbances lead to heat release fluctuations are shown schematically in Figure 12.3. Pressure disturbances cause disturbances in the heat of reaction (route 1), unburned reactant density (route 2), and burning rate (route 3), which directly cause the heat release to oscillate. Additionally, burning rate oscillations in premixed flames cause the burning area to oscillate, similar to the velocity and equivalence ratio mechanism (route 3b), causing heat release oscillations indirectly.

Entropy coupling is similar, in that the flame is disturbed by convecting density and temperature perturbations. The coupling diagram is the same as the acoustic pressure shown previously, although the quantitative sensitivities relating disturbances and the resulting heat release fluctuations are different from the acoustic pressure because of the different disturbance phase speeds.

Lastly, consider acceleration coupling. Acceleration coupling refers to the sensitivity of the flame to acceleration of the flame normal to itself, and is closely associated with the Rayleigh–Taylor instability discussed in Section 3.8. As described in Section

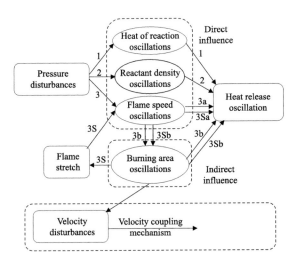

Figure 12.3 Physical mechanisms through which acoustic pressure fluctuations lead to heat release oscillations [2]. Velocity coupling mechanism referred to at the bottom of the figure is detailed in Figure 12.1.

11.6, high-amplitude acoustic fluctuations lead to the formation of fluctuation cells on flames that oscillate at half the forcing frequency. In addition, studies in laminar Bunsen flames have shown that the time-averaged flame shape becomes hemispherical at high amplitudes and forcing frequencies [36]. This mechanism is important at large excitation amplitudes and grows in significance as the density ratio across the flame, σ_p, grows. More work is needed to analyze its significance relative to the other mechanisms noted above.

The rest of this section considers the first three mechanisms in more detail. First, we define the transfer functions F_V, F_p, F_ϕ as

$$F_p = \frac{\hat{\dot{Q}}_1(\omega)/\dot{Q}_0}{\hat{p}_1/p_0}, \tag{12.6}$$

$$F_V = \frac{\hat{\dot{Q}}_1(\omega)/\dot{Q}_0}{\hat{u}_1/u_0}, \tag{12.7}$$

$$F_\phi = \frac{\hat{\dot{Q}}_1(\omega)/\dot{Q}_0}{\hat{\phi}_1/\phi_0}. \tag{12.8}$$

Note that these transfer functions are complex numbers, whose magnitude and phase indicate the relative magnitude ratios and phase difference between the heat release and disturbance quantity. The total unsteady heat release in the linear approximation is then the superposition,

$$\frac{\hat{\dot{Q}}_1(\omega)}{\dot{Q}_0} = F_V \frac{\hat{u}_1(\omega)}{u_0} + F_P \frac{\hat{p}_1(\omega)}{p_0} + F_\phi \frac{\hat{\phi}_1(\omega)}{\phi_0}. \tag{12.9}$$

The flow disturbances are defined at some reference position, such as the flame base. We next consider the relative importance of the different coupling mechanisms as a function of frequency. To do this, it is necessary to assume relationships between pressure, velocity, and fuel/air ratio disturbances. We will assume that acoustic pressure and velocity are related through a traveling wave formulation, $p_1 = \rho_0 c_0 u_1$. As discussed in Exercise 12.1, we will also assume that the velocity, pressure, and fuel/air ratio disturbances are related as

$$\frac{1}{\gamma M_0}\frac{p_1}{p_0} = \frac{u_1}{u_0}, \tag{12.10}$$

$$\frac{\phi_1}{\phi_0} = -\frac{u_1}{u_0}. \tag{12.11}$$

Since the ratio of the amplitudes of velocity perturbations and pressure perturbations is $1/\gamma M_0$, we multiply the pressure-coupled response by γM_0 in order to compare it with velocity and equivalence ratio coupled transfer functions. Consider first the flame response at very low frequencies. In premixed flames, analyses for these limits show [2, 37]:

$$\lim_{St_2 \to 0} F_V = 1, \tag{12.12}$$

$$\lim_{St_2 \to 0} \gamma M_0 F_P = M_0, \qquad (12.13)$$

$$\lim_{St_2 \to 0} F_\phi = k_{R,\phi} = \left.\frac{\partial(k_R/k_{R,0})}{\partial(\phi/\phi_0)}\right|_{\phi=\phi_0}. \qquad (12.14)$$

The same velocity coupling limit applies for high Peclet number nonpremixed flames. In both premixed and nonpremixed flames, this unity transfer function magnitude follows from the fact that in the quasi-steady limit, the instantaneous heat release is proportional to the instantaneous mass flow rate of fuel and oxidizer into the combustion domain; in other words, a 1% fluctuation in velocity induces a 1% fluctuation in heat release. For the same reason, the low St_2 limit for premixed or nonpremixed flames to transverse bulk forcing is $F_{V,transverse} = 0$. The reason this limit can differ from unity in low Peclet number nonpremixed flames is that axial diffusion also contributes to the instantaneous mass flow, while the normalization in Eq. (12.7) only accounts for the convective mass flow portion [38].

These expressions show that the equivalence ratio and velocity coupled responses are similar, differing only by the factor $k_{R,\phi}$, which is of $O(1)$ for lean flames (for example, $k_{R,\phi} = 0.61$ and 0.67 for $\phi = 0.85$ methane–air flames with reactant temperature of 700 K at 1 atm and 5 atm, respectively). In contrast, the pressure-coupled response is $O(M_0)$, and hence is negligible in low Mach number flows.

Conclusions about the relative significance of these mechanisms in premixed flames at high frequencies are more tentative. This limit involves consideration of interference (see Section 11.2.2.3), non-quasi-steady internal flame response (see Section 9.9), and stretch (see Section 11.4) effects. Some studies based on single-step kinetics suggest that pressure coupling may become a significant mechanism for attached flames with tangential flows at high frequencies [2], but more work is needed here. The pressure-coupled mechanism is not considered further in this chapter as it is generally negligible relative to velocity and fuel/air ratio coupling effects at lower frequencies. The exception to this occurs in instances where velocity and fuel/air ratio coupling processes are negligible, such as flat flames propagating through homogeneous mixtures [5, 39–42].

Having discussed the key physics controlling the flame's heat release response, we next consider these characteristics more quantitatively, focusing first on how flame configurations affect its sensitivity to disturbances in Section 12.2. We then detail linear velocity and fuel/air ratio response processes in Sections 12.3.1.1 and 12.3.2, respectively, for premixed flames. Velocity-coupled nonpremixed flame response is discussed in Section 12.3.1.2. Nonlinear premixed flame dynamics are discussed in Section 12.3.3.

12.2 Flame Configuration Effects on Response Sensitivities

The sensitivity of flames to disturbances, whether turbulent or time harmonic, is strongly configuration dependent, due to both variations in mass burning rate and area along the flame. This has significant influences on burning rates and the linear and nonlinear sensitivities of forced flames. For example, the turbulent flame speed in the same turbulent flow field can differ considerably in a turbulent jet flame relative to one anchored by an axisymmetric bluff body. The phase response of these flames excited by the same harmonic disturbance similarly differs significantly.

This point can be seen by expanding Eq. (12.5) to first order:

$$\dot{Q}_1(t) = \int_A \dot{m}''_{F,1} h_{R,0} dA_0 + \int_A \dot{m}''_{F,0} h_{R,0} dA_1 + \int_A \dot{m}''_{F,0} h_{R,1} dA_0. \tag{12.15}$$

Note how the fluctuating mass burning rate is weighted by the nominal area distribution, or the fluctuating area term is weighted by the nominal burning rate distribution. These two effects are discussed further in the following two sections.

12.2.1 Geometry and Flame Area Distribution Effects

Flame area oscillations are important drivers of unsteady and time-averaged heat release (or, equivalently, the turbulent flame consumption rate in turbulent flows). However, the same flame position disturbance, $\xi(y,t)$, can lead to appreciably different effects on flame area. For example, for a two-dimensional or axisymmetric nominal flame position, its position, ξ, is related to flame area by the relation

$$\frac{A(t)}{A_o} = \int_{flame} W(y) \sin \theta(y) \sqrt{1 + \left(\frac{\partial \xi}{\partial y}\right)^2} \, dy. \tag{12.16}$$

The leading-order area fluctuation is then

$$\frac{A_1(t)}{A_o} = \int_0^{W_f} W(y) \sin \theta(y) \cos \theta(y) \frac{\partial \xi_1}{\partial y} \, dy, \tag{12.17}$$

where $W(y)$ is a weighting factor that depends on the nominal flame geometry, to be discussed further later. Consider the axisymmetric cone, two-dimensional, and axisymmetric wedge flame geometries sketched in Figure 12.4. Note that flame area is concentrated near the base and tip for the axisymmetric cone and wedge geometries, respectively, while it is uniformly distributed for the two-dimensional configuration. For example, the first half of the surface area of an axisymmetric cone and wedge are located at $L_F/3$ and $2L_F/3$, respectively.

For premixed flames, a useful limit is a constant angle flame, θ. Defining the width of the flame as $W_f = L_F \tan \theta$, this weighting factor is then given by $W(y) = 2\pi y/(\pi W_f^2)$, $2\pi(W_f - y)/(\pi W_f^2)$, and $1/W_f$ for the axisymmetric wedge, axisymmetric cone, and two-dimensional flames, respectively. This effect of area distribution on global flame response will be returned to at several points in this chapter.

12.2 Flame Configuration: Response Sensitivities

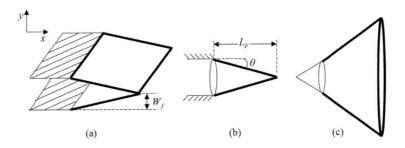

Figure 12.4 Sketch of (a) two-dimensional, (b) axisymmetric cone, and (c) axisymmetric wedge flame geometries.

12.2.2 Burning Rate Distribution Effects

In constant flame speed premixed flames, the burning rate per unit area is constant. This is why premixed flame studies often focus on flame area, as instantaneous area and heat release rate are closely correlated. Of course, in a flow with stratified fuel/air ratio, this need not be the case. Nonpremixed flames usually have completely different behavior: as the burning rate is controlled by gradients in fuel/oxidizer concentration (see Eq. (12.3)), these gradients can vary by orders of magnitude over the flame. In fact, the step mixture fraction inflow profile utilized in Eq. (11.112) to analyze nonpremixed flame sheet dynamics implies an infinite gradient and infinite burning rate at $x = 0$. For this reason, nonpremixed heat release dynamics studies require careful consideration of the inflow profiles, as these details in the highest burning rate region of the flame also control the heat release dynamics. Moreover, axial diffusion, which is commonly neglected in understanding the position of high-Pe flames, plays an important role in controlling the burning rate in this high-gradient region and must be accounted for.

To illustrate, Figure 12.5 shows results from a calculation where axial diffusion is accounted for and the inflow fuel/oxidizer profiles are computed, rather than artificially specified at the inflow. The figures show the steady-state distribution of heat release rate per unit axial distance, $\dot{q}_0(x)$, evaluated along the flame sheet, $\xi_0(x)$, for various Pe values. The cumulative heat release rate distribution, $\dot{q}_{0,c}$, is also plotted, defined as

$$\dot{q}_{0,c}(x) = \frac{\int_0^x \dot{q}_0(s)ds}{\dot{Q}_0}, \qquad (12.18)$$

Where $\dot{Q}_0$ is the total steady-state heat release rate, defined as

$$\dot{Q}_0 = \int_0^{L_{F,0}} \dot{q}_0(x)dx. \qquad (12.19)$$

Also plotted for reference is the axial distribution for a constant burning velocity premixed flame.

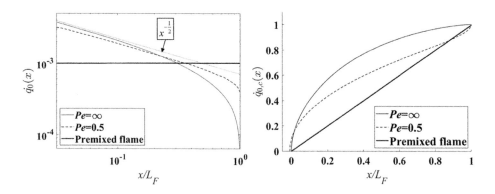

Figure 12.5 Local heat release solutions for two-dimensional nonpremixed jet flame, showing axial dependence of steady-state heat release, both local (left) and cumulative (right, defined in Eq. (12.18)), for $Z_{st} = 0.3$ and two Pe values [38]. Constant burning velocity premixed flame result shown for reference.

For the steady-state distribution, the $Pe \to \infty$ limiting case has a singularity, scaling as $x^{-1/2}$ as $x \to 0$ (denoted by the reference line), at the inlet due to the discontinuity in mixture fraction at the burner exit. With finite Pe values, the inlet profile is smoothed and the peak value at $x = 0$ is reduced. In addition, the heat release rate contribution from the tip (where only axial diffusion contributes) increases with decreasing Pe value. Comparing the premixed and nonpremixed flame curves shows that half of the average heat release rate occurs in roughly the first 15%–20% of the nonpremixed flame, while it occurs at the 50% flame midpoint for the premixed flame. This result clearly shows the need for particular care in accounting for inflow conditions and burner exit details for nonpremixed flame problems. Moreover, they also show that the nonpremixed flame heat release rate dynamics will be very sensitive to flame liftoff from the burner, with a consequent elimination of heat release near the very high gradient burning lip.

12.3 Harmonic Flame Excitation

12.3.1 Linear Dynamics: Velocity-Coupled Response

12.3.1.1 Premixed Flame Response

This section analyzes the response of constant burning velocity flames to velocity fluctuations. In this case, heat release fluctuations are entirely due to area fluctuations, route 1 in Figure 12.1. It focuses on axial forcing of the flame. The flame response to transverse and three-dimensional disturbances is covered in Section 12.3.1.3. To illustrate the solution characteristics, we present results for the flame response using the model velocity profile from Eq. (11.47). The fluctuations in flame area can be determined by taking the solution for the fluctuating flame position in Eq. (11.50) and substituting into Eq. (12.17) for flame area. Setting the decay term to zero for simplicity, $\zeta = 0$, the solution for the axisymmetric wedge geometry is (see Exercise 12.2):

Axisymmetric wedge flame :

$$F_V(St_2, k_c) = \frac{k_c}{2\pi^2(1-k_c)St_2^2}\left(1 - k_c - (1 - 2\pi i St_2)e^{2\pi i St_2} + (k_c - 2\pi i St_2)\,e^{2\pi i St_2/k_c}\right).$$

(12.20)

The corresponding solutions for two-dimensional and axisymmetric conical flames are derived in Exercises 12.3 and 12.4. It can be seen that the linear flame transfer functions depend only on the two parameters, St_2 and k_c, that were introduced in Section 11.1.2.

The transfer function gain, $F_V(St_2, k_c)$, is plotted in Figure 12.6. Note that the gain values tend toward unity at low St_2; i.e., a 1% fluctuation in velocity induces a 1% fluctuation in heat release rate, for reasons discussed in the context of Eq. (12.12). In the bulk forcing (i.e., spatially uniform) velocity case, $k_c = \infty$, the gain decreases nearly monotonically with increases in St_2. This decrease implies that the corresponding 1% velocity fluctuation induces a progressively smaller heat release rate fluctuation. The high-frequency limit for this $k_c = \infty$ case is analyzed in Exercise 12.5 and compared to the nonpremixed flame response in Section 12.3.1.2. In all other cases, the gain increases to values that are greater than unity in certain Strouhal number ranges, due to constructive interference between wrinkles excited by the convecting flow disturbance and those propagating along the flame that are induced by upstream disturbances and the boundary condition. Similarly, the gain curves exhibit "nodes" of zero heat release response. As discussed later, these "nodes" do not imply that the flame is not responding locally – in fact, the reader can verify from the solutions for the flame position in Eq. (11.49) that the flame is wrinkling significantly. Rather, this nodal response is due to destructive interference between oscillations at different parts of the flame which have different phases with respect to each other.

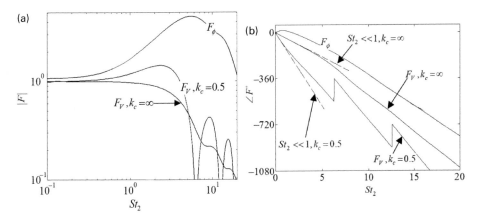

Figure 12.6 Dependence of the velocity and equivalence ratio coupled linear flame transfer function on the Strouhal number for axisymmetric wedge flames showing (a) gain and (b) phase (methane–air, $\phi_0 = 0.9$, $\theta = 30$ degrees).

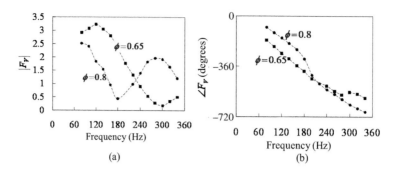

Figure 12.7 Measured (a) gain and (b) phase response of a flame to upstream flow velocity perturbations; reproduced from Jones et al. [43].

The corresponding phase rolls off nearly linearly with Strouhal number, with 180 degree phase jumps at nodal locations in the gain. The slope of this curve has important physical significance. To illustrate, suppose that the velocity and area disturbances are related by a time-delayed relation of the form $\frac{A_1(t)}{A_0} = n_V \frac{u_1(t-\tau)}{u_0}$. The slope of the transfer function phase and the time delay are related by

$$\tau = \frac{1}{2\pi} \frac{d(\angle F_V)}{df}. \tag{12.21}$$

In general, the phase variation with frequency is not linear, although it is for low Strouhal numbers as discussed in the context of Eq. (12.25).

Experiments show transfer functions with similar characteristics. For example, Figure 12.7 plots a measured transfer function for a turbulent, swirling flame. Note the gain values exceeding unity,[1] as well as the near-zero "nodal" values. Similarly, the phase varies almost linearly with frequency.

Consider the high-frequency asymptotic character of these solutions. They predict that the gain curves decay as $1/St_2$ at high Strouhal numbers, as shown in Exercise 12.6. This decay characteristic is due to tangential convection of wrinkles, and therefore heat release fluctuations, leading to phase variations between heat release rate oscillations at different parts of the flame. With increasing Strouhal numbers, the degree of phase variation increases, and consequently so do cancellation effects. To illustrate this point, consider an example where the unsteady heat release rate is convected downstream with a constant axial magnitude and phase velocity as

$$\dot{q}(x,t) = \dot{q}(t - x/u_{t,x,0}). \tag{12.22}$$

The spectrum of these fluctuations is then given by

$$\hat{\dot{q}}(x,f) = \hat{\dot{q}} e^{i 2\pi f x / u_{t,x,0}}. \tag{12.23}$$

---

[1] The reference velocity used for these plots is the acoustic velocity, measured in the nozzle using a two-microphone technique. As such, additional vortical disturbances excited at the separating shear layer may also be contributing to the flame response.

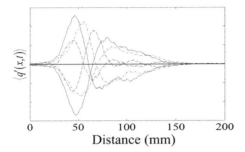

Figure 12.8 Axial variation of the ensemble-averaged unsteady chemiluminescence at several instances of time, measured in a 180 Hz forced flame ($\bar{u} = 25$ m/s, $\phi = 0.75$), a condition corresponding to a node in F_V. Data courtesy of D. Santavicca.

The spectrum of the spatially integrated fluctuations is

$$\hat{Q} = A_0 \int_{x=0}^{L_F} \hat{q}(x,f)dx = \hat{q} A_0 L_F \frac{e^{2\pi i St_2} - 1}{2\pi i St_2}. \qquad (12.24)$$

This expression shows the $1/St_2$ dependence at high frequencies. Moreover, this same asymptotic dependence can be derived for a broad class of more general spatial variation in heat release magnitude using high-frequency asymptotic methods.[2] This cancellation of the unsteady heat release over different parts of the flame is clearly evident in data showing the time variation of the axial heat release distribution of forced flames, such as shown in Figure 12.8. These data were taken at a condition corresponding to a "node" in the gain response, and reiterate the point made earlier, that a zero global heat release gain does not imply that the flame is not responding locally. Rather, this behavior is due to phase cancellation phenomena.

These frequency domain expressions can be converted into time domain expressions using the inverse Fourier transform. For $St_2 \ll 1$ (i.e., a convectively compact flame), the flame area–velocity relationship can be put in terms of an n–τ model, as shown in Exercise 12.7:

$$\frac{A_1(t)}{A_0} = n_V \frac{u_1(t-\tau)}{u_0}, \qquad (12.25)$$

where $n_V = 1$ and $\tau = \frac{2(1+k_c^{-1})L_F}{3u_0 \cos^2\theta}$. For reference, the corresponding time delay for the axisymmetric conical flame (obtained from Exercise 12.3) is half of the wedge flame result, for the reasons discussed in Section 12.2.1.

---

[2] More generally, the power, n, in the asymptotic dependence of St_2^{-n} is related to the character of the derivatives of the integrand at the boundaries. By integrating by parts, it can be shown that the high-frequency asymptotic character of integrals of the form $\int_{x=0}^{L_F} \hat{q}(x) e^{i\omega x/u_{0,x}} dx$ scale as $1/f$ if $\hat{q}(x)$ is not identically zero at both boundaries. If they are zero, then the decay exponent $n = 2$ (implying a faster rolloff), assuming that $d\hat{q}(x,f)/dx$ is not zero, and so forth.

Recall that this velocity-coupled n–τ heat release model was used in Section 6.5.1 for a thermoacoustic instability model problem which demonstrated the significance of the time delay, τ, in controlling combustion instability limits. The above analysis shows that the n–τ heat release model is a rigorous approximation of the heat release rate dynamics in the low-St_2 limit, and that the retarded time equals the time taken for the mean flow to convect some fractional distance of the flame length. This time delay is equivalent to replacing the distributed flame by a concentrated source at this location; e.g., for the axisymmetric wedge flame, this effective position of concentrated heat release rate is $2(1+k_c^{-1})L_F/3\cos^2\theta$.

12.3.1.2 Nonpremixed Flame Response

This section considers the same geometry discussed in Section 11.3.4, analyzing the sensitivity of nonpremixed flames to axial forcing. The flame response to transverse and three-dimensional disturbances is covered in Section 12.3.1.3. As noted in Section 12.2.2, the largest mixture fraction gradients exist near the burner exit. Consequently, this is the region that dominates the unsteady heat release rate. Moreover, axial diffusion effects and the mixture fraction profiles at the burner outlet have important influences on the flame response. For this reason, the analysis is more involved than the premixed flame case and we primarily summarize key results, closely following Magina et al. [38].

Figure 12.9 plots the computed amplitude and phase of the velocity as a function of St_{L_F} for the bulk forced ($k_c = \infty$) case. Results are shown for various Pe values which, as shown in Figure 12.5, influences the axial distribution of mass burning rate, with higher Pe values corresponding to more of the burning happening near the burner exit. There are essentially three zones in Strouhal number space: the quasi-steady, $St_{L_F} \ll 1$, regime, an intermediate regime where $|F_V| \sim 1/St_{L_F}^{1/2}$, and a high-frequency, $1/St_{L_F}$, regime. We discuss each of these next.

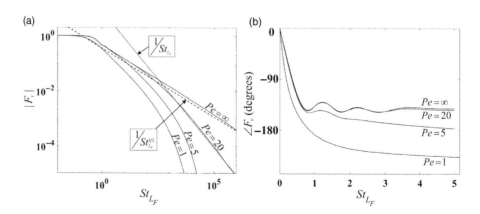

Figure 12.9 Transfer function results for a two-dimensional, bulk forced, nonpremixed flame, showing gain (a) and phase (b) as a function of St_{L_F} for $Z_{st} = 0.3$ for various Pe values [38].

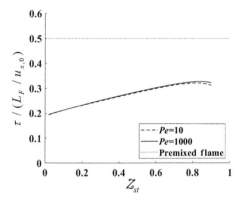

Figure 12.10 Dependence of normalized time delay parameter, $\tau/(L_F/u_{x,0})$, on $\mathcal{Z}_{st}$ for various Pe values for two-dimensional, bulk forced nonpremixed flame.

The low-St transfer function gain, $|F_V|$, equals unity for high Peclet number flames. This value of unity can be understood physically from the fact that in the quasi-steady limit, the instantaneous heat release rate is proportional to the instantaneous mass flow rate of fuel and oxidizer into the combustion domain, which is directly proportional to the fluctuating velocity in the absence of axial diffusion.

The low-St phase varies linearly with frequency, showing that the transfer function can be written as an n–τ model. The time delay, τ, is equivalent to replacing the distributed flame by a concentrated source at some fractional distance of the flame length. For example, the value of τ for the velocity and equivalence ratio coupled premixed flames are discussed in Sections 12.3.1.1 and 12.3.2, respectively. For nonpremixed flames, these dependencies of τ cannot be calculated analytically but must be extracted from the computations, due to the implicit nature of the solution for mean flame position, at which these integrand values are evaluated. Figure 12.10 plots calculations of $\tau/(L_F/u_{x,0})$ as a function of Pe and $\mathcal{Z}_{st}$ for this same case. The corresponding constant burning velocity premixed flame value is also plotted for reference.

Note that the normalized time delay is around 0.2 for a methane–air, nonpremixed system, which has a $\mathcal{Z}_{st}$ value of 0.055. For reference, the normalized time delay for a constant burning velocity premixed flame with the same forcing is 0.5, showing that the nonpremixed flame time delay is about a factor of two smaller than a premixed flame with the same length. This result directly follows from the concentration of heat release near the burner outlet discussed in Section 12.2.2.

Consider next the higher-St behaviors, i.e., where the convective wavelength of the wrinkle is much smaller than the flame length. As noted in the context of Eq. (12.24), highly oscillatory integrals are controlled by the values of the integrands at the integration limits. An approximate expression for this limit for bulk forced, two-dimensional flames is

$$St \gg 1, \ Pe \gg 1: \quad F_V \approx \frac{-(1+i)}{4\sqrt{St_{L_F}}}. \quad (12.26)$$

This shows that the phase between the velocity and heat release has a constant value of 135 degrees and that the gain rolls off as $1/St_{L_F}^{1/2}$. This $1/St_{L_F}^{1/2}$ behavior of the heat release rate transfer function is a much slower roll off than the $1/St_{L_F}$ roll off that occurs to leading order in premixed flames, causing the heat release response of nonpremixed flames to exceed that of premixed flames at high Strouhal numbers. This different asymptotic tendency arises from the fact that, at the scale of the convective wavelength, the burning rate appears to have a singularity at the $x = 0$ boundary; see Figure 12.5.

However, there exists a third, even higher, St asymptotic zone, which takes over when the convective wavelength is of the same order of magnitude as the adjustment zone near the nozzle exit. The transfer function exhibits a $1/St_{L_F}$ roll off for higher frequencies. The Strouhal number at which the $1/St_{L_F}^{1/2}$ to $1/St_{L_F}$ transition occurs is a function of the length of this adjustment region, which, in turn, is a function of Pe.

Comparing the gain results for the premixed and nonpremixed flames (Figure 12.6 (a), $k_c = \infty$ and Figure 12.9) shows they have some similarities in overall shape. However, the key factors controlling these features are completely different. The heat release rate fluctuations for the premixed flame are controlled largely by area fluctuations, with a small contribution from mass burning rate perturbations for stretch-sensitive flames at higher St values (effects that are not included in this particular calculation). In contrast, the nonpremixed flame gain is dominated by mass burning rate oscillations, with only a small contribution from the weighted fluctuating area term.

12.3.1.3 Three-Dimensional Disturbances

This section is the analogue to Section 11.2.2.4, which analyzed how three-dimensional forcing influences the space–time characteristics of flame wrinkling. This section considers how they influence the spatially integrated heat release rate fluctuations. Two types of disturbances are of interest: transverse acoustic waves and helical vortical modes.

First, it is important to note that, from a local point of view, the flame does not differentiate between axial, transverse, helical, or other types of velocity disturbances. It is sensitive to the scalar velocity component normal to it, $u_{n,1}$; e.g., see Eq. (11.37). For example, the local space–time wrinkling of a two-dimensional transversely or axially excited flame is identical, if the dependence of $u_{n,1}$ along the flame is the same.

From a spatially integrated viewpoint, however, the flame's sensitivity is completely different. For example, as noted in the context of Eq. (12.12), the low-St_2 transfer function of the flame's response to bulk axial velocity disturbances is equal to unity for an axially excited flame, and equal to zero for a bulk transversely excited one. In the latter case, this occurs because the flame wrinkling of one side of the flame is exactly out of phase with the other. Thus, while there are local heat release fluctuations on each half, the spatially integrated value is zero. This result could be anticipated, at least in the low Strouhal number limit, as purely transverse forcing causes no fluctuation in reactants into the domain. Since the heat release in the quasi-steady limit is directly proportional to the reactant flow rate into the domain, transverse fluctuations lead to no heat release oscillations [44]. Significantly, this point emphasizes the important distinction between local and global heat release fluctuations. Even though there are no global heat release rate fluctuations, the local heat release rate by each segment of flame is fluctuating, as is the instantaneous flame position.

For helical disturbances exciting premixed or nonpremixed flames this same cancellation can also occur in the azimuthal direction. For example, the fluctuating flame area induced by a helical mode of the form shown in Eq. (11.55) has the form

$$A_1(t) \propto \frac{1}{2\pi} \int_{\theta=0}^{\theta=2\pi} F_0(\theta) e^{im\theta} d\theta, \qquad (12.27)$$

where the function $F_0(\theta)$ is a function of the unforced flame position. If the unforced flame is axisymmetric, then F_0 is independent of θ and can be pulled ouside of the integral, so the integral is only nonzero for the $m = 0$, axisymmetric mode:

$$\frac{F_0}{2\pi} \int_{\theta=0}^{\theta=2\pi} e^{im\theta} d\theta = \begin{cases} F_0 & m = 0, \\ 0 & m \neq 0. \end{cases} \qquad (12.28)$$

To restate, although a variety of helical modes may be present, and often dominant in a swirling flow, the $m = 0$ mode is the sole contributor to the heat release rate oscillations if the flame is asymmetric. In this case, the disturbance convection speed, k_c, is the key parameter influencing the transfer functions [45]. We will refer to this mode leading to maximum heat release fluctuations as $m_{\dot{Q}_{max}}$, in analogy with $m_{\xi max}$ and m_{pmax}, which are the modes leading to maximum flame wrinkling (see Section 11.2.2.4) and sound emissions (see Section 12.4.3.2), respectively. Given the two assumptions previously noted (that the nominal flame is axisymmetric and that the flame dynamics are linear), $m_{\dot{Q}_{max}} = 0$. Finally, note that nonaxisymmetric and/or nonlinear nominal flames can cause some helical modes to excite nonzero $\dot{Q}$ [46, 47].

12.3.2 Linear Dynamics: Equivalence Ratio Coupling [24, 25]

This section describes premixed flame response to convected equivalence ratio, ϕ, disturbances. This is an important case study for a broader understanding of unsteady heat release rate sensitivities to disturbances, as it includes the superposition of additional effects beyond the flame area fluctuations previously considered, as shown in Figure 12.2. In order to illustrate the solution characteristics, we will consider a model fuel/air ratio profile that is radially uniform and occurs at constant density:

$$\phi = \phi_0 \left[1 + \varepsilon \cos \left(2\pi f \left(t - \frac{x}{u_c} \right) \right) \right]. \qquad (12.29)$$

Although the fuel/air ratio disturbance is convected with the flow, $u_c = u_0$, we will write the convection speed explicitly as u_c throughout this analysis so that the reader can see the terms in the solution arising from the forcing (St_c terms) and the natural flame response (St_2 terms). The analysis is quite similar to that described in Section 12.3.1.1, except now the flame speed fluctuation terms are retained and velocity disturbances are neglected in the expression for $v_{F,n1} = u^u_{n,1} - s^u_{d,1}$, where

$$s_{d,1}^u = \left.\frac{\partial(s^u/s_0^u)}{\partial(\phi/\phi_0)}\right|_{\phi=\phi_0} s_0^u \frac{\phi_1}{\phi_0}.$$ Neglecting unsteady stretch effects, the heat release rate response occurs through routes 1, 2a, and 2b in Figure 12.2. Route 1 dominates at low Strouhal numbers, as shown in Exercise 12.8. The linearized transfer function, F_ϕ, can be written as a sum of three contributions due to flame area, burning velocity, and heat of reaction oscillations:

$$F_\phi = F_{\phi,A} + F_{\phi,s^u} + F_{\phi,h_R}. \tag{12.30}$$

For an axisymmetric wedge flame, these terms are given by [34]:

$$F_{\phi,A} = -2\left.\frac{\partial(s^u/s_0^u)}{\partial(\phi/\phi_0)}\right|_{\phi=\phi_0} \left(\frac{\sin^2\theta + (2\pi i St_c - 1)\exp(2\pi i St_c) - \cos^2\theta(2\pi i St_2 - 1)\exp(2\pi i St_2)}{4\pi^2 St_2^2 \sin^2\theta}\right), \tag{12.31}$$

$$F_{\phi,s^u} = -2\left.\frac{\partial(s^u/s_0^u)}{\partial(\phi/\phi_0)}\right|_{\phi=\phi_0} \frac{(1 + 2\pi i St_c e^{2\pi i St_c} - e^{2\pi i St_c})}{4\pi^2 St_c^2}, \tag{12.32}$$

$$F_{\phi,h_R} = -2\left.\frac{\partial(h_R/h_{R,0})}{\partial(\phi/\phi_0)}\right|_{\phi=\phi_0} \frac{(1 + 2\pi i St_c e^{2\pi i St_c} - e^{2\pi i St_c})}{4\pi^2 St_c^2}. \tag{12.33}$$

Note that we have equated the displacement and consumption speeds needed to evaluate the flame dynamics and unsteady heat release rate, respectively. Analogous axisymmetric conical flame results are derived in Exercise 12.9. Note how the area term arising from route 2b, $F_{\phi,A}$, is a function of both St_c and St_2, reflecting the spatially nonlocal dependence of flame wrinkling on burning velocity fluctuations. In contrast, the heat of reaction term from route 1, F_{ϕ,h_R}, and the flame speed term from route 2a, F_{ϕ,s^u}, are only functions of St_c.

Figure 12.6 shows the variation of the phase and magnitude of the linear transfer function with St_2. The results are similar in many ways to the velocity results described earlier, with the gain values oscillating between values that exceed unity to zero (or near zero) through constructive/destructive interference. Similarly, the phase rolls off nearly linearly. One point to note is that the peak gain values can substantially exceed unity – calculations have predicted values on the order of 10. These peak values are a strong function of $\partial s^u/\partial \phi$, which becomes quite large for lean fuel/air ratios. This indicates that lean fuel/air mixtures are particularly sensitive to fuel/air ratio disturbances [48] [49].

As in the velocity disturbance case, it is useful to examine the time domain relationship between equivalence ratio and heat release rate disturbances. For a wedge flame, these low-St dynamics are given by [25]:

$$\frac{\dot{Q}_1(t)}{\dot{Q}_0} = n_H \frac{\phi_1(t-\tau)}{\phi_0} + n_S \frac{d(\phi_1(t)/\phi_0)}{d(t/(L_F/(u_0\cos^2\theta)))}, \tag{12.34}$$

where $n_H = \left.\frac{\partial(h_R/h_{R,0})}{\partial(\phi/\phi_0)}\right|_{\phi=\phi_0}$, $\tau = \frac{2}{3}\frac{L_F}{u_c}$, and $n_s = \frac{2}{3}\left.\frac{\partial(s^u/s_0^u)}{\partial(\phi/\phi_0)}\right|_{\phi=\phi_0}$.

The corresponding conical flame result is derived in Exercise 12.10. This result is interesting as it shows that the flame response does not limit to an n–τ model, but also includes an additional derivative term.

12.3.3 Nonlinear Dynamics

In order to predict the conditions under which combustion instabilities occur or the forced response of linearly stable systems, the linear flame response (such as the bulk parameters n and τ, see Section 6.5.1), must be predicted. In contrast, predicting the amplitude of self-excited oscillations under linearly unstable conditions, or the possibility of destabilizing a linearly stable system by large-amplitude forcing, requires characterization of nonlinearities. As discussed in Section 6.6.2.2, combustion process nonlinearities are important contributors to the combustor system dynamics. This section discusses key processes controlling these heat release nonlinearities.

Combustion process nonlinearities are introduced by the nonlinear dependence of the heat release oscillations on the disturbance amplitude. There are a variety of factors that can cause this nonlinear response, as described next using measurements and illustrative calculations. In this section, the flame transfer function is generalized to the amplitude-dependent "flame describing function," which we will refer to as F_{NL} [50]. The describing function is the ratio of the response only at the forcing frequency (since higher harmonics are also excited) and the reference disturbance. Sections 12.5 and 2.5.3.3 also discuss the effects of external forcing on the nonlinear dynamics of self-excited systems.

12.3.3.1 Kinematic Restoration

As noted in Section 11.2.3, kinematic restoration is a nonlinear effect leading to the destruction of flame area, at a rate that is a function of the amplitude and wavelength of the flame wrinkle. As such, frequency and amplitude effects are strongly coupled.

The nonlinear dependence of the flame area on flame wrinkle perturbation amplitude causes heat release nonlinearities to exhibit some differences from the nonlinear characteristics of the flame position itself. This can be seen explicitly by writing the expression for differential flame surface area, $dA = \sqrt{1 + \left(\frac{\partial \xi}{\partial y}\right)^2} \cdot dy$, using Eq. (11.19) as

$$dA = \sqrt{1 + \left(\frac{\partial \xi}{\partial y}\right)^2} \cdot dy = \frac{1}{s_d^u} \left(\frac{u_n^u}{\sin \theta} - \frac{\partial \xi}{\partial t}\right) \cdot dy. \tag{12.35}$$

The explicitly nonlinear character of the kinematic restoration disappears in Eq. (12.35). Significantly, Eq. (12.35) shows that the flame area fluctuations exhibit a linear dependence on the perturbation velocity amplitude, u_n^u, if $\partial \xi / \partial t$ is either negligible or itself depends linearly on u_n^u. In the low-frequency limit, it is clear that this unsteady term goes to zero, leading to a linear relationship between heat release

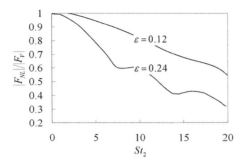

Figure 12.11 Strouhal number dependence of the ratio of the magnitude of the flame area velocity describing function to its linear counterpart for the axisymmetric wedge flame, $\theta = 45$ degrees [51].

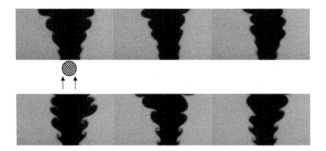

Figure 12.12 Illustration of flames forced with low (top) and high (bottom) axial disturbance amplitudes, showing the manifestation of nonlinear effects through vortex rollup. Images courtesy of S. Shanbhogue.

and axial disturbance amplitude – in other words, nonlinearity in flame area response is intrinsically a non-quasi-steady effect.

To illustrate, Figure 12.11 plots calculations of the Strouhal number dependence of the describing function gain of a wedge flame subjected to the same disturbance field in Eq. (11.47), with $k_c = \infty$ and $\zeta = 0$. The gain describing functions for the nonlinear case, F_{NL}, are normalized by their linear values, F_V. As discussed above, the figure shows that the gain tends to its linear value at low St_2. Note the substantial reduction in flame area relative to its linear value with increases in frequency and/or amplitude; i.e., there is a substantial degree of gain saturation. The wave interference phenomenon causes ripples in the curves with variation in St_2 [51]. Although not shown, the phase can also exhibit substantial amplitude dependence [52].

Calculations like these based on simple model flowfields illustrate important conceptual points, but in practice these kinematic nonlinear effects usually manifest themselves through vortex rollup [53–59]. As the strength of the unsteady vortical disturbances increases with disturbance amplitude, the flame exhibits larger and larger wrinkles on the surface, as shown in Figure 12.12. This figure shows several instantaneous images of the flame at a low (top) and high (bottom) disturbance amplitude.

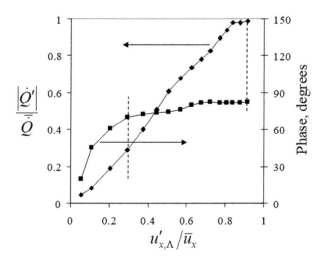

Figure 12.13 Dependence of the unsteady heat release magnitude and phase on the velocity disturbance amplitude from Bellows et al. [60]. Vertical dashed lines denote OH-PLIF data points shown in Figure 12.14.

Note the rollup of the flame in the higher-amplitude forcing case. Kinematic restoration effects become increasingly efficient at destroying flame area as the disturbance amplitude increases, causing the amplitude of flame area fluctuation to increase slower than the amplitude of excitation (see also Section 11.3.3).

To illustrate this point further in a high-velocity, highly turbulent, swirling flame, Figure 12.13 shows the measured unsteady heat release rate response of a forced flame at the forcing frequency $\hat{Q}(f_0)$ (as estimated by a chemiluminescence measurement) as a function of acoustic disturbance amplitude, $u'_{x,\Lambda}/\bar{u}_x$.

Note the linear dependence of the heat release rate gain on velocity for low velocity disturbance amplitudes. The phase exhibits amplitude dependence at lower amplitudes, probably due to changes in time-averaged flame length. At lower amplitudes, the flame is modulated in length because of the oscillating mass flow in the nozzle, as shown by the OH PLIF image in Figure 12.14(a). Above about 80% velocity amplitude, the gain saturates. This gain saturation is clearly coincident with the strong rollup of the flame by a vortex, which leads to rapid destruction of flame surface area. This rollup of the flame is clearly shown in the first image in Figure 12.14(b). Note that the low- and high-forcing cases shown in these images correspond to the amplitudes indicated by the dashed vertical lines in Figure 12.13.

12.3.3.2 Stabilization Point Dynamics

In attached flames, the dynamics of the edge flame near the attachment point play an important role in the overall nonlinear character of the heat release dynamics. If the flame remains anchored at the attachment point, then $\partial \xi / \partial t$ is identically zero at this point for all time. For premixed flames, the flame area perturbations in the vicinity of

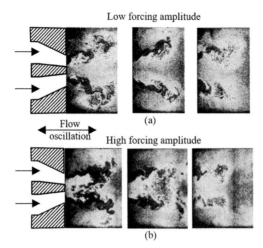

Figure 12.14 Instantaneous OH-PLIF images of an acoustically forced flame at three different phases of the forcing cycle at (a) low and (b) high forcing amplitudes. Images courtesy of B. Bellows.

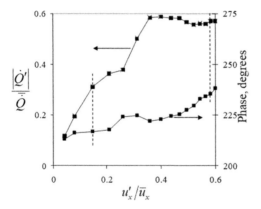

Figure 12.15 Dependence of the unsteady heat release magnitude and phase on the velocity disturbance amplitude, from Bellows et al. [60]. Vertical dashed lines denote OH-PLIF data points shown in Figure 12.16.

the attachment point exhibit a linear dependence on velocity amplitude, as explained in the context of Eq. (12.35). For nonpremixed flames, recall from Section 12.3.1.2 the major influence of near-burner exit details, which would include flame liftoff, on flame response characteristics.

In general, the degree of flame motion near the anchoring point is amplitude dependent. The flame may remain anchored for low amplitudes of excitation, but will either move substantially at higher amplitudes of excitation (but remain attached for part of the cycle), or completely change shape and stabilize at a different time-averaged location. Describing function gain and phase results from a swirl combustor manifesting strong nonlinear behavior due to this effect are shown in Figure 12.15.

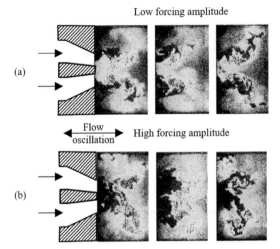

Figure 12.16 Instantaneous OH-PLIF images of an acoustically forced flame at three different phases of the forcing cycle at (a) low and (b) high forcing amplitudes [60]. Images courtesy of B. Bellows.

These data show an approximately linear heat release–velocity gain and phase relationship for disturbance amplitudes below about 35%, but a sharp change in gain relationship at higher amplitudes.

This sharp change in gain is associated with a bifurcation in flame stabilization location at high amplitudes. To illustrate, Figure 12.16 shows OH-PLIF images of the flame at two forcing amplitudes corresponding to the dashed vertical lines in Figure 12.15. The low forcing amplitude case shows a flame stabilized in the inner and outer shear layers of the annular nozzle that is oscillating in length with the fluctuating mass flow. The image at the high forcing amplitude clearly shows that the flame has "blown off" the inner shear layer. The central part of the flame is oscillating back and forth with the vortex breakdown bubble, as opposed to being anchored as it was for low amplitudes. The perturbation amplitude at which local "blow off" of the inner shear layer occurs directly coincides with the point of saturation in the describing function gain [60].

12.3.3.3 Time-Averaged Flame Position and Flowfield

As explained in Section 2.1, nonlinear effects influence not only the nature of the fluctuations, but also the time averages. This section describes nonlinear effects associated with amplitude-dependent time-averaged flowfields and flame positions. The time-averaged flame position is amplitude dependent, through both nonlinear kinematic restoration processes and changes in the underlying fluid mechanics. Indeed, experiments often show a dependence of time-averaged flame position, such as flame spreading angle, on disturbance amplitude [61–63]. From the simplest n–τ framework for understanding flame response, this leads to an amplitude-dependent τ value. This effect is likely responsible for the amplitude-dependent phase response

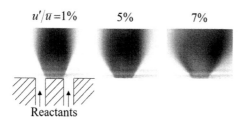

Figure 12.17 Time-averaged, line-of-sight images of a transversely forced swirl flame showing changes in flame shape with increasing disturbance amplitudes [63]. Images courtesy of J. O'Connor.

shown in Figure 12.13. Some processes leading to this change in time-averaged flame length can be understood from quasi-steady arguments using Eq. (11.26). The extent to which the flame length increases is different than it decreases, due to the nonlinear nature of $\sin(\theta)$. For example, consider a flame of length L_F that spreads at an angle of 15 degrees. Quasi-steady variations in flow velocity of 10% and 50% cause the flame length to increase/decrease to $1.1/0.9L_F$ and $2.05/0.63L_F$, respectively. Thus, the flame length fluctuations about the nominal value are not symmetric.

Moreover, time-averaged flowfield characteristics are often themselves amplitude dependent. For example, flows with globally unstable hydrodynamic features (see Section 3.2), such as swirl flows or low density ratio bluff body wakes, can exhibit monotonic or discontinuous dependencies (and the associated hysteresis) of the time-averaged flowfield on disturbance amplitude, a topic also discussed in Section 4.4.4. Figure 12.17 illustrates several images showing this change in time-averaged flame shape with increasing disturbance amplitude in a swirling flow.

12.3.3.4 Geometry and Spatial Flame Area Distribution

This section extends the discussion from Section 12.2. Contributing regions to nonlinearities in heat release rate response are not uniformly distributed along the flame. For example, kinematic restoration effects accumulate with distance downstream from the flameholder, as explained in Section 11.2.3.2. For this reason, the spatial distribution of flame area significantly influences the degree to which local nonlinear effects exert an effect on the global nonlinearities in $\dot{Q}(t)$ [51, 64]. To illustrate, compare anchored, axisymmetric wedge flames and conical Bunsen flames, as shown in Figure 12.4. From purely geometric considerations, note that the latter half of these flames, i.e., $L_F/2 < x < L_F$, contributes 3/4 and 1/4 of the total area of a wedge and conical flame, respectively. Thus, if nonlinear effects are most prominent downstream, they are amplified in wedge flames and reduced in significance in conical flames. The amplitude response characteristics of these two flames are, consequently, quite different, as shown in the describing function data of velocity-forced flames in Figure 12.18. These data show clear gain saturation in the axisymmetric wedge flame, while the phase exhibits little amplitude dependence. In contrast, the conical flame exhibits far less gain sensitivity.

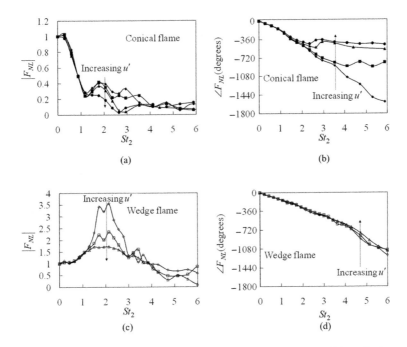

Figure 12.18 Strouhal number dependence of the gain and phase of the velocity-coupled flame describing function for axisymmetric wedge and conical flames at different forcing amplitudes (conical flame, $u'/\bar{u} = 3.6\%, 7.1\%, 14.3\%, 21.4\%$; wedge flame, $u'/\bar{u} = 6.6\%, 13.3\%, 20.0\%$). Figure adapted from Durox et al. [65].

12.3.3.5 Mass Burning Rate and Heat of Reaction Variations

Up to this point, our discussion of heat release rate nonlinearities has primarily focused on flame area effects in Eq. (12.5). In addition to area, the mass burning rate, $\dot{m}_F''$, and heat of reaction, h_R, exhibit a nonlinear dependence on disturbance amplitude, such as fuel–air ratio as shown in Figure 9.5. These nonlinearities appear even in the quasi-steady limit. In particular, excursions in equivalence ratio that cause the instantaneous equivalence ratio to swing from lean to rich, or across lean flammability limits, lead to very strong nonlinear effects [34].

In addition, the velocity and fuel/air ratio relationship is amplitude dependent [66]. This can be seen from the relation

$$\phi = \frac{\dot{m}_{Fuel}/\dot{m}_{Ox}}{(\dot{m}_{Fuel}/\dot{m}_{Ox})_{stoichiometric}}. \tag{12.36}$$

This reduces to the following nonlinear relationship between fuel/air ratio and velocity, derived assumed a low Mach number flow without fuel flow oscillations:

$$\frac{\phi(t)}{\phi_0} = \left(1 + \frac{u_1(t)}{u_0}\right)^{-1}. \tag{12.37}$$

12.4 Broadband Excitation and Turbulent Flame Speeds

In this section we shift from harmonically excited flames and consider the response of flames to broadband random disturbances. As already discussed in Section 11.2.2.6, the finite space–time correlation of the random disturbances leads to significant differences in response relative to harmonic disturbances. Moreover, the quantities of practical interest differ between the harmonic and random forcing cases. While most attention was given in earlier sections of this chapter to the heat release response at the forcing frequency (driven by the thermoacoustic instability application), the dominant item of interest for the broadband excitation problem is the time-averaged burning rate, driven by the turbulent flame speed problem. These time-averaged burning rate characteristics are discussed in Section 12.4.1. Also of interest for the broadband flame excitation problem is the spectrum of heat release fluctuations, which has important applications in turbulent combustion noise [67], discussed in Section 12.4.3.

12.4.1 Time-Averaged Burning Rates

Section 12.3.3 focused on impacts of nonlinearity on the response of forced flames at the forcing frequency. This section considers nonlinear effects of broadband, random forcing. In particular, we analyze the effects of nonlinearity on the time-averaged burning rate of the turbulent flame. Note that the topic of turbulent combustion is a whole field in itself and the reader is referred to focused treatments of the topic for more in-depth discussions of turbulent burning rates [68–71].

We start with a few basic definitions. The thermal progress variable, $\tilde{c}$, is defined as

$$\tilde{c} = \frac{\bar{T} - T^u}{T^b - T^u}, \tag{12.38}$$

whose physical interpretation is illustrated in Figure 12.19, which compares instantaneous and time-averaged features of the turbulent flame. The first two images of the figure show several notional instantaneous planar cuts of a turbulent flame, showing its space–time irregularity. Time-averaged views of these flames show a diffuse brush with a thickness given by δ_T. In addition, time-averaged temperature or fuel concentration measurements show a monotonic variation between unburned and burned values, as shown in (c). In contrast, instantaneous measurements of these quantities show an erratic jumping of these values between burned and unburned values, as the flame flaps back and forth across the probe, as shown in (d).

The time-averaged burning rate is usually quantified via a turbulent flame speed, s_T. It is often conceptually desirable to define these turbulent flame speeds with respect to the unburned flow, i.e., at the $\tilde{c} = 0$ surface, analogous to s^u. While such an approach can be used for analytic calculations, the fact that the number of experimental/computational realizations of the flame tends toward zero at this surface makes it generally necessary to use some other value in practice. Moreover, it is important to distinguish between the different potential definitions of the turbulent flame speed,

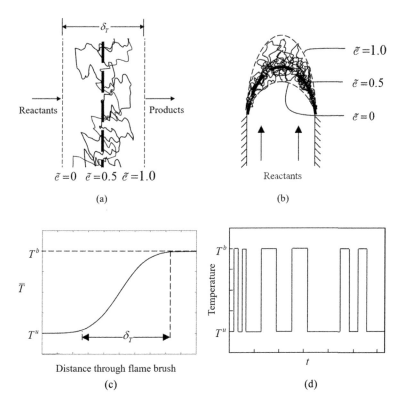

Figure 12.19 Illustration of several realizations of the turbulent flame position for (a) a flat flame normal to the flow and (b) an anchored flame with tangential flow. (c) Spatial dependence of the time-averaged temperature through the flame brush. (d) Temporal dependence of the temperature at a spatial location corresponding to $\tilde{c} = 0.5$.

which can be based on average consumption rates of reactants (the consumption speed) or velocity with respect to the approach flow (the displacement speed). Note the analogies here to the discussion in Section 9.3.4 for stretched laminar flames. We next discuss the turbulent flame speed further, and define and differentiate between different definitions.

The consumption speed definition arises from consideration of the mass burning rate of the flame. It equals the equivalent burning velocity of some reference progress variable surface, $\tilde{c}$, with this same burning rate, i.e.,

$$s_{T,c}^{\tilde{c}} = \frac{-\iiint_{CV} \overline{\dot{w}}_{reactants} dV}{\bar{\rho}^u A_{\tilde{c}}}, \qquad (12.39)$$

where $\iiint_{CV} \overline{\dot{w}}_{reactants} dV$ denotes the time-averaged consumption rate of reactants by the flame in the control volume and $A_{\tilde{c}}$ denotes the area of the reference progress variable surface with respect to which the consumption speed is defined. The consumption speed can also be defined as a "global" or "local" consumption speed, depending on whether this definition is applied locally or is a global average over the entire flame. Note also that the consumption speed is defined with respect to a progress variable surface. Only in special cases, such as a flat flame with normally incident mean flow as considered in Section 11.2.2.5, is A or $s_{T,c}$ independent of the $\tilde{e}$ surface.

The turbulent consumption speed is much larger than the laminar consumption speed. For a constant laminar burning velocity flamelet, the two are directly related by the ratio of the ensemble-averaged, instantaneous turbulent flame area, $\langle A_T \rangle$, to the reference progress variable surface area, $\langle A_T \rangle / A_{\tilde{e}}$. This can be seen from the following expression:

$$\frac{-\iiint_{CV} \overline{\dot{w}_{reactants}} dV}{\bar{\rho}^u} = s_{T,c}^{\tilde{e}} A_{\tilde{e}} = s_c^u \langle A_T \rangle \Rightarrow \frac{s_{T,c}^{\tilde{e}}}{s_c^u} = \frac{\langle A_T \rangle}{A_{\tilde{e}}}. \qquad (12.40)$$

This area ratio, $\langle A_T \rangle / A_{\tilde{e}}$, is essentially an indicator of the degree of flame wrinkling. This is often quantified using the flame surface density, defined as the average instantaneous flame surface area per unit volume [70].

The turbulent displacement speed is equal to the time-averaged normal velocity of the flow with respect to a progress variable surface, i.e., for a statistically stationary flame,

$$s_{T,d}^{\tilde{e}} = (\vec{u} \cdot \vec{n})\big|_{\tilde{e}}, \qquad (12.41)$$

where $\vec{n}$ denotes the vector normal to the progress variable surface.

Similar to laminar flames, the consumption speed and displacement speed yield identical values when the flame and flow are statistically one dimensional. However, they differ in general because of mass flow tangential to time-averaged progress variable contours. To illustrate this point, consider the consumption and displacement speeds with respect to the $\tilde{e} = 0$ contour. Construct a control volume defined by the normal to this $\tilde{e} = 0$ surface on the sides and the $\tilde{e} = 0$ and $\tilde{e} = 1$ surfaces on the front and back, as shown in Figure 12.20. Also, define the time-averaged flow rates of reactants (i.e., not the total mass flow rate) into each surface of the control volume as $\dot{m}_a$, $\dot{m}_b$, $\dot{m}_c$, and $\dot{m}_d$, and assume that all reactants are consumed by the flame, $\dot{m}_d = 0$. These mass flow rates can be related to the reactant mass consumption rate in the control volume as:

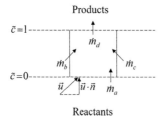

Figure 12.20 Control volume around a turbulent flame brush, showing how tangential mass flux causes differences between the consumption and displacement speeds.

12.4 Broadband Excitation and Turbulent Flame Speeds

$$\dot{m}_a + \dot{m}_b + \dot{m}_c = -\iiint_{CV} \overline{\dot{w}_{reactants}} dV = \bar{\rho}^u A_a s_{T,c}, \qquad (12.42)$$

where A_a denotes the surface area of the front edge of the control volume, defined by the $\tilde{e} = 0$ contour. The mass flow rate of reactants entering the control volume through surface a is given by

$$\dot{m}_a = \bar{\rho}^u \overline{\vec{u} \cdot \vec{n}_a} A_a = \bar{\rho}^u s_{T,d} A_a. \qquad (12.43)$$

Combining Eqs. (12.42) and (12.43) leads to

$$s_{T,c} - s_{T,d} = \frac{\dot{m}_b + \dot{m}_c}{\bar{\rho}^u A_a}. \qquad (12.44)$$

Thus, only if there is no net reactant mass flux through the two sides of the control volume, i.e., $\dot{m}_b + \dot{m}_c = 0$, are the turbulent consumption and displacement speeds equal. Most flows with tangential convection, such as turbulent stagnation flames or anchored flames with spatially developing flame brushes (e.g., Bunsen flames, bluff body stabilized flames), do have net mass flux through the sides, and consequently have different consumption and displacement speeds.

With this background, we next consider the dependence of the turbulent burning rate on flow quantities. Recall that Section 11.2.2.6 analyzed the response of flat and inclined flames to small random fluctuations and showed that the turbulent flame speed has a quadratic dependence on velocity fluctuation amplitude, ε_T, for $\varepsilon_T \ll 1$. This scaling quickly loses accuracy for $\varepsilon_T \sim O(1)$ and cannot be used for typical applications where $\varepsilon_T \gg 1$. There are several reasons for this. First, as shown in Eq. (11.59), the increase in turbulent flame speed for constant burning velocity flames is directly proportional to the increase in flame area, $\langle A_T \rangle$, which is related to the flame position through a term of the form of $\left\langle \sqrt{1 + |\nabla \xi|^2} \right\rangle$. As shown in Eq. (11.60), for weakly wrinkled flames where $|\nabla \xi|^2 \ll 1$, this term scales as $1 + \langle |\nabla \xi|^2 \rangle / 2$. In contrast, for strongly wrinkled flames where $|\nabla \xi|^2 \gg 1$, this term scales as $\nabla \xi|_{rms}$ [72]. Although ε_T and $\nabla \xi|_{rms}$ are not directly proportional in the $\varepsilon_T \gg 1$ limit, this discussion does show the change in scaling that would be expected between the $\varepsilon_T \ll 1$ and $\varepsilon_T \gg 1$ limits. More detailed analyses based on renormalization techniques suggest a high turbulence intensity limit that grows slower than ε_T, such as shown in the following implicit expression [73–76]:

$$\left(\frac{s_T}{s_L}\right)^2 \ln \left(\frac{s_T}{s_L}\right)^2 = 2\varepsilon_T^2. \qquad (12.45)$$

In addition, the linearized flame dynamics calculations in Section 11.2.2.6 only capture the increase of flame surface area by velocity fluctuations. They neglect the nonlinear kinematic restoration effects that cause destruction of flame surface area. Similar to the discussion of harmonically forced flames, kinematic restoration

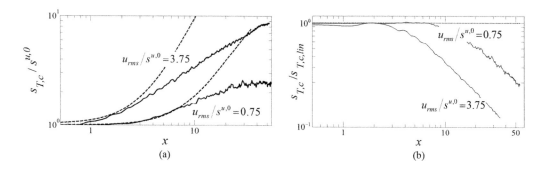

Figure 12.21 (a) Axial variation of the turbulent consumption speed, $S_{T,c}$, based on computations (solid) and linearized theory (dashed). (b) Axial variation of the ratio of computed to linearized turbulent consumption speed. Adapted from Hemchandra et al. [24, 77].

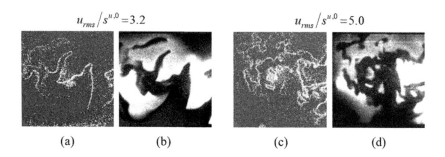

Figure 12.22 CH (a, c) and OH (b,d) PLIF images of turbulent flames at two turbulence intensities, showing increased degree of flame wrinkling with increased turbulence intensity. Adapted from Tanahashi et al. [78].

causes the turbulent wrinkling on the flame to not increase proportionally to disturbance amplitude. To illustrate, Figure 12.21(a) plots a computation of the axial dependence of the turbulent consumption speed for an anchored flame, comparing the $\varepsilon_T \ll 1$ estimate in Eq. (11.82) with a computation of the fully nonlinear level set equation [24, 77]. The turbulent flame speed grows with downstream distance but the computed result grows slower than the low-amplitude ε_T^2 dependence. This can also be seen from Figure 12.21(b), which normalizes the computed result by the asymptotic calculation. The figure shows that the two results are identical near the attachment point, but then the computed value of $s_{T,c}$ is much smaller than the asymptotic one farther downstream. This reflects the growing significance of nonlinear processes with axial distance from the attachment point discussed previously.

Analogous to the discussion in Section 12.3.3.1, the flame becomes increasingly distorted and rolled up around the vortices in real turbulent flow fields with a range of scale sizes, so that kinematic area destruction processes depend nonlinearly on disturbance amplitude. This point is illustrated in Figure 12.22, which shows

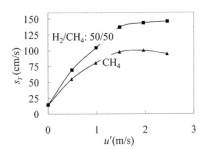

Figure 12.23 Measured dependence of the turbulent consumption speed on turbulence intensity. Fuel burned at $\phi = 0.8$ in a N_2/O_2 blend that was adjusted to keep $s^{u,0}$ constant. Adapted from Nakahara and Kido [88].

simultaneous OH and CH PLIF cuts of turbulent flames at two different turbulence intensities and turbulent Reynolds numbers. Note the increased levels of wrinkling and broader range of wrinkling length scales in the higher turbulence intensity/Reynolds number case.

Data similarly indicate that s_T increases roughly linearly with increasing turbulence intensity and possibly saturates at very high turbulence intensities, such as shown in Figure 12.23. However, the nature of these curves at $\varepsilon_T \gg 1$ is still a subject of active investigation, for reasons that include their dependence on the turbulent flame speed definition used and the type of device [68, 71, 79–87].

Finally, the preceding discussion largely considered turbulent flames with constant laminar burning velocities. However, stretch effects cause the burning velocity to vary significantly along flames whose Markstein number departs appreciably from unity. These cause additional mechanisms of flame area increase through the thermodiffusive stabilizing or destabilizing effects [68, 89]. These effects can be large and are likely the reason for the difference in the two curves shown in Figure 12.23. For example, comparisons of CH_4 and 90/10 H_2/CO blends operated with the same $s^{u,0}$ show differences in $s_{T,c}$ of a factor of 3 at $u_{rms}/s^u = 25$ [90].

12.4.2 Fluctuating Burning and Heat Release Rates

Turbulent combustion processes are inherently unsteady. While turbulence effects on time-averaged burning rates have been extensively studied, the temporal characteristics of the burning rate and heat release rate are less understood. In addition to being of fundamental interest, these temporal characteristics are important for understanding broadband combustion noise. A typical power spectrum of heat release rate fluctuations (inferred from chemiluminescence measurements) for a turbulent Bunsen flame is shown in Figure 12.24. In this particular data set, the power spectrum has a fairly flat frequency dependence up to about 200 Hz, then rolls off at higher frequencies as $f^{-2.2}$.

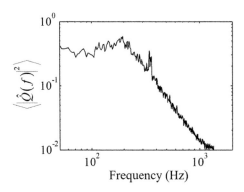

Figure 12.24 Measured spectrum of chemilumenescence fluctuations from a turbulent Bunsen flame. Adapted from Rajaram and Lieuwen [91].

There are two important problems to be addressed in order to understand the heat release spectrum: (1) the relationship between the spectrum of the velocity fluctuations, $\vec{u}(\vec{x},f)$, and the local heat release fluctuations, $\hat{q}(\vec{x},f)$, at each position, $\vec{x}$, and (2) the relationship between the spectrum of the local and spatially integrated (global) heat release rates, $\hat{q}(\vec{x},f)$ and $\hat{Q}(f)$.

The latter question is discussed first. To illustrate some basic ideas, consider the limiting case where the combustion region can be discretized into two regions, whose unsteady heat release rates are given by $\dot{Q}_a'(t)$ and $\dot{Q}_b'(t)$ so that $\dot{Q}'(t) = \dot{Q}_a'(t) + \dot{Q}_b'(t)$. The ensemble-averaged root mean square of $\dot{Q}'(t)$, given by $\langle \dot{Q}'^2(t) \rangle$, is given by

$$\langle \dot{Q}'^2(t) \rangle = \langle \dot{Q}_a'^2(t) \rangle + \langle \dot{Q}_b'^2(t) \rangle + 2\langle \dot{Q}_a'(t)\dot{Q}_b'(t) \rangle. \tag{12.46}$$

For simplicity, if we assume that the mean squared fluctuation level of each region is the same, i.e., $\langle \dot{Q}_a'^2(t) \rangle = \langle \dot{Q}_b'^2(t) \rangle$, then

$$\langle \dot{Q}'^2(t) \rangle = 2\langle \dot{Q}_a'^2(t) \rangle(1 + r_{corr}), \tag{12.47}$$

where r_{corr} is the correlation coefficient between the two time series:

$$r_{corr} = \frac{\langle \dot{Q}_a'(t)\dot{Q}_b'(t) \rangle}{\left(\langle \dot{Q}_a'^2(t) \rangle\right)^{1/2}\left(\langle \dot{Q}_b'^2(t) \rangle\right)^{1/2}}. \tag{12.48}$$

The correlation coefficient is bounded by $-1 < r_{corr} < 1$, showing that $\langle \dot{Q}'^2(t) \rangle$ can take a range of values, $0 < \langle \dot{Q}'^2(t) \rangle < 4\langle \dot{Q}_a'^2(t) \rangle$. In other words, the value of $\langle \dot{Q}'^2(t) \rangle$ is not only a function of the root mean square values of the two contributing regions, but also their correlation. This example is a very rough approximation of a distributed combustion region, which can be modeled with increasing fidelity as a larger and larger number of such regions, each with some degree of correlation to the other regions.

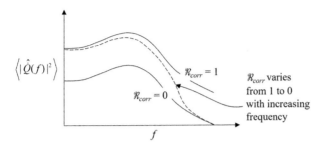

Figure 12.25 Illustration of the power spectrum of the sum of heat release fluctuations from two regions, showing the manner in which the power spectrum of a spatially integrated quantity differs from its local values.

Moving from the time domain to the frequency domain, the correlation between regions a and b is a function of frequency. The correlation coefficient, r_{corr}, then generalizes to the coherence, $\mathcal{R}_{corr}$, which is the correlation between the fluctuations at each frequency [92]:

$$\mathcal{R}_{corr} = \frac{\left\langle \hat{Q}_a(f)\hat{Q}_b^*(f) \right\rangle}{\left(\left\langle |\hat{Q}_a^2(f)| \right\rangle^{1/2}\right)\left(\left\langle |\hat{Q}_b^2(f)| \right\rangle^{1/2}\right)}. \quad (12.49)$$

At low frequencies, turbulent fluctuations are associated with larger eddies correlated over larger length scales, while at higher frequencies they are excited by smaller eddies with shorter correlation lengths. Thus, the coherence between regions a and b decreases with increasing frequency. Three examples of possible spectra of $\hat{Q}(f)$ are shown in Figure 12.25, where we assume that the spectra of the two regions are identical, $\left\langle |\hat{Q}_a(f)| \right\rangle = \left\langle |\hat{Q}_b(f)| \right\rangle$. The figure shows the resulting spectrum for cases where $\mathcal{R}_{corr} = 0$ and 1, which are identical to the individual spectra rescaled by a factor of 2 or 4, respectively. The figure also shows that the power spectrum is different from that of its constituent components when $\mathcal{R}_{corr}$ is a function of frequency.

We now leave this two-zone model and consider a one-dimensional distribution of heat release rate fluctuations, whose spectra is given by $\hat{q}(x,f)$. The power spectrum of $\hat{Q}(f)$ is given by

$$\left\langle |\hat{Q}(f)|^2 \right\rangle = \left\langle \hat{Q}\hat{Q}^* \right\rangle = \left\langle \left(A\int_0^{L_F}\hat{q}(x,f)dx\right)\cdot\left(A\int_0^{L_F}\hat{q}^*(x',f)dx'\right) \right\rangle$$

$$= A^2 \left\langle \int_0^{L_F}\int_0^{L_F}\hat{q}(x,f)\hat{q}^*(x',f)dx'dx \right\rangle. \quad (12.50)$$

Consider an example problem where the power spectrum of $\hat{q}(x,f)$ is spatially uniform (which may not be the case in anchored flames with developing flame brushes) and that the spatial correlation function of the heat release fluctuations is given by

$$\frac{\langle \hat{q}(x,f)\hat{q}^*(x',f)\rangle}{|\hat{q}(f)|^2} = \exp\left(-\frac{\pi}{4}\left(\frac{x-x'}{L_{corr}(f)}\right)^2\right). \tag{12.51}$$

Then, the power spectrum of $\hat{Q}(f)$ is given by

$$\frac{\langle |\hat{Q}(f)|^2\rangle}{A^2} = 2|\hat{q}(f)|^2 L_F^2 \frac{L_{corr}}{L_F} \cdot \left[\mathrm{erf}\left(\frac{\sqrt{\pi}}{2}\frac{L_F}{L_{corr}}\right) + \frac{2}{\pi}\frac{L_{corr}}{L_F}\left(\exp\left(-\frac{\pi}{4}\left(\frac{L_F}{L_{corr}}\right)^2\right)-1\right)\right]. \tag{12.52}$$

Two limits are of interest. If the correlation length of heat release rate fluctuations is much longer than the flame, $L_{corr} \gg L_F$, then

$$\frac{\langle |\hat{Q}(f)|^2\rangle}{(L_F A)^2} = \langle |\hat{q}(f)|^2\rangle. \tag{12.53}$$

In this limit, the spectrum of the normalized global heat release rate fluctuations is identical to that of the local fluctuations. This approximation may be reasonable at low frequencies, if the turbulent eddies are larger than the flame. In contrast, the $L_{corr} \ll L_F$ approximation is more suitable at high frequencies. In this limit, Eq. (12.52) may be expanded to first order as

$$\frac{\langle |\hat{Q}(f)|^2\rangle}{(L_F A)^2} = 2\langle |\hat{q}(f)|^2\rangle \frac{L_{corr}(f)}{L_F} + O\left(\frac{L_{corr}}{L_F}\right)^2. \tag{12.54}$$

This result shows that the power spectrum is reduced in magnitude by the ratio L_{corr}/L_F relative to the perfectly correlated result. It also shows that the frequency dependence of the correlation length, $L_{corr}(f)$, influences $\langle |\hat{Q}(f)|^2\rangle$. Figure 12.26 plots a measurement of the axial dependence of $L_{corr}(f)$ for a turbulent Bunsen flame. This data shows that the size of the coherent region decreases with frequency as expected based on the preceding discussions.[3] This characteristic of randomly forced flames is a key difference from harmonically forced flames, where disturbances are well correlated over the entire length of the flame.

Convection of wrinkles by tangential flow introduces additional phase cancellation effects in the relationship between local and spatially integrated heat release rate fluctuations. Following the discussion in Eq. (12.22), consider an example where the heat release rate fluctuations are given by

---

[3] However, note how slowly the correlation length rolls off. This is likely a manifestation of tangential convection effects that cause the correlation between fluctuations at different time instants to be quite high even at large separation distances.

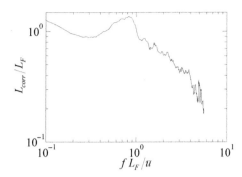

Figure 12.26 Measured dependence of the ratio of axial heat release correlation length, L_{corr}, to the flame length, L_F, on frequency. Data courtesy of R. Rajaram [93].

$$\dot{q}(x,t) = \dot{q}(t - x/u_o). \quad (12.55)$$

It should be emphasized that the axial convection of heat release rate fluctuations described by this equation occurs in the randomly forced problem just as much as it does for the harmonically forced one, and its effects are clearly manifested in data from turbulent flames [24, 77, 91]. Following the treatment in Eq. (12.24), the relationship between local and global heat release rate fluctuations is

$$St \ll 1: \quad \frac{\left\langle \left|\hat{\dot{Q}}\right|^2 \right\rangle}{(AL_F)^2} \approx \left\langle \left|\hat{\dot{q}}\right|^2 \right\rangle, \quad (12.56)$$

$$St \gg 1: \quad \frac{\left\langle \left|\hat{\dot{Q}}\right|^2 \right\rangle}{(AL_F)^2} \approx \frac{\left\langle \left|\hat{\dot{q}}\right|^2 \right\rangle}{St^2}. \quad (12.57)$$

To summarize this discussion, at least two parameters are of interest in relating the local and global heat release rate spectra: the dimensionless correlation length and the Strouhal number.

We next turn to the first question posed at the beginning of this section, which is the relationship between the spectra of the local heat release rate fluctuations and the approach flow disturbances. In the low turbulence intensity limit, the relationship between flame area and turbulent velocity fluctuations is linear [95]. As discussed in the context of Eq. (12.17), the local flame surface area per unit distance is given by $\frac{\partial \xi_1}{\partial y}$, which, in turn, is directly related to the perturbation field through expressions such as Eqs. (11.37) and (11.40). An important implication of this expression is that the relationship between their spectra is localized in frequency space – i.e., velocity fluctuations at a frequency, f, influence the area spectrum at that same frequency, but nowhere else.

In general, however, the dynamics relating the velocity and flame area spectrum are nonlinear and, therefore, nonlocal in frequency–wavenumber space [74, 94–96]; i.e., the heat release generated at some frequency, f, is a convolution of turbulent kinetic energy over a range of frequencies and length scales. The practical implication of this is that variations in the spectrum of the turbulent velocity fluctuations do not cause an analogous change in the spectrum of heat release rate fluctuations [97]. This problem has been analyzed theoretically for flames without tangential convection effects [98], leading to a predicted power-law dependence for the time derivative of the flame area of $f^{-5/2}$ for frequencies corresponding to the inertial subrange in the turbulent velocity field, assuming Kolmogorov scaling and a $k^{-5/3}$ inertial subrange velocity wavenumber spectrum.

12.4.3 Combustion Noise

We next discuss the combustion noise problem [67, 99–105]. This is sometimes referred to as "direct" combustion noise, to differentiate it from "indirect" noise, which is the acceleration of entropy inhomogeneities through the nozzle, see Section 6.3.2. As also discussed in Section 6.5, the far-field noise emitted by isomolar, unsteady combustion in the free field (i.e., in the absence of reflecting walls) is given by [67]

$$p_1(\vec{x}, t) = \frac{1}{4\pi a} \frac{\partial}{\partial t} \int_{\vec{y}} \left[\frac{\rho_0(\gamma - 1)\dot{q}_1(\vec{y}, t - |\vec{y} - \vec{x}|/c_0)}{\rho c^2} \right] d\vec{y}, \quad (12.58)$$

where a denotes the distance from the flame to the observer. In addition, ρ and c denote the density and speed of sound in the combustion zone, while ρ_0 and c_0 denote their free-field values. Also noted in that section was that if the flame is acoustically compact (i.e., much smaller than an acoustic wavelength), then the sound emissions are controlled by the spatially integrated unsteady heat release rate, $\dot{Q}_1(t)$ (see Eq. (6.68)). However, flames still generate sound, even if $\dot{Q}_1(t) = 0$, because of imperfect cancellation of sound emitted from one region of the flame and another, due to retarded time variations, the $t - |\vec{y} - \vec{x}|/c_0$ term in the above equation. Indeed, many combustion noise problems are controlled by the size of the flame with respect to the acoustic wavelength, and the size of correlated regions of unsteady heat release. We will consider these both further in the next sections that consider sound radiation from turbulence and deterministic, convecting disturbances.

12.4.3.1 Turbulent Combustion Noise

A typical combustion noise spectrum obtained from a turbulent Bunsen flame is shown in Figure 12.27.

In contrast to the heat release spectrum shown in Figure 12.24, the acoustic spectrum rises with frequency at lower frequencies. This is due to the $\partial \dot{q}/\partial t$ term in Eq. (12.58), which leads to a roughly f^2 scaling of the low-frequency power spectrum [91]. The spectrum peaks at a frequency f_{peak}. In flames with tangential flow, this peaking is due to phase cancellation effects of convecting heat release disturbances, as

12.4 Broadband Excitation and Turbulent Flame Speeds

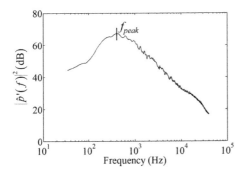

Figure 12.27 Typical combustion noise spectrum, obtained from a turbulent Bunsen flame [93]. Data courtesy of R. Rajaram.

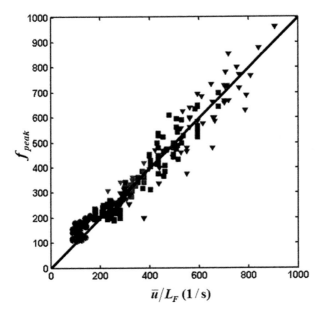

Figure 12.28 Dependence of f_{peak} on the ratio $\bar{u}/L_F$, showing its Strouhal number dependence, obtained with a range of fuels, equivalence ratios, burner diameters, and flow velocities [93]. Data courtesy of R. Rajaram.

discussed in Section 12.4.2. This causes f_{peak} to scale with Strouhal number as $St = f L_F/\bar{u}$ [91, 106, 107], as shown in Figure 12.28.

In addition to the spectrum, the total sound power emitted by the flame is of practical significance [108–111]. An approximate expression for the mean squared pressure fluctuations is [67, 100]:

$$\overline{(p_1^2)} = \frac{(\gamma-1)^2}{16\pi^2 a^2 c_0^4} \int \overline{\frac{\partial \dot{q}}{\partial t}(\vec{x},t) \frac{\partial \dot{q}}{\partial t}(\vec{x}+\vec{\Delta},t)} d^3 \vec{\Delta} \, d^3 \vec{x} \, . \qquad (12.59)$$

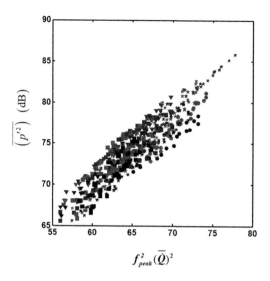

Figure 12.29 Dependence of the measured overall sound pressure level with scaling parameter for a range of fuels, flow velocities, and burner diameters [93]. Data courtesy of R. Rajaram.

This expression can be scaled as

$$\overline{(p_1^2)} \propto \frac{(\gamma-1)^2}{16\pi^2 a^2 c_0^4} f_{peak}^2 \overline{(\dot{Q})}^2 (V_{corr}/V_F). \qquad (12.60)$$

Note that this scaling captures the time derivatives with a characteristic frequency, and the volume integrals of the unsteady heat release rate by the total integrated steady heat release, $\overline{\dot{Q}}$. The correlation volume, V_{corr}, is the three-dimensional analogue of the correlation length, L_{corr}, discussed in Section 12.4.2 and describes the physical volume in space over which heat release rate fluctuations are correlated [112]. Here, V_F denotes the actual volume of the heat release region. Analogous to Eq. (12.54), the total unsteady heat release rate fluctuations and therefore the sound power emitted by the flame, scale with this ratio, V_{corr}/V_F. Figure 12.29 plots overall sound power data obtained with a range of fuels and conditions from a turbulent Bunsen flame using the $f_{peak}\overline{\dot{Q}}$ product as a scaling parameter, showing that it captures the measured trends.

The dependence of the acoustic power emitted by the flame on operational and geometric parameters can be seen by writing $f_{peak} \sim L_F/\bar{u}$ and

$$\overline{\dot{Q}} = \rho^u \bar{u} \frac{\pi D^2}{4} c_p (T^b - T^u). \qquad (12.61)$$

Assuming for simplicity that the unburned and ambient gas properties are the same, e.g., $\rho^u = \rho_0$, then Eq. (12.60) can be written as

$$\frac{\overline{(p_1^2)}}{p_0^2} \propto \frac{\gamma^2}{256 a^2} \frac{M^4 D^4}{L_F^2} \frac{V_{corr}}{V_{flame}} (T^b/T^u - 1)^2. \qquad (12.62)$$

This integral shows a M^4 scaling, typical of monopole sources [67]. It is also proportional to the temperature rise across the flame and the correlation volume, V_{corr}, of the heat release fluctuations.

12.4.3.2 Combustion Noise From Deterministic, Convecting Flow Disturbances [113]

This section considers the sound emissions from flames disturbed by convecting, helical disturbances, returning to the same problem considered in Sections 11.2.2.4 and 12.3.1.3. Because of the coherence of the disturbance source, the sound emissions can exhibit significant directivity, as well as multiple frequencies at which emissions are maximum or zero, due to phase cancellation. While the basic Mach number and mean heat release rate scalings remain the same as in Eq. (12.62), the frequency scaling can be quite different and so is the focus of the section.

Previously, it was noted that that there was a particular helical mode that lead to maximum flame wrinkling, $m_{\xi max}$, and spatially integrated unsteady heat release rate, $m_{\dot{Q}max}$. For nominally axisymmetric flames, $m_{\dot{Q}max} = 0$; i.e., an axisymmetric disturbance. We can similarly identify the helical mode that leads to maximum sound emissions, m_{pmax}. As might be anticipated from the discussion below Eq. (6.68), $m_{pmax} = m_{\dot{Q}max}$ if the flame is acoustically compact, i.e., $kL_F \ll 1$.

While the reader is referred to the reference for details, the pressure radiated to the far field can be directly related to weighted moments of the local unsteady heat release rate. Thus, even if a particular helical mode does not lead to global heat release rate oscillations, it does contribute to the far-field pressure due to imperfect phase cancellation effects that are a function of the flame acoustic compactness parameter, kL_F. If $kL_F \ll 1$, it is convenient to decompose the far-field pressure using a multipole expansion into a monopole, dipole, or quadrupole source, and so forth.

We will briefly discuss the local sound amplitude at a given location, as well as the total sound power (i.e., flux of acoustic power integrated over a far-field surface). While there are many similarities between the total sound power emitted by the flame and root mean square pressure at a given far-field location, there are also some significant differences as well, because of the strong directivity of the sound emissions. Consider the magnitude of the pressure at a given far-field location first. The symmetric mode, $m = 0$, is the sole contributor to the leading-order, monopole contribution. The $|m| = 1$ and $|m| = 2$ modes contribute to the dipole and quadrupole moments, respectively, whose far-field pressure magnitudes scale as $(kL_F)^{|m|}$ when $kL_F \ll 1$. In the $kL_F \gg 1$ limit, the pressure magnitude associated with the $m = 0$ and $|m| > 0$ helical modes scale as $(kL_F)^{-1}$ and $(kL_F)^{-3/2}$, respectively.

Consider next the normalized overall sound power, defined as

$$\mathscr{P}_m = \left(\frac{2a^2 c^2 p_0^2 \sin^2 \psi}{\pi \omega^2 \dot{q}_0^2 (\gamma - 1)^2} \right) \frac{\overline{p_1^2}}{p_0^2}. \qquad (12.63)$$

An example calculation is shown in Figure 12.30 for a case where $m_{\xi max} = +1$. For $kL_F \ll 1$, the peak power comes from the $m_{\dot{Q}max} = 0$ mode, followed by the $|m| = 1$,

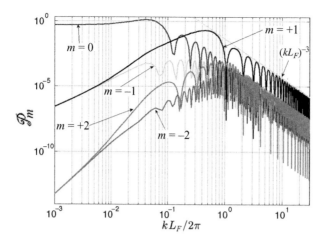

Figure 12.30 Dependence of sound power, $\mathscr{P}_m$, emitted by flame disturbed by convecting, helical flow disturbance, m, on dimensionless flame length kL_F ($\psi = 24°$, $k_c = 0.5$, $M = 0.1$, $\Omega/2\pi f = 0.6$, $m_{\xi max} = +1$).

then $|m| = 2$ modes, and so forth. This follows from the overall sound power scaling of $(kL_F)^{2|m|}$ in the $kL_F \ll 1$ limit. All modes scale equally as $(kL_F)^{-3}$ in the $kL_F \gg 1$ limit. Finally, note how the dominant emitter transitions to $m = +1 = m_{\xi max}$ when kL_F exceeds a value of about 0.2, which is the helical mode leading to maximum local flame wrinkling and local heat release rate oscillations.

12.5 Aside: Effect of External Forcing on Limit Cycle Oscillations

As discussed in Section 2.5.3.3, imposing forcing on a self-excited system undergoing limit cycle oscillations introduces nonlinear interactions between the forced and free oscillations. In particular, this introduces the possibility of frequency locking and suppression of oscillations at the natural system frequency. This behavior is observed in self-excited combustors. For example, Figure 12.31(a) compares two spectra, that of the nominal, limit cycling system and that of the forced system. In the latter case, note the presence of two peaks, corresponding to the forced and self-excited oscillations at the frequencies f_f and f_m, respectively.

Figure 12.31(b) plots the dependence of $\hat{p}'(f_m)$ at the instability frequency and the overall root mean square pressure as a function of disturbance velocity, $u'/\bar{u}$, at the forcing frequency, f_f. Note the monotonic decrease and the disappearance of the self-excited oscillations with increased forcing amplitude. The overall root mean square pressure levels first decrease as the instability amplitude decreases at f_m, but then increase due to the increased pressure amplitudes at the forcing frequency, f_f. Recall also that analogous behavior was shown for the nonreacting, forced bluff body wake in Figure 4.19.

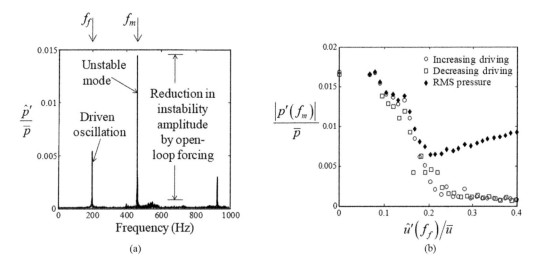

Figure 12.31 (a) Spectra of the combustor pressure at two driving amplitudes showing a decrease in self-excited oscillation amplitude as the driving amplitude is increased. (b) Dependence of the instability amplitude on the driving velocity amplitude [112].

Exercises

12.1 In order to compare the relative roles of pressure, velocity, and fuel/air ratio coupling, it is necessary to prescribe relationships between pressure, velocity, and fuel/air ratio disturbances. Derive the relations in Eq. (12.10) and (12.11) assuming a traveling, planar acoustic wave for pressure and velocity. Derive the fuel/air ratio expression from the definition of equivalence ratio, assuming a low Mach number flow and that only the oxidizer flow rate is oscillating.

12.2 Velocity-coupled response of an axisymmetric wedge flame with zero radius of flameholder and flame width of W_f. Derive Eq. (12.20), which relates the unsteady area fluctuations of the flame to the unsteady velocity.

12.3 Velocity-coupled response of a conical, axisymmetric flame. Expressions for the flame response to a convecting velocity disturbance were derived in Eq. (12.20) for an axisymmetric wedge flame. Derive the analogous result for an axisymmetric conical flame and compare and contrast the resulting flame response characteristics.

12.4 Velocity-coupled response of a two-dimensional flame. Expressions for the flame response to a convecting velocity disturbance were derived in Eq. (12.20) for an axisymmetric wedge flame. Derive the analogous result for a two-dimensional flame and compare and contrast the resulting flame response characteristics.

12.5 Velocity-coupled response of an axisymmetric wedge flame. Write out the flame transfer function for the uniform velocity case, $1/k_c = 0$, from Exercise 12.2. Derive asymptotic results for the gain and phase tendencies for high and low St_2 values and comment your results.

12.6 Asymptotic characteristics of velocity-coupled and fuel/air ratio coupled flame response. From the equations for the velocity and fuel/air ratio coupled flame responses in conical, axisymmetric, and two-dimensional geometries, derive asymptotic results for the high-St_2 flame response. Discuss.

12.7 Asymptotic character of velocity-coupled flame response. From the equations for the velocity-coupled flame response in conical, axisymmetric, and two-dimensional geometries in Eq. (12.20) and Exercises 12.2–12.4, derive asymptotic results for the low-St_2 flame response. Also, convert these results into the time domain.

12.8 Fuel/air ratio coupled flame response. Consider the linear response of flames to fuel/air ratio fluctuations, discussed in Section 12.3.2. For the model problem analyzed in this section, determine the dominant route from Figure 12.2 controlling the flame response in the $St_2 \ll 1$ limit.

12.9 Fuel/air ratio coupled response of an axisymmetric conical flame. The response of an axisymmetric wedge flame was derived in Eqs. (12.31)–(12.33). Derive the corresponding result for axisymmetric conical and two-dimensional flames. Compare and contrast the results.

12.10 Asymptotic character of fuel/air ratio coupled flame response. From the equations for the fuel/air ratio coupled flame response in conical, axisymmetric, and two-dimensional geometries in Eqs. (12.31)–(12.33) and Exercise 12.9, derive asymptotic results for the low-St_2 flame response. Also, convert these results into the time domain.

References

1. Magina N., Shin D.-H., Acharya V., and Lieuwen T., Response of non-premixed flames to bulk flow perturbations. *Proceedings of the Combustion Institute*, 2013, **34**(1): pp. 963–971.
2. Shreekrishna and Lieuwen T., High frequency response of premixed flames to acoustic disturbances, in *15th AIAA/CEAS Aeroacoustics Conference*. 2009, Miami, FL.
3. Clanet C., Searby G., and Clavin P., Primary acoustic instability of flames propagating in tubes: cases of spray and premixed combustion. *Journal of Fluid Mechanics*, 1999, **385**: pp. 157–197.
4. Malik N.A. and Lindstedt R.P., The response of transient inhomogeneous flames to pressure fluctuations and stretch: Planar and outwardly propagating hydrogen/air flames. *Combustion Science and Technology*, 2010, **182**(7–9): pp. 1171–1192.
5. McIntosh A.C., The linearized response of the mass burning rate of a premixed flame to rapid pressure changes. *Combustion Science and Technology*, 1993, **91**: pp. 329–346.
6. McIntosh A.C., Batley G., and Brindley J., Short length scale pressure pulse interactions with premixed flames. *Combustion Science and Technology*, 1993, **91**: pp. 1–13.
7. T'ien J., Sirignano W., and Summerfield M., Theory of L-star combustion instability with temperature oscillations. *AIAA Journal*, 1970, **8**: p. 120–126.
8. Richecoeur F., Ducruix S., Scouflaire P., and Candel S., Effect of temperature fluctuations on high frequency acoustic coupling. *Proceedings of the Combustion Institute*, 2009, **32**(2): pp. 1663–1670.

9. Ayoola B., Hartung G., Armitage C., Hult J., Cant R., and Kaminski C., Temperature response of turbulent premixed flames to inlet velocity oscillations. *Experiments in Fluids*, 2009, **46**(1): pp. 27–41.
10. Kim D., Lee J.G., Quay B.D., Santavicca D.A., Kim K., and Srinivasan S., Effect of flame structure on the flame transfer function in a premixed gas turbine combustor. *Journal of Engineering for Gas Turbines and Power*, 2010, **132**: p. 021502.
11. Fleifil M., Annaswamy A.M., Ghoneim Z.A., and Ghoneim A.F., Response of a laminar premixed flame to flow oscillations: A kinematic model and thermoacoustic instability results. *Combustion and Flame*, 1996, **106**: pp. 487–510.
12. Lieuwen T., Modeling premixed combustion–acoustic wave interactions: A review. *Journal of Propulsion and Power*, 2003, **19**(5): pp. 765–781.
13. Palies P., Durox D., Schuller T., and Candel S., The combined dynamics of swirler and turbulent premixed swirling flames. *Combustion and Flame*, 2010, **157**(9): pp. 1698–1717.
14. Palies P., Schuller T., Durox D., and Candel S., Modeling of premixed swirling flames transfer functions. *Proceedings of the Combustion Institute*, 2010, **33**(2): pp. 107–113.
15. Kim K., Lee J.G., and Santavicca D.A., Spatial and temporal distribution of secondary fuel for suppression of combustion dynamics. *Journal of Propulsion and Power*, 2006, **22**(2): pp. 433–439.
16. Lee J.G., Kim K., and Santavicca D., Measurement of equivalence ratio fluctuation and its effect on heat release during unstable combustion. *Proceedings of the Combustion Institute*, 2000, **28**(1): pp. 415–421.
17. Li H., Wehe S.D., and McManus K.R., Real-time equivalence ratio measurements in gas turbine combustors with a near-infrared diode laser sensor. *Proceedings of the Combustion Institute*, 2010, **33**(1): pp. 717–724.
18. Kang D., Culick F., and Ratner A., Coupling between combustor pressure and fuel/oxidizer mixing and the consequences in flame behavior. *Combustion Science and Technology*, 2008, **180**(1–3): pp. 127–142.
19. Lieuwen T. and Zinn B.T., The role of equivalence ratio oscillations in driving combustion instabilities in low NO_x gas turbines. *Proceedings of the Combustion Institute*, 1998, **27**: pp. 1809–1816.
20. Straub D.L. and Richards G.A., Effect of fuel nozzle configuration on premix combustion dynamics, in *The 1998 International Gas Turbine and Aeroengine Congress and Exhibition*. 1998, Stockholm, Sweden, pp. 1–16.
21. Lieuwen T., Torres H., Johnson C., and Zinn B.T., A mechanism of combustion instability in lean premixed gas turbine combustor. *Journal of Engineering in Gas Turbines and Power*, 1998, **120**: pp. 294–302.
22. Richards G.A. and Janus M.C., Characterization of oscillations during premix gas turbine combustion. *Journal of Engineering for Gas Turbines and Power*, 1998, **121**: pp. 294–302.
23. Kendrick D.W., Anderson T.J., Sowa W.A., and Snyder T.S., Acoustic sensitivities of lean-premixed fuel injector in a single nozzle rig. *Journal of Engineering for Gas Turbines and Power*, 1999, **121**: pp. 429–436.
24. Hemachandra S., Shreekrishna, and Lieuwen T., Premixed flames response to equivalence ratio perturbations, in *Joint Propulsion Conference*. 2007, Cincinnatti, OH.
25. Cho J.H. and Lieuwen T., Laminar premixed flame response to equivalence ratio oscillations. *Combustion and Flame*, 2005, **140**: pp. 116–129.

26. Stufflebeam J.H., Kendrick D.W., Sowa W.A., and Snyder T.S., Quantifying fuel/air unmixedness in premixing nozzles using an acetone fluorescence technique. *Journal of Engineering for Gas Turbines and Power*, 2002, **124**(1): pp. 39–45.
27. Sandrowitz A.K., Cooke J.M., and Glumac N.G., Flame emission spectroscopy for equivalence ratio monitoring. *Applied Spectroscopy*, 1998, **52**(5): pp. 658–662.
28. Fureby C., A computational study of combustion instabilities due to vortex shedding. *Proceedings of the Combustion Institute*, 2000, **28**(1): pp. 783–791.
29. Bai T., Cheng X.C., Daniel B.R., Jagoda J.I., and Zinn B.T., Vortex shedding and periodic combustion processes in a Rijke type pulse combustor. *Combustion Science and Technology*, 1993, **94**(1–6): pp. 245–258.
30. Law C.K. and Sung C.J., Structure, aerodynamics, and geometry of premixed flamelets. *Progress in Energy and Combustion Science*, 2000, **26**(4–6): pp. 459–505.
31. Preetham T., Sai K., Santosh H., and Lieuwen T., Linear response of laminar premixed flames to flow oscillations: Unsteady stretch effects. *Journal of Propulsion and Power*, 2010, **26**(3): pp. 524–532.
32. Wang H.Y., Law C.K., and Lieuwen T., Linear response of stretch-affected premixed flames to flow oscillations. *Combustion and Flame*, 2009, **156**(4): pp. 889–895.
33. Komarek T. and Polifke W., Impact of swirl fluctuations on the flame response of a perfectly premixed swirl burner. *Journal of Engineering for Gas Turbines and Power*, 2010, **132**: p. 061503.
34. Shreekrishna, Hemchandra S., and Lieuwen T., Premixed flame response to equivalence ratio perturbations. *Combustion Theory and Modelling*, 2010, **14**(5): pp. 681–714.
35. Hemchandra S., Direct numerical simulation study of premixed flame response to fuel-air ratio oscillations. in *ASME Turbo Expo 2011*. 2011, Vancouver, British Columbia.
36. Durox D., Baillot F., Searby G., and Boyer L., On the shape of flames under strong acoustic forcing: A mean flow controlled by an oscillating flow. *Journal of Fluid Mechanics*, 1997, **350**: pp. 295–310.
37. Polifke W. and Lawn C., On the low-frequency limit of flame transfer functions. *Combustion and Flame*, 2007, **151**(3): pp. 437–451.
38. Magina N., Acharya V., and Lieuwen T., Forced response of laminar non-premixed jet flames. *Progress in Energy and Combustion Science*, 2019, **70**: pp. 89–118.
39. McIntosh A.C., The interaction of high frequency, low amplitude acoustic waves with premixed flames, in *Nonlinear Waves in Active Media*, J. Engelbrecht, ed. 1989, Heidelberg: Springer.
40. McIntosh A.C., Pressure disturbances of different length scales interacting with conventional flames. *Combustion Science and Technology*, 1991, **75**: pp. 287–309.
41. van Harten A., Kapila A., and Matkowsky B.J., Acoustic coupling of flames. *SIAM Journal of Applied Mathematics*, 1984, **44**(5): pp. 982–995.
42. Ledder G. and Kapila A., The response of premixed flame to pressure perturbations. *Combustion Science and Technology*, 1991, **76**: pp. 21–44.
43. Jones B., Lee J.G., Quay B.D., Santavicca D.A., Kim K., and Srinivasan S., Flame response mechanisms due to velocity perturbations in a lean premixed gas turbine combustor, in *Proceedings of ASME Turbo Expo 2010: Power for Land, Sea and Air*. 2010, Glasgow, UK.
44. Acharya V., Shin D.-H., and Lieuwen T., Swirl effects on harmonically excited, premixed flame kinematics. *Combustion and Flame*, 2012, **159**(3): pp. 1139–1150.

45. Acharya V.S., Shin D.-H., and Lieuwen T., Premixed flames excited by helical disturbances: flame wrinkling and heat release oscillations. *Journal of Propulsion and Power*, 2013, **29**(6): pp. 1282–1291.
46. Acharya V. and Lieuwen T., Premixed flame response to helical disturbances: Mean flame non-axisymmetry effects. *Combustion and Flame*, 2016, **165**(C): pp. 188–197.
47. Acharya V. and Lieuwen T., Nonlinear response of swirling premixed flames to helical flow disturbances. *Journal of Fluid Mechanics*, 2020, **896**: p. A6.
48. Lieuwen T., Neumeier Y., and Zinn B.T., The role of unmixedness and chemical kinetics in driving combustion instabilities in lean premixed combustors. *Combustion Science and Technology*, 1998, **135**: pp. 193–211.
49. Orawannukul P., Lee J.G., Quay B.D., and Santavicca D.A., Fuel-forced flame response of a lean-premixed combustor, in *ASME Turbo Expo*. 2011, Vancouver, British Columbia, Canada.
50. Levine W.S., *Control System Fundamentals*. 2nd ed. 2011, Boca Raton FL: CRC Press.
51. Preetham, Hemchandra S. and Lieuwen T., Dynamics of laminar premixed flames forced by harmonic velocity disturbances. *Journal of Propulsion and Power*, 2008, **24**(6): pp. 1390–1402.
52. Noiray N., Durox D., Schuller T., and Candel S., A unified framework for nonlinear combustion instability analysis based on the flame describing function. *Journal of Fluid Mechanics*, 2008, **615**(1): pp. 139–167.
53. Balachandran R., Ayoola B., Kaminski C., Dowling A., and Mastorakos E., Experimental investigation of the nonlinear response of turbulent premixed flames to imposed inlet velocity oscillations. *Combustion and Flame*, 2005, **143**(1–2): pp. 37–55.
54. Külsheimer C. and Büchner H., Combustion dynamics of turbulent swirling flames. *Combustion and Flame*, 2002, **131**(1–2): pp. 70–84.
55. Poinsot T.J., Trouve A.C., Veynante D.P., Candel S.M., and Esposito E.J., Vortex-driven acoustically coupled combustion instabilities. *Journal of Fluid Mechanics*, 1987, **177**(1): pp. 265–292.
56. Schadow K. and Gutmark E., Combustion instability related to vortex shedding in dump combustors and their passive control. *Progress in Energy and Combustion Science*, 1992, **18**(2): pp. 117–132.
57. Reuter D., Daniel B., Jagoda J., and Zinn B., Periodic mixing and combustion processes in gas fired pulsating combustors. *Combustion and Flame*, 1986, **65**(3): pp. 281–290.
58. Pang B., Cipolla S., and Yu K., Effect of Damköhler number on vortex-combustion interaction, in *Combustion and Noise Control*, G.D. Roy, ed. 2003, Cranfield University Press, pp. 103–115.
59. Hedge U.G., Reuter D., Daniel B.R., and Zinn B.T., Flame driving of longitudinal instabilities in dump type ramjet combustors. *Combustion Science and Technology*, 1987, **55**: pp. 125–138.
60. Bellows B.D., Bobba M.K., Forte A., Seitzman J.M., and Lieuwen T., Flame transfer function saturation mechanisms in a swirl-stabilized combustor. *Proceedings of the Combustion Institute*, 2007, **31**(2): pp. 3181–3188.
61. Dawson J.R., Rodriguez-Martinez V.M., Syred N., and O'Doherty T., The effect of combustion instability on the structure of recirculation zones in confined swirling flames. *Combustion Science and Technology*, 2005, **177**(12): pp. 2349–2371.
62. Lacarelle A., Faustmann T., Greenblatt D., Paschereit C., Lehmann O., Luchtenburg D., and Noack B., Spatiotemporal characterization of a conical swirler flow field under strong forcing. *Journal of Engineering for Gas Turbines and Power*, 2009, **131**: p. 031504.

63. O'Connor J., Kolb M., and Lieuwen T., Visualization of shear layer dynamics in a transversely excited, annular premixing nozzle, in *49th AIAA Aerospace Sciences Meeting*. 2011, Orlando, FL: AIAA.
64. Schuller T., Durox D., and Candel S., A unified model for the prediction of laminar flame transfer functions: Comparisons between conical and V-flame dynamics. *Combustion and Flame*, 2003, **134**(1–2): pp. 21–34.
65. Durox D., Schuller T., Noiray N., and Candel S., Experimental analysis of nonlinear flame transfer functions for different flame geometries. *Proceedings of the Combustion Institute*, 2009, **32**(1): pp. 1391–1398.
66. Peracchio A. and Proscia W.M., Nonlinear heat-release/acoustic model for thermoacoustic instability in lean premixed combustors. *Journal of Engineering for Gas Turbines and Power*, 1999, **121**: pp. 415–421.
67. Dowling A.P., Thermoacoustic sources and instabilities, in *Modern Methods in Analytical Acoustics*, D.G. Crighton, A.P. Dowling, J.E. Ffowcs Williams, M. Heckl, and F.G. Leppington, eds. 1994, Springer. pp. 378–405.
68. Driscoll J.F., Turbulent premixed combustion: Flamelet structure and its effect on turbulent burning velocities. *Progress in Energy and Combustion Science*, 2008, **34**(1): pp. 91–134.
69. Peters N., *Turbulent Combustion*. 1st ed. 2000, Cambridge University Press.
70. Poinsot T. and Veynante D., *Theoretical and Numerical Combustion*. 2nd ed. 2005, R.T. Edwards, Inc.
71. Cheng R.K., Turbulent combustion properties of premixed syngas, in *Synthesis Gas Combustion*, T. Lieuwen, V. Yang, and R. Yetter, eds. 2011, CRC Press, pp. 129–168.
72. Law C.K., *Combustion Physics*. 2006, Cambridge University Press.
73. Kerstein A.R., Simple derivation of Yakhot's turbulent premixed flamespeed formula. *Combustion Science and Technology*, 1988, **60**(1–3): pp. 163–165.
74. Kerstein A.R., Fractal dimension of turbulent premixed flame. *Combustion Science and Technology*, 1988, **60**: pp. 441–445.
75. Sivashinsky G.I., Cascade-renormalization theory of turbulent flame speed. *Combustion Science and Technology*, 1988, **62**(1–3): pp. 77–96.
76. Yakhot V., Propagation velocity of premixed turbulent flames. *Combustion Science and Technology*, 1988, **60**(1–3): pp. 191–214.
77. Hemchandra S. and Lieuwen T., Local consumption speed of turbulent premixed flames: An analysis of "memory effects." *Combustion and Flame*, 2010, **157**(5): pp. 955–965.
78. Tanahashi M., Murakami S., Choi G.M., Fukuchi Y., and Miyauchi T., Simultaneous CH-OH PLIF and stereoscopic PIV measurements of turbulent premixed flames. *Proceedings of the Combustion Institute*, 2005, **30**(1): pp. 1665–1672.
79. Yuen F.T.C. and Gülder Ö.L., Dynamics of lean-premixed turbulent combustion at high turbulence intensities. *Combustion Science and Technology*, 2010, **182**(4–6): pp. 544–558.
80. Yuen F.T.C. and Gülder Ö.L., Premixed turbulent flame front structure investigation by Rayleigh scattering in the thin reaction zone regime. *Proceedings of the Combustion Institute*, 2009, **32**(2): pp. 1747–1754.
81. Daniele S., Jansohn P., Mantzaras J., and Boulouchos K., Turbulent flame speed for syngas at gas turbine relevant conditions. *Proceedings of the Combustion Institute*, 2010, **33**: p. 2937–2944.
82. Griebel P., Siewert P., and Jansohn P., Flame characteristics of turbulent lean premixed methane/air flames at high pressure: Turbulent flame speed and flame brush thickness. *Proceedings of the Combustion Institute*, 2007, **31**(2): pp. 3083–3090.

83. Bastiaans R.J.M., Martin S.M., Pitsch H., Van Oijen J.A., and De Goey L.P.H., Flamelet analysis of turbulent combustion, in *International Conference on Computational Science*, 2005, pp. 64–71.
84. Raman V., Pitsch H., and Fox R.O., Eulerian transported probability density function subfilter model for large-eddy simulations of turbulent combustion. *Combustion Theory and Modelling*, 2006, **10**(3): pp. 439–458.
85. Pettit M.W.A., Coriton B., Gomez A., and Kempf A.M., Large-eddy simulation and experiments on non-premixed highly turbulent opposed jet flows. *Proceedings of the Combustion Institute*, 2010, **33**: pp. 1391–1399.
86. Halter F., Chauveau C., Gökalp I., and Veynante D., Analysis of flame surface density measurements in turbulent premixed combustion. *Combustion and Flame*, 2009, **156**(3): pp. 657–664.
87. Filatyev S.A., Driscoll J.F., Carter C.D., and Donbar J.M., Measured properties of turbulent premixed flames for model assessment, including burning velocities, stretch rates, and surface densities. *Combustion and Flame*, 2005, **141**(1–2): pp. 1–21.
88. Nakahara M. and Kido H., Study on the turbulent burning velocity of hydrogen mixtures including hydrocarbons. *AIAA Journal*, 2008, **46**(7): pp. 1569–1575.
89. Lipatnikov A. and Chomiak J., Molecular transport effects on turbulent flame propagation and structure. *Progress in Energy and Combustion Science*, 2005, **31**(1): pp. 1–73.
90. Venkateswaran P., Marshall A., Shin D.H., Noble D., Seitzman J., and Lieuwen T., Measurements and analysis of turbulent consumption speeds of H_2/CO mixtures. *Combustion and Flame*, 2011, **158**: pp. 1602–1614.
91. Rajaram R. and Lieuwen T., Acoustic radiation from turbulent premixed flames. *Journal of Fluid Mechanics*, 2009, **637**: pp. 357–385.
92. Bendat J.S. and Piersol A.G., *Random Data: Analysis and Measurement Procedures*. 2000, John Wiley & Sons, Inc.
93. Rajaram R., *Characteristics of Sound Radiation from Turbulent Premixed Flames*. 2007, Atlanta: Georgia Institute of Technology.
94. Clavin P. and Siggia E.D., Turbulent premixed flames and sound generation. *Combustion Science and Technology*, 1991, **78**(1–3): pp. 147–155.
95. Lieuwen T., Mohan S., Rajaram R., and Preetham, Acoustic radiation from weakly wrinkled premixed flames. *Combustion and Flame*, 2006, **144**(1–2): pp. 360–369.
96. Aldredge R. and Williams F., Influence of wrinkled premixed flame dynamics on large scale, low intensity turbulent flow. *Journal of Fluid Mechanics*, 1991, **228**: pp. 487–511.
97. Rajaram R. and Lieuwen T., *Effect of Approach Flow Turbulence Characteristics on Sound Generation from Premixed Flames*. AIAA-2004-461, 2004.
98. Clavin P. and Siggia E.D., Turbulent premixed flames and sound generation. *Combustion Science and Technology*, 1991, **78**(1–3): pp. 147–155.
99. Bragg S.L., Combustion noise. *Journal of the Institute of Fuel*, 1963, **36**(264): pp. 12–16.
100. Strahle W.C., On combustion generated noise. *Journal of Fluid Mechanics*, 1971, **66**(3): pp. 445–453.
101. Strahle W.C., Combustion noise. *Progress in Energy and Combustion Science*, 1978, **4**(3): pp. 157–176.
102. Hassan H.A., Scaling of combustion-generated noise. *Journal of Fluid Mechanics*, 1974, **66**(3): pp. 445–453.
103. Chiu H.H. and Summerfield M., Theory of combustion noise. *Acta Astronautica*, 1973, **1**(7–8): pp. 967–984.

104. Hegde U., Reuter D., and Zinn B., Sound generation by ducted flames. *AIAA Journal*, 1988, **26**(5): p. 532.
105. Ihme M., Pitsch H., and Bodony D., Radiation of noise in turbulent non-premixed flames. *Proceedings of the Combustion Institute*, 2009, **32**(1): pp. 1545–1553.
106. Wäsle J., Winkler A., and Sattelmayer T., Spatial coherence of the heat release fluctuations in turbulent jet and swirl flames. *Flow, Turbulence and Combustion*, 2005, **75**(1): pp. 29–50.
107. Hirsch C., Wasle J., Winkler A., and Sattelmayer T., A spectral model for the sound pressure from turbulent premixed combustion. *Proceedings of the Combustion Institute*, 2007, **31**(1): pp. 1435–1441.
108. Roberts J.P. and Leventhall H.G., Noise sources in turbulent gaseous premixed flames. *Applied Acoustics*, 1973, **6**(4): pp. 301–308.
109. Strahle W.C. and Shivashankara B.N., A rational correlation of combustion noise results from open turbulent premixed flames. *Symposium (International) on Combustion*, 1974, **15**(1): pp. 1379–1385.
110. Putnam A.A., Combustion roar of seven industrial gas burners. *Journal of the Institute of Fuel*, 1976, **49**(400): pp. 135–138.
111. Giammar R. and Putnam A., Guide for the design of low noise level combustion systems, in *American Gas Institute Report*. 1971.
112. Bellows B.D., Hreiz A., and Lieuwen T., Nonlinear interactions between forced and self-excited acoustic oscillations in premixed combustor. *Journal of Propulsion and Power*, 2008, **24**(3): p. 628.
113. Acharya V.S. and Lieuwen T.C., Sound generation from swirling, premixed flames excited by helical flow disturbances. *Combustion Science and Technology*, 2015, **187**(1–2): pp. 206–229.

Index

absolute instability, 82, 94, 99–100, 127, 152, 158
acceleration effects on flames, 361, 456, 463, 465
acoustic
    damping, 40, 46, 228, 231
    intensity flux. *See* disturbance, energy flux
acoustic modes
    annular ducts, 205–207
    circular ducts, 192–196
    entropy coupling, 216
    mixed convective modes, 198
    one-dimensional, 188–191
    rectangular duct, 191–192
    sector geometries, 196, 206
acoustic wave equation, 34, 177
    with advection effects, 214–215
    with area change, 210–211
    with temperature gradients, 41
acoustic wave propagation
    flow effects, 213–215
    gas dynamic nonlinearities, 248–250
    interactions with injectors, 225–228
    slowly varying properties, 252–253
    unsteady heat release effects, 233–235
    variable area effects, nozzles, diffusers, 223–225
    variable temperature/density effects, 217–218
    vortical–acoustic coupling, 41, 228
activation energy, 304–305, 322–324, 349, 365, 368–369
aeroacoustics, 40, 43
alkanes, flame properties, 299–300, 314, 327
amplitude equations, 48–54, 244–248
annular geometries. *See* acoustic modes, annular ducts
attractor, 49–50, 207–208
autocorrelation, 55–56, 67
autoignition. *See* ignition, auto
    waves, 310–311, 328–329
axisymmetric breakdown, 151

backflow, effects on hydrodynamic stability, 99, 117, 152, 157
backward-facing step, 87, 156–157, 159, 390
    combustion/nonuniformity effects, 158
    stability analysis, 157–161
baroclinic torque, 18–19, 120, 275, 277–278

base flow, 29, 33, 40, 81, 84, 88, 90, 100–103, 115–116
    swirl rate, 428
    velocity, 90, 94, 449
Bénard/von Kármán instability, 81, 125, 128, 130
bifurcation, 51, 149, 441, 485
    diagram, 51, 54
    subcritical, 52
    supercritical, 51
blowoff, 350, 379–402, 416
    effects on flame response to forcing, 485
    induced combustion instabilities, 198
    kinematic balance effects on, 390–392
    nonpremixed flames, 400–402
    of shear layer stabilized flame, 392–396
    product recirculation effects on, 388–400
bluff body flow field. *See* wakes
Borghi combustion regime diagram, 407
boundary conditions
    acoustic, 181, 185
    choked nozzles, 250, 255
    flow effects, 215
    nonlinearities, 250
boundary layer, 106, 113, 116–117, 124, 134, 140, 145, 231–232, 315, 379, 382–387, 395, 406
    heating effects, 162, 388
    pressure gradient, 165
burning velocity, 270, 288, 301, 324, 328, 335–337, 341–342, 361, 363
    laminar, 267, 331, 351, 361, 390–392, 406–407, 429, 490
    turbulent, 428–432, 488–493
bypass transition, 88

Cabra burner, 351
canonical disturbance modes, 34–35, 39
cavity flows. *See* backward-facing steps
cavity tones, 157, 198
cellular flame structures, 361–362, 456
centrifugal instability mechanisms, 106–107, 145
circular geometries. *See* acoustic modes, circular ducts
circulation equation, 19
coherence, 495, 501

coherent structures, 68, 101, 113, 137, 385
combustion-induced vortex breakdown, 381, 387
combustion instability. *See* thermoacoustic
    instability
combustion noise, 2, 29, 463, 493, 498
compact
    acoustic, 182, 218, 227, 233–235, 414–415, 501
    convective/fluid-mechanic, 414–415, *See*
        Strouhal number
confined flows, stability of, 163–164
conservation
    equations, 9–17, 44
    relations across a discontinuity/interface.
        See jump conditions
conserved scalar, 13
constant-property surface. *See* surface dynamics,
    constant-property surfaces
consumption speed, 262, 397
    laminar, 339–342
    turbulent, 429, 490
convective instability, 86, 94, 136
correlation function, 69, 430–432, 496
counter-rotating vortex pair, 139–140, 278
coupled oscillators, 59
Crocco's equation, 275
cusp formation, 436
cutoff frequency, 202, 204

Damköhler number, 283, 304–306, 352, 354–357,
    359
Darrieus–Landau instability, 360, 456–457
describing function, flame response, 481–482,
    484–487
differential diffusion, 314, 333, 336–337, 363
diffusion velocity, 12
dilatation rate, 1, 10, 46, 275, 278, 289, 366
dilution, with exhaust gas, 359, 397
direct combustion noise. *See* combustion noise
dispersion relation, 90, 97
dispersive wave propagation, 89, 97, 136, 202,
    453
displacement speed, 262, 338, 386, 397, 420, 434
    laminar, 21, 267, 390, 439
    turbulent, 380, 392, 455, 490
disturbance
    energy density, 44
    energy flux, 27, 44
    source terms, 16, 40, 42, 44–45, 47, 247–248, 288
DuFour effect, 333
dynamic mode decomposition, 70–71

eddy viscosity, 117
edge flame, 349–359
    heat loss effects, 355–357
    local stretch rate, 357–359
    overview, 349–351
    velocities, 354–357

eigenvalue problems, 67, 70–71, 82–83, 193, 239,
    250–251
empirical basis function, 62, 66–68
end correction, acoustic, 216
energy equation, 9, 13
    in terms of enthalpy, 14
    in terms of entropy, 15
    in terms of temperature, pressure, 15
    kinetic, 14
    linearized, 33
ensemble average, 72, 429, 454–455, 490, 494
enthalpy, 10
    chemical equation for, 14
    sensible equation for, 15
    total equation for, 14
entrained mass, 186, 197
entrainment of oscillations. *See* frequency locking
entropy–acoustic coupling, 216–217
equivalence ratio coupling. *See* heat release
    dynamics
Euler equations, 30, 61
evanescent disturbances, 202
explosion limit, 300
extinction, 305, 321, 349, *See* quenching line
    stretch rate, 340

failure wave, 349
Fickian diffusion, 332–333, 369
flame speed. *See* burning velocity
flame spread, after ignition, 359
flame stabilization, 388
    in shear layers, 392–396
    product recirculation effects, 396–400
    vortex breakdown/jet core, 380, 387, 392
flame strain. *See* stretch, flame
flame stretch. *See* stretch, flame
flame wrinkle velocity, 413
flameholding, 140, 380
flashback, 379–388
    boundary layer, 382–385
    gas expansion effects, 386–387
    in core flow, 380–381
    stretch effects, 385–388
forced response, 54, 56, 73, 86, 118, 120, 131–132,
    156, 160, 199, 239, 250, 286, 406, 414, 422, 432,
    437, 463, 475, 481, 483, 486, 488, 491, 496
Fourier transform, 62
    prescribed basis function, 62–66
frequency locking, 57, 131, 502
fuel decomposition, 324
fuel/air ratio coupling. *See* heat release dynamics,
    equivalence ratio coupling
Galerkin method, 244

gas expansion
    effects on approach flow, 269–272, 385
    effects on flashback, 385–386

Index

effects on hydrodynamic stability, 120, 124, 151
   flow acceleration, 272–273
Gaster's transformation, 90, 109
G-equation, 266
global mode, 86, 100–101, 138, 159, 164
global stability, 99–100, 125, 131–132, 156, 164
globally unstable, 84–88
group velocity, 90, 94

harmonic generation, 28, 57, 122, 131, 138, 232, 448
heat flux vector, 14, 333
heat release dynamics
   acceleration coupling, 467
   entropy coupling, 467
   equivalence ratio coupling, 466
   linearized analysis, 472
   nonlinear response to harmonic forcing, 481
   pressure coupling, 467
   response mechanism, 463–465
   response to turbulent forcing, 469
   velocity coupling, 465
heat release rate, 2, 14, 235, 309, 324, 463, 493
helical disturbance mode, 127, 134, 147, 151–153
Helmholtz
   equation, 187, 192, 244
   number, 414
   resonator, 197
high activation energy. *See* activation energy
Hopf–Lax formula, 434
horseshoe vortex, 140
Howard's semicircle theorem, 93, 109
Huygens propagation, 434
hydrodynamic stretch. *See* stretch, hydrodynamic
hydrogen, 307, 310, 383, 409
hysteresis, 54, 148, 305, 311, 382, 400

ignition wave, 310–311, 329, 349
ignition, auto, 298–311
   chain branching, 298–300
   in nonpremixed flames, 307–310
   scalar dissipation effects, 345
   stretch effects, 335–341
   temperature, 299
   time, 300
ignition, forced, 311–316
   hot surfaces, 315–316
   spark, 296–297, 312–315
impedance, acoustic, 177, 182, 197, 199–200, 224–225
indirect combustion noise, 498
instability
   combustor/combustion. *See* thermoacoustic instability
   general concepts, 49–52
   intrinsic flame, 360–363
   rotating/swirling flow, 106

integral length/time scales, 406–407, 429, 431
intensity. *See* disturbance, energy flux
interference effects, 74, 181, 220, 425, 428, 433, 438, 451, 473, 480, 482
internal energy, 10, 14, 45, 47, 75

jet in crossflow, 138–144
   nonreacting, 139
   reacting, 142
jets, 132
   forcing effects, 137–138
   noncircular, 164
   stability analysis, 133–136
jump conditions
   acoustic, 218–219, 229, 234
   across shear layer, 96
   finite flame thickness effects, 288
   nonpremixed flame, 269
   premixed flame, 267
   vorticity, 269, 275

Karlovitz number, 283, 336–337, 367, 383–384, 391, 395, 409, 411
Kelvin–Helmholtz instability, 81, 116, 125, 136, 153, 393
kidney vortex, 140
kinematic restoration, 433–437, 481–483, 485–486, 491
Kolmogorov length/time scales, 406, 408, 410–411, 498
Kolmogorov–Petrovsky–Piskunov solutions, 369
Krylov–Bogolyubov decomposition, 246–247

Langevin's equation, 420
level set equation. *See* G-equation
Lewis number, 333, 336–337, 345, 357, 363
liftoff of flames, 400
limit cycle, 50–61, 71, 86, 101, 159, 207, 242, 247, 249–250, 502
linearized Navier–Stokes and energy equations, 29
lumped elements, 196

Markstein number/length, 336, 365, 493
material line
   stretching. *See* stretch, of material line
   vortex interaction, 446–448
mean flow stability, 100
memory, effects on flame dynamics, 423
methane, flame properties, 17, 299, 301, 312, 314, 324, 328–329, 342, 346–347, 359, 368, 383, 400, 456, 477
method of averaging, 247–248
minimum ignition energy, 313–315
mixing layer. *See* shear layer
mixing transition, 118
mixture fraction, 13, 264, 266, 307, 315, 344, 348, 357, 445, 449
   coordinate, 345

modal coupling, 39
    boundary condition coupling, 39
    flow inhomogeneities, 40
    nonlinearities, 42
Moffatt eddies, 157
momentum equation, 9
    linearized form of, 33
momentum ratio, 139
most reacting mixture fraction, 307
multielement canonical flows, 163
multipole expansion, 501

narrowband response, 54–55, 61, 73, 86, 131, 157, 176, 244, 406, 412, 454
natural frequencies, 54, 188, 191, 194, 199, 215, 220, 222
negative flame speed, 338
nodal lines, 192–194
nominal flow conditions, 28
nonlinear effects of acoustic fields
    gas dynamic, 248
nonlinear effects on acoustic fields
    combustion effects, 250
nonlinear interaction, 43, 56, 59, 64, 103, 502
nonlinear oscillators, 56–59
nonpremixed flame
    dynamics, 442–452
    edges. *See* edge flames
    extinction, 350
    finite-rate effects, 349
    harmonic excitation, 448
    intrinsic instabilities, 360
    overview, 343
    scalar dissipation rate. *See* scalar dissipation rate
    surface dynamics. *See* surface dynamics, constant-property surfaces
    transient dynamics, 442–445
    vortex interaction, 445–448
normal mode solution, 89–91
nozzles, wave propagation in, 223–225
    entropy–acoustic coupling, 224
n–τ model, 236, 476

orthogonality, 201, 245, 251
oxycombustion, 16

parallel flow, hydrodynamic stability of, 88–90, 94, 100, 158
passive scalar surface. *See* surface dynamics, passive scalar surfaces
perfect gas, 9, 18, 75, 265
phase space, 53, 60
phase velocity, 89, 92, 97, 134, 179, 202–204, 214, 412–413, 423–424, 426–427, 474
pilot flame, 311, 329, 389
post-flame chemistry zone, 325, 407
potential mode, 36

preferred mode, 137
preheat zone, 322–323, 325, 364, 366, 411, 414
premixed flame
    broadband excitation, response to, 428–432, *See* heat release dynamics
    burning velocity, laminar/turbulent. *See* burning velocity
    edges. *See* edge flames
    extinction, 338, *See* quenching line
    extinction by vorticies, 366–367
    harmonic excitation, response to, 427–428, *See* heat release dynamics
    heat release. *See* heat release dynamics
    intrinsic instabilities, forcing effects, 456–457
    linearized response, 419–433
    nonlinear dynamics, 433–442
    sheet dynamics, 415–442
    stabilization, 388–400
    structure, 322–326
    thickness, 326–328
    transient response, 417–419
    unsteady effects, 341–343
    vortex interaction, 445–448
pressure release boundary condition, 181–187, 192, 215, 220, 223, 235, 244, 251
progress variable, 367, 488–490
propagating surface. *See* surface dynamics, propagating surfaces
propane, 299–300, 310, 313, 326, 336, 384
proper orthogonal decomposition, 66–68

quasi one-dimensional forms of conservation equations, 26
quasi-steady
    acoustic, 229
    flame configuration/heat release response, 413, 486–487
    fluid mechanic, 229
    internal flame response/kinetics, 309, 363, 409, 411, 413–415
quenching line, 367, 411
Q-vortex, 148

Rankine–Hugoniot relations, 269
Rayleigh
    criterion, 45–46, 162, 237, 239
    equation, 93, 95
    inflection point theorem, 90, 92
Rayleigh–Taylor instability, 106, 361, 467
reaction zone, 3, 323, 332–333, 343, 364–365, 398, 408, 414
reaction–diffusion equation, 13
    wave, 311
    wave solutions, 367
receptivity, 106
reflection coefficient, 39–40, 181–187, 215–216, 224, 255

reflection, acoustic wave, 37, 39, 181, 184, 210, 218, 233
refraction, 37, 210, 270, 274
repelling point, 50
resonance, 199–200
rigid-wall boundary condition, 39, 181–183, 188, 191, 193, 195, 205, 220, 222, 225, 235, 244, 251
RQL combustors, 138, 296, 308

saturation, 49, 57, 250, 482–483, 485–486
scalar dissipation rate, 309, 345–346, 357, 401–402, 412, 445, 464
 flame stretch, relation to, 345
scattering, acoustic wave, 43, 232
S-curve, 340, 359
secondary flow instability, 118, 137
self-excited oscillations, 27, 55–58, 86, 156–157, 176, 200, 235, 237–238, 481, 502
self-similar flow evolution, 116, 123, 133, 445
shear layer, 92, 95, 113, 116, 119–156, 279, 321, 350, 379, 386, 392, 474, 485
 nonuniformity/flame effects, 119
 response to harmonic forcing, 120
shocks, 41, 43, 107, 418
sinuous disturbance modes, 105, 125–129, 134
Soret effects, 13
spark ignition. *See* ignition, forced
spatial stability, 90, 97–98
spatiotemporal POD, 68–70
species equation, 13, 302, 304
spectral diagram for turbulent combustion, 410
spectral POD, 70
Squire's transformation, 108
stability of disturbances, 48–49, 81–84, 99–103, 113, 152, 236, 321, 362, 456, 463
stabilization point of a flame, 379, 389–390, 399, 483
stagnating flows, 117, 296, 310, 333, 335–336, 339–341, 346, 348, 397, 445, 491
standing waves, 74, 180, 189, 194, 206–207, 215, 227
strain rate, flow, 11, 282, 333, 335, 346, 351, 393–394, 408–409, 446
strain, flame. *See* stretch, flame
stress tensor, 11
stretch, flame, 282
 curvature, 334–335
 flashback, 385–388
 harmonic forcing effects, 456
 hydrodynamic, 334–335
 in shear layer, 392–396
 relation to scalar dissipation rate, 345–349
 strong stretch effects, 338–341
 weak stretch effects, 335–338
stretch, of material lines/surfaces, 282–285
Strouhal number, 230, 450
 flame, 473, 482, 499

jet, 137
shear layer, 99
swirling flow, 151
wake, 128
Sturm–Liouville problems, 250
subharmonic generation, 118, 121–123, 232
substantial derivative, 33
sum and difference frequencies, 29, 131
surface dynamics, 261
 constant-property surfaces, 262
 material surface, 262–263
 nondiffusive passive scalar, 282, *See* surfaces, material surfaces
 propagating surface, 262–263
swirl flow
 breakdown bubble, 145, 149–150, 152–153, 155
 combustion/nonuniformity effects, 151
 forcing effects, 153
 stability analysis, 153
 subcritical, 145, 151
 supercritical, 145, 151
 vortex breakdown, 147
swirl fluctuations, 41, 466
swirl number, 41, 144, 146–147, 149–153

tangential flow, effects on flame dynamics, 281–283, 350, 358, 386, 394, 420, 428, 432, 496, 498
temporal stability, 90–95
thermoacoustic instability, 122, 132, 235, 389
 active control, 239
three-dimensional disturbances
 flame position, 427
 heat release response, 478
 sound radiation from flame, 498
time average, 29
T-junction, 227, 253
transfer functions, flame response, 468, 473–474, 480
transmission coefficient, 183
trapped vortex combustor, 159
traveling wave, 177–178, 180, 202, 206–208, 212, 468
triple decomposition, 71
turbulent combustion diagram. *See* Borghi combustion regime diagram
turbulent flames
 brush, 2, 233, 368, 429, 442
 burning velocity. *See* burning velocity, turbulent
 harmonic forcing, 454

unflanged pipe, acoustic boundary conditions, 186

van der Pol decomposition, 245
varicose disturbance modes, 105, 125–127, 134
velocity coupling. *See* heat release dynamics, velocity coupling

Index

volume production, by reactions, 16
vortex
 bending, 17–18, 44, 92, 140, 210, 274, 289
 braids, 104, 117, 122, 136–137, 402
 flame extinction by, 278, 366
 induction, 103, 121, 137, 164
 pairing, 118–119, 122–123, 136–138
 stretching, 17, 41, 274, 289
 stretching of material lines, 282–283
 stretching of premixed/nonpremixed flames, 445–448
vorticity
 equation, 11–12, 17, 40–42
 linearized form of, 33
vorticity mode, canonical decomposition, 35

wakes, 123–125
 combustion/nonuniformity effects, 128–131
 forcing effects, 131–132
 shear layer characteristics, 130–131
 stability analysis, 133–136
wave equation. *See* acoustic wave equation
wavelength
 acoustic, 37–38, 212, 233
 convective, 37, 414–415, 478
wavelets, 65
well-stirred reactor, 3, 302

Zeldovich number, 323